SCOTT W. STARRATT

TREATISE ON
INVERTEBRATE PALEONTOLOGY

Prepared under Sponsorship of
The Geological Society of America, Inc.

The Paleontological Society   *The Society of Economic Paleontologists and Mineralogists*
The Palaeontographical Society   *The Palaeontological Association*

RAYMOND C. MOORE
Founder

CURT TEICHERT
Director and Editor

LAVON McCORMICK, ROGER B. WILLIAMS
Assistant Editors

Part W
MISCELLANEA
SUPPLEMENT 1
TRACE FOSSILS AND PROBLEMATICA
Second Edition (Revised and Enlarged)

By †WALTER HÄNTZSCHEL

SCOTT W. STARRATT

THE GEOLOGICAL SOCIETY OF AMERICA, INC.
and
THE UNIVERSITY OF KANSAS
BOULDER, COLORADO, and LAWRENCE, KANSAS

1975

Library of Congress Catalogue Card Number 53–12913

ISBN 0–8137–3027–9

Text Composed by
THE UNIVERSITY OF KANSAS PRINTING SERVICE
Lawrence, Kansas

Illustrations and Offset Lithography
THE MERIDEN GRAVURE COMPANY
Meriden, Connecticut

Binding
TAPLEY-RUTTER COMPANY
Moonachie, N.J.

*Distributed by The Geological Society of America, Inc., 3300 Penrose Place, Boulder, Colorado 80301, from which current price lists of Parts in print may be obtained and to which all orders and related correspondence should be directed. Editorial office for the *Treatise*: Paleontological Institute, 115 Lindley Hall, The University of Kansas, Lawrence, Kansas 66045.

The *Treatise on Invertebrate Paleontology* has been made possible by (1) grants of funds from The Geological Society of America through the bequest of Richard Alexander Fullerton Penrose, Jr., for initial preparation of illustrations, and partial defrayment of organizational expenses in 1948-1957, and again since 1971, and from the United States National Science Foundation, awarded annually since 1959, for continuation of the *Treatise* project; (2) contribution of the knowledge and labor of specialists throughout the world, working in cooperation under sponsorship of The Geological Society of America, The Paleontological Society, The Society of Economic Paleontologists and Mineralogists, The Palaeontographical Society, and The Palaeontological Association; and (3) acceptance by The University of Kansas of publication without any financial gain to the University.

TREATISE ON INVERTEBRATE PALEONTOLOGY

RAYMOND C. MOORE, Founder

Directed and Edited by CURT TEICHERT

LAVON MCCORMICK, ROGER B. WILLIAMS, Assistant Editors

Advisers: EDWIN B. ECKEL, A. L. MCALESTER (The Geological Society of America), ADOLF SEILACHER, N. F. SOHL (The Paleontological Society), W. M. FURNISH, R. S. BOARDMAN (The Society of Economic Paleontologists and Mineralogists), R. V. MELVILLE, M. K. HOWARTH (The Palaeontographical Society), W. D. IAN ROLFE, ALWYN WILLIAMS (The Palaeontological Association).

PARTS

Parts of the *Treatise* are distinguished by assigned letters with a view to indicating their systematic sequence while allowing publication of units in whatever order each may be made ready for the press. The volumes are cloth-bound with title in gold on the cover. Copies are available on orders sent to the Publication Sales Department, The Geological Society of America, P.O. Box 1719, Boulder, Colorado 80302. The prices quoted very incompletely cover costs of producing and distributing the several volumes, but on receipt of payment the Society will ship copies without additional charge to any address in the world. Special discounts are available to members of sponsoring societies under arrangements made by appropriate officers of these societies, to whom inquiries should be addressed.

VOLUMES ALREADY PUBLISHED
(Previous to 1975)

Part C. PROTISTA 2 (Sarcodina, chiefly "Thecamoebians" and Foraminiferida), xxxi+900 p., 5311 fig., 1964.

Part D. PROTISTA 3 (chiefly Radiolaria, Tintinnina), xii+195 p., 1050 fig., 1954.

Part E. ARCHAEOCYATHA, PORIFERA, xviii+122 p., 728 fig., 1955.

Part E, Volume 1. ARCHAEOCYATHA, second edition (revised and enlarged), xxx+158 p., 871 fig., 1972.

Part F. COELENTERATA, xvii+498 p., 2700 fig., 1956.

Part G. BRYOZOA, xii+253 p., 2000 fig., 1953.

Part H. BRACHIOPODA, xxxii +927 p., 5198 fig., 1965.

Part I. MOLLUSCA 1 (Mollusca General Features, Scaphopoda, Amphineura, Monoplacophora, Gastropoda General Features, Archaeogastropoda, mainly Paleozoic Caenogastropoda and Opisthobranchia), xxiii+351 p., 1732 fig., 1960.

Part K. MOLLUSCA 3 (Cephalopoda General Features, Endoceratoidea, Actinoceratoidea, Nautiloidea, Bactritoidea), xxviii+519 p., 2382 fig., 1964.

Part L. MOLLUSCA 4 (Ammonoidea), xxii+490 p., 3800 fig., 1957.

Part N. MOLLUSCA 6 (Bivalvia), Volumes 1 and 2 (of 3), xxxviii+952 p., 6198 fig., 1969; Volume 3, iv+272 p., 742 fig., 1971.

Part O. ARTHROPODA 1 (Arthropoda General Features, Protarthropoda, Euarthropoda General Features, Trilobitomorpha), xix+560 p., 2880 fig., 1959.

Part P. ARTHROPODA 2 (Chelicerata, Pycnogonida, Palaeoisopus), xvii+181 p., 565 fig., 1955.

Part Q. ARTHROPODA 3 (Crustacea, Ostracoda), xxiii+442 p., 3476 fig., 1961.

Part R. ARTHROPODA 4 (Crustacea exclusive of Ostracoda, Myriapoda, Hexapoda), Volumes 1 and 2 (of 3), xxxvi+651 p., 1762 fig., 1969.

Part S. ECHINODERMATA 1 (Echinodermata General Features, Homalozoa, Crinozoa, exclusive of Crinoidea), xxx+650 p., 2868 fig., 1967 [1968].

Part U. ECHINODERMATA 3 (Asterozoans, Echinozoans), xxx+695 p., 3485 fig., 1966.

Part V. GRAPTOLITHINA, xvii+101 p., 358 fig., 1955.

Part V. GRAPTOLITHINA, second edition (revised and enlarged), xxxii+163 p., 507 fig., 1970.

Part W. Miscellanea (Conodonts, Conoidal Shells of Uncertain Affinities, Worms, Trace Fossils, Problematica), xxv+259 p., 1058 fig., 1962.

THIS VOLUME

Part W. Miscellanea (Supplement 1). Trace Fossils, second edition (revised and enlarged), xxi+269 p., 912 fig., 1975.

VOLUMES IN PREPARATION (1975)

EDITORIAL PREFACE

INTRODUCTION

Individual volumes of the *Treatise on Invertebrate Paleontology,* like other similar compilatory works, begin to be out-of-date while they are still in the press. That does not mean that they should not be published, because they incorporate the sum of knowledge at a point in time. Therefore, the idea was developed at an early stage to enhance the value of the *Treatise* as a permanent reference work by publication of revised editions and two such revised editions have already been published. They are of Part V (Graptolithina) by O. M. B. Bulman, published in 1970, and of Part E, volume 1 (Archaeocyatha), by Dorothy Hill, published in 1972.

Research on fossil groups summarized in the various published volumes of the *Treatise* progresses at an uneven pace. Therefore, some sections become outdated sooner than others, and some will retain their value as up-to-date sources for many years. The *Treatise* volume with the most heterogeneous content is probably Part W (Miscellanea), published in 1962, which deals with conodonts, tentaculitids, hyolithids, "worms," trace fossils, and problematica.

Among these groups, increase in knowledge and understanding since publication of the volume in 1962 has been most rapid in the field of conodonts and trace fossils, and already in 1966, need was felt for publication of a small paper containing supplemental information accumulated since 1962.[1] Almost half of this paper was taken up by a contribution by Walter Häntzschel, entitled *Recent contributions to knowledge of trace fossils and problematica.*

Two and one-half years later, in December, 1968, Professor Häntzschel suggested to me the possibility of publication of a second supplementary paper on trace fossils only, because "new ichnogenera were being established all the time." When this suggestion was discussed in an exchange of letters during 1969, it became apparent that new knowledge was being amassed at such a rapid rate that a "supplement" would soon reach the size of the entire original chapter. Professor Häntzschel then offered to completely revise the entire

[1] Rhodes, F. H. T., Häntzschel, Walter, Müller, K. J., Fisher, D. W., & Teichert, Curt, 1966, Treatise on Invertebrate Paleontology, Part W, conodonts, conoidal shells, worms, trace fossils: comments and additions: Univ. Kansas Paleont. Contrib. Paper 9, 17 p., 20 text-fig.

contribution on *Trace fossils and problematica* and it was decided to publish this manuscript as a Supplement to Part W. It was judged that the time was ripe for publication of a comprehensive, thoroughly updated taxonomic text for trace fossils, especially in view of the fairly recent upsurge of interest in this group, particularly in Great Britain and North America, but also in such countries as the Soviet Union, Poland, and France. In Germany, of course, a great volume of work on this group of fossils had been produced at a steady rate since the end of World War II, as is described in the text below.

After his retirement late in 1969, HÄNTZSCHEL spent much of his time on this new task and increased the pace during 1971. He continued to work vigorously up to the beginning of March, 1972, when he suddenly succumbed to an illness from which he did not recover. He died on May 10, 1972.

Fortunately for the work, almost all major problems of policy and general format had been discussed by us before WALTER HÄNTZSCHEL's death and the author's wishes have been persistently respected during the long months of laborious editorial work that followed. Already during the winter 1971-72, HÄNTZSCHEL had submitted long lists of required illustrations, and photography and other art work was proceeding in the editorial office in Lawrence. Also, HÄNTZSCHEL had completed the chapter Introduction, partly in German, and this had been translated. In June, 1972, I spent some days in Hamburg to obtain first-hand knowledge of the degree of completion of the Häntzschel manuscript. Up to that time I could not even be sure that there was enough manuscript or background material available to make it possible for us to complete the job. What I found in Hamburg was a very comprehensive card file, consisting of a separate card for each genus, with descriptions and discussions written for *Treatise*-style publication. On these cards, references to illustrations were only sketchily indicated. Completed, also,

were the introductory parts to individual sections of the manuscript and these were mostly in German. I soon learned that Mrs. MARIANNE HÄNTZSCHEL, Professor HÄNTZSCHEL's widow, was very knowledgeable in her husband's affairs, having assisted him in his work, and she, with great fortitude, guided me in the sifting of these materials.

We were immediately faced with the task of having to prepare captions for nearly 1,000 individual text-figures and integrating these with the descriptive text. Mrs. HÄNTZSCHEL undertook to provide drafts for the captions, working mainly from books and reprints available in her husband's library. Meanwhile, facsimile copies of the entire card file were made and airmailed to the editorial office in Lawrence, where the job of meshing text and figure captions was begun late in 1972. At the same time, photography continued and assembly of illustrations commenced. The bulk of this work was carried out, under my general supervision, by assistant editors LAVON McCORMICK (text) and ROGER B. WILLIAMS (illustrations), ably assisted by special research assistant WILLIAM G. HAKES who was responsible for a great portion of the library research that had to be done and who also made many contributions to the text. All photographic work was done by MICHAEL FREDERICK.

Since Professor HÄNTZSCHEL died before he had completed, or even begun, composition of the manuscript, it was only to be expected that many loose ends would have to be tied up, and such was, in fact, the case. Our greatest headache was references, mainly because the German format of citations differs greatly from the American one, as used in the *Treatise* and other earth science publications in this country. Most Russian references existed only in German, French, or English translation, whereas the *Treatise* requires transliteration of the Russian titles. Innumerable dates, page references, spellings, and occurrence data had to be obtained or verified, some necessitating interlibrary loans, but finally, by the summer of 1973, the first draft of a more or less complete manuscript could be put together. After another five or six months of checks and rechecks, corrections, changes, and additions, the first portions of the manuscript were finally sent to the press in November, 1973. Completed illustrations were sent to the engraver in December.

Anticipating a long gestation period from the arrival of the cardfile copy from Hamburg in August, 1972, until production of a press-ready manuscript, we continued to incorporate into the manuscript file new information, especially taxonomic, as it came in. Many colleagues assisted us in this task. Their names are mentioned below, but I wish to single out ROBERT W. FREY, University of Georgia, for special acknowledgment. A cutoff date for addition of new information was finally set at about April, 1973, but exceptionally important data, especially new taxa, were incorporated in the text up to the time the manuscript went to press, and some even in galley proof. However, no additional illustrations could be added at that late stage.

A matter of special concern was that of availability of names of trace fossil taxa established after 1930. It was discovered that, apparently due to some oversight of the deliberating body, strict application of the rules of the *International Code of Zoological Nomenclature* adopted by the 15th International Zoological Congress and published in 1961, led to the conclusion that names for trace fossil taxa established before 1931 are available, but those published after 1930 are not. Details are given by WALTER HÄNTZSCHEL below. It was immediately obvious that adherence to such a dichotomy would lead to utter chaos in any monographic treatment of trace fossils. Names of trace fossil taxa published before 1931 would have to be printed in italics and all provisions of the *Code* would have to be applied to them, whereas names published after 1930 would have no standing of any kind and would have to be treated as vernaculars to which the laws of priority and synonymy did not apply.

Professor HÄNTZSCHEL and I, at an early stage, refused to be faced with such a chaotic situation and agreed on an arbitrary decision to deal with names of trace fossil taxa in such a manner as if all of them—not only those published before 1931—were coming under the provisions of the *International Code of Zoological Nomenclature*. In consequence, in this volume the criteria of availability, the laws of homonymy and synonymy, and all other provisions of the *Code* are applied equally to taxa established before 1931 and after 1930. Suggestions to this effect are the essence of a recommendation to the ICZN made by HÄNTZSCHEL & KRAUS (1972). We have deviated from this recommendation only in setting trace fossil names in italics, partly because Professor HÄNTZSCHEL preferred this style (written communication, March 1, 1971), and partly in order to preserve compatibility with other *Treatise* volumes. In fact, in this volume all taxonomic names are italicized, except those of pseudofossils.

The editorial work on this volume proceeded as part of the larger *Treatise* project, supported by National Science Foundation Grant GB-31331X with payments of $66,600 in 1972, $58,900 in 1973, and $67,100 in 1974. The Geological Society of America supported the editorial office with grants of $6,000 in 1972, and $7,000 each in 1973 and 1974. Additional support was received through payment of salaries by the University of Kansas Endowment Association ($3,230 in 1973, $1,710 in 1974), and the Wallace Everette Pratt Research Fund in the University of Kansas Department of Geology ($5,130 in 1972-73).

ACKNOWLEDGMENTS

The editorial staff is indebted to many persons for help in a variety of ways. The important role of Mrs. HÄNTZSCHEL, especially in the initial stages of the editing process, has already been mentioned. She was most helpful at all times in supplying data from her husband's files and loaning rare books and reprints from her husband's library.

It is not possible to indicate in detail for each person the nature of his contribution. Assistance came to the author and later to the editors in the form of gifts of reprints, advance manuscripts, photographs and other illustrative materials; as loans of rare books and reprints; as help in tracing hard-to-find references; in the form of discussions of intricate nomenclatural problems, mostly in correspondence. In a few cases, descriptions of individual genera were supplied and these are acknowledged in appropriate places in the text. Because of Prof. HÄNTZSCHEL's death, some contributors may have been overlooked, for which we apologize. Our thanks for their cooperation are extended to the following colleagues: H. ALBERTI (Universität, Göttingen), H.-J. ANDERSON (Universität Marburg a.L.), KLAUS BANDEL (Universität, Bonn), K. W. BARTHEL (Technische Universität, Berlin), M. BEAUVAIS (Université de Paris), JAN BERGSTRÖM (Universitet Lund), LASZLÓ BOGSCH (Univ. Budapest), R. G. BROMLEY (Universitet, København), P. BRÖNNIMAN (Université de Genève), GEORGE CALLISON (California State College, Long Beach, Calif.), BARRY CAMERON (Boston University), RAYMOND CASEY (Institute of Geological Sciences, London), C. K. CHAMBERLAIN (Ohio University, Athens, Ohio), T. P. CRIMES (University of Liverpool), L. DANGEARD (Institut Océanographique, Paris), ALFRED EISENACK (Reutlingen), D. W. ELSTON (U. S. Geological Survey, Flagstaff, Ariz.), R. W. FREY (University of Georgia), T. W. GEVERS (University of the Witwatersrand), O. GIROTTI (Università di Roma), M. F. GLAESSNER (University of Adelaide, South Australia), J. GLAZEK (Instytut Geologiczny, Warszawa), W. C. GUSSOW (Japanese Petroleum Development Corporation, Tokyo), ANTHONY HALLAM (University of Oxford), R. F. HECKER (Akademiya Nauk, Moscow), GERO HILLMER and M. G. SCHULZ (Universität, Hamburg), HEINRICH HILTERMANN and H.-H. SCHMITZ (Bundesanstalt für Bodenforschung, Hannover), H. J. HOFMANN (Université de Montréal), W. KARASZEWSKI (Instytut Geologiczny, Warszawa), E. KEMPER (Bundesanstalt für Bodenforschung, Hannover), W. J. KENNEDY (University of Oxford), ROBERT KNOX (Sedgwick Museum, Cambridge, England), M. KSIĄŻKIEWICZ (Jagellonian University, Cracow, Poland), ULRICH LEHMANN (Universität, Hamburg), OTTO LINCK (Basel), I. R. McLACHLAN (University of the Witwatersrand), OLIVER MACSOTAY (Instituto Oceanografico, Cumaná, Venezuela), ANDERS MARTINSSON (Universitet Uppsala), A. H. MÜLLER (Bergakademie, Freiberg), H. PFEIFFER (Zentrales Geologisches Institut, Berlin), MIROSLAV PLIČKA (Ústředni Ústav Geologicka, Brno, Czechoslovakia), A. RADWAŃSKI (Instytut Geologiczny, Warszawa), E. S. RICHARDSON, Jr. (Field Museum, Chicago), the late O. H. SCHINDEWOLF (Universität, Tübingen), WILHELM SCHLOZ (Bundesanstalt für Gewässerkunde, Koblenz), ADOLF SEILACHER (Universität, Tübingen), FRANK SIMPSON (Subsurface Geological Laboratory, Regina, Saskatchewan), EHRHARD VOIGT (Universität, Hamburg), O. S. VYALOV (Akademiya 'Nauk Ukrainskoi SSR, Lvov), B. D. WEBBY (University of Sydney), and E. L. YOCHELSON (U. S. Geological Survey, Washington, D. C.).

ABBREVIATIONS

Abbreviations used in this division of the *Treatise* are explained in the following alphabetically arranged list.

Abhandl., *Abhandlung(en)*
Abt., *Abteilung*
aff., *affinis* (related to)
Afr., Africa, -an
Ala., Alabama
Alb., Albian
Alg., Algeria
Alta., Alberta
Am., America, -n
Amtl., *Amtlicher*
Anatol., Anatolia
Anis., Anisian
Ann., *Annñaes, Annali,*
    Annal(es), *Annuaire,* Annual
ant., anterior
Antarct., Antarctic
append., appendix
approx., approximately
Apt., Aptian
Arbeit., *Arbeit(en)*
Arch., Archives, *Archivos*
Arenig., Arenigian
Ariz., Arizona
Ark., Arkansas
Arg., Argentina
Årsskr., *Årsskrift*
art., Art., article
AsiaM., Asia Minor
Atl., Atlantic
auctt., *auctorum* (of authors)
Aus., Austria

Bajoc., Bajocian
Barrem., Barremian
B.C., British Columbia
Bd., *Band*
Belg., *Belgique,* Belgium
Ber., *Bericht*
Berrias., Berriasian
Biol., Biological, *Biologicheskaya*
Birrim., Birrimian
Biulet., *Biuletyn*
Bol., *Boletim, Boletín,* Bolivia
Boll., *Bolletino*
Briovér., Briovérian
Brit., Britain, British
Bull., Bulletin

C., Centigrade, Central
ca., *circa*
Calif., California
Callov., Callovian
Cam., Cambrian
Campan., Campanian
Can., Canada
Carb., Carboniferous

Carn., Carnian
Carpath., Carpathians
cat., catalog, catalogue
Cenoman., Cenomanian
Cenoz., Cenozoic
cf., *confer* (compare)
Cincinnat., Cincinnatian
cm., centimeter(s)
Co., County
Coll., Collection(s)
Colo., Colorado
Colom., Colombia
Comanch., Comanchean
commun., communication
Commun., *Communicações*
Comun., *Comunicaciones*
Conn., Connecticut
Contrib., Contribution(s)
cosmop., cosmopolitan
Cr., Creek
Cret., Cretaceous
Czech., Czechoslovakia

Dan., Danian
Denkschr., *Denkschrift(en)*
Denm., Denmark
Dept., Department
Dev., Devonian
diag., diagr., diagrammatical
diam., diameter
diss., dissertation
Distr., District
Dol., Dolomite

E., East
Ecuad., Ecuador
ed., editor
edit., edition
e.g., *exempli gratia*
    (for example)
Eifel., Eifelian
emend., *emendatus(-a)*
Ems., Emsian
Eng., England
enl., enlarged
Eoc., Eocene
Esmark., Esmarkian
est., estimated
Est., Estonia
et al., *et alii*
    (and others, persons)
etc., *et cetera*
    (and others, objects)
Eu., Europe
exfol., exfoliated
expl., explan., explanation

F., Formation
fam., family
Festb., *Festband*
fig., figure(s)
Finl., Finland
Förhandl., *Förhandling(ar)*
Forhandl., *Forhandling(er)*

G.Brit., Great Britain
Ga., Georgia
gen., genus
Geogr., Geographical
Geol., Geological,
    Geologicheskikh, Geologische
Ger., German, Germany
Gotl., Gotland
Gr., Great, Group
Greenl., Greenland
Guatem., Guatemala

Handl., *Handling(ar)*
Hauteriv., Hauterivian
Helvet., Helvetian
Herts., Hertfordshire
Hettang., Hettangian
hom., homonym
horiz., horizontal
Hung., Hungarica, Hungary

ICZN, International Commission
    of Zoological Nomenclature
I.G.P., *Institut für Geologie*
    und Paläontologie
i.e., *id est* (that is)
ichnogen., ichnogenus
ichnosp., ichnospecies
Ill., Illinois
illus., illustrated, -ions
Inaug., Inaugural
inc., incl., including
inc. sed., *incertae sedis*
ind., *indeterminata*
Ind.O., Indian Ocean
Ind., Indiana
Industr., Industrial, Industry
Inst., *Institut,* Institute
Internatl., International
Ire., Ireland
irreg., irregular
Is., Island(s)

Jahrb., *Jahrbuch*
Jahresber., *Jahresbericht*
Jahrg., *Jahrgang*
Jour., Journal
jun., jr., junior
Jur., Jurassic

Kans., Kansas
Kimmeridg., Kimmeridgian
km., kilometer
Ky., Kentucky

L., Lower
Ladin., Ladinian
lat., lateral
Lias., Liassic
loc., locality, location
long., longitudinal
low., lower
Ls., Limestone
Ludlov., Ludlovian

m., meter(s)
M., Middle
M, Monotypy
m.y., million years
Maastricht., Maastrichtian
mag., magnification
Mass., Massachusetts
max., maximum
Md., Maryland
Medd., *Meddelanden, Meddelelser*
Meded., *Mededeelingen*
Medit., Mediterranean
Mem., Memoir, *Memoria, -e*
Mém., *Mémoire(s)*
Merxem., Merxemian
Mesoz., Mesozoic
mid., middle
Mio., Miocene
Misc., Miscellaneous
Miss., Mississippi, Mississippian
Mitteil., *Mitteilungen*
Mittheil., *Mittheilungen*
mm., millimeter(s)
Mo., Missouri
mod., modified
Mon., Monograph, *Monographia, Monographie*
Mon., Monument
Monatsber., *Monatsberichte*
Monatsh., *Monatshefte*
Mont., Montana
MS., manuscript
Mt., Mount, Mountain
Mts., Mtns., Mountains
Mus., Museum

n, new
N., New, North
Nachricht., *Nachrichten*
N.Afr., North Africa
N.Am., North America(n)
Namur., Namurian
Nat., Natural
Natl., National

Naturhist., *Naturhistorische(s)*
N.B., New Brunswick
N.Car., North Carolina
N.Dak., North Dakota
NE., Northeast
Neocom., Neocomian
Neth., Netherlands
Nev., Nevada
Newf., Newfoundland
n.f., *nova forma*
N.J., New Jersey
N.Mex., New Mexico
no., number
nom. correct., *nomen correctum* (corrected or intentionally altered name)
nom. inval., *nomen invalidum*
nom. nov., *nomen novum* (new name)
nom. nud., *nomen nudum* (naked name)
nom. null., *nomen nullum* (null, void name)
nom. oblit., *nomen oblitum* (forgotten name)
nom. subst., *nomen substitutum* (substitute name)
nom. transl., *nomen translatum* (transferred name)
nom. van., *nomen vanum* (vain, void name)
Nomencl., Nomenclature
Nor., Norian, Norway
Notizbl., *Notizblatt*
Nouv., *Nouveaux, Nouvelle*
nov., *novum*
N.S., Nova Scotia
NW., Northwest
N.Y., New York
N.Z., New Zealand

O., Ocean
obj., objective
Occas., Occasional
OD, original designation
Okla., Oklahoma
Oligo., Oligocene
Ont., Ontario
Op., Opinion
Ord., Ordovician
Ore., Oregon
Oxford., Oxfordian

p., page(s)
Pa., Pennsylvania
Pac., Pacific
Pak., Pakistan
Paleoc., Paleocene
Paleont., Paleontological, *Paleontologicheskiy*

Penin., Peninsula
Penn., Pennsylvanian
Perm., Permian
pers., personal
Philos., Philosophical
pl., plate(s), plural
Plat., Platform
Pliensbach., Pliensbachian
Pleist., Pleistocene
Plio., Pliocene
Pol., Poland
Port., Portugal
Precam., Precambrian
prob., probably
Proc., Proceeding(s)
Prof., Professional
Proteroz., Proterozoic
Prov., Providence
pt., part(s)
publ., publication, published
Publ., *Publicacion,* Publication

Quart., Quarterly
Que., Quebec

Rec., Recent, Record(s)
reconstr., reconstructed, -ion
Repos., Repository
Rept., Report
Res., Research, Resources
Rev., Review, *Revista, Revue*
Revin., Revinian
Rhaet., Rhaetian
Rotl., Rotlandian
Rupel., Rupelian
Russ., Russian

S., Sea, South
S.Am., South America
Santon., Santonian
Sax., Saxony
schem., schematic
Schilfsandst., Schilfsandstein
Sci., Science, Scientific
Scot., Scotland
SD, subsequent designation
S.Dak., South Dakota
SE., Southeast
sec., section(s)
Senon., Senonian
ser., serial, series, etc.
sér., *séries*
Sh., Shale
Sib., Siberia
Siegen., Siegenian
Sil., Silurian
Sinemur., Sinemurian
Sitzungsber., *Sitzungsberichte*
Skrift., *Skrift(er)*

REFERENCES TO LITERATURE

Earlier volumes of the *Treatise* were accompanied by selected lists of references to paleontological literature consisting primarily of recent and comprehensive monographs, but also including some older works recognized as outstanding in importance. Publications listed in the *Treatise* were then not regarded as original sources of information concerning taxonomic units, but rather as guides to tell the reader where he may find them.

A departure from this policy occurred with publication of Part C of the *Treatise* in 1964. In these volumes, for the first time, all citations of authors and years in the text were fully documented in the list of references which were well in excess of 2,000. In *Treatise* parts published since 1964 the tendency has been toward fuller bibliographic documentation which is especially evident in Part H, published in 1965, and Part N, published in 1969 and 1971.

Following the wishes of the author, the list of references in the present volume is very comprehensive, comprising 1,720 titles. The editors have endeavored to check the accuracies of all entries, but this proved not to be possible in some cases. Such titles

are indicated by the addition of: [not seen by the editors]. Aiming at something as close to completeness as possible, the author has included in the list a number of references to which no reference is made in the text.

The following is a statement of the full names of serial publications which are cited in abbreviated form in the lists of references in the present volume. The information thus provided should be useful in library research work. The list is alphabetized according to the serial titles which were employed at the time of original publication. Those following in brackets are those under which the publication may be found currently in the *Union List of Serials,* the United States Library of Congress listing, and most library card catalogues. In some instances the current title is followed by the original one in parentheses. The names of serials published in Cyrillic are transliterated; in the reference lists these titles, which may be abbreviated, are accompanied by transliterated authors' names and titles, with English translation of the title. The place of publication is added (if not included in the serial title).

SCOTT W. STARRATT

The method of transliterating Cyrillic letters that is adopted as "official" in the *Treatise* is that suggested by the Geographical Society of London and the U. S. Board on Geographic Names. It follows that names of some Russian authors in transliterated form derived in this way differ from other forms, possibly including one used by the author himself. In *Treatise* reference lists the alternative (unaccepted) form is given enclosed by square brackets (e.g., Chernyshev [Tschernyschew], T.N.).

List of Serial Publications

Académie Polonaise des Sciences, Série des Sciences Techniques, Bulletin. Warszawa.

Académie Royale des Sciences Coloniales, Classe des Sciences naturelles et médicales, Bulletin seances, Mémoires. Bruxelles.

Académie des Sciences [Paris], Comptes Rendus, Mémoires.

Académie Tchèque des Sciences, Bulletin International, Classe des Sciences Mathématiques, Naturelles et de la Médecine. Prague.

Academy of Natural Sciences of Philadelphia, Journal; Proceedings.

Academy of Science of St. Louis, Bulletin; Memoirs; Transactions.

Accademia Gioenia delle Scienze Naturali di Catania, Atti; Bollettino.

[R.] Accademia dei Lincei, Classe di Scienze Fisiche, Matematiche e Naturali, Atti; Memorie; Rendiconti. Roma.

Accademia Pontificia dei Nuovi Lincei, Atti; Memorie. Roma.

[R.] Accademia delle Scienze dell'Istituto di Bologna, Memorie.

[R.] Accademia delle Scienze di Torino, Atti; Memorie.

Acta Geologica Polonica. Warszawa.

Acta Geologica Taiwanica. Series 1 of National Taiwan University, Science Reports. Taipei.

Acta Palaeontologica Polonica [Polska Akademia Nauk, Komitet Geologiczny]. Warszawa.

Acta Universitatis Lundensis (Lund Universitet, Årsskrift).

Akademie der Wissenschaften, physikalisch-mathematische Klasse, Abhandlungen; Monatsberichte. Berlin.

Akademie der Wissenschaften in Göttingen, mathematisch-physikalische Klasse, Nachrichten.

Akademie der Wissenschaften und der Literatur zu Mainz, mathematisch-naturwissenschaftliche Klasse, Abhandlungen. Wiesbaden.

[K.] Akademie der Wissenschaften zu Wien, mathematisch-naturwissenschaftliche Klasse, Denkschriften; Sitzungsberichte.

Akademija Umiejętnosci, Krakow. Komisja fizyograficzna Sprawozdania.

Akademiya Nauk Azerbaydzhan SSR, Doklady. Moskva.

Akademiya Nauk SSSR, Doklady; Izvestiya; Trudy. Moskva.

Akademiya Nauk SSSR, Geologicheskii Institut, Trudy. Moskva.

Akademiya Nauk SSSR, Izvestiya, Seriya Biologicheskaya; Seriya Geologicheskaya, Byulletin. Moskva.

Akademiya Nauk SSSR, Sibirskoe Otdelenie, Institut Geologii i Geofiziki, Trudy. Novosibirsk.

Akademiya Nauk Ukrainskoi SSR, Institut Geologicheskikh Nauk, Trudy. Kiev.

Albany Institute, Proceedings; Transactions.

Algérie, Publications du Service de la Carte Géologique, Bulletin; Mémoires. Alger.

Allgemeine Deutsche Naturhistorische Zeitung. Dresden.

Allgemeine Schweizerische Gesellschaft für die gesamten Naturwissenschaften, Neue Denkschriften. Zürich.

American Academy of Arts and Sciences, Memoirs; Proceedings. Boston.

American Association for the Advancement of Science, Proceedings; Publications. Washington, D. C.

American Association of Petroleum Geologists, Bulletin. Tulsa, Okla.

American Geographical Society, Special Publications, Folio. New York.

American Geologist. Minneapolis, Minn.

American Journal of Science. New Haven, Conn.

American Midland Naturalist. Notre Dame, Ind.

American Naturalist. Lancaster, Pa.

American Philosophical Society, Memoirs; Proceedings; Transactions.

American Zoologist. Utica, N. Y.

Annales des Sciences Naturelles, Zoologie. Paris.

Annals and Magazine of Natural History. London.

Archiv für Fischerei-Wissenschaft. Hamburg.

Archiv für Molluskenkunde. Frankfurt.

Archives de Musée Teyler. Haarlem.

Archives des Sciences. Genève.

Arkiv för Zoologi. Uppsala.

Arquivos do Museu Paranaense. Curitiba, Brazil.

Asociación Mexicana Geólogos Petroleros, Boletín. Mexico D. F.

Association Française pour l'Avancement des Sciences, Compte Rendu. Reims, Paris.

Der Aufschluss. Zeitschrift für die Freunde der Mineralogie und Geologie. Vereinigung der Freunde der Mineralogie und Geologie. Göttingen.

Aus der Heimat. Öhringen.

Australia Bureau of Mineral Resources, Geology and Geophysics, Bulletin: Explanatory Note; Report. Canberra.

Australian Journal of Science. Sydney.

Australian Museum, Memoirs; Records. Sydney.

Azerbaydzhanskoy Neftianoi Khoziaistvo. Baku.

[K.] Bayerische Akademie der Wissenschaften, mathematisch-physikalische Klasse, Abhandlungen, Sitzungsberichte. München.

Bayerische Staatssammlung für Paläontologie und Historische, Geologie, Mitteilungen. München.

Beiträge zur Geologie und Paläontologie. Jena.

Beiträge zur Geologie von Thüringen. Jena.

Beiträge zur Naturkundlichen Forschung in Südwestdeutschland. Karlsruhe.

Biological Bulletin. Woods Hole, Mass.

Biologisches Zentralblatt. Erlangen, Leipzig.

[K.] Böhmische Gesellschaft der Wissenschaften, mathematisch-naturwissenschaftliche Klasse, Sitzungsberichte. Prag.

Boston Society of Natural History, Memoirs; Occasional Papers; Proceedings.

Botanische Zeitung. Berlin, Leipzig.

Brigham Young University, Geology Studies. Provo, Utah.

British Antarctic Survey, Bulletin; Scientific Reports (Faulkland Islands, Dependencies Scientific Bureau). London.

British Association for the Advancement of Science, Reports. London.

British Museum (Natural History), Geology, Bulletin. London.

British Museum, Natural History Magazine. London.

Bulletin of American Paleontology. Ithaca, N. Y.

Bulletin of Zoological Nomenclature. London.

Bureau de Recherches Géologiques et Minières, Bulletin. Paris.

Canada, Geological Survey of, Department of Mines and Resources, Mines and Geology Branch, Bulletin; Memoir; Museum Bulletin; Victoria Memorial Museum Bulletin. Ottawa.

Canadian Journal of Earth Sciences. National Research Council, Canada. Ottawa.

Canadian Journal of Science, Literature, and History. Toronto.

Canadian Naturalist and Geologist. Montreal.

Carnegie Institution of Washington, Papers; Publications. Washington, D. C.

Časopis pro Mineralogii a Geologii. Praha.

Centralblatt für Mineralogie, Geologie, Paläontologie. Stuttgart.

China, Geological Survey of, Palaeontologia Sinica, Bulletin, Memoirs. Peking.

Cincinnati, Quarterly Journal of Science.

Cincinnati Society of Natural History, Journal.

Colorado School of Mines, Professional Contribution; Quarterly. Golden, Colo.

Comitato Geologico d'Italia, Bollettino. Roma.

Commission Géologique du Portugal, Travaux. Lisbon.

Connecticut Academy of Arts and Sciences, Memoirs, Transactions. New Haven, Conn.

Connecticut, State of, Geological and Natural History Survey, Bulletin. Hartford.

Cuerpo de Ingenieros de Minas del Perú. Boletín. Lima.

Current Science. Bangalore, India.

Dansk Geologisk Forening, Meddelelser (Geological Society of Denmark, Bulletin). København.

[K.] Danske Videnskabernes Selskabs, Matematisk-Fysiske Forhandlinger. Skrifter.

Deep-Sea Research. London.

Deutsch Akademie der Wissenschaften zu Berlin, Abhandlungen; Monatsberichte, Geologie und Mineralogie.

Deutsche Botanische Gesellschaft, Berichte. Berlin.

Deutsche Geologische Gesellschaft, Zeitschrift. Berlin, Hannover.

Deutsche Zoologische Gesellschaft Wilhelmshaven, Verhandlungen.

Dictionnaire Universel Histoire Naturelles. Paris.

Direçao dos Trabalhos Geologicos (do Serviço geologico) de Portugal, Comunicações. Lisboa.

Divisão de Geologia e Mineralogia do Brasil, Ministério da Agricultura, Departamento Nacional da Produção Mineral, Boletim. Rio de Janeiro.

Dublin Geological Society, Journal.

Eclogae Geologicae Helvetiae (see Schweizerische Geologische Gesellschaft). Basel.

Edinburgh New Philosophical Journal.

Endeavour. London.

Engenharia, Mineração e Metalurgia. Rio de Janeiro.

Erdöl-Zeitschrift. Wien, Hamburg.

Erlanger Geologische Abhandlungen. Erlangen, Ger.

Evolution. Lancaster, Pa.

Ezhegodnik po Geologii i Mineralogii Rossii (Annuaire Géologique et Minéralogique de la Russie). Novo-Alexandria.

Fieldiana, Geology. Chicago.

Finlande, Commission Géologique, Bulletin. Helsinki.

Firgenwald. Vierteljahrsschrift für Geologie und Erdkunde der Sudentenländer. Reichenberg.

Flora oder Allgemeine Botanische Zeitung. Jena, Regensburg.

Földtani Közlöny (Magyaroni Földtani Társulat Folyóirata). Budapest.

Fortschritte der Geologie und Paläontologie. Berlin.

Fortschritte in der Geologie von Rheinland und Westfalens. Krefeld.

Fossils and Strata. Oslo.

Frankfurter Beiträge zur Geschiebeforschung (Beiheft zur Zeitschrift Geschiebeforschung). Frankfurt.

Freiberger Forschungshefte. Berlin.

Geognostische Jahreshefte. München.

Geologia Romana. Roma.

Geologica Carpathica, Geologický Zborník. Bratislava.

Geological Journal. Liverpool.

Geological Magazine. London, Hertford.

Geological Society of America, Bulletin; Memoir; Special Paper. Boulder, Colo.

Geological Society of Australia, Journal. Adelaide.

Geological Society of Dublin, Journal.

Geological Society of Glasgow, Transactions.

Geological Society of Japan, Journal. Tokyo.

Geological Society of London, Memoir; Proceedings; Quarterly Journal; Transactions. [Now The Geological Society.]

Geologicke Práce, Zprávy. Bratislava.

Geologie (Zeitschrift für das Gesamtgebiet der Geologie und Mineralogie sowie der angewandten Geophysik). Berlin.

Geologie der Meere und Binnengewässer. Berlin.

Geologie en Mijnbouw. Den Haag.

Geologija i Razwedka. Moskva.

Geologische Blätter für Nordost-Bayern und angrenzende Gebiete. Erlangen.

[K. K.] Geologische Bundesanstalt Wien, Abhandlungen; Jahrbuch; Verhandlungen.

Geologische Gesellschaft in der Deutschen Demokratischen Republik für das Gesamtgebiet der Geologischen Wissenschaften, Berichte. Berlin.

Geologische Gesellschaft in Wien, Mitteilungen.

Geologisches Jahrbuch, Beihefte. Hannover.

Geologische Jahresberichte. Berlin.

[K. K.] Geologische Reichsanstalt Wien (see Geologische Bundesanstalt Wien).

Geologische Rundschau. Geologische Vereinigung, Stuttgart.

Geologisches Staatsinstitut in Hamburg, Mitteilungen.

Geologiska Föreningens i Stockholm, Förhandlingar.

Geologist. London.

Geologists' Association, Proceedings. London.

Geology. Geological Society of America, Boulder, Colo.

Geološki Vjesnik. Zagreb.

Georgia, Geological Survey of, Bulletin. Atlanta.

GEOS. Universidad Central de Venezuela, Escuela de Geologia, Minas y Metalurgia. Caracas, Venezuela.

Geotimes. American Geological Institute, Washington, D. C.

Germania. Frankfurt.

Gesellschaft zur Beförderung der Gesammten Naturwissenschaften zu Marburg, Sitzungsberichte.

Gesellschaft Deutscher Naturforscher und Aerzte, Amtlicher Bericht, Verhandlungen. Berlin, Leipzig.

Gesellschaft von Freunden der Naturwissenschaften in Gera, Jahresbericht.

Gesellschaft der Geologie- und Bergbaustudenten, Mitteilungen. Wien.

Gesellschaft Naturforschender Freunde Berlin, Magazin; Sitzungsberichte.

Glückauf, Berg- und Hüttenmännische Zeitschrift. Essen.

Great Britain, Geological Survey of, Palaeontology, Bulletin; Memoirs. London.

Greifswald, Geologisch-palaeontologisches Institut der Universität, Abhandlungen.

Grønlands Geologiske Undersøgelse, Rapport. København.

Gulf Coast Association of Geological Societies, Transactions. Houston, Texas.

Hallesches Jahrbuch für Mitteldeutsche Erdgeschichte. Leipzig.

Harvard University, Museum of Comparative Zoology, Breviora; Bulletin; Memoirs; Special Publications. Cambridge, Mass.

Hebrew University, Geology Department, Bulletin. Jerusalem.

Heidelberger Akademie der Wissenschaften, Abhandlungen. Heidelberg.

Hessenland. Zeitschrift für hessische Geschichte und Literatur. Kassel.

Hessisches Landesamt für Bodenforschung, Abhandlungen; Notizblatt.

Himmel und Erde. Leipzig, Berlin.

Ichnology Newsletter. Savannah, Ga.

India, Geological Survey of, Bulletin; Memoirs; Records.

Indian Botanical Society, Journal. Madras.

Indiana Department of Geology and Natural History, Annual Report. Bloomington, Ind.

Indiana, Geological Survey, Annual Report. Bloomington, Ind.

Institut Géologique de Roumanie, Comptes Rendus des Séances. Bucuresti.

Institution of Mining Engineers, Transactions. Newcastle-on-Tyne.

Instituto Ecuatoriano de Ciencias Naturales, Universidad Central, Boletín. Quito, Ecuador.

Instituto Geológico Barcelona, Publicatiónes.

Instituto Geológico y Minero de España, Boletín; Memorias; Notas y Comunicaciónes. Madrid.

Instytut Geologiczny, Biuletyn; Kwartalnik Geologiczny. Warszawa.

International Sedimentary Petrographical Series. Leiden.

International Union Geological Sciences, Commission on Stratigraphy, Second Gondwana Symposium, Proceedings and Papers. Pretoria, South Africa.

Iran, Geological Survey of, Report. Tehran.

Irish Naturalist. Dublin.

Jahrbuch für Geologie. Zentrales Geologisches Institut, Berlin.

Japan Academy, Proceedings. Tokyo.

Japan, Geological Survey of, Bulletin; Report. Kawasaki City.

Japanese Journal of Geology and Geography. Tokyo.

Jenaische Zeitschrift für Naturwissenschaft. Jena.

Johns Hopkins University, Oceanographic Studies. Baltimore, Md.

Journal of Geological Education. Columbus, Ohio.

Journal of Geology. Chicago.

Journal of Paleontology. Tulsa, Okla.

Journal of Protozoology. Washington, D. C.

Journal of Sedimentary Petrology. Tulsa, Okla.

Julius Klaus-stiftung für Vererbungsforschung,

Sozialanthropologie und Rassen-Hygiene, Archiv. Zürich.

Kansas Academy of Science, Transactions. Topeka, Kans.

Kansas City Scientist. Kansas City, Mo.

Kansas, The University of, Paleontological Contributions, Article; Paper. Lawrence, Kans.

Komisja Fizyograficzna oraz Materyaly do fizyografii Krajowej, Sprawozdania. Kraków.

Korrespondenzblatt des Naturforschervereins zu Riga.

[K.] České Společnost Nauk, Třída Matematicko-Přírodovědecká, Rozpravy; Věstník. Práce.

Leidse Geologische Mededeelingen. Leiden.

Leningrad Universitet, Vestnik.

Lethaia. Oslo.

Limnology and Oceanography. Woods Hole, Mass.

Liverpool and Manchester Geological Journal.

Louisiana State University, Miscellaneous Publications. Baton Rouge, La.

Lunds Universitet, Årsskrift.

Lvovskoe Geologicheskoe Obshchestvo pri Gosudarstvennyi Universitet Ivana Franko, Geologicheskiy Sbornik, Mineralogicheskiy Sbornik, Paleontologicheskiy Sbornik, Trudy. Lvov.

Lyceum of Natural History of New York, Annals.

Manchester Literary and Philosophical Society, Memoirs and Proceedings. Manchester, Eng.

Manchurian Science Museum, Bulletin. Mukden.

Marine Geology. International Journal of Marine Geology, Geochemistry, and Geophysics. Amsterdam.

Maroc, Service Géologique du Division des Mines et de la Géologie, Notes et Mémories. Rabat.

Maryland, Geological Survey of. Baltimore.

Meddelelser om Grønland. København.

Medizinisch-naturwissenschaftlichen Gesellschaft zu Jena, Denkschriften.

The Mercian Geologist (East Midlands Geological Society). Nottingham.

"Meteor" Forschungsergebnisse. Berlin.

Meyniana. Kiel Universitaet Geologisches Institut.

Michigan, University of, Museum of Paleontology, Contributions. Ann Arbor.

Micropaleontology. American Museum of Natural History. New York.

Mikroskopie. Zentralblatt für Microscopische Forschung und Methodik. Wien.

Mineralogisch-geologisches (Staats-) Institut Hamburg, Mitteilungen.

Minnesota Academy of Natural Sciences, Bulletin. Minneapolis.

Moskovskoe Obshchestvo Ispytatelei Prirody, Byulletin (formerly Société Impériale des Naturalistes de Moscou). Moskva.

Mountain Geologist. Rocky Mountain Association of Geologists. Denver, Colo.

Musée Royal d'Histoire Naturelle de Belgique, Annales; Bulletin; Mémoires (continued as Institut Royal des Sciences Naturelles de Belgique). Bruxelles.

Museo Civico di Storia Naturale di Trieste, Atti.

Museo Civico di Storia Naturale di Verona, Memorie.

Museo de Historia Natural de Mendoza, Revista.

Museo Libico Storia Naturale, Annali. Tripoli.

Muséum d'Histoire Naturelle, Annales; Nouvelles Archives. Paris.

Museum des Königlich Bayerischen Staates, Paläontologische Mittheilungen. Stuttgart.

Museum Senckenbergianum. Frankfurt. (See Senckenbergische Naturforschende Gesellschaft.)

Mycologia. Lancaster, Pa.

Nassauischer Verein für Naturkunde, Jahrbuch. Wiesbaden.

National Academy of Sciences, Memoirs; Proceedings. Washington, D. C.

Natur und Museum. Senckenbergische Naturforschende Gesellschaft. Frankfurt.

Natur und Volk. Senckenbergische Naturforschende Gesellschaft. Frankfurt. (Temporary name for Natur und Museum.)

Natura. Milano.

Natural History Society of New Brunswick, Bulletins. St. John.

The Naturalist. London.

Le Naturaliste, Annales. Paris.

Nature. London.

Nature. Paris.

Naturforschende Gesellschaft in Basel, Bericht; Verhandlungen.

Naturforschende Gesellschaft zu Freiburg im Breisgau, Berichte.

Naturforschende Gesellschaft Graubündens, Jahresbericht. Chur.

Naturforschende Gesellschaft zu Leipzig, Sitzungsberichte.

Naturforschende Gesellschaft in Zürich, Vierteljahrschrift.

Naturhistorische Gesellschaft Nürnberg, Abhandlungen.

Naturhistorischer Verein der Preussischen Rheinlande und Westfalens, Sitzungsberichte; Verhandlungen. Bonn.

[K. K.] Naturhistorisches Hofmuseum, Annalen. Wien.

Naturhistorisch-medizinischer Verein zu Heidelberg, Verhandlungen.

Die Naturwissenschaften. Berlin.

Naturwissenschaftlicher Verein zu Bremen, Abhandlungen.

Naturwissenschaftlicher Verein für Neu-Vorpommern und Rügen, Greifswald, Mitteilungen.

Naturwissenschaftlicher Verein zu Troppau, Mitteilungen.

Naturwissenschaftlicher Verein für Sachsen und Thüringen, Jahresbericht. Berlin.

Naturwissenschaftliche Wochenschrift. Jena.

Natuurhistorisch Maandblad. Maastricht.

Nebraska Geological Survey, Bulletin; University Studies. Lincoln.

[K.] Nederlandse Akademie van Wetenschappen,

Afdeeling Natuurkunde, Verhandelingen. Amsterdam.

Neues Jahrbuch für Geologie und Paläontologie (Before 1950, Neues Jahrbuch für Mineralogie, Geologie, und Paläontologie), Abhandlungen; Beilage-Bände; Monatshefte. Stuttgart.

Neues Jahrbuch für Mineralogie (Before 1950, Neues Jahrbuch für Mineralogie, Geologie, und Paläontologie), Abhandlungen; Beilage-Bände; Monatshefte. Stuttgart.

New Mexico State Bureau of Mines and Mineral Resources, Bulletin; Circular; Memoir. Socorro, New Mexico.

New South Wales, Geological Survey of, Ethnology, Memoirs; Geology, Memoirs; Paleontology, Memoirs; Records. Sydney.

New York Academy of Sciences, Annals.

New York State Agricultural Society, Proceedings; Transactions. Albany, N. Y.

New York State Geological Survey, Natural History of New York; Palaeontology of New York; Annual Report. Albany.

New York State Museum of Natural History, Annual Report; Bulletin. Albany.

New Zealand Journal of Geology and Geophysics. Wellington.

Norges Geologiske Undersökelse, Skrifter. Oslo, Norway.

Norsk Geologisk Tidsskrift. Norsk Geologisk Forening, Oslo.

Norsk Polarinstitutt Skrifter. Oslo.

Norsk Videnskaps-Akademi i Oslo, Skrifter. Oslo.

Nova Acta Leopoldina. Halle.

Nova Scotian Institute of Science, Proceedings and Transactions. Halifax.

Oberrheinische Geologische Abhandlungen. Karlsruhe.

Oberrheinischer Geologischer Verein, Jahresbericht und Mitteilungen. Stuttgart.

Offenbacher Verein für Naturkunde, Bericht über die Tätigkeit. Offenbach.

Ohio Journal of Science. Columbus.

Oklahoma Geology Notes. Norman.

Ontario Department of Mines, Annual Report. Toronto.

The Ore Bin. Oregon Department of Geology and Mineral Industries. Portland.

Österreichische Akademie der Wissenschaften, mathematisch-naturwissenschaftliche Klasse, Denkschriften; Sitzungsberichte. Wien.

Pacific Science. University of Hawaii Press. Honolulu.

Pakistan, Geological Survey of, Palaeontologia Pakistanica, Records. Karachi.

Palaeobiologica. Wien.

Palaeogeography, Palaeoclimatology, Palaeoecology. Amsterdam.

Palaeontographia Italica. Pisa.

Palaeontographica. Stuttgart, Kassel.

Palaeontographica Americana. Ithaca, N. Y.

Palaeontographica Bohemiae. Praha.

Palaeontographical Society, Monograph. London.

Palaeontologia Africana. Johannesburg.

Palaeontologia Sinica, Geological Survey of China, Peking.

Palaeontological Society of India, Journal. Lucknow.

Palaeontological Society of Japan, Transactions and Proceedings. Tokyo.

Paläontologische Zeitschrift. Berlin, Stuttgart.

Palaeontology. Palaeontological Association, London.

Paleontologicheskiy Sbornik, Vsesoyuznyy Nauchno Issledovatel'skiy Geologo-Razvedochnyi Neftianoi Institut. Moskva.

The Paleontologist. Cincinnati, O.

Pan-American Geologist. Des Moines, Ia.

Państwowy Instytut Geologiczny, Biuletyn. Warszawa.

Pennsylvania Academy of Science, Proceedings. Harrisburg.

Pennsylvania, Geological Survey of, Annual Report; Report of Progress. Harrisburg.

Plateau. Northern Arizona Society of Science and Art, Flagstaff.

Polskiego Towarzystwa Geologicznego w Krakówie, Rocznik. Kraków.

Praktika. Athens.

[K.] Preussische Geologische Landesanstalt, Abhandlungen; Jahrbuch. Berlin.

[K.] Preussische Geologische Landesanstalt und Bergakademie, Abhandlungen; Jahrbuch. Berlin.

Priroda. Akademiya Nauk SSSR, Moskva.

Provincia de Buenos Aires, Gobernación. Comisión Investigaciónes Cientifica, Notas. La Plata, Argentina.

Provinzialstelle für Naturdenkmalpflege Hannover, Mitteilungen. Hildescheim, Ger.

Przegląd Geologiczny, Wydawnictwa Geologczne. Warszawa.

Queensland Museum, Memoirs. Brisbane.

Reichsstelle (Reichsamt) für Bodenforschung, Bericht; Jahrbuch. Wien.

Research Council of Israel, Bulletin; Special Publication. Jerusalem.

Revista Española Micropaleontologia. Madrid.

La Revue de Géographie. Montreal.

Revue de Micropaléontologie. Paris.

Revue des Sciences Naturelles de l'Ouest. Paris.

Rivista Italiana di Paleontologia. Milano.

Rochester Academy of Science, Proceedings. Rochester, N. Y.

Royal Dublin Society, Scientific Proceedings.

Royal Physical Society of Edinburgh, Proceedings.

Royal Geological Society of Cornwall, Transactions. Penzance.

Royal Geological Society of Ireland, Journal. Dublin.

Royal Society of Canada, Proceedings and Transactions. Ottawa.

Royal Society of Edinburgh, Memoirs; Proceedings; Transactions.

Royal Society of Ireland, Journal. Dublin, Edinburgh.

Royal Society of London, Philosophical Transactions; Proceedings.

Royal Society of New Zealand, Proceedings. Transactions and Proceedings. Wellington.

Royal Society of South Africa, Transactions. Capetown.

Royal Society of South Australia, Memoirs; Transactions and Proceedings. Adelaide.

Royal Society of Tasmania, Papers and Proceedings. Melbourne.

Royal Society of Victoria, Proceedings. Melbourne.

Saarbrücken, Universität des Saarlandes (Annales Universitatis Saraviensis), Scientia.

Saito Ho-on Kai Museum, Research Bulletin. Sendai, Japan.

Science. New York.

Science and Culture. Calcutta.

Science Record. Chunking.

Scientific American. New York.

Scottish Journal of Geology. Edinburgh.

Sedimentologija (Yugoslavia, Zavod za Geološka i Geofizička Istraživanja). Belgrade.

Sedimentology. Journal of the International Association of Sedimentology. Amsterdam.

Senckenbergiana Lethaea (Senckenbergische Naturforschende Gesellschaft, Wissenschaftliche Mitteilungen). Frankfurt.

Senckenbergiana Maritima. Frankfurt.

Senckenbergische Naturforschende Gesellschaft, Abhandlungen; Aufsätze und Reden. Frankfurt.

Sigma Gamma Epsilon, The Compass. Provo, Utah.

Skandinaviske Naturforskeres Møte, Forhandlinger. Kristiania.

Slovensko Akademie Znanosti in Umetnosti Razred za Prirodoslovne Vede, Razprave. Ljubljana.

Smithsonian Miscellaneous Collections. Washington, D. C.

[R.] Sociedad Española de Historia Natural, Boletín. Madrid.

Società Geológica Italiana, Bolletino; Memorie. Roma.

Società Italiana di Scienze Naturali e del Museo civile di storia naturale, Atti; Memorie. Milano.

Società Ligustica di Scienze e Lettere, Atti. Pavia (Univ. Genova).

Società dei Naturaliste e Matematici, Atti. Modena.

Società Toscana di Scienze Naturali Residente in Pisa, Atti; Memorie.

Société Belge de Géologie, de Paléontologie et d'Hydrologie, Bulletin. Bruxelles.

Société des Études Scientifiques d'Angers, Bulletin.

Société Geologique de Belgique, Annales, Mémoires. Liège.

Société Géologique de France, Compte Rendu des Séances; Bulletin; Mémoires. Paris.

Société Geologique et Minéralogique de Bretagne, Mémoires. Rennes.

Société Géologique du Nord, Annales; Mémoires. Lille.

Société Géologique de Pologne, Annales. Kraków.

See Polskiego Towarzystwa Geologicznego w Krakówie;

Société d'Histoire Naturelle de Paris, Mémoires.

Société d'Histoire Naturelle de Toulouse, Bulletin.

Société Linnéenne de Normandie, Bulletin; Mémoires. Caen.

Société de Naturalistes Luxembourgeois, Bulletin. (See Verein Luxemburger Naturfreunde, Mitteilungen.)

Société Paléontologique de la Suisse, Mémoires (see Schweizerische Paläontologische Gesellschaft). Zürich.

Société de Physique et d'Histoire Naturelle de Genève, Mémoires.

Société des Sciences de Nancy et de la Réunion Biologique, Bulletin; Mémoires.

South Africa, Geological Survey of, Annals. Pretoria.

South African Geographical Journal. Johannesburg.

South African Museum, Annals. Capetown.

South Australian Museum, Records. Adelaide.

Southern California Academy of Sciences, Bulletin. Los Angeles.

Spisanie na B'lgarskoto Geologichesko Druzhestvo. Sofia.

Sprawozdania z Posiedzén Komisji Oddzialu Pan w Krakówie.

Státního Geologického Ústavu Československe Republiky, Věstník. Praha.

Stuttgart Universität, Geologisch-Paläontologisches Institut, Arbeiten.

Stuttgarter Beiträge zur Naturkunde. Stuttgart.

[K.] Svenska Vetenskapsakademien, Arkiv för Mineralogi och Geologi; Arkiv för Zoologi; Handlingar. Stockholm.

Sveriges Geologiska Undersökning, Afhandlingar; Årsbok. Stockholm.

Tartu Ülikooli Geoloogia-Instituudi Toimetused (Acta et Commentationes Universitatis Tartuensis, Dorpatensis). Tartu.

Tartu, Ülikooli juures oleva Loodusuurijate Seltsi Araunded.

Természetrajzi Füzetek. Budapest.

Texas Agricultural and Mechanical University, Oceanographic Studies. College Station.

Texas Geological Survey, Annual Report. Austin.

Texas, University of, Bulletin; Publications. Austin.

Tohoku University, Science Reports. Sendai, Japan.

Tunisia, Direction des Travaux Publics, Annales des Mines et de la Géologie. Tunis.

United States Geological Survey, Annual Report; Bulletin; Monographs; Professional Paper. Washington, D. C.

United States Geological and Geographical Survey of the Territories, Annual Report. Washington, D. C.

United States National Museum, Bulletin; Proceedings. Washington, D. C.

Universum. Natur und Technik. (Gesellschaft für Natur und Technik). Vienna.

Uppsala, University of, Geological Institution, Bulletin.

Ústředního Ústavu Geologickeho, Rozpravy; Sborník; Věstník. Praha.

Utah Geological and Mineralogical Survey, Bulletin. Salt Lake City, Utah.

The Veliger. Berkeley, Calif.

Verein der Freunde der Naturgeschichte in Mecklenburg, Archiv. Güstrow.

Verein Luxemburger Naturfreunde, Mitteilungen. Luxembourg.

Verein für Vaterländische Naturkunde in Württemberg; Jahreshefte. Stuttgart.

Versammlung Deutscher Naturforscher und Aerzte, Bericht. Mainz.

Victorian Naturalist (Journal and Magazine of the Field Naturalists' Club of Victoria). Melbourne.

Videnskabs-Selskabet i Kristiania, Forhandlinger; Skrifter. Oslo.

Vlastivédného Ústavu v Olomouci, Zpravy. Olomouc, Moravia.

Vsesoyuznyy Geologo-Razvedochnyy Ob'edineniya SSSR, Trudy. Moskva.

Vsesoyuznyy Neftianoi Nauchno-issledovatelskyi Geologo-razvedochyni Institut (VNIGRI), Trudy. Leningrad.

Vsesoyuznyy Paleontologicheskiy Obshchestvo, Ezhegodnik, Trudy. Moskva. (Formerly Russkoe Paleontologicheskiy Obshchestvo, Ezhegodnik; Société Paléontologique de Russie, Annuaire.)

Wagner Free Institute of Science of Philadelphia, Bulletin.

Washington Academy of Sciences, Journal; Proceedings. Washington, D. C.

Wyoming, University of, Contributions to Geology. Laramie, Wyo.

Yokohoma Kokuritsu Daigaku Science Reports, Section II, Biological and Geological Sciences. Kamakura, Japan.

Yorkshire Geological Society, Proceedings. Manchester, Leeds.

Zavod za Geološka i Geofizička Istraživanja, Vesnik Primenjena Geofizika. Belgrad.

Zeitschrift für Angewandte Mineralogie. Berlin.

Zeitschrift für die Gesamte Naturwissenschaft. Braunschweig.

Zeitschrift für Geschiebeforschung. Berlin. (Formerly Zeitschrift für Geschiebeforschung und Flachlandsgeologie.)

Zeitschrift für Morphologie und Ökologie der Tiere. Berlin.

Zeitschrift für Naturwissenschaften. Halle.

Zeitschrift für Wissenschaftliche Zoologie. Leipzig.

Zeitschrift für Zellforschung und Mikroskopische Anatomie. Berlin.

Zentralblatt für Geologie und Paläontologie (Before 1950, Neues Jahrbuch für Mineralogie, Geologie, und Paläontologie). Stuttgart.

Zoological Society of London, Proceedings.

[K. K.] Zoologisch-botanische Gesellschaft in Wien, Verhandlungen.

SOURCES OF ILLUSTRATIONS

At the end of figure captions a name and date are given to supply record of the author of illustrations used in the *Treatise,* reference being made either (1) to publications cited in reference lists or (2) to the names of authors with or without indication of individual publications concerned. Previously unpublished illustrations are marked by the letter "n" (signifying "new") with the name of the author.

Classification of rocks forming the geologic column as commonly cited in the *Treatise* in terms of units defined by concepts of time is reasonably uniform and firm throughout most of the world as regards major divisions (e.g., series, systems, and rocks representing eras) but it is variable and unfirm as regards smaller divisions (e.g., substages, stages, and subseries), which are provincial in application. Users of the *Treatise* have suggested the desirability of publishing reference lists showing the stratigraphic arrangement of at least the most commonly cited divisions. Accordingly, a tabulation of European and North American units, which broadly is applicable also to some other continents, is given here.

Generally Recognized Divisions of Geologic Column

EUROPE	NORTH AMERICA
CAINOZOIC ERATHEM	**CENOZOIC ERATHEM**
QUATERNARY SYSTEM	**QUATERNARY SYSTEM**
Holocene (Recent) Series	Holocene (Recent) Series
Pleistocene Series	Pleistocene Series
TERTIARY SYSTEM[1]	**TERTIARY SYSTEM**[1,2]
Pliocene Series	**Pliocene Series**
Astian Stage	Foley
Pontian Stage	
Miocene Series	**Miocene Series**
Sarmatian Stage	Clovelly
Tortonian Stage	Duck Lake
Helvetian Stage	Napoleonville
Burdigalian Stage	Anahuac
Oligocene Series	**Oligocene Series**
Aquitanian Stage	Chackasaway
Chattian Stage	
Rupelian Stage	Vicksburgian Stage
Lattorfian Stage	
Eocene Series	**Eocene Series**
Ludian Stage	
Bartonian Stage	Jacksonian Stage
Auversian Stage	Claibornian Stage
Lutetian Stage	Sabinian Stage
Ypresian Stage	
Paleocene Series	**Paleocene Series**
Sparnacian Stage	
Thanetian Stage	Midwayan Stage
Montian Stage (includes Danian)	
MESOZOIC ERATHEM	**MESOZOIC ERATHEM**
CRETACEOUS SYSTEM	**CRETACEOUS SYSTEM**
Upper Cretaceous Series	**Gulfian Series (Upper Cretaceous)**
Maastrichtian Stage[3]	Navarroan Stage
Campanian Stage[3]	Tayloran Stage
Santonian Stage[3]	Austinian Stage
Coniacian Stage[3]	Eaglefordian Stage
Turonian Stage	Woodbinian (Tuscaloosan) Stage
Cenomanian Stage	
Lower Cretaceous Series	**Comanchean Series**
	(Lower Cretaceous)
Albian Stage (Gault)	Washitan Stage
Aptian Stage	Fredericksburgian Stage
Barremian Stage[4]	Trinitian Stage
Hauterivian Stage[4]	
Valanginian Stage[4]	**Coahuilan Series (Lower Cretaceous)**
Berriasian Stage[4]	Nuevoleonian Stage
	Durangoan Stage

Lower Devonian Series
   Emsian Stage
   Siegenian Stage
   Gedinnian Stage
SILURIAN SYSTEM
  Upper Silurian Series
   Pridolian Stage
   Ludlovian Stage[11]
   Wenlockian Stage[11]
  Lower Silurian Series[11]
   Llandoverian Stage[11]
ORDOVICIAN SYSTEM
  Upper Ordovician Series

   Ashgillian Stage[11]

   Caradocian Stage[11]

  Lower Ordovician Series
   Llandeilian Stage[11]
   Llanvirnian Stage[11]
   Arenigian Stage[11]
   Tremadocian Stage[11]
CAMBRIAN SYSTEM
  Upper Cambrian Series (Merioneth)

  Middle Cambrian Series (St. David)
  Lower Cambrian Series (Comley)

ROCKS OF PRECAMBRIAN ERAS
PROTEROZOIC ERATHEM
   Dalradian, Eocambrian,
   Vendian, Riphean,
   and equivalents

Ulsterian Series (Lower Devonian)
   Onesquethawan Stage[10]
   Deerparkian Stage[10]
   Helderbergian Stage[10]
SILURIAN SYSTEM

Cayugan Series[12] (Upper Silurian)
Niagaran Series[12] (Middle Silurian)

Alexandrian Series[12] (Lower Silurian)

ORDOVICIAN SYSTEM
Cincinnatian Series
(Upper Ordovician)
   Richmondian Stage
   Maysvillian Stage
   Edenian Stage
Champlainian Series
(Middle Ordovician)
   Mohawkian Stage
    Trentonian Substage
    Blackriveran Substage
   Chazyan Stage
   Whiterockian Stage
Canadian Series (Lower Ordovician)

CAMBRIAN SYSTEM
Croixian Series (Upper Cambrian)
   Trempealeauan Stage
   Franconian Stage
   Dresbachian Stage
Albertan Series (Middle Cambrian)
Waucoban Series (Lower Cambrian)

ROCKS OF PRECAMBRIAN ERAS
PROTEROZOIC ERATHEM
   Algonkian, Beltian,
   Hadrynian, Helikian,
   Aphebian, and equivalents

CURT TEICHERT

---

[1] For convenience Miocene and Pliocene are often grouped as Neogene, Paleocene, Eocene, and Oligocene as Paleogene subsystems.

[2] Follows essentially Gulf Coast usage.

[3] Classed as division of Senonian Subseries.

[4] Classed as division of Neocomian Subseries.

[5] Included in Upper Jurassic by some authors.

[6] Equivalent to upper Thuringian (Zechstein) deposits.

[7] Equivalent to lower Thuringian (Zechstein) deposits.

[8] Equivalent to upper Autunian and part of Rotliegend deposits.

[9] Also known as Eifelian.

[10] Applies essentially to eastern United States; in western North America European stage terminology is used.

[11] Classified as Series by many English geologists; Tremadocian placed in Cambrian by some authors.

[12] Applies essentially to eastern North America only. BERRY and BOUCOT have advocated use of the English standard scale everywhere in North America (Geol. Soc. America, Spec. Paper 102, 1970).

PART W SUPPLEMENT 1

TRACE FOSSILS AND PROBLEMATICA

By †Walter Häntzschel

[Hamburg, West Germany]

———

CONTENTS

INTRODUCTION

When the manuscript of the first edition of Part W of the *Treatise* (1962) was completed, it was the first of a very few such general compilations to be published. Since its appearance, not only have numerous new trace fossils been described and new ichnogenera named, but also, the results of many new investigations in general ichnology have been published. The significance of trace fossils for sedimentology, facies interpretation, and paleontology is becoming more and more recognized, and this branch of paleontology arouses worldwide interest. Thus, it has become necessary to revise and expand the entire edition.

It is the primary purpose of this revision not only to give complete descriptions of the increasing number of important ichnogenera but also to increase the number and improve the quality of the illustrations selected from new literature.

This introduction, which was likewise revised and expanded, cannot be an extensive treatment of general ichnology. Instead, one may refer to a complete discussion of this general subject given recently by FREY (1971). Presently, an exhaustive book on ichnology is in preparation under the editorship of FREY (1974, in press) with the collaboration of many paleontologists. The materials in this edition of the *Treatise* have been divided into many sections, each with an expanded introduction. Within each section, the generic names are listed in alphabetical order as in the first edition.

A criticism of the 1962 edition was that unidentified trace fossils were not included. This has been practically impossible to correct as such descriptions are generally incomplete and are hidden and scattered in the world literature.

In the present volume, an attempt has been made to take into consideration all the trace fossil literature of the world published until about the beginning of 1973. As a result, the bibliography of the earlier edition has been extensively enlarged. Because of the extraordinarily scattered trace fossil literature, this reference list was necessary, especially since the last detailed list in *Fossilium Catalogus* (HÄNTZSCHEL, 1965) had only limited distribution.

ACKNOWLEDGMENTS

Numerous paleontologists, in all parts of the world, have assisted me in the preparation of this second edition of my contribution to the *Treatise,* Part W. Their kind assistance has made available to me specimens, literature, illustrations, and other information. It is not possible to name individually these people, and my thanks to them are expressed collectively. I would also like to thank Professor CURT TEICHERT for granting all my requests in regard to the illustrations and the increased number of references. Similar thanks go to the *Treatise* editorial staff at the University of Kansas for the very careful preparation of manuscripts and numerous illustrations for printing.

GLOSSARY OF TERMS

ichnocoenosis, ichnocoenose (DAVITASHVILI, 1945; again proposed independently by LESSERTISSEUR, 1955, p. 10). Association of trace fossils, corresponding to biocoenosis; ichnocoenosis used by DAVITASHVILI only for Recent assemblages of traces; a fossil association regarded by him as an oryctocoenosis EFREMOV (*see* RADWAŃSKI & RONIEWICZ, 1970).

ichnofossil (SEILACHER, 1956a, p. 158) (German, Spuren-Fossil, KREJCI-GRAF, 1932, p. 21). Trace fossil.

Ichnolites (HITCHCOCK, 1841, p. 476). Name proposed for a "class" including all sorts of tracks, divided into "orders" (depending on number of feet of animal that made the tracks): Polypodichnites, Tetrapodichnites, Dipodichnites.

ichnolithology (HITCHCOCK, 1841, p. 770). "History of fossil footmarks"; same as ichnology, term not widely adopted.

ichnology (BUCKLAND, about 1830). Entire field of lebensspuren (all tracks, trails, burrows, and borings); in fossil state, paleoichnology or palichnology; Recent, neoichnology.

lebensspur (ABEL, 1912, p. 65) [Synonymous German terms: *biogene Spur, organogene Spur* (KREJCI-GRAF, 1932); French, *trace physiologique* (D'ORBIGNY, 1849); *vestige fossile de vie* (VAN STRAELEN, 1938); *trace de vie* (ROGER, 1962); *trace d'activité animale* (LESSERTISSEUR, 1955); Italian, *impronte fisiologiche* (DESIO, 1940); Spanish, *huella problematica* (MACSOTAY, 1967); Russian, sled, bioglyph (VASSOEVICH, 1953); *International Code of Zoological Nomenclature* (1964) refers to "work of an animal"]. Used for fossil and Recent tracks, trails, burrows, and borings; fossil *Lebensspur* =trace fossil, ichnofossil (German, *Spuren-Fossil* KREJCI-GRAF, 1932); ABEL (1912) did not define term, but using it in a wide sense he (ABEL, 1912, 1935) included under this heading not only tracks, trails, burrows, borings, coprolites, but also death agony, pathological phenomena, symbiosis, parasitism, gastroliths, etc. Shortest definition (preferred here) was given by HAAS (1954, p. 379): "Lebensspuren are structures in the sediment left by living organisms"; in my opinion the words "or in hard substrates" should be added behind "in the sediment," thus including borings. New definition given by OSGOOD (1970, p. 282): "Evidence of the activity of an organism in or on the sediment, produced by some voluntary action of that organism." FREY (1971, p. 94) included coprolites, fecal castings, and similar features and excluded biostratification structures as stromatolites, byssal mats, biogenic graded bedding, and related phenomena. SIMPSON (1957, p. 477) restricted the term trace fossil to activity of an animal moving on or in the sediment at time of its accumulation, which excludes borings in shells or in consolidated sediment. There is still some discussion on the best definition of this term. (Also for discussion, *see* MARTINSSON, 1970, p. 323-324.)

nucleocavia (RICHTER & RICHTER, 1930, p. 168). General name (not generic) for small, winding canals, which generally occur in form of furrows on surfaces of originating steinkerns; producers are probably worms, small arthropods, or other animal groups. (*See* also RICHTER, 1931, p. 308.)

spreite. German noun, often literally translated as "spread," meaning structures spread between limbs of a U-tube comparable to web of duck's foot and representing a transverse zone of disturbed sediment appearing as series of concentric arcs between limbs of U-tube, and generally parallel to base of tube; produced by shifting tube transversely through sediment. Protrusive and retrusive spreiten are to be distinguished, indicating deepening or elevation of bottom of tube respectively, according to erosion or accumulation of sediment. *Spreite* plus U-tube=spreite burrow (German, *Spreitenbau*); observed as early as in Lower Cambrian sand-

stones, fossil spreite burrows may be horizontal, oblique, or perpendicular to bedding, bladelike or spiral-shaped. Recent spreite burrows are very difficult to observe in unconsolidated sediment, but are known in various environments, and are made by animals of very different systematic position (SEILACHER, 1967b, p. 414, fig. 1).

track, trackway. Impression left in sediment by feet of animals; term sometimes used for isolated impressions left by individual feet, but also used for the "trackway," or assemblages of tracks reflecting directional locomotion.

trace fossil. Fossil lebensspur.

trails. More or less continuous grooves left by (mostly creeping) animals as they move over bottom and have part of their bodies in contact with substrate or sediment surface. PACKARD (1900), CASTER (1938), NIELSEN (1949), and OSGOOD (1970, p. 351) used "track" for "the whole record of walk" of an arthropod (see also CASTER, 1938, p. 5, footnote 2).

vestigiofossil (R. C. MOORE, written commun., 1956). Unpublished suggestion to replace term "ichnofossil" because of its bilinguistic derivation from both Latin and Greek.

For terms on arthropod (especially trilobite) tracks, see OSGOOD (1970, p. 351), for terms on U-tubes with and without spreite, see OSGOOD (1970, p. 314), and for further terms and their definitions see the following chapters: Introduction, Nomenclature, Position of Traces in the Sediment, and, particularly, Classification.

Until recently, the majority of the world's literature on trace fossils had been published in either German or French. Because of this, Table 1 has been included to facilitate the translation of foreign terms into English. In addition, the Russian language is well represented by a book by VYALOV (1966), which describes many different types of trace fossils.

GEOLOGICAL OCCURRENCE AND SIGNIFICANCE OF TRACE FOSSILS FOR STRATIGRAPHY AND TECTONICS

GENERAL REMARKS

Trace fossils occur in marine, lacustrine, and continental sedimentary rocks of all

TABLE 1.—*Equivalent Terms in English, German, and French** (after Frey, 1973, append. 1, mod.).

(List of German terms prepared by H.-E. Reineck and G. Hertweck; list of French terms prepared by J. Lessertisseur)

ENGLISH	GERMAN	FRENCH
active fill	aktive Verfüllung	remplissage actif
back fill	Versatzbauten; Versatzgefüge	terrier (*or* galerie) remblayé
biodeformational structure	Verformungswühlgefüge	structure de biodéformation
bioerosion structure	Bioerosion	structure de bioérosion
biogenic sedimentary structure	biogenes Sedimentgefüge	structure sédimentaire biogène
biogenic structure	biogenes Gefüge	structure biogène
biostratification structure	biogenes Schichtgefüge	structure de biostratification
bioturbate texture	Verwühlung	texture bioturbée
bioturbation	Verwühlung; Bioturbation	bioturbation
bioturbation structure	Wühlgefüge; Bioturbationsgefüge	structure de bioturbation
body fossil	Körperfossil	corps fossile; fossile corporel
boundary relief	Grenzrelief	relief limite
burrow	Gang	terrier
burrow cast	Gangverfüllung	moulage (du terrier)
burrow lining	Gangwandung	paroi (du terrier)
burrow mottle	durch Gänge erzeugte Flecken	amas (*or* agglomérat) de terriers
burrow system	Gangsystem	terrier composé
cleavage relief	Spaltrelief	relief sur clivage (sur délit)
configuration	Konfiguration	configuration
crawling trace	Kriechspur	trace de locomotion (*or* de reptation, *in a restricted sense*)
dwelling burrow	Wohngang	terrier d'habitation
dwelling structure	Wohnbau	structure d'habitation (*or,* logement)
dwelling tube	Wohnröhre	tube d'habitation
epirelief	Epirelief	épirelief
escape structure	Fluchtspur	structure d'évitement
ethology	Verhaltensforschung; Ethologie	éthologie
feeding structure	Fresspur	structure de nutrition
full relief	Vollrelief	plein relief
grazing trace	Weidespur	trace de pacage
groove	Furche	sillon
hyporelief	Hyporelief	hyporelief
ichnocoenose	Ichnocoenose	ichnocénose
ichnofauna	Ichnofauna	ichnofaune
ichnoflora	Ichnoflora	ichnoflore
ichnology	Ichnologie; Spurenkunde	ichnologie
lebensspur; spoor	Lebensspur	trace d'activité; trace de vie
neoichnology	Neo-Ichnologie	néoichnologie
palichnology	Palichnologie	palichnologie
passive fill	passive Verfüllung	remplissage passif
resting trace	Ruhespur	trace de station
ridge	Kamm; Grat; Rücken	bourrelet
semirelief	Halbrelief	demirelief
shaft	Schacht	tube; tuyau
spreite	Spreite	traverse
stuffed burrows	Stopfbauten; Stopfgefüge; Stopftunnel	
trace; spoor	Spur	trace
trace fossil; ichnofossil	Spurenfossil; Ichnofossil	trace fossile; fossile de trace; ichnofossile
track	Trittsiegel; (*in a strict sense,* Fusspur)	empreinte
trackway	Fährte	piste; (*at depth,* galerie)
trail	Kriechspur	″ ″ ″ (de reptation)
toponomy	Toponomie	toponomie
tunnel	waagerechter Gang	tunnel

* Not all of these terms have exact counterparts in English, German, and French, but an attempt was made to approximate a common meaning as closely as possible. Several ichnological terms derived directly from classical words, such as *pascichnion* and *endichnion*, are cognates in all three languages, and are not listed here.

geologic systems from the Precambrian to the Recent (Fig. 1). Trace fossils are most abundant and best preserved in clastic rocks with alternating sandy and shaly beds.

Trace fossils found in the Late Precambrian are particularly significant for the investigation of the development of life before the Cambrian, especially that of metazoans. Also important is the comparison of lebensspuren in Late Precambrian sediments with those of undoubted Early Cambrian age. Such investigations have been made by SEILACHER (1956a) and GLAESSNER (1969) in the United States and Australia and have proven that trace fossils are scarce in Late Precambrian rocks when compared with their occurrences in lowest Cambrian rocks. In the Ediacara fauna of South Australia, there are perhaps six different ichnofossils produced by soft-bodied organisms creating grazing trails and ingesting sediment (GLAESSNER, 1971, p. 1337). GLAESSNER (1969, p. 381) has assigned one of these trace fossils to *Margaritichnus* BANDEL [=*Cylindrichnus* BANDEL], and the others remain unknown.

In general, the oldest lebensspuren are somewhat uncertain finds in the Grand Canyon Series (Hakatai Shale) and the Belt Series of the United States. These occurrences are both about 1,000 m.y. old, but whether or not they are genuine trace fossils must be verified. A trace fossil that is certainly of Late Precambrian age is *Bunyerichnus* GLAESSNER, 1969, which was discovered in South Australia (Brachina Formation, Wilpena Group) (see Fig. 30,3). *Bunyerichnus* is a crawling trail, 2 to 3 cm. wide, produced by a bilaterally symmetrical animal undoubtedly related to primitive mollusks. Precambrian lebensspuren cannot always be definitely identified when a distinction between body fossils and inorganic pseudofossils is difficult. This is shown by old and new discoveries of such fossils in the Precambrian from Canada, most recently discussed by HOFMANN (1971).

In several Paleozoic rocks, trace fossils are so characteristic and numerous that they have furnished the names of stratigraphic units, e.g., the *Skolithos* Sandstone, Fucoid Sandstone, and *Diplocraterion* Sandstone of the Lower Cambrian in Sweden, the *Phycodes* beds of the Lower Ordovician in Germany, the "Grès à *Harlania*" in the Paleozoic of North Africa, and others (see Fig. 37,2; 59,2; 64,2). In these types of sediments, contemporaneous body fossils are usually absent, but the trace fossils inform us of the existence of large numbers of bottom-dwelling animals. SEILACHER (1970) has pointed out that trace fossils can be considered to be a useful aid in the age determination and the stratigraphic correlation of such "unfossiliferous" sediments.

Trace fossils found in flysch facies are numerous and morphologically diverse. These synorogenic geosynclinal sediments have worldwide distribution and are generally deposited during orogenic times of the earth's history. Petrographically, flysch deposits are characterized by rhythmic alternations of coarser clastic sediments intercalated with pelitic sediments. Such rocks are especially favorable for the preservation of trace fossils. Since body fossils are rare in flysch deposits, the only paleontological evidence in these sediments are the ichnocoenoses, composed of traces of sediment ingestion, *Fressbauten*, and predominantly grazing trails, *Weidespuren* (see p. W32).

Also, many marine epicontinental sediments of all geological ages are rich in lebensspuren. However, these trace fossil associations are of different composition and show less diversity than those in flysch facies.

In sediments not entirely marine in origin, for example, the Lower Triassic Buntsandstein, which was deposited under essentially continental conditions, trace fossils are also present. However, in contrast to the ichnocoenoses of marine environments, the number of different types of nonmarine trace fossils is considerably less.

Sediments without lebensspuren are rare. There are also sediments in which some

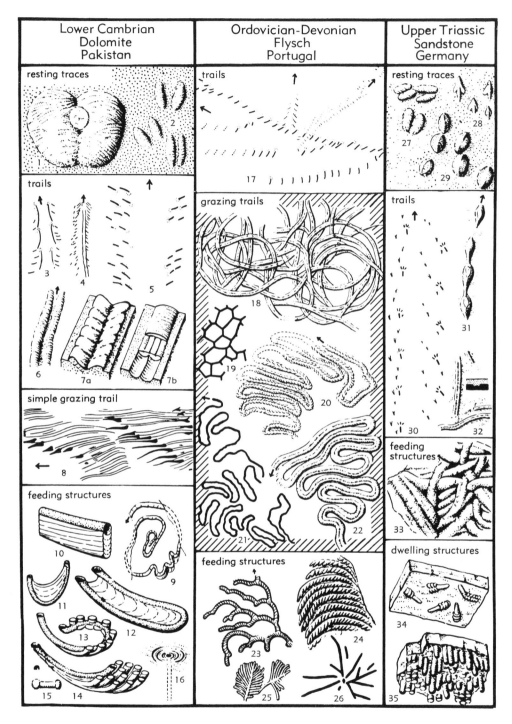

F

Fɪɢ. 1. Examples of different trace fossil assemblages (modified from Seilacher, 1955). (For explanation see p. *W*8.)

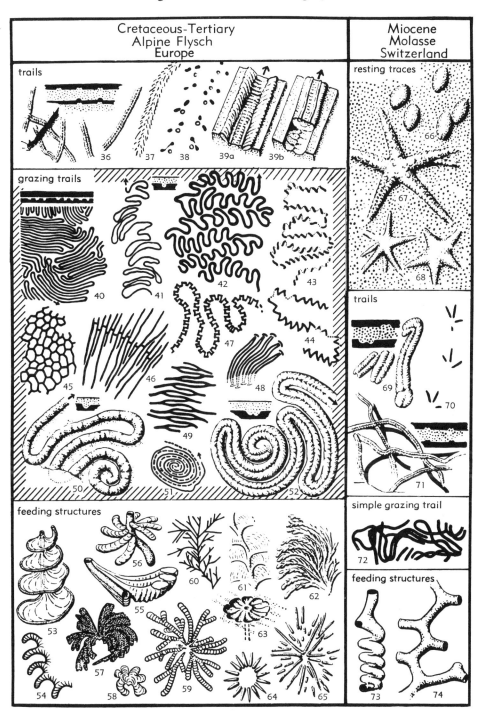

FIG. 1. (*Continued from facing page;* for explanation see p. *W*8.)

exogenic traces are preserved, whereas endogenic burrows are absent, due to ecologically unfavorable substrates. An example of such sediments is the Solnhofen Limestone (ABEL, 1927).

Homogeneous sediments may appear completely devoid of lebensspuren, but this is often only due to the fact that the lebensspuren are not visible to the unaided eye. HAMBLIN (1962, 1965) was the first to recognize distinct burrows in homogeneous sediments by the use of X-ray photography. X-radiography has also revealed elaborate boring networks in shell material (Fig. 2).

FIG. 2. Radiograph of *Pecten maximus* with camerate boring of *Cliona vastifica* (Bromley, 1970, p. 75, in: *Trace Fossils*, edited by T. P. Crimes & J. C. Harper, Geol. Jour. Spec. Issue 3, Seel House Press, Liverpool).

FIG. 1. (*Continued from page W6, 7.*)

1,2. Rusophycus, ×0.3, ×0.75.	40. Helminthoida, ×0.5.
3,4. Protichnites, ×0.75.	41. "Helminthoida," ×0.25.
5. Diplichnites, ×0.75.	42. Cosmorhaphe, ×0.16.
6. Crossopodia, ×0.75.	43. Helicolithus, ×0.75.
7a,b. Scolicia.	44. Belorhaphe, ×0.75.
8. Dimorphichnus, ×0.3.	45. Paleodictyon, ×0.5.
9. Dictyodora, ×0.3.	46. Desmograpton, ×0.5.
10. Teichichnus, ×0.3.	47. Paleomeandron, ×0.75.
11. Corophioides, ×0.3.	48. "Unnamed form," ×0.3.
12. Rhizocorallium, ×0.3.	49. Helminthoida, ×0.25.
13,14. Phycodes, ×0.7, ×0.3.	50. Spirophycus, ×0.3.
15. Bifungites, ×0.75.	51. Spirorhaphe, ×0.3.
16. Laevicyclus, ×1.3.	52. Taphrhelminthopsis, ×0.16.
17. "Trilobite trails," ×0.3.	53. Zoophycos, ×0.25.
18. "Irregularly circular bilobate trails," ×0.5.	54. Phycosiphon, ×0.75.
19. Paleodictyon, ×0.3.	55. Pennatulites, ×0.1.
20. Nereites, ×0.3.	56. "Gyrophyllites," ×1.
21. ?Nereites, ×0.3.	57. "Chondrites," ×0.25.
22. Crossopodia, ×0.3.	58. Hydrancylus, ×0.5.
23. Phycosiphon, ×0.75.	59. Taenidium, ×0.2.
24. Lophoctenium, ×0.5.	60. Chondrites, ×0.3.
25. "Undescribed trail similar to Oldhamia," ×1.	61. "Unnamed form," ×0.3.
26. Chondrites, ×0.5.	62. Lophoctenium, ×0.5.
27. Rusophycus, ×0.75.	63. Gyrophyllites, ×0.3.
28. Sagittichnus, ×1.5.	64. Lorenzinia, ×0.3.
29. Lockeia, ×0.75.	65. "Unnamed star-shaped feeding structure," ×0.16.
30. Kouphichnium, ×0.3.	
31. "Unnamed bivalve trail," ×0.3.	66. Lockeia, ×0.75.
32. "Bilobate worm trail."	67,68. Asteriacites, ×0.25, ×0.75.
33. "Unilobate feeding structures," ×0.3.	69. "Isopodichnus," ×0.16.
34. Biformites, ×0.5.	70. "Bird tracks," ×0.25.
35. Cylindricum, ×0.5.	71. Gyrochorte, ×0.5.
36. Gyrochorte, ×0.5.	72. Helminthoida, ×0.5.
37. "Undetermined articulated trail," ×0.2.	73. Gyrolithes, ×0.16.
38. "Large tetrapod striding trail," ×0.05.	74. "Spongites," ×0.05.
39a,b. Scolicia, ×0.3.	

STRATIGRAPHIC USE

Lebensspuren usually have little importance in stratigraphy. In restricted areas, however, they may attain the rank of index fossils. A burrow, *Arenicolites franconicus* TRUSHEIM, 1934, from the Muschelkalk of southern Germany may serve as an example: this fossil occurs abundantly in a layer only 3 to 4 cm. thick and may be followed for a horizontal distance of 26 km. (see Fig. 24,2). Another example is a track-bearing horizon in the Eocene Green River Formation of Utah, which is traceable laterally for about 40 km. (MOUSSA, 1968, p. 1434). It consists of three beds containing bird and mammal tracks associated with invertebrate trails some of which are of very regular wave-like shape.

A long time-range is one of the characteristics of most biogenic structures, the vast majority of which remain unchanged throughout geologic time. This is true for nondescript, smooth, furrowlike crawling trails and cylindrical burrows, as well as for more distinctive U-shaped burrows with spreite and even for the honeycomb-like networks named *Paleodictyon* by MENEGHINI (in MURCHISON, 1850), which are known from Silurian to Tertiary.

In some cases, ichnospecies of widely distributed and "long-lived" ichnogenera have been proven to be useful guide fossils for age determinations. Species of the ichnogenus *Cruziana* D'ORBIGNY have been proven to be useful guide fossils for lower Paleozoic rocks in Wales (*Cruziana semiplicata* for Upper Cambrian, *C. furcifera* for Lower Ordovician). In homogeneous rocks of uncertain age in which body fossils are absent, the generally abundant trace fossils may be used for stratigraphic correlation (CRIMES, 1968, 1969, 1970; SEILACHER, 1960, 1970). CRIMES distinguished between Cambrian and Ordovician rocks by determining the differences in morphological characteristics between certain mo-

tion trails *(Laufspuren)* and grazing trails *(Weidespuren)* of trilobites. SEILACHER (1970) established an elaborate stratigraphic succession for *Cruziana* in lower Paleozoic rocks (Fig. 3). Some other trace fossils have also proven themselves to be useful for age determination, such as *Oldhamia* for the Cambrian and *Phycodes circinnatum* for the Ordovician. Another example is the beaded coprolite *Tomaculum* GROOM, which so far has been found only in Ordovician strata of England, France, Germany, and Czechoslovakia.

USE IN STRUCTURAL GEOLOGY

In structurally complicated areas where inverted beds may be expected to occur, burrows and trails may be useful for distinguishing top and bottom of strata as has been rather extensively discussed by SHROCK (1948, p. 175-188) and more recently by FREY (1971). Especially well suited for this purpose are U-shaped burrows, which are invariably built either horizontally or with the curved part toward the bottom. Burrows of the *Skolithos* type are usually excavated vertical to the bedding in undisturbed beds. If they are inclined strongly in one direction in disturbed beds they may serve to determine direction and amount of the tectonic movement. Burrows or borings of pelecypods that are enlarged and rounded at the bottom may be used as reliable top and bottom criteria by their shape.

By observing vertical and horizontal burrows that originally had tunnels with circular cross sections and now are elliptical, the amount of lateral and vertical compression may be quantitatively determined. PLESSMANN (1966) has measured the vertical diagenetic "contraction" and the lateral compressional forces on sediments in the flat Upper Cretaceous deposits at the northern margin of the Harz Mountains in Germany and in the flysch deposits of Sanremo in the Maritime Alps of Italy.

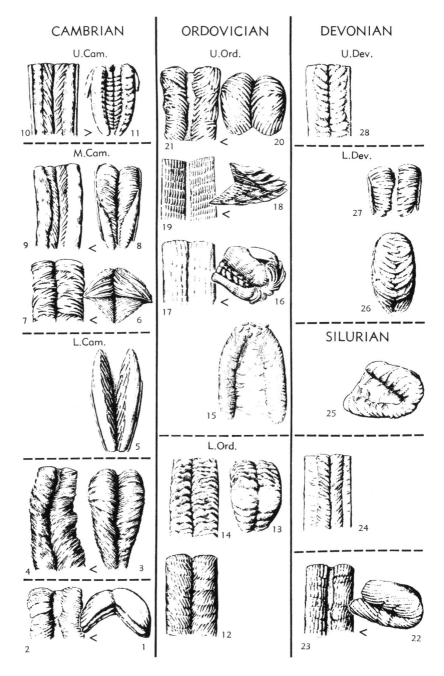

Fig. 3. *Cruziana* stratigraphy of Paleozoic sandstone of Europe, North Africa, and Southwest Asia (after Seilacher, 1970, p. 458, in: *Trace Fossils,* edited by T. P. Crimes & J. C. Harper, Geol. Jour. Spec. Issue 3, Seel House Press, Liverpool). < and > signs indicate whether the furrow (left) or the resting track expression (right) is more common. Forms not separated by dashed line may occur in the same unit.

POSITION OF TRACES IN THE SEDIMENT, THEIR FOSSILIZATION AND PRESERVATION

EXOGENIC TRACES

The most remarkable forms of traces observable in Recent sediments are lebensspuren made on the surface of sediments. They originate on the sediment surface at the bottom of flowing or stationary water at all depths or subaerially on the land (HERSEY, 1967; HEEZEN & HOLLISTER, 1971). Such lebensspuren are called surface, or surficial, trails, which is the same as exogene epirelief of SEILACHER (1953a) (Fig. 4). They belong to the group of semi-, or demi-reliefs.

It has often been noted that surficial trails produced in marine environments, especially in shallow water with tidal currents, have only a very small chance of preservation. Such trails can be destroyed by currents or wave action, especially on tidal flats. There is, however, a chance of preservation under certain favorable conditions, such as 1) rapid drying-up of the sea bottom during ebb tide especially near the shore, 2) cementation of the sediment by mucus, or 3) by infilling of the trail by wind-blown sand or by rapidly accumulating sediment. Preservation of trace fossils may also be expected to be more common in quiet, current-free, deep water. Here grain size and consistency of the sediment play an important role. In Recent clayey sediments of some coherency, trails are distinctly preserved under water. Preservation of such features as small ripples and microripples, and especially very thin, linearly striated groove casts and similar marks frequently found on bedding planes show that not all such features are easily destroyed. In pelitic freshwater sediments, as, for example, in the Lower Permian of Germany, delicate arthropod tracks have been preserved on the bedding planes of claystone. Such trails also have been discovered in Pleistocene varves in Germany and in Upper Paleozoic varves in Natal (SAVAGE, 1971), and surface trails have been preserved in ancient terrestrial sandstones. An example of this would be vertebrate tracks in the eolian Permian Coconino Sandstone of Arizona (United States), described by McKEE (1947). McKEE also performed experiments with several types of lizards moving on Recent sand dunes and determined that preservation of tracks was likely to occur as the sand surface, moistened by dew or mist, was consolidated and attached to dry eolian sand that covered it.

Ethologically considered, surface trails are either movement traces (running or crawling traces, more seldom swimming trails), resting traces, or sediment-ingesting trails.

When surface trails are normally epichnial grooves (MARTINSSON, 1965) or concave epireliefs (SEILACHER, 1964a), they can later become epichnial ridges or convex epireliefs, respectively. These "relief-tracks" may be formed from vertebrate trails (WASMUND, 1936) when the footprints are more resistant to the wind than the surrounding sediment. They have been

---

FIG. 3. *(Continued from facing page.)*

1,2. *C. cantabrica*, Spain.
3,4. *C. fasciculata*, Spain.
5. *C. carinata*, Spain.
6,7. *C. barbata*, Spain.
8,9. *C. arizonensis*, USA(Mont.-Ariz.).
10. *C. semiplicata*, North Wales.
11. *C. polonica*, Poland.
12. *C. rugosa*, Northern Iraq.
13,14. *C. imbricata*, Portugal.
15. *C. lineata*, South Jordan.

16,17. *C. almadenensis*, Spain.
18,19. *C. flammosa*, South Jordan.
20,21. *C. petraea*, South Jordan.
22,23. *C. acacensis*, Libya.
24. *C. quadrata*, Libya.
25. *C. pedroana*, Spain.
26. *C. uniloba*, Algeria.
27. *C. rhenana*, Germany.
28. *C. lobosa*, Libya.

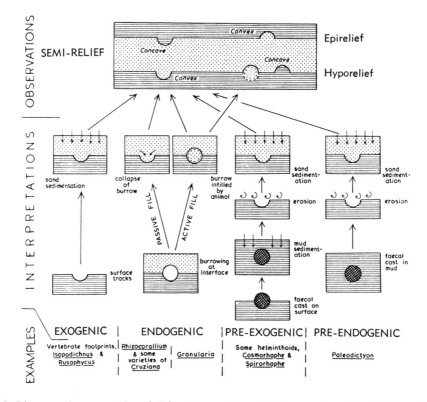

FIG. 4. Diagrammatic representation of different types of trace fossil preservation (after Webby, 1969a).

observed in snow as well as in terrestrial and marine sediments (TEICHERT, 1934; LINKE, 1954; SCHÄFER, 1951).

ENDOGENIC TRACES

Lebensspuren originating within sediment layers are designated as endostratal or endogenic. They are produced by animals that either move constantly in the sediment or live more or less permanently in structures within the sediment. There is also a transition between endostratal and surface trails. It is not always discernible whether a crawling surface trail has originated on an exposed sandy layer or whether the sedimentary surface was covered by a layer of sediment and endostratal lebensspuren were produced by the mixing and digging of an animal at the sediment interface in the sand beneath. If clay is overlain by sand, a distinct endo-

stratal resting trace is produced in the clay, and an indistinct concave form is produced in the sandstone. Running arthropods, especially limulids and trilobites, leave behind in the sediment surface trackways of different appearance, varying according to which part of the animal's extremities were impressed to different depths on the sediment surface (undertracks, GOLDRING & SEILACHER, 1971, p. 424; cleavage relief type, OSGOOD, 1970, p. 292) (Fig. 5; FREY, 1973a, fig. 5). Another transitional form between surficial and endostratal trails are tunnel trails *(Tunnelfährten)*.

Very many trace fossils occur at sedimentary interfaces where sand is underlain by mud. They are then found on the underside of the sandstone beds and generally are well preserved. They have been described as convex hyporeliefs (SEILACHER, 1964a) or hypichnia (MARTINSSON, 1965).

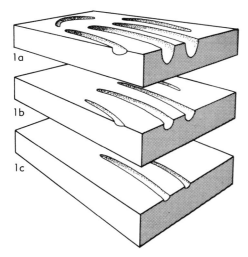

FIG. 5. Differential preservation of a hypothetical arthropod track (after Osgood, 1970; mod.). Each block is 1 mm. thick.——*1a*. Concave epirelief at depositional interface; quadrifid track with an arcuate posterior fringe.——*1b*. Cleavage relief 1 mm. below depositional interface; arcuate fringe not preserved.——*1c*. Cleavage relief 2 mm. below depositional interface; only two imprints preserved.

SEILACHER especially called attention to this kind of trace fossil, and employed the German word *Innenspuren*.

All trace fossils on lithologic bedding planes are semi- (demi-) reliefs. It is possible to distinguish between "cleavage reliefs" in a uniform sediment and "boundary reliefs" between petrographically different layers, especially between sandstone and shale (SEILACHER, 1953a, p. 438). However, in practice, this distinction may be difficult to make.

Clearly delineated burrows within one stratum that were originally formed as hollows (endogene full reliefs) have been named endichnia by MARTINSSON (1965) (=fossitextura figurativa, SCHÄFER, 1956a; 1972). Such burrows can be actively or passively filled. Burrowing textures (*Wühlgefüge*) are bioturbate shapes without sharp outlines, which may be filled in from above. MARTINSSON has named these structures exichnia (=fossitextura deformativa, SCHÄFER, 1956a).

There are still more complex endogenic

burrows, especially in flysch sediments, which have been described as pre-exogene or pre-endogene. Their origin is shown in Fig. 4 (see also p. *W20*).

Endostratal lebensspuren also include dwelling burrows in the sediment having very different morphological features, such as vertical shafts, J- or U-shaped tubes with or without spreite, Y- or W-shaped tunnels, irregular and complicated tunnel systems that may be arranged horizontally, vertically, or in netlike forms, or a combination of all three.

The walls of such Recent burrows are usually compacted by mucus and many animals press infiltrating sand grains against the walls, which are thereby strengthened. Burrows constructed in this manner have a good chance of being preserved as fossils. This is seen in the tidal flats of the North Sea where the upper end of *Arenicola* U-tubes may be solidified and thus escape being washed away. The tubes may protrude several centimeters above the sediment surface (HÄNTZSCHEL, 1938). In Recent lime muds from Florida and the Bahamas, SHINN (1968) has observed unoccupied decapod burrows that were still open. Covered by sediment, such burrows could possibly remain open for centuries. That such burrows can become indurated relatively rapidly is shown by the sedimentation of the U-shaped *Spreitenbauten (Rhizocorallium)* in the Lower Jurassic of southern Germany (SCHLOZ, 1968).

In Recent sediments, complex forms of endostratal burrows *(Innenspuren)* are more difficult to observe than in the fossil record. Especially fine structures of backfill origin *(Versatzbauten)* or *Spreitenbauten*, for example, in the sandy mud flats of the North Sea, are difficult to recognize. Thus, little is known about Recent spreiten structures, although they are common as fossils. Diagenetic processes greatly enhance the preservation and recognition of trace fossils in the sediment (SEILACHER, 1957). In order to study and observe endostratal burrows in Recent sediments, special methods must be used (HERTWECK & REINECK, 1966).

HISTORICAL REVIEW

No complete history of paleoichnological investigations has been written. WINKLER's *"Histoire de l'ichnologie"* (1886) represents only a chronologically arranged, annotated bibliography covering paleoichnological publications (mainly on vertebrate tracks) for the period 1828 to 1886. The following section briefly describes only a few stages of the rather discontinuous development of this branch of paleontology.

OSGOOD (1970, p. 286-291) has published a comprehensive survey of the historical development of ichnology, to which reference may be made. He divided the history of ichnology into three parts: 1) the "age of the fucoids" and 2) the "period of reaction," followed by 3) rapid advances in paleoichnology and neoichnology since the 1920's and continuing to the present time. The development of ichnology is important for paleontology and sedimentology, because it is a "development of ethological and paleoecological approaches."

In the early years of paleontology, many fossils, especially cylindrical and U-shaped burrows, now identified as lebensspuren, were considered to be remains of marine algae. This is apparent in names such as *Fucoides, Algacites, Chondrites,* and the many generic names having the ending *-phycus.* Ramification of the burrows was considered the most conclusive evidence for their interpretation as plants. In publications of these "algae," Recent Thallophyta were commonly figured in order to show the identity or relationship of the fossil forms with them. Occasionally, even the drawings of the fossils were modified so as to make them look more like algae.

According to OSGOOD, the "age of the fucoids" began in 1828, the year that *Fucoides* BRONGNIART, 1822, was divided into *"sectiones,"* and it ended in 1881. Nevertheless, in the nineteenth century,

many *"Fucoiden"* were described as marine algae. Most were labeled *incertae sedis,* although a few paleontologists recognized and named traces produced by invertebrates. One of these paleontologists was E. HITCHCOCK (1792-1864), geologist, astronomer, minister, and pedagogue. He named the first ichnogenus with the characteristic ending *-ichnus,* i.e., *Cochlichnus* HITCHCOCK, 1858, an invertebrate meander trail. In the same year, JARDINE established many genera with the same ending. Most of these were vertebrate tracks. The oldest established names for invertebrate trace fossils are *Harpagopus* HITCHCOCK, 1848, and *Herpystozeum* HITCHCOCK, 1848. HITCHCOCK was the first to publish a detailed description of a trace fossil assemblage consisting of numerous trails from Triassic sandstone of the Connecticut Valley (HITCHCOCK, 1858, 1865).

DAWSON (1864, p. 367) recognized that the traces named *Rusophycus* HALL, 1853, especially *R. grenvillensis* BILLINGS, were produced by trilobites as resting impressions, or as cavities made for shelter. He suggested, therefore, that the name *Rusophycus* should be changed to the more descriptive name *Rusichnites.*

Astonishingly, some ethological or general genetic interpretations of certain trace fossils have remained valid for nearly a century. NICHOLSON (1873, p. 288-289) regarded *Skolithos*-structures as true burrows of habitation, whereas he explained horizontal burrows as wandering tunnels excavated by worms in search for food. NICHOLSON also declared that forms combined by him under the name *Planolites* were "not the actual burrows themselves but the burrows filled up with sand or mud which the worm has passed through its alimentary canal." His interpretations were repeated, independently, decades later by subsequent authors. These early contributions must be recognized again, today.

Often, in the "age of the fucoids," forms such as *Nereites* MACLEAY (1839) were not considered to be trace fossils but body fossils. *Nereites* was claimed to be a *Nereis-*

type worm. Other grazing trails, such as *Helminthoida,* puzzled paleontologists, but it, too, was explained as being of plant origin. Some of the best examples of botanical interpretation of many trace fossils are found in the important, voluminous monograph, *"Flora fossilis Helvetiae"* (HEER, 1877), in which numerous flysch lebensspuren are described in great detail as plants.

The next forty years, from 1881 to about 1921, is OSGOOD's second period in the development of ichnology, the "period of reaction." This period should be expanded to begin with the publication of the classic works by the Swedish paleobotanist NATHORST (1873, 1881a,b). On the basis of systematic neoichnological observations and experiments on traces of marine animals, he pointed out the striking similarity of many "fucoids" and problematica to the tracks and trails of marine invertebrates. This evidence, together with the information that animal trails may ramify, permitted NATHORST to challenge the doctrine of plant origin for these fossils. The years between 1881 and 1885 were characterized by the violent controversy between NATHORST and his opponents DELGADO, LEBESCONTE, and DE SAPORTA, who tenaciously defended the botanical origin of these doubtful fossils. These arguments also dealt with the origin of the genera *Cruziana* and *Rusophycus,* which are today recognized as definite trilobite lebensspuren, at least in the majority of Paleozoic sediments. However, specimens of *Cruziana* and *Rusophycus* have been recognized in Triassic sediments in East Greenland and questionably attributed to notostracans or conchostracans (BROMLEY & ASGAARD, 1972). Since the recounting of this embittered controversy would take up too much space and because it has only historical significance, the reader is referred to OSGOOD (1970, p. 287-288) for a more detailed account.

Independently of NATHORST and without knowledge of his publications, J. F. JAMES (1857-97) in the United States published numerous and often overlooked works protesting the plant interpretation of most fucoids of the Cincinnatian. He explained their origin as animal trails, marks, or body fossils, and cautioned against many hasty publications and the assignment of names to poorly preserved and uncertain "fucoids." Attention must be called to his warning, which was long ignored but is still valid: "When every turn made by a worm or shell, and every print left by the claw of a Crustacean is described as a new addition to science, it is time to call halt and eliminate some of the old before making any more new species."

Only gradually did NATHORST's interpretation of many fossil "algae" as lebensspuren become accepted. Even today several "genera" of lebensspuren (e.g., *Chondrites, Fucoides*) are sometimes interpreted as algae. Canadian and Indian papers from 1938 and 1949 refer typical trace fossils to algae. FUCINI (1936, 1938), in extensive publications, described Problematica from the Cretaceous "Verrucano" of Toscana, Italy, mainly inorganic markings, as plant fossils.

Even in the beginning of this century many forms of lebensspuren were not recognized as trace fossils, including all grazing trails in Cretaceous or Tertiary flysch sediments in Europe called hieroglyphs or graphoglyphs. A number of these especially peculiar forms such as the ichnogenera *Paleodictyon, Urohelminthoida* [=*Hercorhaphe*], and *Spirorhaphe* were assumed by FUCHS (1895) to be spawn, presumably of gastropods. Similar interpretations are still being discussed for similar forms (e.g., *Spirodesmos*).

After several decades of stagnation following the turn of the century, substantial progress was made in lebensspuren studies by ABEL and his pupils, and especially in the course of "actuopaleontologic" investigations in marine biology of the North Sea tidal flats by RUDOLF RICHTER. His studies included 1) a survey of Recent and fossil worm trails and burrows, 2) an elucidation of general questions of palichnology, and utilization of lebensspuren for paleogeographic interpretation, and 3) an interpretation of many problematica, as well as an analysis of numerous arthropod trails and Recent and fossil U-shaped burrows. Until World War II, the efforts and results of RICHTER and his collaborators at the marine-geologic Forschunganstalt "Senckenberg" in Wilhelmshaven (HÄNTZSCHEL, SCHÄFER, SCHWARZ, TRUSHEIM) were focused in the same general direction.

Since the end of World War II, paleontologists and geologists, especially those from Europe and North America, have developed a tremendous interest in neoichnology and even more in paleoichnology. This interest was stimulated by the intensive investigations concerning the nature and origin of depositional basins, and the inorganic and biogenic textures of Recent and fossil sediments. It has been shown by trace fossil investigations that there are types of ichnocoenoses with characteristic elements having worldwide distribution independent of sediment age. Single lebensspuren, and especially ichnocoenoses, are good facies indicators, and they give reference to paleoenvironments. Trace fossils are usually not rare in rocks containing them, but are the most common fossils. Trace fossils and trace fossil associations are of great value for sedimentology and paleontology owing to their facies range. This significance of trace fossils is becoming more and more recognized in paleoecology because they furnish direct evidence of autochthonous life in the sediment, and thanatocoenoses do not exist. Many types of trace fossils remain unchanged and can be recognized during very long periods of time in the stratigraphic record. Such forms, therefore, permit the evaluation of ichnofacies.

CLASSIFICATION

The possible diversity of lebensspuren made by an individual animal, dependent on its activity (crawling, eating, running, burrowing, swimming), and the dependence of traces on fortuitous preservational properties of the sediment, make it impossible to clarify lebensspuren in a manner corresponding to a zoological pattern.

Classifications, or at least categorizing, of similar forms into groups have been attempted from many different viewpoints based on either: 1) the shape (morphological arrangement) of the trace fossil, 2) the kind of preservation and occurrence in the sediment, specifically the position of the boundary between calcareous and arenaceous sediments (stratinomic or toponomic arrangement), 3) ethological interpretations, or 4) a combination of the taxonomic, morphologic, and stratinomic bases (VYALOV, 1968b). In addition, an attempt has been made to arrange lebensspuren by taxonomic rank of the producer of the trace. HITCHCOCK (1844, p. 318) proposed a "new order including all sorts of footless trails made by worms, molluscs, and fishes," to be called Apodichnites. Lebensspuren produced by animals with more than four feet were called Polypodichnites (HITCHCOCK, 1841, p. 476). SALTER (1857, p. 204) named long, sinuous surface trails or filled-up burrows of marine worms without impressions of lateral appendages *Helminthites* (=*Helmintholites* MURCHISON, 1867, p. 514). Possibly a classification of trails produced by vertebrates will become feasible when footprints prove to be assignable with certainty to a particular taxonomic group of vertebrates.

MORPHOLOGICAL-DESCRIPTIVE CLASSIFICATION

In the early stages of paleontological research, most trace fossils were interpreted as marine algae, and were arranged exclusively according to morphological characters. The shape of the "thallus" was regarded as a determining factor and fucoid species were distinguished according to the angle of divergence of branches. FUCHS (1895), accepting such structures to be trace fossils, tried to arrange them into family-like groups, determined mainly by morphological criteria.

Many excellent, well-preserved examples of trace fossils can be seen in the Cretaceous-Tertiary flysch of southern Europe. FUCHS described the following different types:

1) GRAPHOGLYPTEN (FUCHS, 1895, p. 394; =*Hieroglyphen s.s.,* FUCHS, 1895, p. 394). Trace fossils appearing as reliefs on lower surface of beds (mostly sandstones) and resembling ornaments, or letters (e.g., *Paleodictyon, Paleomeandron,* explained by FUCHS, however, as strings of spawn of gastropods).

2) VERMIGLYPHEN (FUCHS, 1895, p. 390). Collective name for threadlike, straight, or variously winding reliefs on undersurface of sandstone beds in flysch and similar sediments; mostly unbranched; width usually only a few millimeters.

3) RHABDOGLYPHEN (FUCHS, 1895, p. 391). General and informal name for nearly straight bulges, mostly on undersurface of sandstone beds of flysch and similar sediments; greatest diameter several centimeters.

RUDOLF RICHTER presented good examples of a possible simple classification by 1) the distinction of U-shaped burrows with or without spreite (Rhizocorallidae, Arenicolitidae; see RICHTER, 1926, p. 211), and 2) the division of worm trails according to "basic architectural forms" (*bauliche Grund-Formen*) on a mechanical and biological basis (RICHTER, 1927a). Similarly, RICHTER (1941) arranged trails from the Hunsrück Shale morphologically into the following groups:

1) *Ichnia taeniata.* Regularly developed, bandlike grooves and tunnels, not filled by sediment.

2) *Ichnia catenaria.* Strings of pearl-like trails.

3) *Ichnia spicea.* Spike-shaped trails.

4) *Ichnia disserta.* Arthropod trails of separated rows of footprints.

However, this classification has not been generally adopted and has enjoyed very little use in the literature.

KREJCI-GRAF (1932) proposed a very comprehensive classification based on the life activities of the animals. He established three division units: 1) traces of rest, 2) traces of motion, and 3) traces of "existence," and defined these units with extremely detailed subdivisions. However, the number of minor categories makes the application of this elaborate classification difficult.

LESSERTISSEUR (1955) suggested a classification based mainly on morphological criteria which distinguishes 1) *traces exogènes* (simple bilobate and trilobate crawling trails, meanders, spirals, starlike trails, etc.) and 2) *traces endogènes* (burrows and tunnels of various forms, fucoids, resting trails, U-shaped burrows with or without spreite, and screw-shaped burrows) (Table 2).

VASSOEVICH (1953, p. 41) devised a classification that is strictly morphological in content and may be called "Fucoids in a wider sense." Accordingly, lebensspuren have been categorized as to whether they are two-dimensional or three-dimensional. These two major divisions are further subdivided on the basis of similarities of morphology such as meanders, braids, screw shapes, spiral shapes, U- or J-structures, presence or absence of branches, and other characters.

EWING & DAVIS (1967, p. 265-267) developed a very detailed morphological classification of Recent trails and dwelling structures found in the deep sea, arranged in geometric groups. Because the producers of lebensspuren almost always remain unknown, these authors adopted a strictly morphological classification. They distinguished between ridges and sets of ridges, lumps and sets of lumps, grooves and sets of grooves, depressions and sets of depressions and one or more grooves together, and sculptured strips. However, because transitional forms exist and there are problems of definition of the forms, nomenclatural problems arise.

HOROWITZ designed a new descriptive classification of lebensspuren which has been reproduced by FREY (1971, p. 96)

TABLE 2.—*Lessertisseur's (1955) Proposed Classification for Traces of Activity of Invertebrates* (translated from Lessertisseur, 1955).

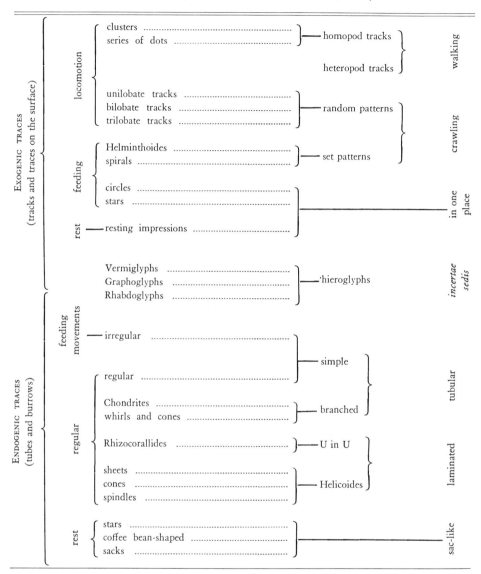

(Fig. 6). This classification is similar to LESSERTISSEUR's in using two main groups, i.e., intrastratal and bedding-surface structures, which then are further subdivided.

PRESERVATIONAL ASPECTS

Most trace fossils are preserved at the interface between clay and coarser-grained clastic sediments. For example, in flysch sediments, trace fossils are found on the underside of the coarse-grained clastic beds.

Therefore, it has also been possible to establish classifications based on the position of the trace fossil relative to the sediment

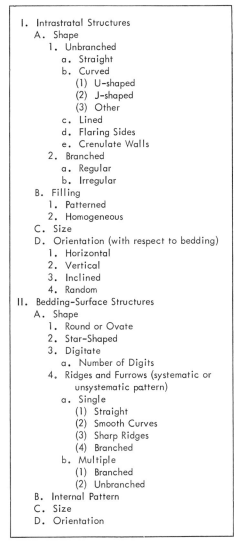

I. Intrastratal Structures
  A. Shape
    1. Unbranched
      a. Straight
      b. Curved
        (1) U-shaped
        (2) J-shaped
        (3) Other
      c. Lined
      d. Flaring Sides
      e. Crenulate Walls
    2. Branched
      a. Regular
      b. Irregular
  B. Filling
    1. Patterned
    2. Homogeneous
  C. Size
  D. Orientation (with respect to bedding)
    1. Horizontal
    2. Vertical
    3. Inclined
    4. Random
II. Bedding-Surface Structures
  A. Shape
    1. Round or Ovate
    2. Star-Shaped
    3. Digitate
      a. Number of Digits
    4. Ridges and Furrows (systematic or
        unsystematic pattern)
      a. Single
        (1) Straight
        (2) Smooth Curves
        (3) Sharp Ridges
        (4) Branched
      b. Multiple
        (1) Branched
        (2) Unbranched
  B. Internal Pattern
  C. Size
  D. Orientation

Fig. 6. Descriptive classification of lebensspuren proposed by Horowitz (Horowitz in Frey, 1971).

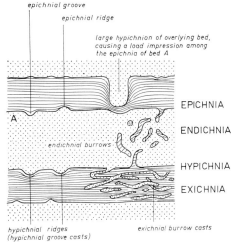

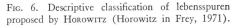

Fig. 7. Diagrammatic representation of toponomic terminology suggested by Martinsson (1970) and shown in cross section (Martinsson, 1970, p. 327, in: *Trace Fossils*, edited by T. P. Crimes & J. C. Harper, Geol. Jour. Spec. Issue 3, Seel House Press, Liverpool). [Stippled areas are siltstones and ruled areas, shales. For descriptive terms at right refer to bed A.]

interface. Martinsson (1965, p. 202-203) created a "stratinomic classification" or, as it has also been called, a "topographic classification." Recently, Martinsson (1970) has given another detailed discussion of his trace fossil classification, which he renamed the "Toponomy of Trace Fossils" (Fig. 7). It is a purely descriptive terminology including no ethological interpretation of the trace or trace producer. Only the position of the trace fossil in the sediment is important and is identified by the following four "toponomic" terms.

  1) Epichnia. Traces on upper surfaces of the main casting medium.
  2) Endichnia. Traces inside sediment within the casting medium (in German, *Innenspuren*).
  3) Hypichnia. Traces in firm primary contact with the lower surface of the clastic bed (sole trails).
  4) Exichnia. Mostly burrows in calcareous sediments but consisting of coarser materials introduced from a coarser bed.

These four terms have the advantage that they can be used either as adjectives (epichnial) or as nouns (epichnion). They may also be combined with simple morphological terms such as ridge, groove, furrow, burrow, or cast (e.g., epichnial ridge).

In the strictest sense, such a descriptive "system" is actually not a classification of lebensspuren, as any descriptive system must be supplemented with an ethological analysis and interpretation of trace fossils in general.

In this connection the classification developed by SEILACHER (1964a, p. 254-255; 1964c, p. 297) must be mentioned, which takes into consideration both the type of preservation and the origin of the trace fossils (but not in an ethological sense). In an expansion of his earlier somewhat schematic, stratinomic terms (SEILACHER, 1953a, p. 437), in his 1964 publications he has further refined previous classification.

1) FULL RELIEFS (Ger., *Vollformen*). Preservation of the entire structure ("fills" comparable to internal molds, "cavities"=open burrows).

2) SEMIRELIEFS (Ger., *Halbformen;* French, *demireliefs*). Sculptures on sand/clay interfaces; two kinds are to be distinguished, a) epireliefs, grooves or ridges on the top surface of a psammitic sediment, and b) hyporeliefs, on the undersurface of psammitic beds (ridges or grooves).

These forms can be produced in different ways, and additional observations are necessary. Thus, endogenic burrows may be exposed on the surface if the overlying sediments are eroded away, after which another layer of sediment may be deposited on the erosional surface, filling the excavated burrow. This burrow will then be preserved as "pseudoexogenic." Therefore, it must be determined if a burrow underwent active or passive filling. WEBBY (1969a, p. 90) felt that the term pseudoexogenic was unsatisfactory, and proposed that forms such as *Paleodictyon* are best named "preendogenic." Ichnogenera *Cosmorhaphe* and *Spirorhaphe*, originally surface fecal casts that have been eroded and later filled with sand, are described as "preexogenic" (Fig. 4).

Lebensspuren from flysch sediments that are generally interpreted as turbidites have been differentiated as either predepositional or postdepositional, based upon their chronologic relation to turbidity currents

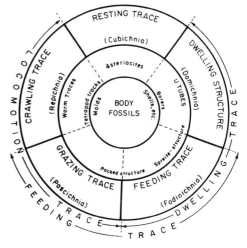

FIG. 8. Ethologic classification of trace fossils proposed by SEILACHER (1953) (from Osgood, 1970).

(KSIĄŻKIEWICZ, 1954, p. 446). A classification of the numerous trace fossils from Polish flysch deposits was made by KSIĄŻKIEWICZ (1970, p. 315-317) according to whether they were predepositional or postdepositional in origin. A discussion of his criteria for division and classification has been included because some forms are impossible to place in either group. SEILACHER (1962, p. 230) discussed a similar arrangement for sole trails in flysch deposits of northern Spain where similar turbidite sequences have been observed. Some sole trails were obviously of endogenic origin, and after weak compaction, were exposed on a bedding plane, eroded and later filled by sediment. Such trails were called "preendogene" by WEBBY (1969a) (see Fig. 4). A comparison of the lists given by KSIĄŻKIEWICZ (1970) and by SEILACHER (1962) of ichnogenera which they regarded as predepositional and postdepositional shows some agreement, but also some uncertainties of such a classification.

ETHOLOGICAL ASPECTS

A classification according to ethological principles proposed by SEILACHER (1953a, p. 432-434) (Fig. 8), is based on the fact that different groups of animals with similar life habits or behavioral patterns produce traces with similar basic characters, even though the animals themselves have quite different body shapes. Working out these common basic characters, SEILACHER distinguished five ethological groups: dwelling structures (domichnia), feeding structures (fodinichnia), grazing traces (pascichnia), resting traces (cubichnia) (=Ger., *Ruhespuren*, RICHTER, 1926, p. 223; repose imprints, KUENEN, 1957, p. 232), and crawling traces (repichnia) (=Herpichnites GÜMBEL, 1897, general term, not used as "genus"). For each of these groups typical features may be characterized as follows:

1) DOMICHNIA. Simple or U-shaped burrows or burrow systems with horizontal and vertical components, or dwelling tubes; perpendicular or oblique to the surface. More or less permanent domiciles for most semisessile suspension-feeding animals.

2) FODINICHNIA. Variously shaped burrows (with or without spreite) and burrow systems, at various angles to the bedding. More or less temporarily by used semisessile sediment-eaters simultaneously as domicile, "mine," or hunting-ground.

3) PASCICHNIA. Highly winding bands or furrows, not crossing each other, with intense utilization of the surface available for grazing or feeding, commonly resulting in surface ornamentation such as meanders or letterlike patterns ("parqueting").

4) CUBICHNIA. Isolated, mostly shallow depressions of troughlike relief, outlines corresponding roughly to the shapes of their producers. Commonly arranged parallel to each other as a result of like orientation (rheotactic rectification) toward currents, vertical and horizontal repetition possible.

5) REPICHNIA. Furrows, trackways, trails, and shallow crawling tunnels of variable direction, linear or sinuous, ramified or unramified, smooth or sculptured.

SEILACHER's system has the advantage of grouping ethologically similar assemblages of lebensspuren. Questions as to identity of their producers may be disregarded here, for these can only rarely be answered unequivocally on the basis of morphological criteria. The characterization of groups is, also, independent of time; for example, the assemblage termed cubichnia is equally valid for extinct arthropods of the Paleozoic (e.g., trilobites), as for Recent arthropods that have a corresponding mode of life. BERGSTRÖM (1972) has observed that the bend in the anterior cephalic margin of the trilobite *Cryptolithus* appears to have the same function in plowing as the limulid prosoma.

Due to its easy application, this system has proved useful for fossil and Recent lebensspuren. In the literature dealing with trace fossil associations, ichnogenera are assigned to one or the other of these groups. The ethological classification makes it possible to compare different ichnocoenoses which are characterized by giving percentage contribution by each group ("trace fossil-spectra"). In this manner, SEILACHER was able to distinguish several ichnofacies (e.g., *Nereites* facies and *Cruziana* facies) characterized by pascichnia in which cubichnia predominate. (For a complete discussion, see p. W32-W33.)

Trace fossils reflect the behavioral patterns of their producers. Therefore, in SEILACHER's ethological classification, it is not possible to assign each trace fossil to a particular group. An example is the vertical dwelling tube (*Wohnröhre*) of a polychaete worm that produces star-shaped grazing trails (*Weidespuren*) in the sediment surface surrounding the opening of the burrow, because such structures can be described as a combination of domichnia and pascichnia (HÄNTZSCHEL, 1970, p. 262). FREY (1971, p. 99) has considered trace fossils produced by two behavioral patterns in giving the name "combined feeding-dwelling burrows" to burrows produced by sediment-ingesting organisms that also double as domiciles for those animals.

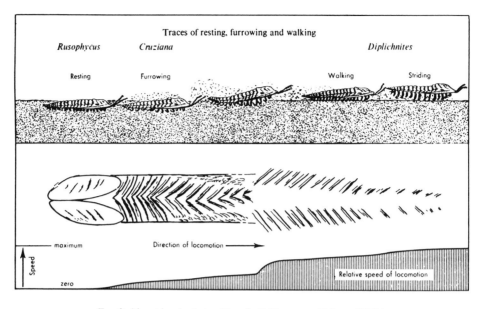

FIG. 9. Transitional relationships of trilobite traces (Crimes, 1970b).

Another example is the transition from resting impressions *(Ruhespuren)* to motion trails *(Bewegungsspuren)* of trilobites observed by CRIMES (1970c, pl. 5, fig. e) (Fig. 9). Nomenclatural problems arise when the two forms have received names, because they are also found singly (e.g., the motion trail *Cruziana* and the resting impression *Rusophycus*), both made by trilobites. One could, of course, consider these names to be synonyms and use only the older one *(Cruziana)* as was done by SEILACHER (1970).

SEILACHER (1953a, p. 434-435) supplemented his classification, especially for Recent lebensspuren, by including swimming trails, hatching structures, and functional structures mostly for the seizure of food (i.e., nets, traps, and others).

MÜLLER (1962, p. 25-28; 1963, p. 167) expanded SEILACHER's classification (see Fig. 10 [from OSGOOD, 1970, p. 290, fig. 3] for a complete English translation) and distinguished four main groups: Quietichnia (resting traces), Cibichnia (feeding structures), Movichnia (movement traces),

and Bioreactions (disease, parasitism, etc.), and four subgroups: Mordichnia (biting and gnawing traces), Cursichnia (running traces), Natichnia (swimming traces), and Volichnia (flying traces).

However, by the use of this expanded

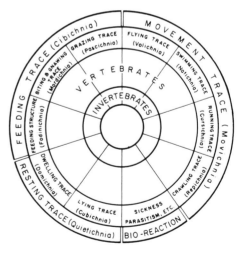

FIG. 10. MÜLLER's (1962) ethologic classification of lebensspuren as an expansion of SEILACHER's (1953) classification (from Osgood, 1970).

system, the application of the German terms can be misunderstood, and it also appears that this system is not entirely correct, as dwelling traces (domichnia SEILACHER) are included as a subgroup of quietichnia MÜLLER (=*Ruhespuren* MÜLLER, 1962; *non Ruhespuren* RICHTER, 1926, *nec* SEILACHER, 1953). By strict definition, bioreactions are not trace fossils. Swimming traces have so far been described from the Culm of Western Germany (FIEGE, 1951) and the Dwyka Group of South Africa (ANDERSON, 1972), but flying trails are, as yet, known only in the Recent and are difficult to identify as such. Therefore, I recommend that in the future, SEILACHER's (1953a) classification be adopted with his original definitions.

TAXONOMIC-STRATINOMIC-MORPHOLOGIC CLASSIFICATION PROPOSED BY VYALOV

VYALOV (1968b, p. 125; 1972) named all lebensspuren zooichnia or vivichnia. Since his classification differentiated between vertebratichnia and invertebratichnia, it was the first to classify trace fossils according to their producers (e.g., piscichnia, amphibipedia, etc.). Lebensspuren produced by invertebrates were divided into two main groups, bioendoglyphia and bioexoglyphia, which respectively correspond to endogenic and exogenic structures. VYALOV named traces produced by the appendages of organisms podichnacea, and all others, apodichnacea. These terms, respectively, correspond to the terms tracks and trails. Lebensspuren produced within a substrate have been named either 1) foroglyphia, produced in solid substrates such as hardgrounds and shells, or 2) fossiglyphia, produced in unconsolidated sediments. VYALOV (1968b, p. 126-127) introduced numerous additional morphological subgroups with so many new names that it is impractical to quote them all here. The names of these groups have endings analogous to those used for higher taxonomic units of the

VIVICHNIA

Invertebratichnia

  Bioendoglyphia (traces within the sediment)

    Foroglyphia (borings in hard substrate)

      Lithoforida (in stones and rocks)

      Coproforida (in organic substrate)

        Conchoforoidea (in shells)

        Arboforoidea (in wood)

    Fossiglyphia (burrows in unconsolidated sediment)

      Endotubida (tubular)

        Rectotubae (straight; Skolithos, Tigillites)

        Arcotubae (U–shaped; Arenicolites)

        Spirotubae (spiral; Gyrolithes, Xenohelix)

        Chondritae (chondrites; Chondrites)

      Crustolithida (branched, unordered; Ophiomorpha, Radomorpha)

      Helicoidida (helicoidal; Zoophycos)

      Cryptoreptida (subsuperficial; Scolicia)

FIG. 11. A portion of VYALOV's (1968b, 1972) classification of trace fossils (after Vyalov, 1968b).

zoological system (i.e., -a and -ae) and are easy to recognize. In 1972 VYALOV summarized and slightly modified his earlier views and presented them in tabular form (Fig. 11).

In this system, it could happen that ethologically and morphologically heterogeneous ichnogenera are placed in the same group. For example VYALOV (1968b, p. 127, table 3) placed the sinusoidal crawling trace *Cochlichnus,* the cylindrical and horizontal burrow *Palaeophycus,* and the meandering, grazing trail, *Cosmorhaphe,* all in the subgroup Vermiglyphidae, a subdivision of the Unipartoidae. I maintain that a classification that unites so many different forms in one and the same group is of little use. Even VYALOV described his classification as "artificial and conditional."

NOMENCLATURE OF TRACE FOSSILS

Since about 1850 it has become customary to use binary nomenclature for trace fossils in the same way that it has been used for body fossils. With trace fossils, however, the terms "genera" and "species" have a meaning different from that which is applicable to body fossils. As may be understood from the history of palichnology, too many finely differentiated genera and species have been established for trace fossils, because they originally were believed to be fossil plants, in particular, marine algae. This is especially true for the host of fucoids, as evidenced by the description of the history of the "genus" *Fucoides* by JAMES (1884).

The numerous, isolated descriptions scattered throughout world literature in paleobotanical, paleozoological, faunistic, stratigraphical, regional geological, and strictly palichnological papers have led to an excessive number of described genera and species. Because of the worldwide distribution and considerable vertical ranges of numerous trace fossils, the "new" forms were often published without knowledge or consideration of earlier literature.

Binary nomenclature has not been accepted universally for lebensspuren. Many authors have declined to give even descriptive informal names to trace fossils, which is an understandable and justifiable procedure, especially with poorly preserved forms. However, experience shows that these unnamed forms usually escape notice in later literature. I agree with OSGOOD (1970, p. 295), who asserts that "a form must be named if it is not to be 'lost' in the literature."

FAUL's (1951) suggestion of a designation by formulas may perhaps be suitable for vertebrate tracks, but it is not applicable to trails of invertebrates.

Repeatedly, the early term *Ichnium* was used as a blanket designation for undifferentiated trails. This was done in connection with species names, especially for Lower Permian vertebrate trails described from Germany (publications by PABST from 1896 to 1908) and later for invertebrate trails from the Lower Permian of Germany (SCHMIDTGEN, 1927, 1928). Some authors preferred HITCHCOCK's general term *Ichnites* for "all footmarks." This served as 1) a collective name, or 2) a special description when accompanied by a specific name describing single trails produced by vertebrates or invertebrates. A few paleontologists have generally opposed the use of names for trace fossils. NATHORST (1883a, p. 34, 287) observed that in view of the great similarity of trails produced by totally different animals, names for fossil forms were nearly worthless.

However, to make possible international discussion about individual forms or components of ichnocoenoses, trace fossils must be formally named. Supposedly new names of ichnogenera and ichnospecies should be based only on well-preserved material with well-defined morphological characteristics. Names should not be given to poorly preserved material or obscure forms. As long ago as 1894, JAMES drew attention to the many useless names which did not represent scientific progress, but were only a burden in the literature.

JARDINE (1853) proposed that the ending *-ichnus* be added to the generic names of vertebrate trails from Scotland so that it would be possible to distinguish names of trace fossils from body fossils by their characteristic endings. Soon after this, invertebrate trails were named in the same manner (i.e., *Cochlichnus* HITCHCOCK, 1858). More recently, SEILACHER (1953a, p. 446) and HÄNTZSCHEL (1962, p. *W*182) have recommended the application of the *-ichnus* ending for new ichnogenera, and this procedure is, at present, often employed.

When describing new ichnogenera or ichnospecies, it is suggested that the abbreviations *nov. ichnogen.* or *nov. ichnosp.* should follow the proposed names, not *nov. gen.* or *nov. sp.*

A survey of ichnogenera shows that quite frequently the name of the animal that produced the trail or structure is incorporated in the name of the ichnogenus. Some examples are *Arenicolites* SALTER and *Annelidichnus* KUHN. Just as often, trace fossils were named because of morphological characteristics (e.g., *Asterichnites* BROWN & VOKES, *Cylindricum* LINCK, and *Monocraterion* TORELL), or because they were originally thought to be of plant origin (e.g., names having the ending *-phycus* and such names as *Fucoides* BRONGNIART and *Hormosiroidea* SCHAFFER). Only occasionally is the age of the trace fossil indicated by its name (i.e., *Archaeichnium* GLAESSNER and *Permichnium* GUTHÖRL) or the locality where it is found (*Steigerwaldichnium* KUHN).

It is unavoidable that trace fossils, which were formerly assumed to be bodily preserved plants or animals and were named accordingly, now carry inconsistent names that have to be retained (e.g., *Fucoides,* for feeding burrows of marine animals).

The question as to whether a previously unknown trace fossil should be named as a new ichnogenus or should be established as a new ichnospecies of an existing "related" ichnogenus, is very difficult to answer. Such judgments are more or less subjective and depend entirely on the personal opinions of the investigator who establishes the new name. The same is true in considerations of questions of synonymy and the establishment of validity of names. When trace fossils are described according to the International Code, as has been common practice, the establishment or designation of a type species is necessary, but

the great variability of forms makes it very difficulty to select an ichnospecies that adequately represents all morphological variations of an ichnogenus. For this reason alone, a large number of monotypic ichnogenera have been established, and the number of trace fossil names is dismayingly large.

In view of these difficulties, it is understandable why MARTINSSON (1965, p. 204; 1970, p. 324) suggested that for trace fossils the practice of formalizing generic descriptions and designating type species should be abandoned. He proposed replacing ichnogeneric and ichnospecific names "by adopting terms which designate ecological types rather than taxia, such as cruzianae, dimorphichnia, and halopoans" (MARTINSSON, 1965, p. 204). Undoubtedly, a loose and unconstrained terminology has merit since these names would not be printed in italics and thus could be distinguished from generic names given to body fossils. Therefore, no diagnosis of new forms would be required. On the other hand, without clear and concise definitions of such terms as "a cruziana" or "a halopoan," they would be impossible to use in practice.

There are two opposing definitions of the meaning of names of trace fossils, which can be considered either 1) for the trace fossil itself, as the "work of an animal" (*Code,* Art. 16,a), or 2) for the producer of the trace fossil. These different points of view have been discussed quite recently, and it is still possible to speak of "two apparently irreconcilable schools" (OSGOOD, 1970, p. 296-297). SEILACHER (1956b, p. 158) stated, *"Ichnofossilien werden nicht in Stellvertretung ihres Urhebers benannt"* [Trace fossils are not to be named as substitutes for their producers] and considered trace fossils to be features independent of their producers. I am of a similar opinion, and believe that a name should describe only the trace fossil and not its producer. It must, however, be taken into consideration that when only behavioral patterns and biogenic sedimentary structures are named, one can only guess as to the identity of the animal that produced a particular trace fossil, particularly if the producer is an invertebrate.

For trace fossils in hard substrates, such as borings, BROMLEY (1970) has emphatically insisted that only the names of the trace and not that of the animal producer of the trace should be valid. Names such as *Cliona* or *Polydora* should not be applied to borings because they apply to the producer of the structure. The name of a boring should suggest no more than that it is a hole in a shell or some other hard substrate. An example of the alternative interpretation of trace fossil names is the description of the genus *Ixalichnus* CALLISON (1970), which by the ending *-ichnus* is clearly established as a trace fossil. However, CALLISON (1970) assigned *Ixalichnus* as a new genus to the subphylum Trilobitomorpha, phylum Arthropoda, adding that *Ixalichnus* "spent much of his time swimming. . . ."

The trace fossil and its producer are rarely found together. This situation has been observed for trilobite lebensspuren when a typical resting impression is found associated with its producer *in situ* (OsGOOD, 1970, p. 296, pl. 57, fig. 1 and pl. 58, fig. 4,5). In a few rare cases, the producer is found at the end of its running or crawling trail and in this manner, a definite producer can clearly be demonstrated (e.g., limulid trails from the Upper Jurassic Solnhofen Limestone) (Fig. 12).

Since the *Code* is inconsistent and contradictory in regard to the naming of ichnotaxa, the nomenclature of trace fossils is in a state bordering on chaos. As regards names established before 1931, Article 12 of the *Code* prescribes that, in order to be available, such a name must be accompanied by a "description, definition, or indication." Article 16 defines "what constitutes an indication" and includes as one of the definitions "the description of the work of an animal, even if not accompanied

Fig. 12. A *Limulus* preserved at the end of its trail (Abel, 1935).

by a description of the animal itself." It is thus perfectly clear that names given to trace fossils before 1931 are available under the *Code* and have to be treated on an equal footing with all other zoological names. This is further clarified by Article 24 (b) (iii) which states that the Law of Priority applies "when, before 1931, a name was founded on the work of an animal before one is founded on the animal itself."

However, for names published after 1930 a different set of rules applies. The critical rule is that stated in Article 13 (a) (i) which requires that such a name must be "accompanied by a statement that purports to give characters differentiating the taxon." This requirement is, of course, impossible to fulfill in the case of trace fossils of which the producer is generally not known. Hence, names for trace fossils established after 1930 are not available under the *Code*.

In order to clarify this situation, HÄNTZSCHEL & KRAUS (1972) submitted an application to the I.C.Z.N. which has been published in Volume 29 of the *Bulletin of*

Zoological Nomenclature. In this application, the authors asked the Commission to issue a Recommendation (Appendix E of the *Code*) that all names of lebensspuren should be treated in the same way as prescribed for categories of names presently governed by the *Code*. They also recommended that names of ichnogenera should not be italicized, but for purpose of conformity with general *Treatise* style, such names are printed in italics here. With this exception, the trace fossil names in the present volume are dealt with in conformity with recommendations made by HÄNTZSCHEL & KRAUS (1972). (See also Editorial Preface, p. vii.)

[As might be expected, the HÄNTZSCHEL & KRAUS proposal has received critical review from scientists in many countries (FREY, 1972; MARTINSSON, 1972; TEICHERT, 1972; VOIGT, 1973; LEMCHE, 1973; YOCHELSON, 1973). All are unanimous in their desire that the problem of the availability of trace fossil names be faced now and settled once and for all, but not everyone has agreed on how this should be accomplished.

FREY, MARTINSSON, TEICHERT, and YOCHELSON agreed basically with the proposal supporting availability of all names for trace fossils and emphasized the need for these names to continue in italic print. YOCHELSON (p. 71) in addition suggested a logical solution for all this confusion: "by removing the post-1930 restriction, the rules will be allowed to operate for the 'indications' of animals. A minimum of problems results from such a course of action."

LEMCHE (p. 70) on the other hand believed that there was excellent justification for the freeing of all post-1930 trace fossil names from the rules of the *Code*, adding that if anybody can propose a better system "than that proposed by the present applicants, he should hasten to do so." Perhaps SARJEANT & KENNEDY (1973) have already answered LEMCHE's plea with their "Proposal of a code for the nomenclature of trace fossils" which would exempt the names of trace fossils from the rules of both the Zoological and Botanical Codes. However, as the title suggests, this is only a proposal, or more properly, a "draft and not a finished product" which "may at least stimulate thought and discussion" (SARJEANT & KENNEDY, 1973, p. 465). It has no legal standing, especially if the HÄNTZSCHEL & KRAUS proposal is accepted.—CURT TEICHERT, W. G. HAKES.]

SIGNIFICANCE OF TRACE FOSSILS FOR SEDIMENTOLOGY

Inorganic sedimentary structures produced by physical processes can be altered or destroyed by burrowing, crawling, agitating, and ingesting the sediment by infaunal elements (Fig. 13). These biological processes produce sedimentary structures that have been described as bioturbation or biogenic sedimentary structures.

Vagile sediment ingestors and the more or less stationary dwelling structures of animals in the sediment interact with the sedimentation processes in their environ-

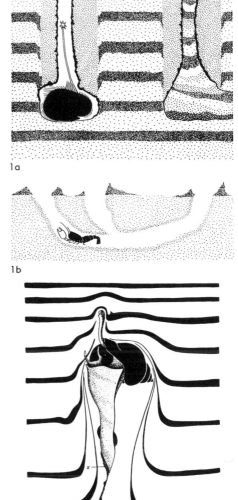

1a

1b

1c

FIG. 13. Some examples of sedimentary structures associated with biogenic activity (Schäfer, 1956). ——*1a.* Left: *Echinocardium* at the bottom of its burrow; right: after sea urchin leaves its burrow, cavity is later filled by inorganic sedimentation. ——*1b.* Cross section of *Callianassa* burrow. Sediment is piled at openings of burrow by the crab. ——*1c.* Deformation of sand layers produced by the upward movement of the gastropod *Buccinum* in the sediment (x = sand mixed with mucus).

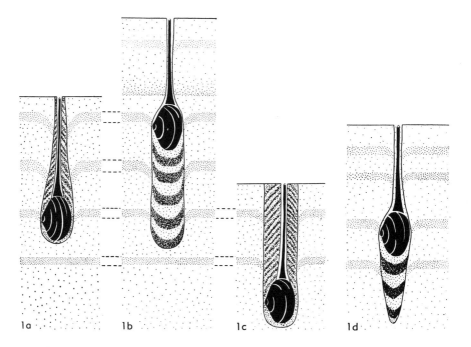

Fig. 14. Relationships of burrowing structures of unisiphonal pelecypods to rates of sedimentation (Reineck, 1958a).——*1a.* No sedimentation: a conical burrow forms above a growing pelecypod.—— *1b.* Rapid sedimentation: as the animal moves upward through the sediment, a burrow is formed below equal to the animal's width.——*1c.* Erosion: animal migrates downward in sediment producing a burrow above it equal to its width.——*1d.* Very slow sedimentation: a growing pelecypod follows the accumulation of sediment upward creating a conical burrow beneath it.

ment. Rapid or slow sedimentation, non-deposition, or the removal and change of sedimentary processes can often be determined by studying trace fossils.

The paleoichnology of marine sediments must be based on detailed knowledge of the relationships of Recent benthonic communities to the sediment. SCHÄFER (1956; 1972) and REINECK (1958a,b; 1972) have studied the influence of different benthonic organisms on the bedding of Recent sediments by observations on the tidal flats of the North Sea and in aquariums. However, little is as yet known about occurrences of lebensspuren in the neritic, bathyal, and abyssal zones of the ocean (HERSEY, 1967; HEEZEN & HOLLISTER, 1971; PEQUEGNAT et al., 1972).

Benthonic organisms live at specific depths in the sediment (Fig. 14). When excessive amounts of sediment accumulate above an animal, it will create an escape structure or tunnel, primarily by digging upward, in order to raise its position in the sediment. This upward motion within the sediment produces a displacement or bending of the sedimentary layers above and below the animal's escape burrow (Fig. 13,*1c*; Fig. 15,*4*). The very vagile *Sipunculus* produces upward warping of the sedimentary layers during the production of escape tunnels (Fig. 15,*3*). In comparison, downward arching of sedimentary layers has been observed mostly in the escape tunnels of polychaetes (Fig. 15,*2*), some bivalves (Fig. 14,*1b*), and the sea anemone, *Cerianthus* (Fig. 15,*1*). Similar sedimentary deformation is produced by the burrowing of many polychaetes, echinoderms, and brachyurans, and such bioturbate sedi-

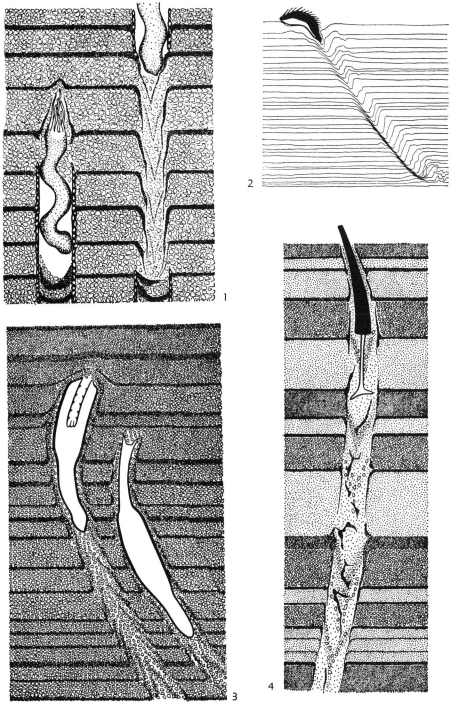

FIG. 15. Examples of escape structures (from Schäfer, 1972).——*1.* Sea anemone, *Cerianthus,* covered by sediment, evacuates its burrow and moves upward in the sediment (schem.).——*2.* As large polychaete, *Aphrodite aculeata,* moves upward, beds sag downward behind it (schem.).——*3.* *Sipunculus* moves upward in the sediment, and beds are pulled upward with the animal, ×0.3.——*4.* Turbate trail of scaphopod moving upward in the sediment (schem.).

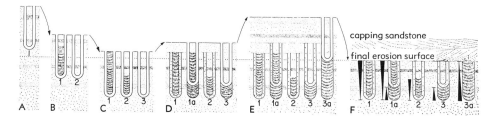

FIG. 16. Movement pattern of *Diplocraterion yoyo* (Goldring, 1964). In the Upper Devonian Baggy Beds, England, this trace occurs in various types shown in *(F)*, where all have been truncated to a common erosion surface. Repeated phases of erosion and sedimentation led to the development of the various types. Stage *(A)*, development of burrow *(1)*: with degradation of surface, this tube migrates downward, and at intervals, new tubes (2 and 3) are constructed *(B and C)*. Sedimentation follows *(D and E)* but some of the tubes are abandoned. Stage *(F)*: all tubes are abandoned and erosion reduces them to a common base.

mentary structures occur around burrowed tubes in Cambrian sandstone and quartzite beds in Europe. However, it appears that such "escape structures" have been recorded only rarely in the literature (FREY, 1973b). Perhaps they have been overlooked.

Erosion can cause infaunal elements to migrate downward through horizontal sedimentary layers in order to reach their required living depth. This is especially true of pelecypods, which also produce similar biogenic structures (Fig. 14,*1c*).

An excellent example of the reaction of sediment-dwellers to sedimentation processes is seen in the Upper Devonian *Diplocraterion* tubes in England studied by GOLDRING (1962) (Fig. 16). Different types of U-shaped tubes, normal protrusive, retrusive, and abandoned, with spreite structures, give an indication of the reaction of the infauna to repeated changes from deposition to erosion. For these occurrences, the appropriate species name *Diplocraterion yoyo* was coined. In the Aptian of England, MIDDLEMISS (1962) concluded that poorly preserved burrows are commonly found in highly turbated beds deposited during periods of slow sedimentation, whereas better preserved burrows indicate rapid sedimentation. In Jurassic sandstones, resting impressions such as *Asteriacites* have been observed to exhibit vertical repetition of impressions within the sediment. These oc-

currences are undoubtedly the result of the upward escape of the animal through the sediment in response to considerable sediment influx (SEILACHER, 1953b) (Fig. 17).

Areas of slow deposition or nondeposition provide favorable substrates for the settlement in the sediment of burrowing organisms and filter-feeders. For the most part, presence of numerous excavated burrows *(Wühlspuren)* indicates stable substrates or slow sedimentation rates.

Occasionally, during temporary nondeposition of sediment the surface of fine-grained sediments may be converted into hardgrounds. Such occurrences are typical for the Upper Cretaceous of western Europe where domiciles *(Wohnbauten)* of crustaceans and echinoderms are found in such rocks in many places. The abutment of such burrows against an obstacle such as a shell, or detour of a tunnel around an obstacle, indicate that the burrow was excavated before the sediment was lithified (RASMUSSEN, 1971).

Many seemingly homogeneous sediments have completely lost their original bedding as a result of intense bioturbation (MOORE & SCRUTON, 1957, p. 2743). However, complete obliteration of bedding features is rare and occurs only if an abundant infauna was present, sedimentation was slow or absent, and if the infaunal animals had enough time to rework the sediment.

These examples show the importance of

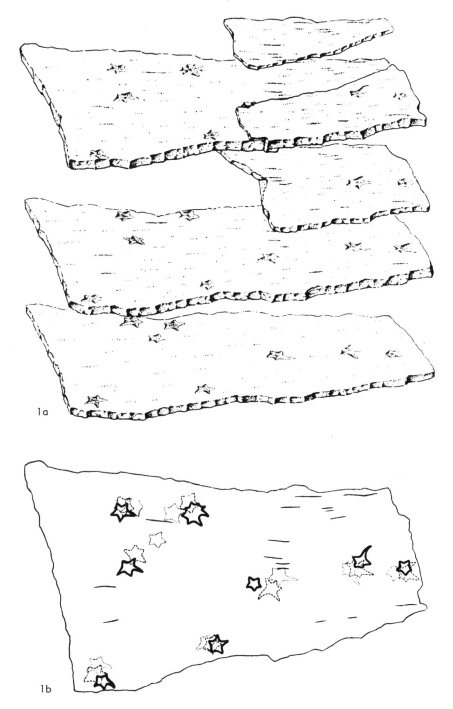

FIG. 17. Starfish impressions, *Asteriacites lumbricalis,* Lower Triassic, southern Tirol (Seilacher, 1953). ——*1a.* Expanded view of bedding planes showing upward migration of starfish as a result of rapid sediment influx.——*1b.* Composite overview of *1a,* solid outlines indicate impressions stratigraphically above dotted outlines.

endogenic traces and burrows for the clarification of sedimentological problems and for interpretation of the depositional history of many sediments. Further investigations on interrelationships between Recent infauna and sediments in different biotypes are necessary to provide a sounder basis for paleoichnological research.

SIGNIFICANCE OF TRACE FOSSILS FOR PALEOENVIRONMENTAL INVESTIGATIONS

For the most part, the paleoenvironment of marine sediments can be interpreted by investigating lithology, primary sedimentary structures, and faunal elements. In recent years, trace fossils and associations of trace fossils, because of their autochthonous nature, have been shown to be particularly useful in paleogeographic investigations. With very few exceptions trace fossils are preserved *in situ*. They cannot be displaced, and, in contrast to many body fossil assemblages, they form no thanatocoenoses. Lebensspuren provide certain evidence of life on and within the sediment. In addition, many trace fossils are good facies indicators.

Through worldwide comparison of ichnocoenoses in marine sediments of different ages, SEILACHER (numerous publications since 1954) has shown that characteristic trace fossil assemblages occur in many places in sediments of different ages. Each such assemblage belongs to a particular marine environment and is composed of specific associations of trace fossils, constituting an ichnofacies. The environment is characterized by the composition and texture of the sediment, and by oceanographic factors such as water depth, salinity, water circulation, and many others.

The contrasts between different ichnofacies are best recognized in the "ichnospectra," which give a quantitative picture of the individual trace fossil associations according to their ethologic classification. As a rough generalization, the differences between trace fossil assemblages in shallow

and deep water can be characterized as follows: In shallow water, vertical tubes, burrowing structures, dwelling burrows, and resting impressions predominate. In deep water, complicated spreitenbauten and many, varied, grazing trails of sediment-ingestors develop. SEILACHER (1954, 1955, 1959) was first to call attention to different ichnocoenoses and their time-independent facies relationships associated with flysch and molasse deposits. The trace fossils associated with geosynclinal flysch sediments contain assemblages of different grazing trails, whereas epicontinental and paralic molasse deposits are characterized by various resting impressions. Both of these examples have been found in Paleozoic, Mesozoic and Cenozoic rocks. The ichnocoenoses in predominantly fluviatile and continental deposits, with only periodic marine inundations, again show a different composition. Here, all ethologic associations are represented, with the exception of grazing trails. These associations have low diversity, but are generally rich in individuals. The ichnocoenoses of the Buntsandstein ("Bunter," Lower Triassic) and the Keuper Sandstone (Upper Triassic) of central Europe are examples.

More recent investigations of ichnocoenoses of different ages and from different geographic areas have shown the necessity to establish additional types of trace fossil assemblages. In some cases, small, local "subassociations" of trace fossils have been established. Every ichnocoenosis corresponds to a defined relatively narrow, facies range. There are no restrictions to certain sediment types and they are named after trace fossils characteristic for them. SEILACHER (1967b) distinguished the following ichnofacies and compared them with their particular environments at different bathymetric levels (Fig. 18):

1) *Scoyenia* facies: nonmarine; commonly redbeds.

2) *Skolithos* facies: littoral; rapid sedimentation and frequent transportation.

3) *Glossifungites* facies: littoral; ero-

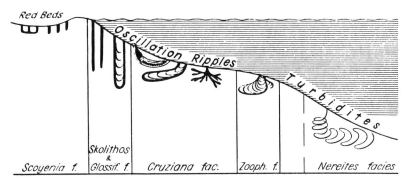

Fig. 18. Bathymetric zonation of trace fossil assemblages [*f* = facies] (Seilacher, 1967b).

sional surfaces, restricted to single bedding planes (erosion surfaces).

4) *Cruziana* facies (formerly: resting-impression facies): deeper shallow water, below the true littoral zone.

5) *Zoophycos* facies: transitional to bathyal zone.

6) *Nereites* facies (formerly: grazing-trail facies): bathyal to abyssal; pelagic sediments and turbidites.

CHAMBERLAIN (1971c) established a *Chondrites* assemblage in the Upper Paleozoic of Oklahoma (United States) which is a bathymetric zone transitional between the *Nereites* and *Zoophycos* associations.

Almost certainly, marine trace fossil assemblages are not solely depth-dependent. SEILACHER and, more recently, OSGOOD (1970, p. 403) and FREY (1971, p. 110-111) have pointed out that in addition to oceanographic conditions, factors such as nutrient supply may influence the composition of biologic ichnocoenoses, independent of bathymetry. Future investigations probably will introduce additional subassociations of trace fossils, or the boundaries between ichnofacies will be less distinct. OSGOOD (1970, p. 403) believes that, for example, a coexistence of pascichnia and cubichnia "at some intermediate depth" is possible and that a sharp distinction between the *Cruziana* facies and *Nereites* facies cannot be made. He also doubted that the *Zoophycos* facies was anything but a transitional facies, because it seems that

in the United States *Zoophycos* occurs in both deep and shallow water sedimentary deposits. [See OSGOOD & SZMUC (1972) for a more detailed discussion.] FREY & MAYOU (1971) have studied the distribution of Recent decapod burrows from Holocene barrier island beaches along the Georgia coast, and according to these authors, burrow orientation and morphology reflects distance from shore (Fig. 19).

On the other hand, similarities exist between Recent lebensspuren produced at great depths and trace fossils that were probably produced in a similar environment. Thus, spiral lebensspuren have been observed in the abyssal zone of the present seas which are similar to many grazing trails found in flysch deposits (BOURNE & HEEZEN, 1965; EWING & DAVIS in HERSEY, 1967; HEEZEN & HOLLISTER, 1971). Also, very large star-shaped lebensspuren have been found on the deep sea bottom which resemble similar forms found in Polish and Spanish flysch sediments. SEILACHER (1967b) compared the cross section of horizonal spreite structures found in Recent deep sea muds to *Zoophycos,* which is found in many flysch deposits.

As might have been expected, regional geological investigations have shown that as the depositional environment changes with time, trace fossil assemblages vary in vertical succession through the rock sequence. They reflect accurately the geological development, especially in geosynclinal

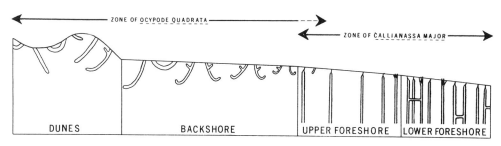

Fig. 19. Zonation of decapod burrows in Holocene barrier island beaches, Georgia. Diagram stresses form and configuration, rather than size and relative abundance, of ghost crab burrows (Frey & Mayou, 1971).

areas. Successive stages are also reflected in the lithology of the sediments and their primary structures. Such investigations make it possible to check paleogeographic conclusions drawn from observation of changes in the ichnocoenoses (see SeiLACHER, 1963; SEILACHER & MEISCHNER, 1965; CHAMBERLAIN, 1971a,c).

Regional comparisons of trace fossil as-

semblages are also possible in the horizontal dimension. If lithologies change from one to another, the trace fossil assemblages associated with them are also different. It is therefore possible by combined ichnologic and sedimentologic studies to reconstruct the paleogeographic development of large areas.

In some instances, the occurrence of just

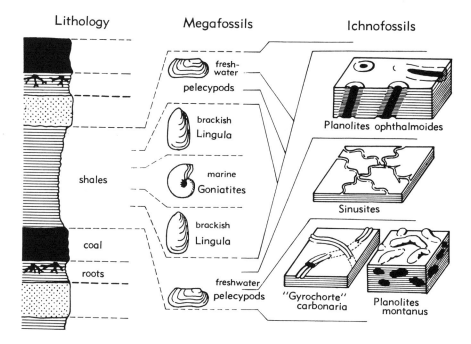

Fig. 20. Within lithologic cyclothems in paralic deposits of Carboniferous age in the Ruhr Basin, as shown above, more members can be recognized with the help of trace fossils. For this purpose it makes no difference that these trace fossils belong to rather insignificant types which in other formations may occur in dissimilar types of facies (Seilacher, 1964c).

a few trace fossils is sufficient to make possible deductions regarding the depositional environment of the sediment. Rudolf Richter (1931) demonstrated that the occurrence of *Chondrites* in the Hunsrück Shale of Germany indicates that the original sediment did possess an infauna and was not an H$_2$S-rich sapropel as had been believed previously. In a genuine euxinic environment, lebensspuren would be entirely absent.

Trace fossils can also help to determine certain characteristics of the depositional environments of sediments, especially in the marine realm. By studying trace fossils, lithologies, and body fossils in paralic Upper Carboniferous cyclothems of western Germany, Seilacher (1963, 1964c, p. 307) (Fig. 20) has been able to distinguish whether a sediment was deposited in freshwater, brackish water, or under marine conditions. Some conclusions as to the strength and direction of currents can be drawn from the study of trace fossils. A few examples are: 1) deviation and obliteration of trilobite running trails, especially by lateral currents across the trails, 2) current orientation of resting impressions parallel to the direction of flow (rheotactic orientation, mostly against current direction), 3) existence of different kinds and varying abundances of lebensspuren in areas with strong, as contrasted with weak currents, and 4) orientation against the current (presumably tidal currents) of some dwelling structures in the Jurassic of England (Farrow, 1966).

TRACE FOSSILS

The definition of the concept "trace fossil" in the Introduction indicates the kind of fossils discussed in this section. As the result of the very numerous trace fossil investigations undertaken since the first edition of this chapter (Häntzschel, 1962), the number of ichnogenera has increased considerably. Unfortunately, many forms lacking definite characters have been given names when only simple morphological descriptions were needed. In some cases, descriptions as well as illustrations were insufficient. Some of the original "generic" diagnoses were changed by some authors, mostly expanded, so that forms that diverged considerably from the early definitions were listed under the old names. Also, many transitional forms between well-defined and well-known ichnogenera have been recognized. This was to be expected and it demonstrates the difficulties of identification and nomenclature of trace fossils. It is not easy to find a compromise between a narrow and a broad definition of trace fossil generic concept. Frequently also, authors have changed their ideas about the definition of an ichnogenus, thus creating synonyms.

I have tried to list all ichnogenera published before the end of 1971. Since good, clear illustrations are very important in the description of trace fossils, the illustrations have been improved and their number has been increased as far as possible. In many recent ichnological publications, ichnocoenoses have been classified according to the well-known "ecological" system of Seilacher discussed above. However, in this volume, for reasons given in the first edition, the arrangement of ichnogenera in alphabetical sequence of names has been preserved. Descriptions of especially widespread and important ichnogenera are given in greater detail, and following them, expanded statements concerning former and present interpretations. Complete references to old and new literature about ichnogenera are found in the reference list.

In a review of the *Treatise* Part W of

1962, SEILACHER (1964b) stated that in the section on trace fossils, about half of the names could have been placed in synonymy. This proportion seems too great to me. As already indicated in the Introduction, the placing of trace fossil names as synonyms depends very much on subjective judgment. In the future, careful research on individual ichnogenera based on abundant and well-preserved material on a worldwide basis is required, and so are fundamental monographs of entire rich ichnocoenoses, such as, for example, the extensive investigations of trace fossils of the Cincinnatian of Ohio by OSGOOD (1970). Such large, regional works are not only necessary for paleoichnology; they also contribute to understanding of the paleoecology and paleogeography of sedimentary basins and of animal-sediment interrelationships generally. This is true for Recent as well as fossil ichnocoenoses.

Acanthichnus HITCHCOCK, 1858, p. 150 [*A. cursorius*; SD LULL, 1953, p. 40] [Includes 9 widely different "species" (HITCHCOCK, 1865, p. 13-15); see also *Pterichnus* HITCHCOCK, 1865, p. 14]. Linear tracks consisting of 2 parallel rows of short straight strokelike impressions mostly slightly turned outward; tracks very different in width, position, and length of impressions of feet. [?Made by insects.] *Trias.*, USA(Mass.).

Acanthorhaphe KSIĄŻKIEWICZ, 1970, p. 301 [*A. incerta*; OD] [=*Acanthoraphe* KSIĄŻKIEWICZ, 1961, p. 883, 888; published as "n.f." without species name]. Thin sole trails, 1 mm. in width; winding in somewhat irregular curves of small "amplitude"; with short lateral thornlike branches, usually on convex side of curves, sometimes on both sides. [See also *Unarites* MACSOTAY, 1967, p. *W*120, and *Protopaleodictyon* KSIĄŻKIEWICZ, 1970, p. *W*97.] *L.Cret.(low.Neocom.,Berrias.); Tert.(low.Eoc.),* Eu.(Pol.).——FIG. 21,*1a. A.* sp., L.Cret., Pol.; ×0.6(Książkiewicz, 1961).—— FIG. 21,*1b-e.* *A. incerta,* L.Cret.(Berrias.), Pol.; ×0.5 (Książkiewicz, M., 1970, p. 302, in: *Trace Fossils* edited by T. P. Crimes & J. C. Harper, Geol. Jour. Spec. Issue 3, Seel House Press, Liverpool).

Aglaspidichnus RADWAŃSKI & RONIEWICZ, 1967, p. 545 [*A. sanctacrucensis*; M]. Hypichnial trace (15.5 cm. long by 12.0 cm. wide, max.) composed of sinuous, longitudinal axial ridge with ovally triangular posterior ending and 8 laterally opposed, posteriorly bent ridges projecting from axial ridge. [Interpreted as cast of rest-

ing place of aglaspid arthropod (very probably of family Beckwithiidae RAASCH, 1939); impression of pygidial shield preserved; no trace of prosoma visible; only one specimen known.] *U.Cam.,* Eu.(C.Pol.).——FIG. 21,*2. *A. santacrucensis; 2a,b,* ×0.5 (Explanation of *2a: a', a''* =two small hieroglyphs, *Rusophycus* sp., deforming ant. part; *b*=small synaeresis crack cutting first three ridges on right side) (Radwański & Roniewicz, 1967, mod.).

Agrichnium PFEIFFER, 1968, p. 671 [*Palaeophycus fimbriatus* LUDWIG, 1869, p. 111; OD]. Groups of small subparallel smooth furrows of unequal moderate length. [Probably grazing trails; see also *Schaderthalis* HUNDT, 1931 *(nom. nud.),* p. *W*000, whose type species *S. bruhmii* has been ascribed to *Agrichnium* by PFEIFFER, 1968, p. 672.] *L.Carb.(Kulm),* Eu.(Ger.,Thuringia).—— FIG. 22,*1. *A. fimbriatum* (LUDWIG); ×1.2 (Pfeiffer, 1968).

Algites SEWARD, 1894, p. 4, *emend.* STOPES, 1913, p. 254 [no type species to be designated]. Seldom used, comprehensive generic name given to replace all older generic names of "algae" which suggest relationship with living forms. [According to PIA (1927, p. 110), SEWARD's "species" belong to algae but other species interpreted as algae (JACOB, 1938; ?Jur., India) represent trace fossils *(Chondrites).*]

Allocotichnus OSGOOD, 1970, p. 358 [*Asaphoidichnus dyeri* MILLER, 1880, p. 219; OD] [=*partim Asaphoidichnus* MILLER, 1880, p. 217 (type, *A. trifidus*); for discussion see OSGOOD, 1970, p. 359]. Wide, bifid, dimorphic track; each set consisting of maximum of 4, occasionally only 3 or 2 pairs of imprints; on one side arranged as 4 long subparallel raking imprints, on other side preserved as *en echelon* support imprints; only first 4 or 5 pairs of walking legs used; body of producer angled to right of direction of movement; detailed morphology varies. [Interpreted as crawling track of large arthropod with relatively small number of walking legs, probably made by multisegmented trilobites *(Isotelus?),* but differing greatly from known trilobite tracks by uniqueness of motion.] *U.Ord.(low.Cincinnat.),* USA(Ohio-Ky.).——FIG. 23,*1. *A. dyeri* (MILLER), Eden beds, repichnia of *?Isotelus,* convex hyporelief; *1a,c* (Ky.), ×0.56, ×0.6; *1b,d*(Ohio), ×0.45, ×0.56; *1e* (Ohio), holotype, ×0.6 (Osgood, 1970).

Amphorichnus MYANNIL, 1966, p. 202 [*A. papillatus*; OD]. Fillings of cylindrical and amphoralike hollows; length (max.) 7 to 8 cm., diameter (max.) 3 to 4 cm.; at lower end distinct peak similar to mamilla; perpendicular to bedding plane. [Dwelling burrow or resting trail.] *Low. M.Ord.,* USSR(Est.).——FIG. 24, *3. *A. papillatus,* Ord. Kalke, Baltic; ×0.75 (Myannil, 1966).

Annelidichnium KUHN, 1937, p. 368 [*A. triassicum*; M]. Tunnel fillings with irregular sculp-

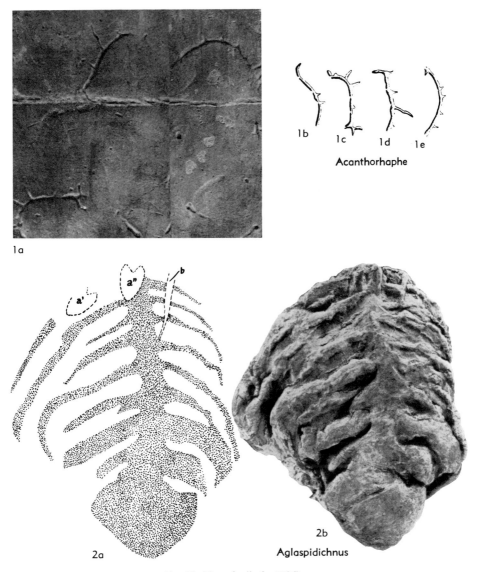

1a

1b 1c 1d 1e

Acanthorhaphe

2a

2b

Aglaspidichnus

FIG. 21. Trace fossils (p. *W*36).

ture; ornamented with sharp or rounded longitudinal ridges or blunt tubercles. [Type or other specimens are lost.] *U. Trias.,* Eu.(S.Ger., Bavaria).

Archaeichnium GLAESSNER, 1963, p. 117 [**A. haughtoni;* M]. Fillings of cylindrical burrows with external longitudinal striation (10 to 12 striae), faint transverse sculpture inside; diameter (max.) about 5 mm., thickness of walls 1 mm. [Erroneously described as Archaeocyatha by HAUGHTON (in HAUGHTON & MARTIN, 1956; 1960), certainly a trace fossil.] *Uppermost Pre-*

cam.(Nama Syst., Kuibis Quartzite), S.Afr.——— FIG. 24,4. **A. haughtoni;* ×1.5 (Glaessner, 1963).

Archaeonassa FENTON & FENTON, 1937, p. 454 [**A. fossulata;* OD]. Crawling trail, 1 to 7 mm. wide, consisting of regularly convex furrow and 2 low narrow lateral ridges; furrow rarely smooth, mostly crossed by rounded wrinkles, which are convex in anterior direction. [Interpreted as gastropod trail, similar to those made by Recent snails on tidal flats (e.g., *Ilyanassa obsoleta* or small species of *Littorina*; trails probably belong-

Agrichnium

Fig. 22. Trace fossils (p. *W*36).

ing to "group" *Scolicia* QUATREFAGES, 1849, p. 265.] *L.Cam.,* Can.(B.C.).

Ardelia CHAMBERLAIN & BAER, 1973, p. 88 [**A. socialia;* OD]. Cylindrical tunnels projecting radially from vertical and/or oblique shafts; may possess either smooth or nodose surfaces; straight or curving; radial bifurcations from central shaft may extend up to 10 cm. and can apparently occur at different levels; no internal lining of burrow system is apparent; diameter of total structure 10 to 20 cm.; depth in sediment 0.3 to 2 meters or more. [Interpreted to have been produced by a thalassinid decapod; judging from original description, it is possible that *Ardelia* is the same as *Ophiomorpha* (see p. *W*85) and *Thalassinoides* (see p. *W*115).] *U.Perm.(Wolfcamp.),* N.Am.(USA,Utah). [Description supplied by CURT TEICHERT & W. G. HAKES.]

Arenicolites SALTER, 1857, p. 204 [**Arenicola carbonaria* BINNEY, 1852, p. 192; SD RICHTER, 1924, p. 137; second SD (rejecting RICHTER's designation) by BATHER, 1925, p. 198: *Arenicolites didymus* SALTER, 1857, p. 200 (=*Arenicola didyma* SALTER, 1856, p. 248)] [=*Arenicolithes* HILDEBRAND, 1924, p. 27 *(nom. null.)*]. Simple U-tubes without spreite, perpendicular to bedding plane; varying in size, tube diameter, distance of limbs, and depth of burrows; limbs rarely somewhat branched, some with funnel-shaped opening; walls commonly smooth, occasionally lined or sculptured; burrows may reach considerable depth. [Certainly made by worms or wormlike

animals; in places widely distributed; TRUSHEIM (1934) described, from German Middle Triassic Muschelkalk, *Arenicolites franconicus* marker-bed, only 2 to 5 cm. thick, which has observed lateral extent of about 25 kilometers.] *Cam.-U.Cret.,* Eu.-USA-Greenl. [Probably worldwide.]——FIG. 24,*2a. A.* sp. SALTER; schematic (Trusheim, 1934).——FIG. 24, *2b. A. franconicus* TRUSHEIM, M.Trias., Ger.; schem. cross sec. of burrow, ×0.8 (Trusheim, 1934).

[Several species should not be placed in genus, e.g., *A. didyma* SALTER seems to be resting trace of the *Rusophycus* type; *A. spiralis* TORELL, 1868, is type species of *Spiroscolex* TORELL, 1870; *A. lunaeformis* KOLESCH, 1922, *A. zimmermanni* KOLESCH, 1922, *A. statheri* BATHER, 1925, and *A.? lymensis* BIGOT, 1941, are U-shaped burrows with spreite.]

Arthraria BILLINGS, 1872, p. 467 [**A. antiquata;* M]. Bars on bedding surfaces with spheroidal expansions at each end, similar to pair of dumbbells. [*Arthraria biclavata* MILLER, 1875 (p. 354), from the Cincinnatian of USA(Ohio) has been interpreted by K. E. CASTER (pers. commun.) and HÄNTZSCHEL (1962, p. *W*184), as U-shaped burrow with spreite, similar or possibly identical with *Corophioides* or *Diplocraterion;* OSGOOD, 1970, p. 323, placed this species in *Corophioides,* regarding it as base of U-tube with spreite "where the arms have been secondarily deepened below the base of the spreite"; type specimen of *A. antiquata* BILLINGS, 1872, from Silurian of Canada has been lost, thus this species is *incertae sedis;* as concerns *A. magna* RUEDEMANN, 1925 (U.Ord., USA), see OSGOOD, 1970, p. 325; *A. renzii* HUNDT, 1929 (Sil., Ger.) = *nom. nud.*] *?Cam., Ord.-Sil., ?Penn.,* N.Am.

Arthrophycus HALL, 1852, p. 4 [**A. harlani;* M (=*Fucoides harlani* CONRAD, 1838, p. 113)] [=*Harlania* GOEPPERT, 1852, p. 98 (no type species designated); *Rauffella palmipes* ULRICH, 1889, p. 235; *Arthrophicus* HERNANDEZ-PACHECO, 1908, p. 83 *(nom. null.);* for synonymy see also BASSLER, 1915, p. 70]. Bundles of annulated curved burrows, simple or branched, subquadrate in cross section, mostly 1 to 2 cm. in diameter, up to 60 cm. long, commonly bilobate with median longitudinal depression; surface showing strong, very regularly spaced transverse ridges; internal chevron-shaped filling. [Feeding burrow; for history of the genus see JAMES (1893); at first regarded as plant (even as late as 1952 by BECKER & DONN), inorganic (tectonic origin advocated by SCHILLER, 1930); trails produced by arthropods or worms; first explanation as lebensspur given by NATHORST (1881a); according to SARLE (1906a), perhaps made by sedentary polychaetes; *Arthrophycus* sometimes considered junior synonym of *Phycodes* RICHTER, 1850 (e.g., by SEILACHER, 1955, p. 386); OSGOOD (1970, p. 342) agrees with author in differentiating the two genera; similar burrows from the Lower and Upper Cretaceous of USA (HOWARD, 1966; FREY

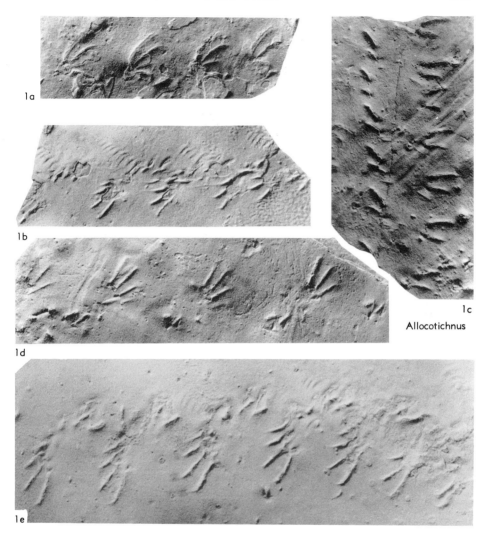

1a

1b

1c
Allocotichnus

1d

1e

FIG. 23. Trace fossils (p. *W*36).

& HOWARD, 1970) and from European Cretaceous and Upper Tertiary deposits (Pol., Aus.) have been compared with *Arthrophycus,* but are not typical.] *Ord.-Sil.,* N.Am.-S.Am.-Eu.-N.Afr.-Asia M.——FIG. 25,4. *A. alleghaniensis* (HARLAN), L.Sil., N.Y., ×0.3 (Häntzschel, 1962).

Arthropodichnus CHIPLONKAR & BADWE, 1970, p. 3 [*A. indicus; OD]. Track 1.8 to 2.0 cm. wide with 2 parallel rows of transverse slitlike opposing depressions separated by central axial region 0.25 to 0.3 cm. wide; distance between consecutive, marginal depressions 0.3 cm.; axial region has serially arranged slitlike depressions, apparently preserved in epirelief. [Probably produced by appendages of arthropod, perhaps a myriapod or chilopod.] *L.Cret.,* India.——FIG. 24,1. *A. indicus,* Nimar Ss., Amba Donger; ×0.7 (Chiplonkar & Badwe, 1970). [Description supplied by W. G. HAKES.]

Asaphoidichnus MILLER, 1880, p. 217 [*A. trifidus;* SD HÄNTZSCHEL, 1962, p. *W*184]. Large tracks, 6 to 15 cm. wide, consisting of 2 rows of mostly trifid imprints, about 2 cm. in length, individually varying in morphology; also combinations of unifid, bifid and trifid impressions observed; average per set, 9 imprints; tracks show both oblique and straight-ahead movement. [Produced by trilobites, most likely *Isotelus; A. dyeri* MILLER, 1880, removed by OSGOOD (1970, p. 359) from genus and placed as type species

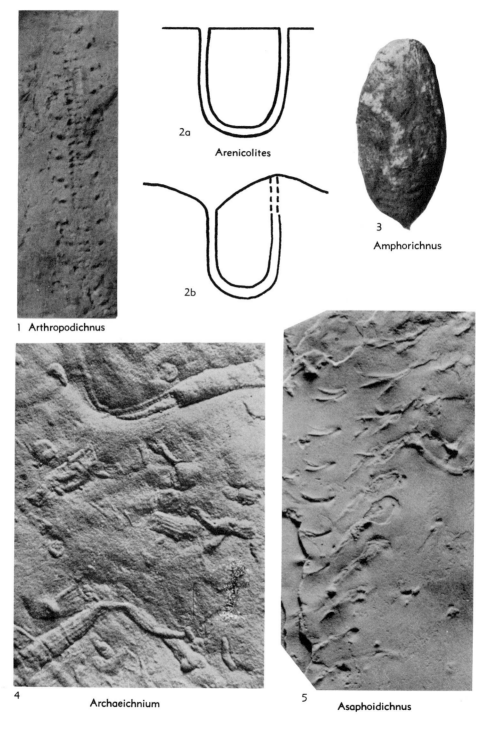

1 Arthropodichnus

2a Arenicolites

2b

3 Amphorichnus

4 Archaeichnium

5 Asaphoidichnus

FIG. 24. Trace fossils (p. *W36-37, 39*).

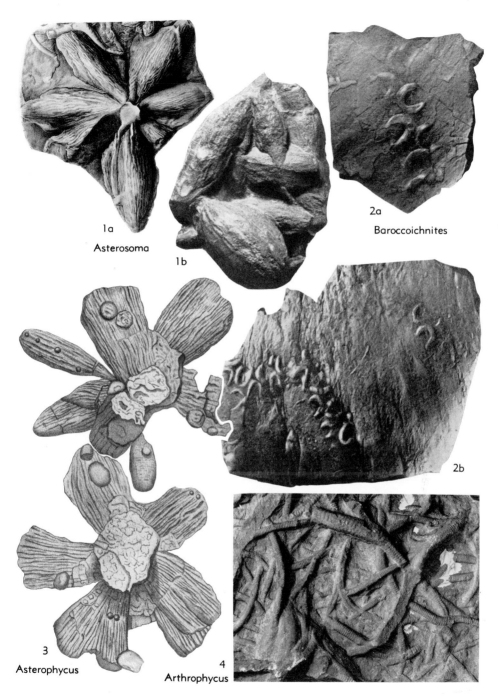

1a
Asterosoma

1b

2a
Baroccoichnites

2b

3
Asterophycus

4
Arthrophycus

FIG. 25. Trace fossils (p. *W*38-39, 43, 45).

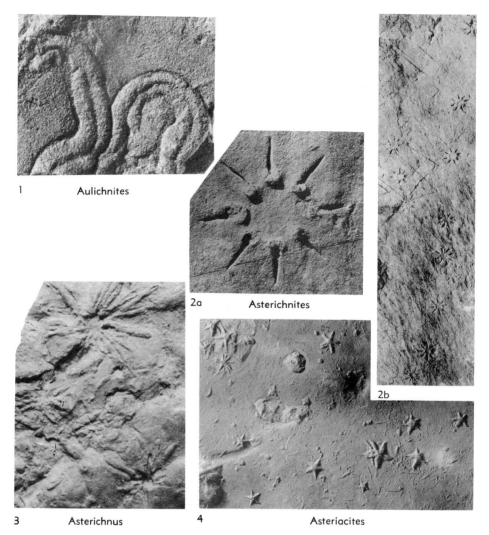

1   Aulichnites

2a   Asterichnites

2b

3   Asterichnus   4   Asteriacites

FIG. 26. Trace fossils (p. W42-43).

of *Allocotichnus* OSGOOD, 1970.] *U.Ord.(Cincinnat.)*, USA(Ohio).——FIG. 24,5. **A. trifidus,* loc. unknown; $\times 1$ (Osgood, 1970).

Asteriacites VON SCHLOTHEIM, 1820, p. 324 [*non* VON SCHLOTHEIM, 1822, p. 71] [**A. lumbricalis;* SD SEILACHER, 1953, p. 93] [=*Heliophycus* MILLER & DYER, 1878b, p. 2 (type, *H. stelliforme); Spongaster* FRITSCH, 1908, p. 9 (*non* EHRENBERG, 1860) *(nom. nud.); Asterocites* MIROSHNIKOV, 1959; *Ateriacites* CHAMBERLAIN, 1971, p. 212; *Asteriachites* CHAMBERLAIN, 1971a, p. 217); *(nom. null.)*]. Impressions in form of asteroids or ophiuroids, with transversely sculptured arms; their striae produced by activity of digging tube feet; often intersected by traces of neighboring animals ("horizontal repetition") or (as reaction to rapid sedimentation; see Fig. 17) "vertical repetition"; morphology dependent on preservation as convex hyporelief or concave epirelief. [Three different conical, subconical, or subcylindrical biogenic structures with pentameral symmetry on sides (ridges coarsely striated or double rows of nodes or rounded radial ridges) from the Pennsylvanian of USA (Okla.) were ascribed to *Asteriacites* by CHAMBERLAIN (1971a, p. 219), who named them "*A, lumbricalis* hiding forms A, B, C" and regarded them as true resting trace fossils; the proposal to expand the diagnosis of *Asteriacites* based on these forms is not accepted here. Regarded as body fossils of asteroids (ventral casts) ("stella lumbricalis") by KNORR & WALCH (1769) (=*"Asterias lumbricalis"* GOLDFUSS, 1833); interpreted by SEILACHER (1953b) as resting traces of Asterozoa such as *A. lumbricalis* SCHLOTHEIM, produced by ophiuroids, and *A. quinquefolius* QUENSTEDT, produced by starfishes. The nomenclatorial status of *Asteriacites* is confused; the name *Asteriacites* VON SCHLOTHEIM, 1820, p. 324, has been interpreted by NEAVE *(Nomenclator Zoologicus)* as *lapsus pro Asteriatites* VON SCHLOTHEIM, 1813 (p. 68, 99, 108; used for at least two different fossils from Trias., Jur., and Cret. rocks). For nomenclatorial discussions see *Treatise*, Part C (1964, p. C796) and Part U (1966, p. U103); however, *Asteriacites* has been used so often in paleoichnological papers that it is the opinion here that *Asteriacites* VON SCHLOTHEIM, 1820, should be preserved for asterozoan resting trace fossils.] *Ord.-Tert.,* Eu.-USA.——FIG. 26,4. **A. lumbricalis,* L.Jur., Ger.; ×0.5 (Seilacher, 1953).

Asterichnites BROWN & VOKES, 1944, p. 658 [**A. octoradiatus;* OD]. Rows of stellate imprints, about 6 cm. in diameter, each consisting of unmarked central disc and 8 radiating grooves 13 to 18 mm. long; arranged in rows on bedding planes. [Probably produced by tentacles of dibranchiate cephalopod as the animal apparently bounced over the bottom of the sea on the tips of its tentacles while its body was in nearly perpendicular position.] *U.Cret.(Mowry Sh.),* USA(Mont.-Wyo.).——FIG. 26,2. **A. octoradiatus,* Mont.; *2a,* ×0.6; *2b,* ×0.1 (Brown & Vokes, 1944).

Asterichnus BANDEL, 1967, p. 2 [*non Asterichnus* NOWAK, 1961, p. 227, *nom. nud.; nec Asterichnus* KSIĄŻKIEWICZ, 1970, p. 310 (type, *A. nowaki*) (=*Subglockeria* KSIĄŻKIEWICZ, 1974, herein, p. *W*112)] [**A. lawrencensis;* OD]. Starlike traces, approximately circular in cross section; diameter 4 to 12 cm., 10 to 30 unbranched "rays" consisting of grooves or tubelike ridges, 5 to 8 mm. wide; center formed by an irregularly oval to round knob. [According to BANDEL (1967a, p. 3) subsurface traces made within sediment along bedding planes in same way as other known Recent and fossil starlike traces on the surface of the sediment; producer probably a relatively large organism of unknown systematic position.] *Penn.,* USA(Kans.-Okla.).——FIG. 26,3. **A. lawrencensis,* Rock Lake Sh., Kans.; ×0.55 (Bandel, 1967a).

Asterophycus LESQUEREUX, 1876, p. 139 [**A. coxii;* M]. Large starlike trace fossil similar to *Asterosoma* VON OTTO, 1854; diameter about 6 to 12 cm.; individual "rays" radiating from central tube, oblong or obovate, 1 to 2 cm. in diameter, cross section irregular, surface longitudinally wrinkled. [At first described by

LESQUEREUX as plant; interpreted by DAWSON (1890, p. 603) as burrows of ?worms; no type or other specimen of occurrences in Indiana could be located.] *Miss.-Penn.,* USA (Kans.-Ind.-Ky.).——FIG. 25,3. **A. coxii,* Penn., Ky.; ×0.3 (Lesquereux, 1876).

Asterosoma VON OTTO, 1854, p. 15 [*non* GRUBE, 1867] [**A. radiciforme;* M]. Big stars diameter about 20 cm., with elevated center; about 3 to 9 rays, bulbous, tapering toward ends, longitudinally wrinkled, of different length, 2 of them mostly lying in same direction and commonly longer than other ones; rays sometimes do not radiate in all directions but form only acute-angled sector; longitudinally wrinkled. [Very probably burrows with radiating feeding trails; the Mesozoic forms suggested by ALTEVOGT (1968a) and HÄNTZSCHEL to have been made by decapod crustaceans. HÄNTZSCHEL agreed with GLAESSNER (1969, p. 375) that the following forms very probably have been incorrectly assigned to *Asterosoma:* FARROW's (1966) stellate structures (M. Jur., Eng.); three "forms" described as *Asterosoma by* FREY & HOWARD (1970) from the Upper Cretaceous of USA; and a starlike trace fossil from the Lower Tertiary (Paleoc.) of England (DURKIN, 1968). Similar starlike trace fossils were described from Paleozoic rocks partly as *Asterosoma* (Sil., Nor., SEILACHER & MEISCHNER, 1965, p. 616; Dev., Libya, SEILACHER, 1969a, p. 122) and partly as *Rosselia* DAHMER, 1937 (L.Cam., Pak., SEILACHER, 1955, p. 389; L.Dev., Ger., DAHMER, 1937, p. 532); for *Asterosoma? canyonensis* (BASSLER) (Precam., USA) see GLAESSNER (1969, p. 375). *Rosselia* has been regarded by SEILACHER (1969a, p. 122) as junior synonym, but as yet, no detailed discussion has been published.] *?Precam.,* USA(Ariz.); *Paleoz.,* Libya-Pak.; *?Paleoz.,* Eu.(Ger.-Nor.)-USA (Okla.); *?M.Jur.,* G.Brit.(Eng.); *?U.Jur.,* Eu. (France); *?L.Cret.,* Eu. (Ger.); *U.Cret.(Turon.),* Eu.(Ger.-Czech.), *?U.Cret.,* USA(Kans.-Utah). ——FIG. 25,*1a. *A. radiciforme,* U.Cret.(Turon.), Ger.; ×0.3 (von Otto, 1854).——FIG. 25,*1b. Asterosoma* assemblage, *Cruziana* facies, Dev., Libya; ×0.67 (Seilacher, 1969a).

Aulichnites FENTON & FENTON, 1937, p. 1079 [**A. parkerensis;* OD]. Trail, 5 to 10 mm. wide, commonly strongly curved; consisting of 2 convex ridges, separated by rather deep median groove in epirelief. [Crawling and/or grazing trail, most probably made by gastropod.] *Sil.,* USA(Ga.); *?M.Dev.,* Eu.(Ger.); *Penn.,* USA (Texas-Kans.); *?Penn.,* USA(Ark.); *?Cret.,* USA (Utah); *?Tert.,* S.Am.(Venez.)——FIG. 26,*1. *A. parkerensis,* Penn., Texas; ×1 (Howell in Häntzschel, 1962).

Balanoglossites HÄNTZSCHEL, 1962, p. *W*185 [**B. triadicus* MÄGDEFRAU, 1932, p. 153; OD] [=*Balanoglossites* MÄGDEFRAU, 1932, p. 153, *nom.*

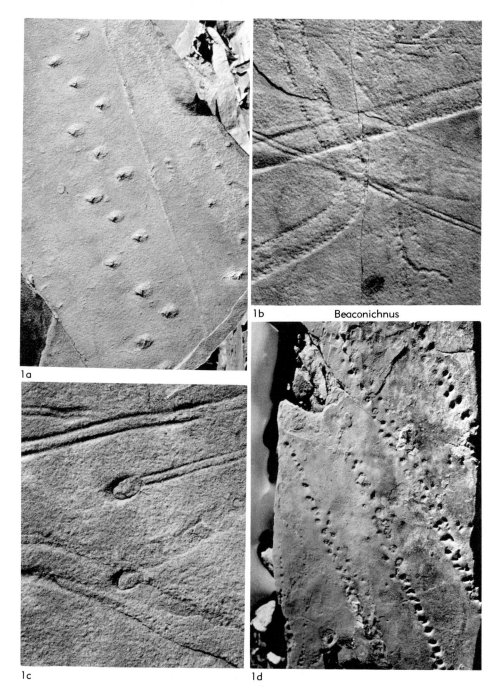

1a

1b   Beaconichnus

1c   1d

FIG. 27. Trace fossils (p. *W*45).

nud., established without designation of type species; *?Unculiferus* HUNDT, 1941, p. 58 (type, *U. transversus*), according to MÄGDEFRAU (1941, p. 526) identical to *Balanoglossites*]. Burrows, 1 to 3 cm. wide and up to 15 cm. deep, irregularly branched, with several openings that are sometimes funnel-shaped (e.g., *B. eurystomus* MÄGDE-FRAU); walls of burrows may be sculptured by transverse ridges and delicate longitudinal striations. [Suggested by MÄGDEFRAU to have been made by polychaetes or enteropneusts; KAZMIERC-ZAK & PSZCZOLKOWSKI (1969, p. 305) compared *Balanoglossites* with very similar burrow systems from the Middle Triassic (low. Muschelkalk) of Poland, which they interpreted as made by enteropneusts.] *?Ord.*, Eu.(Ger.); *M.Trias.* *(Muschelkalk)*, Eu.(Ger.); *?M.Trias. (Muschelkalk)*, Eu.(Pol.).

Baroccoichnites VYALOV, 1971, p. 88 [**B. pamiricus;* OD]. Chain consisting of 2 rows of arched cylinders, each bent in different direction—open to outside, arranged in checkerboard pattern and in contact with each other along their convex lateral sides. *U.Trias.*, C.Asia(Pamir).——FIG. 25,2. **B. pamiricus; 2a,b,* ×0.67 (Vyalov, 1971). [Description supplied by CURT TEICHERT.]

Beaconichnus GEVERS, 1973, p. 1002 (*nom subst. pro Arthropodichnus* GEVERS in GEVERS *et al.*, 1971, p. 92 (*non* CHIPLONKAR & BADWE, 1970)) [**Arthropodichnus darwinum* GEVERS, 1971, p. 93; OD]. Ichnogenus comprising 3 different types: 1) 2 narrow parallel grooves, 9 to 18 mm. apart, absolutely linear or only slightly curving; length up to more than 1 m.; depth and width of grooves 1 to 4 mm.; very small closely spaced foot imprints may be preserved in wider trails (=**B. darwinum* (GEVERS)); 2) paired parallel rows of commonly very closely spaced footprints; rows 2 to 4 cm. apart, usually broadly curving; foot or claw imprints appearing as small circular pits or in larger trails commonly elongated, oblique to trend of tracks (=*B. gouldi* (GE-VERS)); 3) large tracks, about 30 cm. wide, mostly straight, consisting of short parallel rows of foot imprints, up to 3 cm. wide, regularly arranged in sets of 3 or rarely 4, oblique (35°) to median line representing telson drag marking; distance between footprints averages 6 cm.; footprint pits show angular imprints of arrowhead shape indicating bipartite spines (=*B. antarcticum* (GEVERS)). [Epichnial crawling and walking trails; producers of *B. darwinum* probably shovelling and burrowing arthropods (?trilobites); *B. gouldi* possibly made by trilobites, the "species" is comparable to *Diplichnites* DAWSON, 1873 (see p. *W61*); origin of the large track *B. antarcticum* is doubtful (made by eurypterids?), somewhat resembles *Palmichnium* RUDOLF RICHTER, 1954 (L.Dev., Ger.), a smaller track tentatively interpreted as produced by eurypterids (see p. *W91*).] *?Dev.(up.Hatherton Ss.)*, Antarctic

(Victoria Land).——FIG. 27,1a. *B. antarcticum* (GEVERS); single trails, ×0.17 (Gevers *et al.*, 1971).——FIG. 27, 1b. *B. gouldi* (GEVERS); large tracks in center and at left, crossed by **B. darwinum* (GEVERS), also by narrow forms of *Beaconites antarcticus* VYALOV, ×0.11 (Gevers *et al.*, 1971).——FIG. 27,1c. **B. darwinum* (GEVERS); trails and burrows, ×0.3 (Gevers *et al.*, 1971).——FIG. 27,1d. *B. giganteum* GEVERS, low.Beacon sediments; irreg. pattern evident in each tread line (Dept. Geology coll., Univ. Witwatersrand).

Beaconites VYALOV, 1962, p. 728 [**B. antarcticus;* M] [=*?Laminites* GHENT & HENDERSON, 1966, p. 158 (type, *L. kaitiensis*); for description and discussion see p. *W78*]. Large horizontal segmented ("septate") burrows, many of them of giant size; 3 to 13 cm.(max.) wide, 8 to 10 cm. very common width; somewhat sinuous, large forms relatively straight; rather long (up to about 1 m.); commonly crowded; associated with rounded pits of similar cross section; marginal welts 5 to 30 mm. wide; curving "septal" ridges mostly remarkably equidistant; those of giant forms usually markedly crescentic, but size, shape, and spacing may vary considerably. [Originally doubted whether trace or body fossils were represented; interpreted by GEVERS *et al.* (1971, p. 83) as burrows made by unknown animals within the sediment; observed in highly bioturbated layers; for detailed discussion see GEVERS *et al.* (1971, p. 83-85)]. Dev., Antarct. ——FIG. 28,1. **B. antarcticus*, up.Hatherton Ss., Victoria Land; ca. ×0.24 (Gevers *et al.*, 1971).

Belorhaphe FUCHS, 1895, p. 395 [**Cylindrites zickzack* HEER, 1877, p. 159; OD] [=*Beloraphe multorum autorum (nom. null.);* *Helicolithus fabregae* AZPEITIA MOROS, 1933, p. 32 (see SEILACHER, 1959, p. 1068); *Belorapha* DIMIAN & DIMIAN, 1964, pl. 8 *(nom. null.)*]. Sharply zigzag-shaped locomotion trails, commonly ,with short protrusion at corners.[1] [Evidently post-depositional trail; MICHELAU (1955) placed *Sinusia* KRESTEW, 1928 (*nom. inval.*: preoccupied) and *Sinusites* DEMANET & VAN STRAELEN, 1938(U.Carb., Eu.) in *Belorhaphe*, but these two "genera" belong to regularly sinuous trail *Cochlichnus* HITCHCOCK, 1858, resembling sine curve.] [Found in flysch deposits.] *Cret.-L.Tert.*, Eu. ——FIG. 29,2. *B.* sp. FUCHS, Aus.; ×0.6 (Fuchs, 1895).

Bergaueria PRANTL, 1946, p. 50 [**B. perata;* OD] [=*?Palaeactis* DOLLFUS, 1875 (type, *P. vetusta*); see WELLS & HILL (1956), p. *F233*; probably *nom. oblit.*]. Cylindrical or baglike protrusions with smooth walls, length and diameter subequal (2-4 cm.); lower end rounded, with shallow depression which is sometimes sur-

---

[1] For a discussion of the origin of these protrusions, see NOWAK (1970, fig. 3). [W. G. HAKES.]

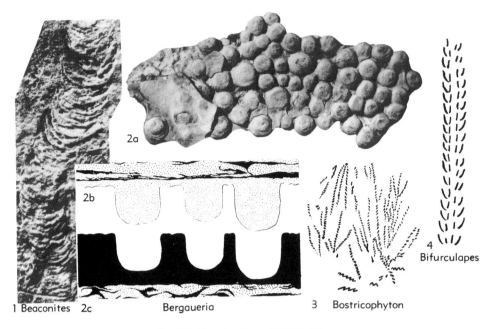

FIG. 28. Trace fossils (p. W45-46, 48).

rounded by 6 to 8 very short radially arranged tubercles; some specimens (L.Cam., Nev.) display biradially symmetrical impressions on ventral surface. [Probably resting burrows of suspension-feeding coelenterate, possibly of actinian anemones (ALPERT, 1973); comparisons have been made with *Edwardsia* or *Phyllactis conguilegia*; for detailed discussion of the origin see ARAI & MCGUGAN (1968, p. 206).] *Cam.-Ord.*, N.Am. (USA-Can.)-Eu.——FIG. 28,2*a*. *B.* sp., ventral surface of colony, L.Cam., Can.(Moraine Lake, Banff Area, Alta.); ×0.17 (Arai & McGugan, 1968).——FIG. 28,2*b,c.* **B. perata,* Ord., Czech.; *2b,* casts in overlying sandstone; *2c,* original burrow-cavities in underlying shale, ×0.3 (Prantl, 1946).

Bifasciculus VOLK, 1960, p. 152 [**B. radiatus*; M]. Starlike trace fossil, consisting of many (up to 40) tunnels 2 to 3 cm. long, radiating from central area and ending blindly, bent slightly upward and downward. [Feeding burrow.] *Ord. (Griffel-Schiefer),* Eu.(Ger., Thuringia), *Ord. (Arenig.),* Eng.-Ire.——FIG. 29,4. **B. radiatus,* Griffel-Schiefer, Ger.; ×1 (Volk, 1960).

Biformites LINCK, 1949, p. 44 [**B. insolitus*; OD]. Bimorphous form, consisting of narrow section, partly divided by longitudinal furrows, continuing into wider section with prominent transverse ribs; resembles shafted hand grenade; fillings visible at lower surface of layers. [According to SEILACHER (1955), dwelling burrow.] *?Penn.,* USA(Okla.); *U.Trias.(M.Keuper),* Eu.(S.Ger.).

——FIG. 29,3. **B. insolitus,* U.Trias.(M.Keuper), Ger.; *3a,* ×0.8; *3b,* ×1 (schem.) (Linck, 1949b).

Bifungites DESIO, 1940, p. 78[**B. fezzanensis* (=*?Buthotrephis impudica* HALL, 1852, p. 20); M]. Structures dumbbell-like or arrow-shaped, 1 to 5 cm. long; ends commonly hemispherical, diameter up to 1 cm.; on bedding planes respectively at erosional interfaces; preserved as positive hyporeliefs or positive epireliefs; similar to *Arthraria biclavata* MILLER (placed in *Corophioides* by OSGOOD, 1970, p. 323). [Interpreted by DESIO as fucoid or colonial animal; according to DUBOIS & LESSERTISSEUR (1965) filling of top of U-shaped burrow perhaps inhabited by small trilobite; regarded by SEILACHER (1955, fig. 5; 1969a, p. 112) as special kind of preservation of protrusive vertical U-tube representing feeding burrow; *Bifungites* predominant ichnogenus of ichnocoenoses in Upper Devonian of USA(Mont.) (RODRIGUEZ & GUTSCHICK, 1970, p. 418).] *L. Cam.,* Pak.; *?Ord.,* Eu.(Czech.); *?Sil.,* USA (N.Y.); *Dev.,* N.Afr.-USA(Mont.).——FIG. 29,1. **B. fezzanensis,* M.Dev.-U.Dev., N. Afr.; ×0.7 (Desio, 1940).

Bifurculapes HITCHCOCK, 1858, p. 152 [**B. laqueatus*; SD LULL, 1953, p. 42] [=*Bifurculipes, Biferculipes, Bifurcalipes* HITCHCOCK, 1865, p. 13, 14 *(nom. null.)*]. Four regular rows of tracks, commonly resembling small forks when united at base; may have 2 additional rows with pairs of opposing tracks; similar to *Permichnium* GUTHÖRL, 1934, and *Triavestigia niningeri* GIL-

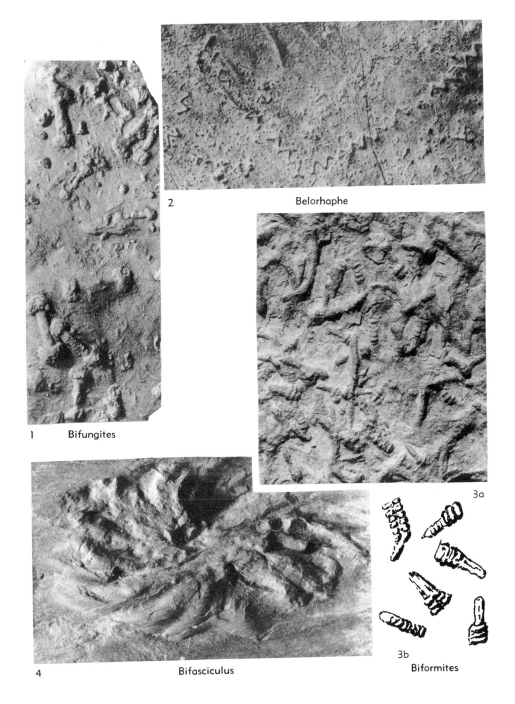

1 Bifungites

2 Belorhaphe

3a

3b Biformites

4 Bifasciculus

Fig. 29. Trace fossils (p. W45-46).

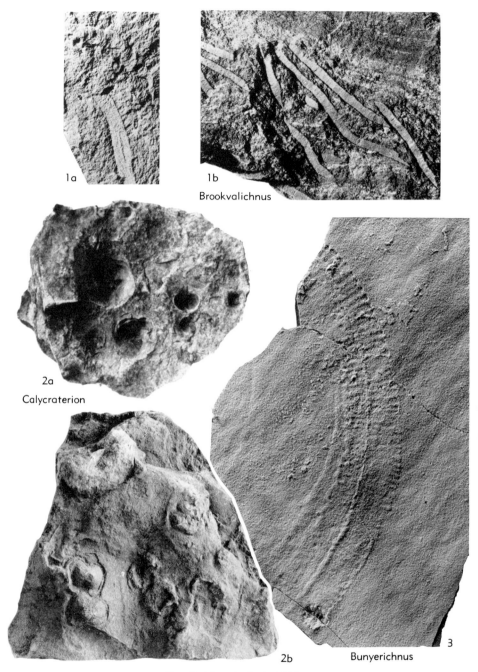

1a

1b

Brookvalichnus

2a

Calycraterion

2b

Bunyerichnus

3

FIG. 30. Trace fossils (p. W49).

MORE, 1927. [Interpreted by HITCHCOCK (1858, 1865) and LULL (1915, 1953) as probably made by insects.] *Trias.,* USA(Mass.).——FIG. 28,4. **B. laqueatus;* ×0.7 (Lull, 1953).

Bostricophyton SQUINABOL, 1890, p. 181 [**B. pantanellii;* SD ANDREWS, 1955] [=*Bostrichophyton* ANDREWS, 1955 *(nom. null.)*]. Very thin threadlike burrows, spirally rolled, ?horizontal, ?branched. [Originally described as threadlike alga with spiral branchlets; according to Fuchs (1905, p. 366), identical to *Chondrites intricatus*; possibly related to *Helicolithus* Azpeitia Moros; not studied for many decades except for brief description of dubious "n. sp." from Precam.(U.Vindhyan) of India (Verma & Prasad, 1968).] [Found in flysch deposits.] *Precam.-Cam., India; Cret.-L.Tert.,* Eu.——Fig. 28,*3*. **B. pantanellii*, L. Tert., Italy; ×0.7 (Fuchs, 1895).

Brancichnus Doughty, 1965, p. 148 [**B. dudleyi*; M]. Horizontal and branching cylindrical structures, maximal length 60 cm., main cylindrical portion ("stem") fairly straight; diminishing in diameter distally. [Burrow systems, according to Doughty, more probably "remains of branching marine alga or some form of branching Porifera"; name rather superfluous; resembling or identical to *Saportia* Squinabol, 1891.] *L.Jur.,* Eu.(Eng.).

Brookvalichnus Webby, 1970, p. 528 [**B. obliquus*; OD]. Flat ribbonlike structures, sometimes in groups, straight to slightly curving, unbranched, normally inclined 10 to 15° to horizontal, up to 9 cm. long, uniform width 3.5 to 4.0 mm.; very thin ribbonlike part exhibits transverse annulations bordered on either side by a thicker structureless layer, consisting of dark shale; structures most likely originated by collapse of tubelike dwelling-burrows. [Perhaps made by freshwater (?) wormlike animal or insect larva.] *M.Trias.(up.Hawkesbury Ss.),* Australia (NewS. Wales, Sydney Basin).——Fig. 30,*1*. **B. obliquus*, shale lens in Hawkesbury Ss., NewS. Wales (Brookvale); *1a*, ×1.7; *1b*, ×0.83 (Webby, B. D., 1970a, p. 529, in: *Trace Fossils* edited by T. P. Crimes & J. C. Harper, Geol. Jour. Spec. Issue 3, Seel House Press Liverpool).

Buchholzbrunnichnus Germs, 1973, p. 69 [**B. kröneri (recte kroeneri)*; OD]. *Precam.(Nama Syst., Kuibis F.),* SW.Afr.

Bunyerichnus Glaessner, 1969, p. 379 [**B. dalgarnoi*; OD]. Curved surface locomotion trail; somewhat variable width, changing throughout observed length from 2 to 3 cm.; submedian ridge about 2 mm. wide, distance from margins slightly variable; distinctly transverse rise-and-groove sculpture: grooves about 2 mm. long, separated by longer straight rises ending in pit-like depressions. [Produced by bilaterally symmetrical animal employing rhythmic muscular contractions, probably related to primitive mollusks without mineralized shells; only single specimens.] *U.Precam.(Brachina F.),* S.Australia (Flinders Ranges).——Fig. 30,*3*. **B. dalgarnoi*; holotype, ×0.67 (Glaessner, 1969).

Calycraterion Karaszewski, 1971, p. 104 [**C. samsonowiczi*; M]. Regular calyx-shaped depressions; smaller ones similar to impression of lower part of very large hazelnut; inner walls smooth; "calyx" 15 to 40 mm. in diameter, 5 to 15 mm. in depth; 2 or 3 small circular depressions on the bottom representing outlets of filled burrows, 2 to 5 mm. in diameter. *L.Jur. (Hettang.),* Eu.(Pol.).——Fig. 30,*2*. **C. samsonowiczi*, Holy Cross Mts.; *2a*, preserved in concave epirelief; ×0.25 (Karaszewski, 1971a); *2b*, bottom of rock slab shown in *2a* with molds of calyces in convex hyporelief, ×0.25 (Karaszewski, n; I. G. 1285.11.2, Geol. Inst. Mus. Warsaw).

Capodistria Vyalov, 1964, p. 113 [**C. vettersi*; OD]. Starlike trace fossil; superfluous name for "genus" based on only one specimen observed in stone wall at Capodistria (Istria) and described by Vetters (1910). [Found in flysch deposits.] *Tert.(Eoc.),* Eu.(Italy-Yugosl., Istria).

Caulerpites von Sternberg, 1833, p. 20 [**Fucoides Targionii* Brongniart, 1828, p. 56 (=*C. targionii* von Sternberg, 1833, p. 25); SD Andrews, 1955, p. 130] [=*Caulerpides* Schimper, 1869, p. 160 *(nom. null.)*]. Very heterogeneous "genus" including plants (even conifers, according to Schimper) as well as trails (e.g., *C. marginatus* Lesquereux, 1869, p. 314 = *Spreitenbau* similar to *"Taonurus"; C. annulatus* Ettinghausen, 1863, p. 462 = stuffed burrow similar to *Keckia* or *Muensteria*); other "species" also classified with Recent genus *Caulerpa* Lamouroux, 1809.

Chomatichnus Donaldson & Simpson, 1962, p. 78 [**C. wegberensis*; OD]. Small circular conical mounds consisting of fecal castings, about 5 to 7 cm. high, connected with a vertical burrow; somewhat similar to piles of fecal castings produced by Recent polychaete *Arenicola*; according to Simpson (1970, p. 510), these castings probably produced by the *Zoophycos* animal. *L.Carb. (Dibunophyllum Z.),* G.Brit.(Eng.); *Cret.,* USA (N.Mex.).——Fig. 31,*6*. **C. wegberensis*, Carb., Eng. (Carnforth, Lancash.); *6a*, vert. sec., based on holotype, ×0.67; *6b*, ×0.4 (Donaldson & Simpson, 1962).

Chondrites von Sternberg, 1833, p. 25 [*non* M'Coy, 1848] [**Fucoides lycopodioides* Brongniart, 1828, p. 72 (=*C. lycopodioides* von Sternberg, 1833, p. 20); SD Andrews, 1955, p. 127] [=*Caulerpites* von Sternberg, 1833, p. 20 *(partim); Sphaerococcites* von Sternberg, 1833, p. 28 *(partim); Buthotrephis* Hall, 1847, p. 8 *(partim); Phymatoderma* Brongniart, 1849, p. 59 *(partim); ?Trevisania* de Zigno, 1856, p. 23; *Phycopsis* von Fischer-Ooster, 1858, p. 64 ("subg."); *Bythotrephis* Eichwald, 1860, p. 56 *(nom null.); Nulliporites* Heer, 1865, p. 140 *(non* Krueger, 1823, *nom. nud.); Chondrides* Schimper, 1869, p. 168 *(nom. null.); Leptochondrites* Schimper, 1869, p. 171 *(partim)* ("subg."); *?Theobaldia* Heer, 1877, p. 114 *(partim); ?Aulacophycus* Heer, 1877, p. 111 (type, *A. sulcatulus); Palaeochondrites* de Saporta, 1882, p. 35; *Chondropogon* Squinabol, 1890, p. 180; *?Prochondrites* Fritsch, 1908, p. 22; *?Labyrinthochorda* Weissenbach, 1931, p. 76; *?Isawaites* Hatai &

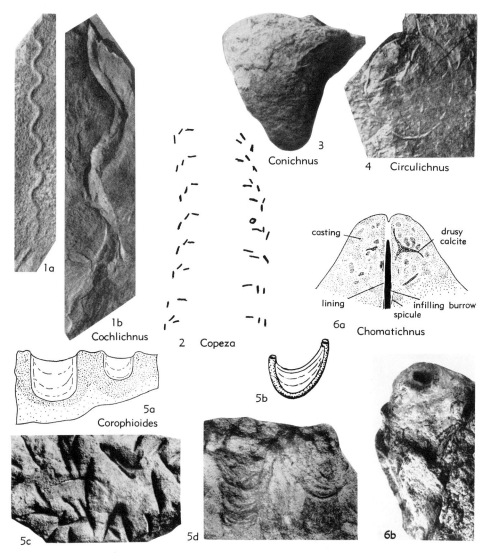

FIG. 31. Trace fossils (p. *W*49, 52-53).

NODA, 1971b, p. 5]. "Form genus" in widest possible sense; plantlike dendritic patterns of small cylindrical ramifying tunnel systems; individual tunnels neither crossing each other nor interpenetrating (perhaps only between tunnels of different systems); one or few main axes open to surface; branching tunnels trending downward across bedding and then (at least their distal portions) mostly lying parallel to bedding planes; may branch in regular or irregular patterns (highly variable); angle of branching may also be variable or constant, between 25 and 40°; branches may be arranged in pinnate or radial patterns or form compact groups; diameter of tunnels 0.5 to 5 mm., remaining constant within entire tunnel system; otherwise varying from large (e.g., *"Buthotrephis"*) to small (most *Chondrites*); some tunnels with transversely built-in ellipsoidal pills (their probable fecal origin doubted); preservation of fillings of tunnels controlled by stratinomic factors; trace fossil nature convincingly proved first by RICHTER (1927a, 1931), though earlier NATHORST (1881a) and FUCHS (1895) had rejected the former interpretation as algae; producer unknown, perhaps worms. SIMPSON (1957) suggested sipunculoid worms working from fixed center on the surface of sediment and producing tunnels by an ex-

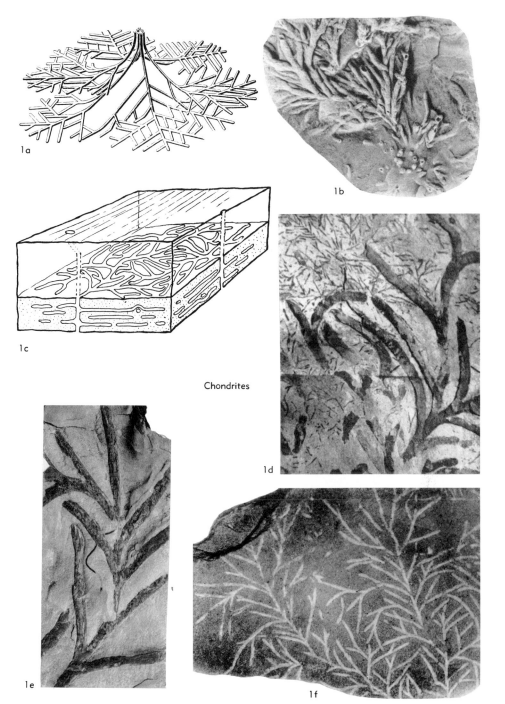

Chondrites

FIG. 32. Trace fossils (p. *W*49-50, 52).

tensible proboscis; branching pattern may be affected by phobotaxis (RICHTER, 1927a, p. 218; 1928, p. 226; 1931, p. 302); ethological interpretation is still discussed but *Chondrites* undoubtedly belongs to Fodinichnia and is to be regarded as feeding structures of sediment-eating animals (RICHTER, 1927; SEILACHER, 1955; OSGOOD, 1970) and not dwelling burrows of filter-feeding annelids (TAUBER, 1949); detailed studies of the ichnogenus would certainly lead to several additional "new ichnogenera"; some dozens of "ichnospecies" have been described but recognition of these within *Chondrites* very difficult (OSGOOD, 1970, p. 489); for historical account of many theories of the nature of *Chondrites*, detailed treatments, and literature, see especially SIMPSON (1957) and OSGOOD (1970, p. 328-331); for discussion and various reconstructions of tunnel systems of this form see RICHTER (1931, p. 301, fig. 2), TAUBER (1949, p. 149-150, fig. 1,2), and SIMPSON (1957, p. 484, fig. 2).] *?Cam., Ord.-Tert.*, cosmop.——FIG. 32,1a,e,f. C. sp., U.Cret.; *1a*, reconstr. of tunnel system (Simpson, 1957); *1e*, large form, Aus.; ×0.9 (Häntzschel, 1955); *1f*, small form, Maastricht., Spain; ×0.9 (Gómez de Llarena, 1946).——FIG. 32,1b. *Chondrites*, type C, U.Ord.(Cincinnat., Whitewater beds), USA(Ind.); ×1.2 (Osgood, 1970).——FIG. 32,1c. C. bollensis ZIETEN, Lias ε Ger.(Holzmaden); schem.(Richter, 1931).——FIG. 32,1d. C. furcatus VON STERNBERG, flysch deposits, ?Tert., Aus.; ×0.9 (Derichs, 1928).

Chondritoides BORRELLO, 1966, p. 15 [*C. insolitus*; OD]. Superfluous name for poorly figured straight burrows, 7 mm. in diameter, bifurcating at various angles up to 60°; somewhat resembling large "species" of *Chondrites* VON STERNBERG. *Ord.*, S.Am.(Arg.).

Circulichnis VYALOV, 1971, p. 91 [*C. montanus*, p. 91; OD]. Ring-shaped trace, almost circular (or oval), formed by some cylindrical object. *U.Trias.*, C.Asia(SW.Pamir).——FIG. 31,4. *C. montanus*; ×0.67 (Vyalov, 1971). [Description supplied by CURT TEICHERT.]

Climactichnites LOGAN, 1860, p. 285 [*C. wilsoni*; M] [=*Climachtichnites* MILLER, 1877, p. 214 *(nom. null.)*; *Climactichnides* CHAPMAN, 1878, p. 490 *(nom. null.)*]. Very large trails, width about 15 cm., maximum length 3 to 4 m., with prominent, slightly arched or V-shaped transverse ridges and very delicate, closely spaced arched rills; dishlike impressions, oval, distinctly bounded at beginning of trail. [Crawling trail of unknown producer; interpreted also as of plant origin (CHAPMAN, 1878); many groups of animals have been proposed as producers: burrowing crustaceans, eurypterids, large trilobites, worms, and mollusks; according to ABEL (1935, p. 242-249) most probably made by gastropods (likely marine nudibranchs); for history of genus see BURLING (1917, p. 390) and ABEL (1935, p. 242).]

U.Cam., USA-Can.——FIG. 33,1a. *C. wilsoni*, Potsdam Ss., USA(N.Y.); ×0.02 (Walcott, 1912). ——FIG. 33,1b,c. C. youngi (CHAMBERLAIN), St. Croix, USA(Wis.); *1b*, ×0.4 (Walcott, 1912); *1c*, ×0.5 (Walcott, 1912, in Malz, 1968).

Cochlichnus HITCHCOCK, 1858, p. 161 [*C. anguineus*; M] [=*Sinusia* KRESTEW, 1928, p. 574 *(non* CARADJA, 1916) *(nom. nud.)*; *Sinusites* DEMANET & VAN STRAELEN, 1938, p. 107; MICHELAU (1956) incorrectly considered *Sinusia* and *Sinusites* to be synonymous with *Belorhaphe* FUCHS, 1895, p. 395 (p. W45)]. Regularly meandering smooth trails, resembling sine curve. [Found in flysch deposits.] *U.Precam.*, Australia (New S.Wales); *U.Precam. or L.Cam.*, Eu.(Nor.); *?L.Cam.-M.Cam.*, Eu.(Eng.); *Ord.(Arenig.)*, Eu. (Eng.); *U.Carb.*, Eu.-USA; *Perm.*, Antarct.; *Trias.*, N.Am.(USA, Mass.); *L.Jur.*, Eu.(Ger.); *L.Jur. (Pliensbach.)*, Greenl.; *L.Cret.*, Eu.(Ger.-Eng.-Pol.); *Tert.(Oligo.)*, Eu.(Pol.).——FIG. 31, 1. C. kochi (LUDWIG), Carb., Ger.; *1a,b*, ×0.67 (Michelau, 1956).

Conichnus MYANNIL, 1966, p. 201 [*C. conicus*; M]. Fillings of conical or conelike hollows; mostly very regular forms with circular cross section; lower end round, without distinct mammilliform peak, thus differing from *Amphorichnus* MYANNIL; length (max.) 12 cm., diameter (max.) 8 cm., perpendicular to bedding plane. [Dwelling burrow or resting trail.] *M.Ord.-U.Ord.*, USSR (Est.).——FIG. 31,3. *C. conicus*, M.Ord. (?Kukruse F.), Est.; ×0.5 (Myannil, 1966).

Conispiron VYALOV, 1969, p. 106 [*Xenohelix babkovi* GEKKER in GEKKER, OSIPOVA, & BELSKAYA, 1962, p. 205; OD]. Dextrally or sinistrally coiled burrows having circular or elliptical cross sections; diameter of spiral possibly decreasing downward; vertical distance between the twist also decreases downward, the entire spiral having a conical outline. *Tert.(?mid. Oligo.)*, USSR(Crimea). [Description supplied by CURT TEICHERT.]

Conopsoides HITCHCOCK, 1858, p. 152 [*C. larvalis*; M]. Tracks in 3 (?4) rows, divergent from median line; foot impression linear, blunt anteriorly; tracks straight or sharply curved. [?Made by insect.] *Trias.*, USA(Conn.-Mass.).

Copeza HITCHCOCK, 1858, p. 159 [*C. triremis*; M]. Three rows of impressions on either side of median line, with main track at right angles to that line; width of trackway 35 mm.; oblique impressions not outside of longitudinal ones as in *Lithographus*, but inside. [?Made by podites of insect.] *Trias.*, USA(Mass.).——FIG. 31,2. *C. triremis*; ×0.7 (Lull, 1915).

Coprinisphaera SAUER, 1955, p. 9 [*C. ecuadoriensis*; M]. Spherical structures with one opening; about 6 cm. in diameter; walls about 1 cm. thick; mostly hollow or filled with consolidated mass similar to argillaceous excrement; found in loess-like tuffs *(cancagua)*. [Probable breeding places

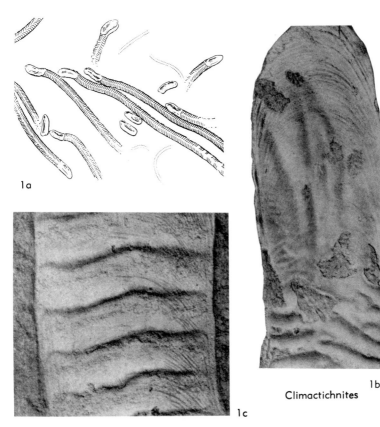

1a

Climactichnites

1b

1c

FIG. 33. Trace fossils (p. *W*52).

of scarabaeid beetles.] *Pleist.* (guide fossil of 3rd interglacial stage), S.Am.(Ecuad.-Colom.).

Corophioides SMITH, 1893, p. 292 [**C. polyupsilon*; M] [=*Arenicoloides* BLANCKENHORN, 1916, p. 39 (type, *A. luniformis*); *Arenicolithes* HILDE-BRAND, 1924, p. 27 *(nom. null.)*; *Corophyoides* ÖPIK, 1956, p. 108 *(nom. null.)*; *Corophiodes* BORRELLO, 1966, p. 11 *(nom. null.)*]. U-shaped spreiten burrows similar to *Rhizocorallium*, but shorter and always perpendicular to bedding plane (Richter, 1926). Both limbs of each successive U-tube typically show lateral displacement from limbs of preceding U-tube (see KNOX, 1973, p. 135, for further discussion). [*Arenicoloides* comprises crescent-shaped grooves in bedding planes produced by erosion of burrows to their basal ends.] *Cam.-U.Cret.*, Eu.-Asia.——FIG. 31,5*a,c,d. C. luniformis* (BLANCKENHORN), L.Trias., Ger.; *5a*, side (somewhat schem.), ×0.67; *5c*, lower ends of U-shaped burrows with spreite, ×0.4; *5d*, side, ×0.4 (Abel, 1935).——FIG. 31,5*b. C.* sp. cf. *C. rosei* DAHMER, L.Cam., Pak.; ×0.4 (Seilacher, 1955).

Cosmorhaphe FUCHS, 1895, p. 395 (misprinted *Cosmoraphe*; correct spelling *Cosmorhaphe* twice on p. 447) [**Helminthopsis sinuosa* AZPEITIA MOROS, 1933, p. 45; SD HÄNTZSCHEL, herein]. "Free meanders" of simple, smooth ridges, of extraordinarily regular form, meanders commonly in 2 orders of size; windings not physically close to each other. [At first compared by FUCHS (1895) with spawn strings of gastropods; however, *Cosmorhaphe* is typical grazing trail. For discussion of the preservation (predepositional, formed along bedding planes, secondary casts of surface trails), see WEBBY (1969a, p. 84). *C. timida* PFEIFFER (*L.Carb.*, Ger.) is not typical *Cosmorhaphe*.] [Found in flysch deposits.] *?Ord.*, Eu.(Nor.); *U.Sil.*, Australia(New S. Wales); *?Dev.*, Eu.(Ger.)-USA(Mont.); *?U.Cret.*, Alaska; *U.Cret.-L.Tert.*, Eu.; *L.Tert.*, S.Am. (Venez.); *M.Tert.*, N.Z.——FIG. 34,*3. C.* sp., low.mid.Eoc., Pol.; ×0.6 (Książkiewicz, 1960).

Crossopodia M'COY, 1851, p. 395 [**C. scotica*; SD HÄNTZSCHEL, 1962, p. *W*189] [=*Crassopodia* TATE, 1859, p. 66 *(nom. null.)*; *Crossochorda* SCHIMPER, 1879, p. 52 (modified name for algal interpretation); *Chrossocorda, Chrossochorda, Chrossocarda* WILLIAMSON, 1887, p. 21, 22, 29

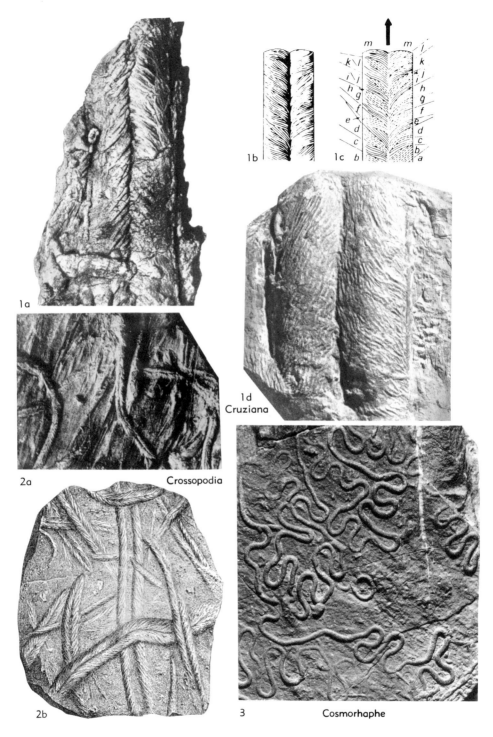

FIG. 34. Trace fossils (p. *W*53, 55).

(nom. null.)]. Meandering, curved, or straight trails, width about 1 cm., with broad dense fringe on each side (formerly regarded as "segments" of supposed worm), mostly with median furrow. [Crawling trail, at first interpreted as worm or algá; name *Crossopodia* should be restricted to the type of crawling trails as figured (e.g., by SCHIMPER or WILLIAMSON); *C. tuvaensis* MASLOV (1956, p. 87) (Sil., USSR) to be excluded from *Crossopodia* (markings?); concerning *C. henrici* (GEINITZ), see *Dictyodora* WEISS.] Ord.-Carb., Eu.-USA(Kans.); ?Ord.-Carb., S.Am.(Arg.-Brazil), ?U.Cret., USA(Kans.-Okla.-Iowa).——FIG. 34,2a. *C. tuberculata* (WILLIAMSON), Carb., Eng.; ×0.3 (Williamson, 1887).——FIG. 34,2b. *C. scotia* (M'COY), Ord., France; ×0.5 (Schimper in Schimper & Schenk, 1879).

Cruziana D'ORBIGNY, 1842, p. 30 [*C. rugosa*; (first?) SD BASSLER, 1915, p. 292; later SD: *C. furcifera* D'ORBIGNY, 1842, p. 30, by SEILACHER, 1953, p. 107] [=Bilobites D'ORBIGNY, 1839, expl. pl. 1, fig. 1-3 *(non* DEKAY, 1824; *nec* RAFINESQUE, 1831; *nec* BRONN, 1848; *nec* QUENSTEDT, 1869), for discussion see SINCLAIR, 1951; *Crusiana* DAWSON, 1888, p. 30 *(nom. null.)*; *Bilobichnium* KREJCI-GRAF, 1932, p. 31 (no formal species name; proposed as *nom. nov.*)]. Elongate bandlike furrows covered by herringbone-shaped ridges, with or without 2 outer smooth or finely longitudinally striated zones outside V-markings occasionally with lateral grooves and/or wisp markings; variability in size and sculpture due to varied behavior of producer and preserved width of trail (0.5 cm. to about 8 cm.); length up to more than 1 m. (RADWAŃSKI & RONIEWICZ, 1972), commonly 10 to 20 cm.; V-angle quite variable, acute to blunt, along length of an individual trail. V-markings are scratch markings made by appendages of producer, certainly mostly by digging activity of endopodites of trilobites; V-markings grouped in sets of distinct parallel claw markings produced by multiple or serrate claws, thus consisting of 2 or more parallel or slightly diverging grooves. [Interpretation was very controversially discussed in many publications from 1881-87; some regarded *Cruziana* as plants or sponges (DELGADO, 1885; LEBESCONTE, 1883a,b; DE SAPORTA 1884), but NATHORST (1881a, 1886) argued for trace fossil nature; for a short account of this controversy see OSGOOD, 1970, p. 287. Occurrences in France were regarded as *"pas de boeuf"* or even as *"monument druidique"* (see DESLONGCHAMPS, 1856, p. 299; FAUVEL, 1868; MORIÈRE, 1879). These forms now are generally regarded as made by furrowing, burrowing, or shoveling trilobites or trilobite-like arthropods, in part perhaps of merostome origin, and have also been found in freshwater deposits, questionably attributed to notostracan branchiopods (BROMLEY & ASGAARD, 1972); originated by simple ploughing

using all or only anterior appendages; lateral ridges may be made by dragging of genal spines; trails may also possess additional impressions of coxae, pleural spines, exopodites and/or carapace edges; produced at mud-sand interface or in muddy sediment by burrowing beneath a sand layer (BIRKENMAJER & BRUTON, 1971, p. 315). For undertrack trails see SEILACHER (1970, p. 448); V-shaped pattern points in opposite direction to that of animal's movement, V's gape forward; for many conclusions from studies of *Cruziana* on morphology of trilobite legs, trilobite motion and behavior, gradients in digging direction, and preservation, see SEILACHER (1962; 1970, fig. 1-6), CRIMES (1970b,c), BIRKENMAJER & BRUTON (1971, p. 314, 317).[1] Intermediate forms between *Cruziana, Rusophycus,* and *Diplichnites* have been observed; *Cruziana* and *Rusophycus* were often regarded as synonyms, but LESSERTISSEUR (1955, p. 45), SEILACHER (1955, p. 366), and particularly OSGOOD (1970, p. 303) recommended restricting *Rusophycus* to the short bilobate resting trails of trilobite origin, naming the longer bilobate forms *Cruziana*; however, SEILACHER (1970) did not follow that suggestion and placed all "resting tracks," "resting nests," and "resting burrows" in *Cruziana*. Owing to the difficulties in separating *Cruziana* and *Rusophycus*, it seems best "to base the names strictly on morphology" (OSGOOD, 1970); for discussion of stratigraphic significance of *Cruziana* see CRIMES (1968, 1969) and SEILACHER (1960, 1970); for detailed discussion of the genus see LEBESCONTE (1883a, p. 59-73), DE SAPORTA (1884, p. 58-89), DELGADO (1885, p. 27-68), DESIO (1940, p. 64-67), LESSERTISSEUR (1955, p. 44-47), SEILACHER (1955, p. 364-366) and other papers quoted above.] U.Precam.-Dev., cosmop., Trias., E.Greenl.——FIG. 34,1a. *C. semiplicata*, U.Cam., North Wales(Snowdonia); ×0.5 (Crimes, 1968).——FIG. 34,1b,c. *Cruziana*, Cam.; *1b*, diag. showing herringbone pattern consisting of sets of scratch marks thought to be produced by backward movement of trilobite appendages; *1c*, detail of the various sets (represented by letters *a-l*), schem. (arrow represents direction of movement of the animal) (Birkenmajer & Bruton, 1971).——FIG. 34, 1d. *C. furcifera*, L.Ord., North Wales; ×0.37 (Crimes, 1968).

Ctenopholeus SEILACHER & HEMLEBEN, 1966, p. 47 [*C. kutscheri*; M]. Long horizontal tunnel-like burrow, straight or somewhat curved, with vertical shafts rising at equal intervals; burrow only rarely branched horizontally; fragments up to 60 cm. in length. [Feeding burrow.] *L.Dev. (Hunsrück Sh.),* Eu.(Ger.).——FIG. 35,1. *C.

---

[1] BERGSTRÖM (1973, p. 52-59) discussed the above mentioned papers and others in an excellent summary of the behavioral patterns of trilobites as they relate to the formation of different species of *Cruziana* and other related trace fossils. [W. G. HAKES.]

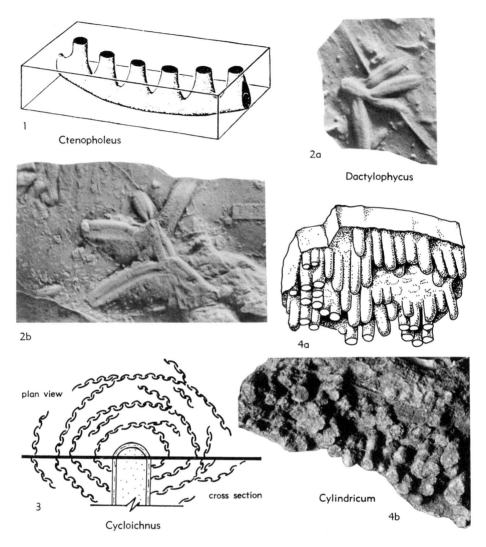

FIG. 35. Trace fossils (p. *W*55-58).

kutscheri; schem., ×1.3 (Seilacher & Hemleben, 1966).

Curvolithus FRITSCH, 1908, p. 13 [**C. multiplex*; SD HÄNTZSCHEL, 1962, p. *W*189]. Ribbonlike trails, more or less straight, flat; consisting of 3 parts: broad, usually smooth central stripe (about 0.5 to 2 cm. wide) and very narrow lateral ridges (1 to 2 mm. wide). [Endichnial crawling trails, also cutting bedding planes and passing over and under each other, probably produced by burrowing gastropods.[1] Two varied types of the genus described by HEINBERG (1970, p. 23); perhaps the grooved pipes described by KEIJ

[1] See also HEINBERG (1973).

(1965, p. 226) (Mio., Borneo) should also be placed in *Curvolithus* as proposed by CHAMBERLAIN (1971a, p. 224). Identity of the Ordovician specimens with the Jurassic ones is not yet proved; *C. gregarius* FRITSCH, 1908, p. 13, differs distinctly from the type species.] ?*Precam.*, Australia; *Ord.*, Eu.(Czech.); *Sil.*, USA(Ga.); *Penn.*, USA(Okla.); *L.Jur.-M.Jur.*, Eu.(Ger.)-Greenl.; *Cret.*, Eu.(Ger.); *Tert.*(?*Mio.*), Borneo.——FIG. 36,*3a*. *C.* sp., M.Jur., Ger.; ×0.7 (Seilacher, 1955).——FIG. 36,*3b*. *Curvolithus* FRITSCH, Low. Lias (Hettang.), Ger. (Helmstedt); ×1.3 (Häntzschel & Reineck, 1968).

Cycloichnus GREGORY, 1969, p. 13 [**C. waitema-*

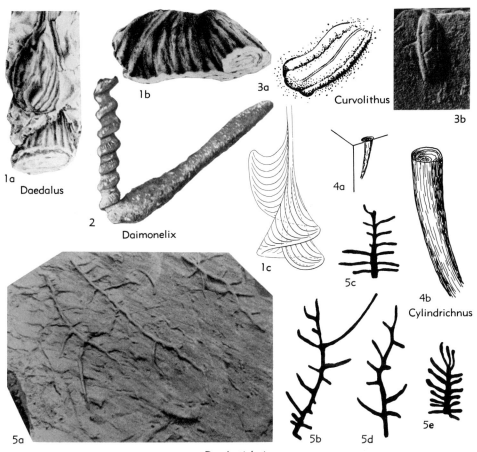

1a Daedalus
1b
2 Daimonelix
1c
3a Curvolithus
3b
4a
4b Cylindrichnus
5c
5a Dendrotichnium
5b 5d
5e

FIG. 36. Trace fossils (p. *W*56-57).

taensis; OD]. Simple central shaft, structureless, diameter about 1 cm., length 2 cm.; wall probably smooth, with several saucer-shaped galleries diverging from it, irregularly constricted to give small leaf-shaped impression, visible on bedding plane with concentric markings surrounding central shaft; galleries somewhat branching but not interconnected. [Tentatively interpreted by GREGORY to be result of proboscis-bearing animal systematically culling sediment about dwelling shaft. An inorganic origin (except central shaft) may be possible.] *Tert.(low.Mio.,Waitemata Gr.),* N.Z.——FIG. 35,*3*. *C. waitemataensis,* Whanga Paroa Penin., Auckland, schem., ×1.1 (Gregory, 1969).

Cylindrichnus TOOTS in HOWARD, 1966, p. 45 [*C. concentricus; M] [=Anemonichnus CHAMBERLAIN & CLARK, 1973; obj.] [non Cylindrichnus BANDEL, 1967a, =Margaritichnus BANDEL, 1973]. Subconical form, weakly curved, circular to oval

in cross section with diameter of 10 to 20 mm., most commonly 12 to 15 mm.; central core 2 to 4 mm.; exterior wall composed of concentric layers; preserved in full relief; orientation from nearly horizontal to vertical. [Interpreted as permanent burrow (domichnia) of filter-feeding organism. Considered by FREY & HOWARD (1970) as a form of *Asterosoma.*] *L.Penn.,* USA(Utah); *U.Cret.,* USA(Utah-Wyo.-Kans.).——FIG. 36,*4*. *C. concentricus,* Utah; *4a,b,* diagram. (Howard, 1966). [Description supplied by W. G. HAKES.]

Cylindricum LINCK, 1949, p. 19 [*Tubifex antiquus PLIENINGER, 1845, p. 159 (=Cylindricum gregarium LINCK, 1949b, p. 19; see LINCK, 1961, p. 9); OD]. Plugs (fillings of tubes) shaped like test tubes, rounded at lower end, not pointed, walls smooth; diameter up to 5 cm., up to several cm. long; preserved in groups in convex hyporelief oriented perpendicular to bedding plane. [Dwelling burrow.] *?Dev.,* Antarct.; *?L.*

Carb.(Kulm), Eu.(Ger.); *L.Trias.(Buntsandstein)*, Eu.(Ger.); *U.Trias.(Keuper)*, Eu.(Ger.); *?M.Jur.*, Eu.(Ger.).——Fig. 35,4. **C. antiquum* (Plieninger), U.Trias.(M.Keuper), Ger.; *4a*, U.Trias. (Schilfsandst.), diagram. (after Seilacher, 1955); *4b*, ✕1 (Linck, 1949b).

Dactylophycus Miller & Dyer, 1878, p. 1 [**D. tridigitatum*; SD Osgood, 1970, p. 345]. Delicately annulated bilobate burrows, small, about 15 mm. long, 2 to 4 mm. in diameter; radiate or randomly branching, number of branches varying. [Originally regarded as plant; considered by James (1884) to be fragments of burrows or of inorganic origin; according to Osgood (1970, p. 346), belongs to *Fodinichnia*, "excavation of *Sedimentfresser*"; as stated by James (1885) and Osgood (1970) possibly identical with *Palaeophycus radiata* Orton, particularly *Phycodes flabellum* (Miller & Dyer); type specimen of *D. tridigitatum* not located.] *U.Ord.(Cincinnat.)*, USA(Ohio).——Fig. 35,2. *D. quadripartitum* Miller & Dyer, Eden beds; *2a*, ✕1; *2b*, ✕1.3 (Osgood, 1970).

Daedalus Rouault, 1850, p. 736 [*non* Redtenbacher, 1891] [**Vexillum desglandi* Rouault, 1850, p. 733; SD Häntzschel, 1961, p. *W191*] [=*Vexillum* Rouault, 1850, p. 733 (*non* Bolten, 1798) *(nom. nud.); Humilis* Rouault, 1850, p. 738 (no type species designated); *Vescillum* Lebesconte, 1892, p. 76 *(nom. null.)*]. Spreiten structures, J-shaped at beginning, later spirally twisted; spreiten surface may cut through itself, as in *Dictyodora* Weiss. [For synonymy of type species see Rouault in Lebesconte (1883a, p. 45-47).] *Ord.-Sil.*, Eu.-Asia(Iraq)-USA.——Fig. 36,1. **D. desglandi* (Rouault); *1a, 1b*, Ord., France, ✕0.25 (Lebesconte, 1892); *1c*, L.Sil., USA, diagram showing gradation from vertical to spiral (Sarle, 1906).

Daimonelix Barbour, 1892, p. 99 [**D. circumaxilis*; SD Häntzschel, herein] [=*Daemonelix* Barbour, 1895, p. 517; *Helicodaemon* Claypole, 1895, p. 113 ("a more appropriate name") (all *nom. van.*); *Daemonhelix* auctt. (*non Daemonhelix krameri* von Ammon, 1900, p. 63 (L. Tert., S.Ger.), see *Gyrolithes* de Saporta); *non Daimonhelix Dusli* Fritsch, 1908, p. 6 (Ord., Czech.)]. Large, vertical, open, spiral structure, regular in form, mostly coiled with strict uniformity; transverse rhizome-like piece at base. [Explained as freshwater sponges, or casts of rodent burrows; some forms also resembling concretions; interpretation of helical burrows as *Daimonelix* and other forms by Toots (1963); history of genus discussed by Schultz (1942) and Lugn (1941). A somewhat comparable, though more tightly coiled, spiral structure was described by Whitehouse (1934) from the Lower Cretaceous of Queensland.] *Tert.(Mio.)*, USA. ——Fig. 36,2. **D. circumaxilis*, USA(Neb.); side view, ✕0.3 (Barbour, 1895).

Delesserites von Sternberg, 1833, p. 32 [**Fucoides Lamourouxii* Brongniart, 1823, p. 312 (=*D. lamurouxii* von Sternberg, 1833, p. 32); SD Andrews, 1955, p. 144] [=*Delessertites* Bronn, 1853, p. 110 (*non* Ruedemann, 1925, p. 8 = *Delesserella* Ruedemann, 1926, p. 156)]. Very heterogeneous "genus," including obvious trace fossils (e.g., *D. sinuosus, D. gracilis, D. foliosus* Ludwig, 1869, from Devonian and Lower Carboniferous of Germany) and equally obvious plants (e.g., probably *D. lamourouxii*, and, according to Pia (1927), *D. salicifolia* Ruedemann, 1925, Ord., N.Y.); Cenozoic "species" are under name of Recent genus *Delesseria* Lamouroux.

Dendrotichnium Häntzschel, herein [**D. llarenai* Farrés, 1967, p. 30; OD] [=*Dendrotichnium* Gómez de Llarena, 1949, p. 123 (*nom. nud.*, without species designation); *Dendrothichnium* Farrés, 1963, p. 105 (*nom. null.*); *Dendrotichnium* Farrés, 1967, p. 30 (*nom. nud.*, established without type species)]. Treelike trail, 7 to 30 cm. long; straight or somewhat curved "main stem," with several "side-branches" on both sides, their length quite variable, branching off perpendicularly in type species, but obliquely in *D. haentzscheli* Farrés. [Found in flysch deposits.] *U.Cret.*, Eu.(Spain).——Fig. 36,5a. *D. haentzscheli* and **D. llarenai*; ✕0.2 (Farrés, 1967).——Fig. 36,5b,d. *D. haentzscheli* Farrés; *5b,d*, ✕0.25 (Farrés, 1967).——Fig. 36,5c,e. **D. llarenai* Farrés; *5c,e*, ✕0.25 (Farrés, 1967).

Desmograpton Fuchs, 1895, p. 394 [no type species named] [=*Pseudodesmograpton* Macsotay, 1967, p. 36 (type, *P. ichthyformis* Macsotay, 1967, p. 36)]. Trail, roughly in form of long and very narrow letter H, single patterns usually lined up in ribbons; form variable; similar to *Paleomeandron* Peruzzi but with long appendices. [Grazing trail. With reference to great similarity of *Pseudodesmograpton* to *Desmograpton* and varying pattern of latter, *Pseudodesmograpton* should not be considered separate genus.] [Found in flysch deposits.] *Cret.-L.Tert.*, Eu.-S. Am.(Venez.).——Fig. 37,4. *D.* sp., *?U.Cret.*, Italy; ✕0.6 (Seilacher, in Häntzschel, 1962, coll. Florence Geol. Dept.).

Dictyodora Weiss, 1884, p. 17 [**Dictyophyton? liebeanum* Geinitz, 1867, p. 288; M] [=*?Nemertites* McLeay, in Murchison, 1839, p. 701 (certainly *N. sudeticus* Roemer, 1870, p. 33; for discussion, see Walter, 1903, p. 76); *Myrianites gracilis* Delgado, 1910, p. 28, and very probably several other "new species" of *Myrianites* McLeay, in Murchison, 1839, p. 700, in Delgado, 1910]. Complicated three-dimensional spreiten structure, irregularly conical, vertical to bedding; apex of cone upward; very thin spreite (=*Dictyodora s.s.*) with exterior surface delicately striated, intensely "folded," may cut through itself, consisting of furrowlike lamellae crescent in cross section; irregular spiral or meandering "band"

FIG. 37. Trace fossils (p. *W*58, 62, 64).

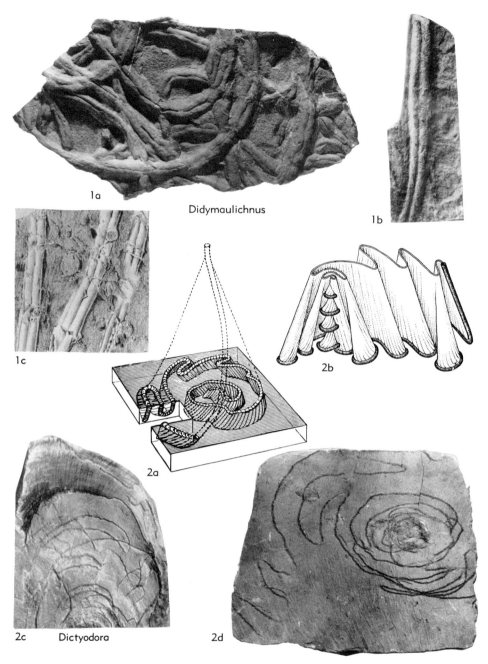

1a

Didymaulichnus

1b

1c

2a

2b

2c   Dictyodora   2d

Fig. 38. Trace fossils (p. *W*58, 60-61).

(=*"Palaeochorda marina"* Geinitz) on cleavage planes parallel to bedding represents line of intersection of vertically spacious spreite; lower margin nonhorizontal, thick, tunnel-shaped, padlike

(=*"Crossopodia henrici"* Geinitz); height of entire structure (in L. Carb. of Ger.) 3 to 18 cm. [Internal meandering foraging trail; producer unknown; Seilacher (1967c, p. 78) explained the different forms of *Dictyodora* from Cambrian to Mississippian by anatomical changes (increasing length of the supposed siphons?) and behavioral evolution of producer; for various former interpretations and for discussion see WEISS (1884a), ZIMMERMANN (1889, 1891, 1892), ABEL (1935, p. 429), SEILACHER (1955, p. 379), PFEIFFER (1959), A. H. MÜLLER (1962, 1971b), SEILACHER (1967c, p. 78).] *L.Cam.,* Asia(Pak.); *Ord.,* Eu.(Ger.-Port.); *Sil.,* Eu.(Eng.)-USA(Ga.); *U.Dev.-L.Carb.,* Eu.(Ger.Aus.).——FIG. 38,*2a. D. simplex* SEILACHER, L.Cam., Pak., drawing of a model, ×0.5 (Seilacher, 1955).——FIG. 38,*2b. Dictyodora* trail, reconstr.: "The animal left the upper sediment, eating its way deeper along a corkscrew-like path and then meandering in a restricted manner" (Seilacher, 1967).——FIG. 38, *2c,d.* **D. liebeana* (GEINITZ), *2c,* L.Carb., Aus.; ×0.3 (Abel, 1935); *2d,* L.Carb.(Kulm facies), Wurzbach (Frankenwald); ×0.5 (Müller, 1962).

Didymaulichnus YOUNG, 1972, p. 10 [*Fraena Lyelli ROUAULT, 1850, p. 732; OD] [=?*Fraena* ROUAULT, 1850, p. 729; *Rouaultia* DE TROMELIN, 1878, p. 501, obj. (*non Rouaultia* BELLARDI, 1878 (?1877), p. 233); *?Cruziana rouaulti* LEBESCONTE, 1883a, p. 67; *Rouaulita* HÄNTZSCHEL, 1962, p. *W212 (nom. null.)*]. Simple, smooth, gently curving bilobate trails (about 2 cm. wide) preserved in convex hyporelief; parallel to bedding; lobes separated by distinct furrow; may have 2 asymmetric "marginal bevels"; trails may overlap and truncate one another. [Origin speculative but possibly crawling trail of molluscan origin; similar to "molluscan trails" of GLAESSNER (1969, fig. 9B-9C); *Rouaultia rouaulti* considered by CRIMES (1970b, p. 56) probably to have been made by trilobites.] *U.Precam.,* N.Am.(Can.)-Australia-C.India; *?U.Precam.(U.Vindhyan),* C. India; *L.Cam.(U.Arumbera F.),* C.Australia; *U. Cam.(Ffestiniog Stage),* Eu.(N.Wales); *L.Ord.,* Eu.(France); *Ord.,* Eu.(France-Port.-Spain); *?Sil.,* N.Afr.-AsiaM.(Jordan).——FIG. 38,*1a. D. miettensis* YOUNG, U.Precam.(Miette Gr.), Can.(B.C., Alta.); ×0.2 (Young, 1972).——FIG. 38,*1b.* **D. lyelli* (ROUAULT), Ord., Port.; ×0.7 (Delgado, 1885).——FIG. 38, *1c. D. rouaulti* (LEBESCONTE), L.Ord.(Arenig.), France; ×0.75 (Lebesconte, 1883a). [Description supplied by W. G. HAKES.]

Dimorphichnus SEILACHER, 1955, p. 346 [**D. obliquus;* M]. Asymmetrical trails with 2 different types of impressions; thin sigmoidal ones, produced by raking movement (*"Hark-Siegel"* of SEILACHER), and blunt ones, similar to impressions of toes ("support imprints," *"Stemm-Siegel"* of SEILACHER); both types arranged in series oblique to direction of movement. [Made by laterally grazing trilobites; for discussion of the paleoecologic significance see OSGOOD (1970, p. 353.] *Cam.,* Eu.(Swed.-Eng.-Pol.)-Asia(Pak.); *Ord.,* S.Am.(Arg.).——FIG. 38A,*1.* **D. obliquus,*

1a   1b   1c
Dimorphichnus

FIG. 38A. Trace fossils (p. *W*61).

Jutana Dol., Pak.; *1a,* ×0.16; *1b,* ×0.8; *1c,* ×0.4 (Seilacher, 1955).

Diplichnites DAWSON, 1873, p. 19 [**D. aenigma;* M] [=*Acripes* MATTHEW, 1910, p. 122 (type, *A. incertipes;* SD HÄNTZSCHEL, 1965, p. 6), for synonymy, see SEILACHER, 1955, p. 343]. Morphologically simple track, width about 1 to 2 cm., consisting of 2 parallel series of fine ridges (1-5 mm. long), individual ridges elongate obliquely to track axis, sometimes apparently occurring in pairs (illustrating two-clawed limbs of animal producing track), anterior ridge then more prominent. [Originally interpreted by DAWSON as traces of large worms or crustaceans or imprints of spines of fish; now considered locomotion tracks of trilobites, walking or striding in straightforward movement across the surface of the sediment; CRIMES (1970b, p. 57) observed transitional forms between *Diplichnites* and *Cruziana;* OSGOOD (1970, p. 352) is skeptical about the trilobite origin and the marine environment of DAWSON's type specimens of ichnogenus; a comparable track from the Devonian of Antarctica is *Beaconichnus gouldi* GEVERS, 1971 (see GEVERS et al., 1971, p. 86, 93).] *L.Cam.,* Can.; *Cam.,* Eu.(Eng.-Swed.-Pol.)-USSR(Sib.)-Greenl.-Asia(Pak.)-Australia; *?Cam.,* Eu.(Nor.); *Ord.,* Eng.; *?Ord.,* Asia (Jordan); *?Dev.-Carb.,* N.Am.(Can.); *L.Perm.(Dwyka Gr.),* S.Afr.——FIG. 39,*4a. D.* sp., L.Cam., Asia(Pak.); schem., ×1.3(Seilacher, 1955).——FIG. 39,*4b. Diplichnites,* U.Cam., N.Wales; ×? (from Crimes, T. P., 1970, p. 120, in *Trace Fossils* edited by T. P. Crimes & J. C. Harper, Geol. Jour. Spec. Issue 3, Seel House Press, Liverpool).

Diplocraterion TORELL, 1870, p. 13 [**D. parallelum*; SD RICHTER, 1926, p. 213] [=*Polyupsilon* HOWELL, 1957, p. 151 (type, *"Tigillites" habichi* LISSON, 1904, p. 41); for discussion see GOLDRING, 1962, p. 238; *Diplocration* GOLDRING, 1964, p. 137 *(nom. null.)*]. U-shaped burrow with spreite; vertical to bedding plane; limbs of U parallel; both limbs of each successive U-tube confluent with limbs of preceding U-tube (see KNOX, 1973, p. 134); openings of tubes mostly funnel-shaped (but apparently often truncated by erosion); commonly protrusive, but also retrusive forms observed; bottom of burrow semi-circular, rarely straight; horizontal cross section on bedding planes dumbbell-shaped; diameter of tubes 5 to 15 mm., distance between limbs 1 to 7 cm. (average, 2-3 cm.), depth of burrows 2 to 15 cm. (max. 35). [Dwelling burrow of suspension feeding animal, probably living in environment of high wave energy; several stages of erosion and sedimentation may be recognized from various levels of tube (e.g., *D. yoyo*; see GOLDRING, 1962, p. 235, and Fig. 16); intermediate forms between *Diplocraterion* and *Rhizocorallium* observed in the Carboniferous of Scotland (CHISHOLM, 1970b, p. 49).] *Cam.,* Eu. (Swed.-Nor.-Pol.-Spain)-N.Am. (USA-Newf.)-Australia; *Cam.,* Pleist. drift, Eu.(N.Ger.); *Ord.,* Eu.(Nor.); *L.Paleoz.,* N.Afr.(Libya); *M.Dev.,* Eu. (Ger.); *Sil.,* USA(Ga.); *U.Dev.,* Eu.(Eng.); *Carb.,* Eu.(Scot.); *Jur.,* Eu.(Eng.-N.France-Pol.)-Greenl.; *Cret.,* N.Am.(USA,Colo.)-S.Am.(Peru); *?Cret.,* Eu.(Ger.).——FIG. 37,*2a.* **D. parallelum,* L. Cam.(Mickwitzia Ss.), Swed.: ×0.7 (Westergård, 1931).——FIG. 37,*2b.* *D. lyelli* TORELL, L.Cam., Swed.; funnel-shaped openings of U-shaped burrow to surface, concave epirelief, ×1.5 (Westergård, 1931).

Diplopodichnus BRADY, 1947, p. 469 [**D. biformis*; OD]. Long, continuous trail, consisting of 2 or 3 parallel grooves, each pair separated by narrow, low ridge; rarely with faint foot impression; somewhat similar to *Gordia* EMMONS [=*Unisulcus* HITCHCOCK]. [Made by arthropods; common in Coconino Sandstone.] *L.Perm.,* USA (Ariz.).

Diplopodomorpha CHIPLONKAR & BADWE, 1970, p. 4 [**D. cretaceca*; OD]. Trail 0.7 to 0.8 cm. wide with clusters of 3 or 4 tubercles separated by smooth axial region 0.35 to 0.4 cm. wide, consecutive clusters spaced 0.15 to 0.2 cm. apart, each cluster of tubercles composed of one large and 2 or 3 subequal smaller impressions. [Probably produced by diplopodous arthropod.] *U. Cret.,* India.——FIG. 37,*3.* **D. cretaceca,* Nimar Ss.; ×0.7 (Chiplonkar & Badwe, 1970). [Description supplied by W. G. HAKES.]

Echinospira GIROTTI, 1970, p. 60 [**E. pauciradiata*; OD]. Similar to *Zoophycos s.l.*; characterized by an "aculeate" edge; unbranched "radioli," 30 to 40 cm. in length; "pinnulae" on only one side of "radioli," disappearing at their end.

[GIROTTI followed PLIČKA's description of these supposed imprints of sabellid prostomia, designating *Echinospira* as ichnofossil, although he was in agreement with PLIČKA's interpretation of *Zoophycos* and similar forms as "anatomical parts of polychaetes."] *U.Tert.(Mio.),* Eu.(C.Italy).

Eugyrichnites AMI, 1905, p. 291 [**E. minutus*; M]. Minute tortuous trail, about 1 mm. wide; with fine annulations (25-30 closely set parallel lines in 1 cm.). [Said to resemble *Gyrichnites* WHITEAVES; never figured, no specimens located in Canadian collections.] *?Sil.,* Can.(N.B.).

Fascifodina OSGOOD, 1970, p. 340 [**F. floweri*; OD]. Vertically bundled shafts; mostly preserved as crescentic or horseshoe-shaped groups of short concave and vermiform markings (epireliefs) surrounding lower part of single shaft; upper portion of original burrow, particularly upper part of master shaft, very probably stripped away by erosion. [Morphology not yet fully understood; first described by FLOWER (1955), but left unnamed and interpreted as the vermicular markings produced by tentacles of orthoconic nautiloid *Orthonybyoceras* grasping sea bottom for feeding and clinging to substratum to resist motion of the water; interpreted by OSGOOD (1970) as feeding burrow.] *U.Ord.(mid.Cincinnat.),* USA(Ohio).——FIG. 39,*3.* **F. floweri*; block diagram, ×0.19 (after Osgood, 1970).

Fascisichnium KSIĄŻKIEWICZ, 1968, p. 10 (Pol.), p. 16 (Eng.) [**F. extendum*; M]. Large central area surrounded by numerous arrowlike ribs arranged like bundle of scattered rods; ribs straight or curved, tapering to point, not diverging from center of inner field, but lying excentrically outside of it; whole trail 8 to 10 cm. long, up to 5 mm. wide. [Found in flysch deposits.] *L.Tert. (Paleoc.-low.Eoc.),* Eu.(Pol.).——FIG. 37,*1.* **F. extendum,* Paleoc.-low.Eoc., Carpathians; ×0.8 (Książkiewicz, 1968).

Felixium DE LAUBENFELS, 1955, p. E36 [*pro Rhizocorallium* FELIX, 1913 *(non* ZENKER, 1836)] [**Rhizocorallium glaseli* FELIX, 1913; OD]. Elaborately sculptured, curved cylinder 5×20 cm. *Cret.,* Eu.(Ger.).

Fraena ROUAULT, 1850, p. 729 [**F. Sancti-Hilairei*; SD PÉNEAU, 1946, p. 77]. Rarely used name for simple trails, unilobate as well as particularly bilobate, some smooth, some striated longitudinally or transversely. [ROUAULT combined in "genus" seven "species" which were subsequently placed in *Cruziana, Rouaultia* (type species, *Fraena lyelli*), and *Rusophycus*; DE TROMELIN & LEBESCONTE (1876, p. 627), MATTHEW (1891, p. 158), PÉNEAU (1946, p. 77), and other authors recommended restricting name *Fraena* to simple smooth unilobate trails of type of first "species" described by ROUAULT, *F. sanctihilairei*; see also *Palaeotenia guilleri* CRIÉ, 1883.] *Ord.,* Eu.(France).

Fucoides BRONGNIART, 1823, p. 308 [**F. strictus*; SD JAMES, 1894, p. 69]. Formerly used as generic

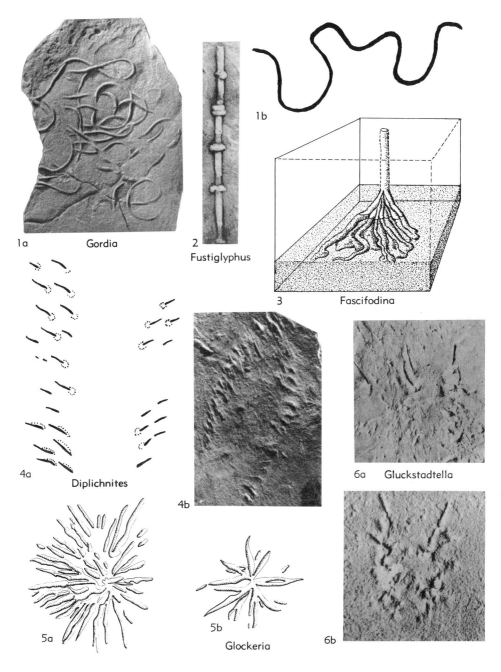

FIG. 39. Trace fossils (p. *W*61-62, 64).

name mostly for regularly branching, plantlike tunnel structures; at present only used informally ("fucoid"), due to too many widely differing "species" descriptions; BRONN's *"Index Palaeon-* *tologicus"* (1849) listed 59 "species," JAMES (1894), describing history of "genus," ascertained 85 "species"; by 1825 *Fucoides* had been divided into "subgenera" and by 1828 into *"sectiones."*

JAMES (1892a, p. 76) wrote "that before many years the genus *(Fucoides)* began to overflow and then, like an overloaded wagon, broke down. . . . Among the debris we find tracks of crustaceans, burrows of worms, trails of mollusks, marks made by trailing tentacles of medusae, markings made by the tide or waves, rills by running water, and holes formed by burrowing worms."; *Fucoides graphica* VANUXEM, 1842, common in the lower Upper Devonian of western New York, has been used for determining trends or directions of paleocurrents in that sequence (COLTON, 1967). [See also *Chondrites.*]

Fucusopsis PALIBIN in VASSOEVICH, 1932, p. 51 [**F. angulatus*; M] [=*Trichophycus sulcatum* MILLER & DYER, 1878, p. 4 (for synonymy see OSGOOD, 1970, p. 380); *Fucopsis* GROSSHEIM, 1946, p. 115 *(nom. null.)*; *?Gyrochorda fraeniformis* FARRÉS, 1963, p. 116]. Stretched tubiform burrows (2-10 mm. in diameter), long, straight, sometimes branching, crossing over and interpenetrating; with typical threadlike sculpture; regarded by SEILACHER (1959, p. 1070) as produced by burrowing activity; interpreted by OSGOOD (1970, p. 380) as "tension faulting" in sole of host rock; appearance depending on kind of preservation. [Originally regarded as marine alga or inorganic; now interpreted as burrows of infaunal origin.] [Found in flysch deposits.] *U.Ord.(Cincinnat.),* USA(Ohio); *Cret.-L.Tert.,* Eu.(Switz.-Spain-Pol.-Italy-?Aus.-USSR); *?Tert.,* S.Am.(Venez.).——FIG. 37,5. **F. angulatus,* U. Cret.(Senon.), USSR; ×0.3 (Gekker, in Häntzschel, 1962).

Fustiglyphus VYALOV, 1971, p. 90 [**F. annulatus* (=*Rhabdoglyphus grossheimi* BOUČEK & ELIÁŠ, 1962, p. 146 *(partim),* non VASSOEVICH, 1951, p. 61); M]. Straight strings or narrow cylinders of varying length encircled by ringlike "knots" or well-defined swellings at regular or varying intervals; rosary-like. [Difficult to interpret as trace fossil; according to OSGOOD, 1970, p. 369-371, a variety of repichnia or fodinichnia; believed by BOUČEK & ELIÁŠ (1962) to be made by amphipods or gastropods or even a holothuroid similar to *Leptosynapta;* for detailed discussion of *Fustiglyphus* see *Rhabdoglyphus* in OSGOOD (1970, p. 369-371) and (in Czech language) BOUČEK & ELIÁŠ (1962). See also *Rhabdoglyphus* BOUČEK & ELIÁŠ, p. W99, for a discussion of the nomenclatural history of *Fustiglyphus.*] *U.Ord.(Cincinnat.),* USA(Ohio); *Tert.(Eoc.),* Eu.(Pol.); *Tert. (Paleog., Magura Gr.),* Eu.(Czech., Carpathians).——FIG. 39,2. **F. annulatus,* Magura Gr., Carpathians; ×0.56 (Bouček & Eliáš, 1962). [Description supplied by CURT TEICHERT and W. G. HAKES.]

Glockeria KSIĄŻKIEWICZ, 1968, p. 9 (Pol.), p. 15 (Eng.) [**G. glockeri;* OD]. Starlike trace fossil with numerous long rays, straight, pointed, commonly dichotomous and radiating from small central area; small ones between main ribs; diameter 6 to 13 cm.; feeding burrow. [Found in flysch deposits.] *L.Cret.,* Japan-Eu.(Pol.); *U. Cret.(Senon.)-L.Tert.(Paleoc.),* Eu.(Pol.-Spain). ——FIG. 39,5a. **G. glockeri,* L.Cret.(Berrias., Cieszyn Ls., Pol.(Goleszów); ×0.3 (Książkiewicz, M., 1970, p. 311, in: *Trace Fossils* edited by T. P. Crimes & J. C. Harper, Geol. Jour. Spec. Issue 3, Seel House Press, Liverpool).——FIG. 39,5b. *G. sparsicostata* KSIĄŻKIEWICZ, U.Cret.(Senon., Inoceramian Beds), Pol.(Zawoja); ×0.3 (from Książkiewicz, M., 1970, p. 311, in: *Trace Fossils* edited by T. P. Crimes & J. C. Harper, Geol. Jour. Spec. Issue 3, Seel House Press, Liverpool).

Gluckstadtella SAVAGE, 1971, p. 231 [**G. cooperi;* OD]. "Arthropod resting impression," 8 to 22 mm. long, 5 to 14 mm. wide; showing 6 pairs of appendage marks; anterior 2 pairs longest, remaining 4 pairs shorter and forming distinct group directed obliquely backwards. [Perhaps producer of trails described by SAVAGE (1971, p. 225) as *Diplichnites* sp.; from freshwater periglacial environment.] *L.Perm.(Dwyka Gr.),* S. Afr.(N.Natal).——FIG. 39,6. **G. cooperi; 6a,* ×1.4; *6b,* ×1.7 (Savage, 1971).

Goniadichnites MATTHEW, 1891, p. 160 [**G. trichiformis;* M]. Small sinuous smooth trails, no larger than slender thread, commonly branching, apparently forking dichotomously; resembling trails of Recent *Goniada* as figured by NATHORST (1881a). *Cam.,* Can.

Gordia EMMONS, 1844, p. 24 [*non* MELICHAR, 1903] [**G. marina;* M] [=*Palaeochorda* M'COY in SEDGWICK, 1848, p. 224 (type, *P. minor* M'COY; SD HÄNTZSCHEL, herein) *(non P. marina* (EMMONS) *sensu* GEINITZ, 1867, p. 14; see *Dictyodora* WEISS, 1884); *Palaeochordia* EICHWALD, 1860, p. 53 *(nom. null.); Herpystozeum* HITCHCOCK, 1848, p. 245 (type, *H. marshii;* SD LULL, 1953, p. 50); *Helminthoidichnites* FITCH, 1850, p. 868 (type, *H. tenuis;* SD HÄNTZSCHEL, 1965, p. 45); *Unisulcus* HITCHCOCK, 1858, p. 160 *(nom. nov. pro Herpystezoum* HITCHCOCK, 1848); *Gordiopsis* HEER, 1865, p. 439 (type, *G. valdensis;* M)]. Long, slender, smooth wormlike trails of uniform thickness throughout; mostly bent but not meandering; resembling hair-worm *Gordius.* *?Precam.,* Can.; *Paleoz.-Cenoz.,* Eu.-N. Am.——FIG. 39,1a,b. *G.* sp.; *1a,* M.Dev.(Eifel.), Ger.(Holzmülheim, Eifel), ×0.5 (Fischer & Paulus, 1969); *1b,* schem. drawing, ×0.5 (Häntzschel, 1962).

Granularia POMEL, 1849, p. 333 [**Algacites granulatus* VON SCHLOTHEIM, 1822, p. 45; "OD"] [*non Granularia* POLETAEVA, ?1936] [=*Alcyonidiopsis* MASSALONGO, 1856, p. 48 (no type species designated)]. Elongated fillings of burrows; long, diameter up to about 15 mm.; twig-shaped, with rather regular branching; walls originally lined with clay particles; burrows observed by SEILACHER (1962, p. 228) in flysch deposits of

Spain which are up to several meters thick. [BATHER (1911, p. 555) wrote erroneously that "*Granularia* was established by POMEL (1847) with *G. repanda* as genotype"; POMEL described six "species" and he founded *Granularia* on *Algacites granulatus,* the same species on which BRONGNIART (1849) founded his "genus" *Phymatoderma*; for discussion of somewhat confused synonymy and nomenclature see also ROTHPLETZ (1896, p. 889).] [Found in flysch deposits.] *Sil.,* Australia; *M.Jur.,* Eng.; *Cret.-L.Tert.,* Eu.——FIG. 40,*3a.* *G.* sp. *cf. G. arcuata* SCHIMPER, L.Tert.Alberese, Italy; ×1.25 (Reis, 1909).——FIG. 40,*3b.* *G. lumbricoides* (HEER), L.Tert. (Alberese), Italy; ×1.25 (Reis, 1909).

Gyrichnites WHITEAVES, 1883, p. 111 [**G. gaspensis*; M]. Trails of large size; undulating, slender, rounded furrows marked transversely by nearly straight, subparallel and subequidistant grooves. [?Annelid trail; name given as "provisional and local," apparently never used since 1883.] *?U. Cam.,* USA(N.Y.); *Dev.,* N.Am.(Can.).——FIG. 40,*4.* **G. gaspensis,* L.Dev., Can.; ×0.3 (Whiteaves, 1883).

Gyrochorte HEER, 1865, p. 142 [**G. comosa*; SD HÄNTZSCHEL, 1962, p. *W*196] [=*Gyrochorda* SCHIMPER in SCHIMPER & SCHENK, 1879, p. 51 *(nom. null.); ?Equihenia* MEUNIER, 1886, p. 567 (type, *E. rugosa*)]. Trace up to 5 (rarely 10) mm. wide; in epirelief preserved as plaited ridges with biserially arranged, obliquely aligned pads of sediment ("*Zopf-fährten*" of German literature); in hyporelief preserved as smooth biserial grooves separated by median ridge; course strongly winding and direction changing sharply; trace may intersect itself or other traces; ridges and their grooves may be separated by vertical distance of 1 cm.; usually preserved in clastic sediments. [Crawling trails, similar to amphipod trails (e.g., *Corophium*); doubtless made by tunnelling through sediment; producer unknown; ?worms or crustaceans; for model of this trail see SEILACHER (1955, p. 380, fig. 2b); for detailed discussion of mechanism of formation of this trail see HALLAM (1970, p. 192-195).[1] *G. bisulcata* GEINITZ, 1883-95 (Eoc., N.Ger.) does not belong to *Gyrochorte, s.s.,* but is similar to *Dreginozoum* VAN DER MARCK; "*Gyrochorte*" *carbonaria* SEILACHER, 1954 (U.Carb., Ger.) is no true *Gyrochorte;* for discussion see SEILACHER (1963, p. 83). MARTINSSON (1965, p. 219) has described the relationship of *Gyrochorte* to *Halopoa* TORELL.] *Sil.,* USA(Ga.), *?Carb., Jur.-Tert.,* Eu.-Greenl.; *?Carb., ?Jur.-Tert.,* USA-S.Am.-Antarct.——FIG. 40,*1.* **G. comosa,* M.Jur., Switz.; ×1 (Heer, 1865).

---

[1] HALLAM's proposed mode of origin for *Gyrochorte* as a collapsed tunnel has been recently rejected by HEINBERG (1973), who described vertical spreite-like structures connecting the epichnial ridges with the hypichnial grooves and felt that *Gyrochorte* was produced by a polychaete-like worm moving obliquely through the sediment. [W. G. HAKES.]

Gyrolithes DE SAPORTA, 1884, p. 27 [**G. davreuxi*; SD HÄNTZSCHEL, 1962, p. *W*200] [="Gyrolithen" DEBEY, 1849, p. 10 (*partim*; not used as "genus"); *Siphodendron* DE SAPORTA, 1884, p. 38 (type, *S. girardoti*); *Syringodendron* FUCHS, 1895, p. 404 (?erroneously *pro Siphodendron*); *Daemonhelix krameri* VON AMMON, 1900, p. 63; *Xenohelix* MANSFIELD, 1927, p. 6 (type, *X. marylandica*)]. Dextrally or sinistrally coiled burrows up to several cm. in diameter, sometimes with rounded or elongate processes which may be branching near upper end; diameter of whorls mostly uniform; vertically oriented; up to several decimeters high. Thin mantle of burrows may be formed by network of small *Chondrites;* "*Xenohelix*" with *Ophiomorpha*-like ornament are also known from Tertiary of Germany (KILPPER, 1962) and Borneo (KEIJ, 1965). [Probably made by decapod crustaceans (with exception of the specimens from L. Cam., Nor.); for discussion see FUCHS, 1894b; UMBGROVE, 1925; HÄNTZSCHEL, 1934; KILPPER, 1962; TOOTS, 1963.] *?L.Cam.,* Eu.(Nor.); *Jur.-Tert.,* Eu.-USA-S.Am. (Venez.)-Borneo.——FIG. 41,*4a.* *G. marylandicus* (MANSFIELD), *?Mio.,* Md.; ×? (Mansfield, 1927).——FIG. 41,*4b.* *G. saxonicus* (HÄNTZSCHEL), U. Cret.(Turon.), Ger.; ×0.4 (Häntzschel, 1934).

Gyrophyllites GLOCKER, 1841, p. 322 [**G. kwassizensis*; M] [=*Sargassites rehsteineri* FISCHER-OOSTER, 1858, p. 34; *?Discophorites* HEER, 1877, p. 145 (no type species designated)]. Vertical or oblique shaft from which 5 to 20 (average 10) club- or leaf-shaped feeding tunnels radiate at different levels in whorled or helical arrangement; rosettes up to several cm. in diameter, becoming larger upward; tunnels may show spreiten structure; shape of whole structure conical. [Definite trace fossil, producer unknown; for description of several "species" and interpretation as algae see LORENZ VON LIBURNAU (1900, p. 568); VONDERBANK (1970, p. 104) reconstructed complete sequence of rosettes of various sizes connected by the central shaft and ending in funnel-shaped aperture above highest rosette (Tert., Spitz.).] [Found in flysch deposits.] *Dev., Jur.-Tert.,* Eu.; *?Jur.-Tert.,* N.Z.——FIG. 40,*2.* *G.* sp., U.Cret., Aus.; *2a,* ×1 (Fuchs, 1895); *2b,* schem. (Seilacher, 1957).

Haentzschelinia VYALOV, 1964, p. 113 [**Spongia ottoi* GEINITZ, 1849, p. 113; OD]. Starlike trail with elevated center, about 5 cm. in diameter; generally 6 to 10 radiating grooves, rather irregularly and often only unilaterally developed. [Originally described as sponge similar to *Peronidella furcata* (GOLDFUSS); obviously a feeding burrow made by crustaceans or worms.] *Trias.,* Asia(USSR, NE.Sib.); *U.Cret.(Cenoman.),* Eu. (Ger., Saxony).——FIG. 42,*3.* **H. ottoi* (GEINITZ), U. Cret., Ger.; *3a,* ×12.5; *3b,* ×0.33 (Häntzschel, 1930).

Halimedides LORENZ VON LIBURNAU, 1902, p. 710

FIG. 40. Trace fossils (p. *W*64-65).

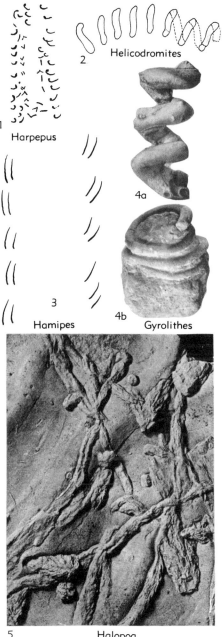

FIG. 41. Trace fossils (p. *W*65, 67).

[**Halimeda juggeri* Lorenz von Liburnau, 1897, p. 177; M]. Burrow with bilaterally ("pinnate') arranged, kidney-shaped extensions. [Morpho-

logically very similar to Recent alga *Halimeda* Lamouroux; *Halimedides* proposed only for *Halimeda juggeri*; *Halimeda saportae* Fuchs (1894c, p. 204) identical to problematical body fossil *Halysium* Swidzinski, 1934.] [Found in flysch deposits.] *Cret.*, Eu.(Aus.).——Fig. 42,*1*. **H. juggeri*; ✕0.3 (Lorenz von Liburnau, 1897).

Halopoa Torell, 1870, p. 7 [**H. imbricata*; SD Häntzschel, herein (not Andrews, 1970, p. 99, which was a proposal rather than a valid designation)] [=*Scotolithus* Linnarsson, 1871, p. 18 (type, *S. mirabilis*); for discussion see Martinsson, 1965, p. 219]. Long, slightly curved trails dug along surface; surface of trail with typical imbricate or lycopodiaceous structure; diameter of burrows about 0.5 to 1 cm. [Probable producers epipsammonts; for the first time since Torell's description in 1870, figured and discussed by Martinsson (1965, p. 219), who grouped "the halopoans" with *Zopffährten* (=*Gyrochorte* Heer) although they show no typical plaitlike structures.] *Cam.*, Eu.(Swed.).——Fig. 41,*5*. **H. imbricata*; L.Cam., Lugnås, Västergötland; ✕0.5 (Martinsson, 1965).

Hamipes Hitchcock, 1858, p. 150 [**H. didactylus*; M]. Two paired, regular, parallel rows of equidistant impressions of steps, curved inward, somewhat hook-shaped; width of trackway 40 mm; foot impressions nearly parallel, may be slightly divergent. [Arthropod trail.] *Trias.*, USA(Mass.).——Fig. 41,*3*. **H. didactylus*; ✕0.7 (Hitchcock, 1858).

Haplotichnus Miller, 1889, p. 578 [**H. indianensis*; OD]. Simple trail, straight or curved, sometimes bent sharply. [Supposed to be made by larva of ?palaeodictyopterid.] *U.Miss.(Kaskaskia Gr.)*, USA(Ind.).

Harpepus Hitchcock, 1865, p. 16 [**H. capillaris*; M]. One or 2 rows of tracks showing slightly curved feet impressions, somewhat sickle-like, one end raised, blunt. *Trias.*, USA(Mass.).——Fig. 41,*1*. **H. capillaris*; ✕0.7 (Lull, 1953).

Helicodromites Berger, 1957, p. 540 [**H. mobilis*; M]. Smooth screw-shaped burrows, horizontal; diameter of tunnels about 2 mm.; interval between spiral turns about 5 mm. [For discussion of similar Recent traces from marine and terrestrial sediments, see A. H. Müller (1971a).] *Oligo.(Rupel.)*, Eu.(S.Ger.).——Fig. 41,*2*. **H. mobilis*; ✕0.7 (Berger, 1957).

Helicolithus Häntzschel, 1962, p. *W*200 [**H. Sampelayoi* Azpeitia Moros, 1933, p. 48; OD] [=*Helicolithus* Azpeitia Moros, 1933, p. 48, *nom. nud.*, established without designation of type species]. Small, meandering, screw-shaped burrows; diameter of tunnels 1 mm.; diameter of spiral up to about 3 mm.; somewhat similar to *Helicodromites* but much smaller; *Helicolithus fabregae* Azpeitia Moros resembling *Belorhaphe* Fuchs, but with sharp turns. [Grazing trails, first

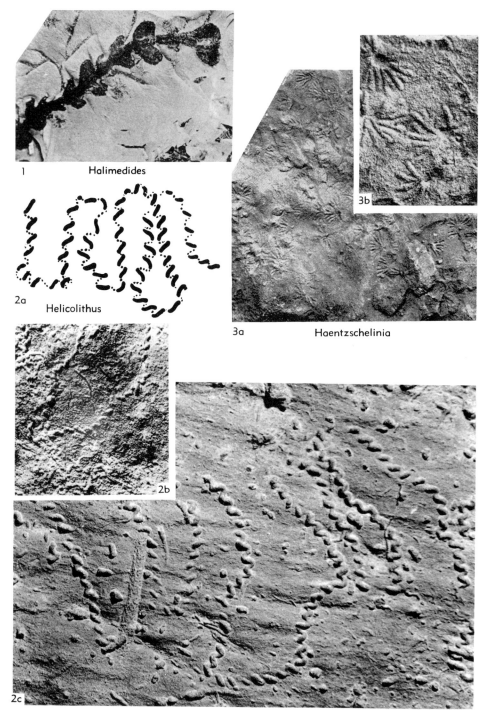

1   Halimedides

2a   Helicolithus

3b

3a   Haentzschelinia

2b

2c

FIG. 42. Trace fossils (p. *W*65, 67, 70).

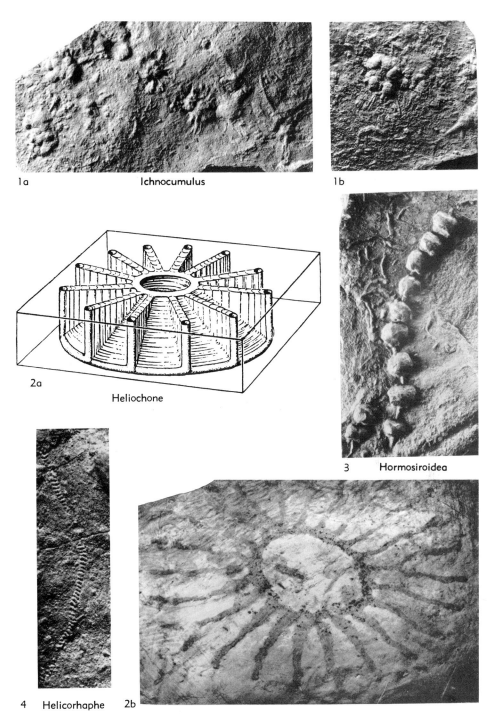

1a   Ichnocumulus   1b

2a   Heliochone

3   Hormosiroidea

4   Helicorhaphe   2b

Fig. 43. Trace fossils (p. *W*70, 74).

interpreted as algae.] [Found in flysch deposits.] *?L.Cam.*, Eu.(Nor.); *Cret.-L.Tert.*, Eu.(Aus.-Pol.-Spain-Italy).——FIG. 42,2. **H. sampelayoi* AZPEITIA MOROS; *2a*, ?Cret., Italy; schem. drawing, ×1.5 (Seilacher, 1955); *2b*, U.Cret., Spain; ×1 (Azpeitia Moros, 1933); *2c*, U.Eoc. (Magura Ss.), Carpathians; ×1.3 (Książkiewicz, M., 1970, p. 297, in: *Trace fossils*, ed. by T. P. Crimes & J. C. Harper, Geol. Jour. Spec. Issue 3, Seel House Press, Liverpool).

Helicorhaphe KSIĄŻKIEWICZ, 1970, p. 286 [**H. tortilis*; OD] [=*Helicoraphe* KSIĄŻKIEWICZ, 1961, p. 885, 889; published as "n.f." without species name]. Very narrow sole trails resembling horizontal spring with narrow turns (15/1 cm.); differing from similar *Helicolithus* AZPEITIA MOROS, 1933, by nearly straight course, not meandering and more tightly twisted. [For discussion of similar Recent traces see A. H. MÜLLER (1971a).] [Found in flysch deposits.] *Tert. (low.Eoc.)*, Eu.(Pol.).——FIG. 43,4. *H.* sp., Pol.; ×1.5 (Książkiewicz, 1961).

Heliochone SEILACHER & HEMLEBEN, 1966, p. 46 [**H. hunsrueckiana*; M]. Large, somewhat complex system of burrows consisting of circular tunnel with numerous (max., 22) vertical shafts proceeding from it at equal intervals; shafts connect tunnel with surface of sediment; whole starlike system originated by congruent enlargement of ring-shaped tunnel in outward and downward direction; diameter of burrow up to 50 cm.; probably feeding burrow. *L.Dev. (Hunsrück shale)*, Eu.(Ger.).——FIG. 43,2. **H. hunsrueckiana; 2a*, schem. drawing; *2b*, ×0.4 (Seilacher & Hemleben, 1966).

Helminthoida SCHAFHÄUTL, 1851, p. 142 [**H. labyrinthica* HEER, 1865, p. 246; SD HÄNTZSCHEL, 1962, p. W200] [=*Elminthoida* SACCO, 1886, p. 940; *Helminthoidea* MAILLARD, 1887, p. 7 (and other authors); *Helminthoides* FUCHS, 1895, p. 385; *Helmintoidea* VINASSA DE REGNY, 1904, p. 318 (all *nom null.*)]. Meandering tunnel trails; meanders numerous, very regular, parallel and closely spaced, but may be irregular, not always parallel, and not closely spaced; about 1 to 3 mm. wide, max. width of meanders about 1 cm., and length 10 cm.; regular meanders (particularly *H. crassa*) are type of the "guided meanders" (RICHTER, 1924, p. 153); species of this "genus" exhibit much variability, thus KSIĄŻKIEWICZ (1970, p. 296) introduced two "groups" and some "formae" (e.g., *H. labyrinthica forma lata*). [Former interpretations: plants (algae), worms, feeding traces of gastropods, strings of spawn (RECH-FROLLO, 1962); now regarded as internal grazing trails of wormlike animals. For behavioral analysis, see RICHTER (1928) and SEILACHER (1967a,c); meanders are probably effected by stimuli (homostrophy, thigmotaxis, phobotaxis); SEILACHER (1967c, p. 76) described areas of disturbance created by churning

of sediment along sides of tunnels.] [Found in flysch deposits.] *Cert.-Tert.*, Eu.-N.Am.(Alaska)-S.Am.(Chile-Venez.-Trinidad)-Asia (Japan)-?N.Z.——FIG. 44,*1a,b. H.* sp.; *1a*, schem. drawing; *1b*, Tert., Toscana, Italy; ×0.75 (Seilacher, 1967a,c).——FIG. 44,*1c. *H. labyrinthica*, U. Cret., Aus.; ×1 (Häntzschel, 1955). [See also Fig. 55,*1b,c.*]

Helminthopsis HEER, 1877, p. 116 [*non* GROUVELLE, 1906] [**H. magna*; SD ULRICH, 1904, p. 144] [=*Elminthopsis* SACCO, 1886, p. 939; *Helmintopsis* VINASSA DE REGNY, 1904, p. 319 *(nom. null.); Magarikune* MINATO & SUYAMA, 1949, p. 277 (type, *M. akkesiensis);* *?Serpentinichnus* MAYER, 1956, p. 8 (type, *S. bruchsaliense); Tosahelminthes* KATTO, 1960, p. 333 (type, *T. curvata); Helmenthiopsis* CHAMBERLAIN, 1971a, p. 216 *(nom. null.)*]. Simple meandering smooth trails, but not as strictly developed as *Helminthoida s.s.* (RICHTER, 1928); in part with marginal ridges. [*Helminthopsis involuta* DE STEFANI, 1895, and *H. ?concentrica* AZPEITIA MOROS, 1933, p. 46, are to be placed in *Spirorhaphe* FUCHS; *H. sinuosa* AZPEITIA MOROS, 1933, p. 45, in *Cosmorhaphe* FUCHS; *H. tenuis* KSIĄŻKIEWICZ, 1968, p. 7, should be ascribed to the "genus" *Gordia* EMMONS.] *Ord.-Tert.*, Eu.-Asia-N.Am.-Antarct.-S.Am.(Venez.).——FIG. 44, *2. H.* sp., U.Cret.; *2a*, Alaska, ×1 (Ulrich, 1904); *2b*, Aus., ×0.75 (Abel, 1935).

Hexapodichnus HITCHCOCK, 1858, p. 158 [**H. magnus*; SD LULL, 1953, p. 45]. Triple rows of tracks on either side of median line; inner impressions parallel, outer tracks also parallel or diverging outward; width 15 to 20 mm. [Probably made by insects.] *Trias.*, USA(Mass.).

Himanthalites VON FISCHER-OOSTER, 1858, p. 54 [**H. taeniatus*; M] [=*?Chondrites taeniatus* KURR, 1845, p. 16; *?Taeniophycus* SCHIMPER, 1869, p. 190 (type, *T. liasicus)*]. Probably only a large *Chondrites*; specimens from Switz. with fewer ramifications. *?Jur.*, Ger.; *Cret.-Tert.*, Eu. (Switz.-Italy), *Tert.(Mio.)*, N.Z.

Histioderma KINAHAN, 1858, p. 70 [**H. hibernicum*; M]. Curved tubes, upper extremities trumpet-shaped, lower turned up at right angle to bedding plane; upper portion of tubes marked by several ridges crossing each other at irregular intervals. [Dwelling burrow.] *Cam.*, Ire.——FIG. 45,2. **H. hibernica; 2a,b, ca.* ×0.7 (Hallissy, 1939).

Hormosiroidea SCHAFFER, 1928, p. 214 [**H. florentina*; OD]. Hemispherical or spherical bodies arranged on thin strings like pearls; diameter of hemispheres 0.5 to 1 cm., of string 1 to 2 mm.; surface of some specimens coarsely granulose. [SCHAFFER regarded *Hormosiroidea* (1928) as alga similar to Recent *Hormosira*, explaining the swellings as spore cases; interpreted by SEILACHER (1959, p. 1068) as a rosary-like trail of unknown origin. It is doubtful whether SCHAFFER

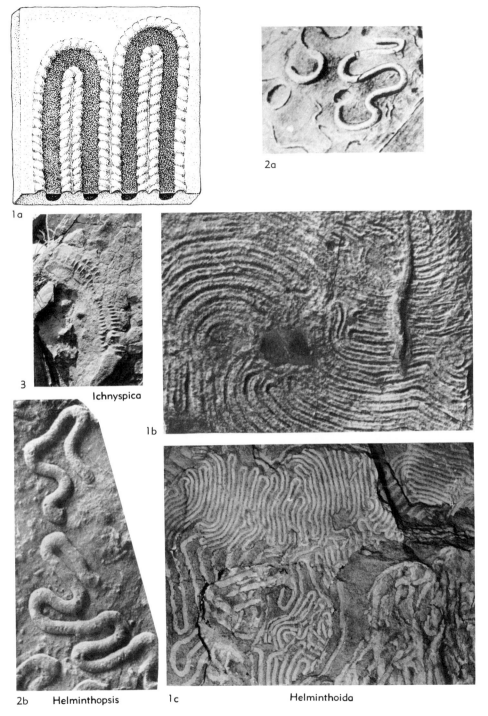

1a

2a

3 Ichnyspica

1b

2b Helminthopsis 1c Helminthoida

Fig. 44. Trace fossils (p. *W*70, 74).

FIG. 45. Trace fossils (p. W70, 74).

(1928) regarded *Hormosira moniliformis* HEER, 1877, p. 161 (flysch, Switz.), as belonging to *Hormosiroidea*; for *Hormosira* see *Halysium* SWIDZINSKI (probably a body fossil). For discussion of possible synonymy of *Hormosiroidea* with *Fustiglyphus* BOUČEK & ELIÁŠ, see OS-GOOD, 1970, p. 369, who placed *Hormosiroidea* in repichnia.[1] [Found in flysch deposits.] *Cret.-L.Tert.,* Eu.(Aus.-Switz.-Spain-Italy).——FIG. 43, *3. *H. florentina,* U.Cret., Italy; ×0.7 (Häntzschel, 1962, courtesy Naturhist. Mus. Wien).
Hydrancylus VON FISCHER-OOSTER, 1858, p. 39 [*Muensteria geniculata* VON STERNBERG, 1833, p.

32; OD] [=*Hydrancilus* NATHORST, 1881, p. 83 *(nom. null)*]. Groups of rounded leaflike impressions arranged irregularly or in lyre shape. [Feeding burrow, proposed as "subgenus" of *Muensteria* VON STERNBERG, originally interpreted as plant; first interpretation as trace fossil was by NATHORST, 1881a, p. 83.] [Found in flysch deposits.] *Cret.-L.Tert.,* Eu.——FIG. 46,*1. H. oosteri* VON FISCHER-OOSTER, ?U.Cret., Switz.; ×1.5 (von Fischer-Ooster, 1858).

FIG. 46. Trace fossils (p. W72, 74-75).

---

[1] OSGOOD (1970) compared his specimens with those figured by BOUČEK & ELIÁŠ (1962); see both *Fustiglyphus* VYALOV, 1971, p. W64 and *Rhabdoglyphus* BOUČEK & ELIÁŠ, 1962, p. W99, for clarification. [W. G. HAKES.]

1a

1b   Imbrichnus

FIG. 46A. Trace fossils (p. *W*74).

Ichnites HITCHCOCK, 1837, p. 175 [=*Ichnites* VINASSA DE REGNY, 1904, p. 320]. Name introduced as general term for all footmarks (i.e., subgroups Tetrapodichnites, Sauroidichnites, Ornithichnites); sometimes also used as generic designation with species name for several tracks and trails of invertebrates and vertebrates (e.g., *Ichnites lithographicus* OPPEL, 1862, a xiphosuran track from the Upper Jurassic Solnhofen Limestone from Bavaria, type species of *Kouphichnium* NOPCSA).

Ichnocumulus SEILACHER, 1956, p. 154 [**I. radiatus*; OD]. Small pustule-shaped bodies possessing straight, radiate projections. [Resting traces made by unknown animals hiding temporarily in sediment.] *L.Jur.-M.Jur.*, Eu.(S.Ger.).——FIG. 43,*1*. **I. radiatus*, L.Lias(Angulaten-Schichten; *1a*, holotype, ×1; *1b*, another specimen with especially thin projections, ×1 (Seilacher, 1956b).

Ichnyspica LINCK, 1949, p. 36 [**I. pectinata*; OD] [=*Ichnispica* LESSERTISSEUR, 1955, p. 35 *(nom. null.)*]. Double track, each composed of numerous "teeth" as in a comb; teeth straight and ending in very sharp points; rows curved, parallel and equidistant. [Type of "ear-shaped" trails (e.g., *Ichnia spicea* RICHTER (1941, p. 229); according to LINCK (1956, p. 50), *Ichnyspica* is sometimes difficult to distinguish from comblike drag marks of *Equisetites*.] *U.Trias.(M.Keuper)*, Eu.(S.Ger.).——FIG. 44,*3*. **I. pectinata*; ×0.3 (Linck, 1949b).

Ichthyoidichnites AMI, 1903, p. 330 [**I. acadiensis*; M] [*?partim* = *Protichnites carbonarius* DAWSON, 1873, p. 16]. Two rows of dashlike impressions with small ridges or monticules at posterior ends. [Believed to be made by fin or finlike appendages of acanthodians (AMI, 1903) or by arthropods (ABEL, 1935, p. 79).] *L.Dev.(Knoydart F.)*, N.Am.(Can., Nova Scotia).

Imbrichnus HALLAM, 1970, p. 197 [**I. wattonensis*; M]. Sediment-filled, winding burrows 0.5 to 1.0 cm. in diameter, commonly parallel to bedding plane, only locally slightly ascending or descending, on lower surfaces of sandstones preserved as semirelief or full relief; characterized by superficial imbricate structure, formed by successive pads of sandy sediment, 1 to 3 mm. thick, inclined at approximately 60° to horizontal. [Produced by movement of an animal along or below sand-mud interface, perhaps by a small bivalve, imbrication formed by periodic extension of the foot, the smooth-walled core by the shell.] *M.Jur.(Bathon., Forest Marble F.)*, Eu.(Eng., Dorset).——FIG. 46A,*1*. **I. wattonensis*, Forest Marble F., Watton Cliff, Dorset; *1a*, holotype, ×0.38; *1b*, ×0.34 (Hallam, A., 1970, p. 197, in: *Trace Fossils* ed. by T. P. Crimes & J. C. Harper, Geol. Jour. Spec. Issue 3, Seel House Press, Liverpool).

Imponoglyphus VYALOV, 1971, p. 89 [**I. torquen-*

dus, p. 89; OD]. Single trace, curved to a greater or lesser degree, a cord with regularly spaced constrictions, being like truncated cones invaginated into one another. *U.Trias.*, C.Asia(SW. Pamir).——FIG. 45,*1*. **I. torquendus*, *1a,b*, ×0.67; *1c*, enl. (Vyalov, 1971). [Description supplied by CURT TEICHERT.]

Incisifex DAHMER, 1937, p. 525 [**I. rhenanus*; M]. Two parallel rows of obliquely arranged notches, stemming from 3-membered extremities; between and outside rows smooth strips of sediment made by sliding ventral side of animal. [Produced by arthropods, perhaps *Homalonotus*.] *L.Dev.*, Eu.(Ger.-Belg.); *?Perm.*, S.Afr.——FIG. 46,*2*. **I. rhenanus*, L.Dev.(Seifener beds), Ger.; ×0.7 (Dahmer, 1937).

Irredictyon VYALOV, 1972, p. 79 [**I. chaos*; OD]. Similar to *Paleodictyon*, but meshwork of burrows more irregular. *Tert.(low.Paleoc.)*, USSR (N. Daghestan). [Description supplied by CURT TEICHERT.]

Isopodichnus BORNEMANN, 1889, p. 25, explan. pl. III [*emend.* SCHINDEWOLF, 1928, p. 27 (*non* BRADY, 1947, p. 470)] [**I. problematicus*; SD SCHINDEWOLF, 1928, p. 27 (=*Ichnium problematicum* SCHINDEWOLF, 1921, p. 21)] [=*?Bipezia* MATTHEW, 1910 (type, *B. bilobata*); for discussion see GLAESSNER, 1957, p. 107]. Dimorphous trace fossil consisting of small, straight, or curved double-ribbon trails, up to about 6 mm. wide, transversely striated by fine furrows; both "ribbons" separated by median ridge; trail may be intermittent; associated with "coffee-bean"-shaped impressions of corresponding size. [Combination of ribbonlike ploughing or raking trails (in German, *Weidespuren*) and coffee-beanlike resting trails, produced by arthropods, possibly by phyllopods or another group of entomostracans. SEILACHER (1970, p. 456) considered *Isopodichnus* to be a facies indicator for the nonmarine environment; LINCK (1942, p. 253) restricted its facies range to brackish water. Most similar trails in marine Paleozoic beds are probably made by trilobites, thus *Isopodichnus* has been regarded as a synonym of *Rusophycus* HALL or even *Cruziana* D'ORBIGNY; for detailed discussion of *Isopodichnus* see LINCK (1942), GLAESSNER (1957), and BIRKENMAJER & BRUTON (1971, p. 311, 317, 318); OSGOOD (1970, p. 303) restricted the name *Isopodichnus* to short *Rusophycus*-like imprints of non-trilobite origin.] *L.Cam.*, Asia(Pak.); *?Ord.*, S.Am.(Arg.); *U.Sil. (?Downton.)*, Spitz.; *?L.Dev.*, Eu.(Ger.); *Carb.*, Australia, Can.(Nova Scotia-N.B.); *L.Perm. (Dwyka Ser.)*, S.Afr.; *Trias.*, Eu.(Ger.)-USA. [The following "species" should be excluded from *Isopodichnus*: *I.* sp. SPECK (1945, p. 411) (Mio., Switz.) (according to SEILACHER, 1953b, p. 115, internal trails of creeping gastropods); *I. raeticus* LINCK, 1942, p. 242 (U.Trias., S.Ger.);

I. sp. Müller, 1955b, p. 483 (L.Trias., Ger.) and probably also *I. tritylotos* Hunger, 1947, p. 419 (M.Trias., Ger.).]——Fig. 46,*3.* **I. problematicus,* L.Trias.(Buntsandstein), Ger.; *3a,* ×0.67 (Seilacher, 1960); *3b,* schem., ×0.3 (Seilacher, 1963); *3c-e,* ×0.5 (Schindewolf, 1928).

Ixalichnus Callison, 1970, p. 20 [**I. enodius*; OD]. Short track of subrectangular shape, formed by 2 rows of 15 to 18 impressions; 5 cm. in length, width decreasing more or less from rear to front. [Made by a vagile trilobite, usually swimming rather than crawling.] *U.Cam.(Deadwood F.),* N.Am.(USA, S.Dak.).——Fig. 47,*4.* **I. enodius,* W.S.Dak.; *4a,* trackway, ×1.1; *4b,* trackway of holotype, ×1.4 (Callison, 1970).

Keckia Glocker, 1841, p. 319 [**K. annulata*; M]. Fillings of cylindrical tunnels with transverse annulation, single "segments" bent; burrows straight or slightly curved, branched, 1 to 2 cm. wide, of varying length, lying in bedding plane; similar to *Taenidium* but much larger; fillings probably fecal material passed through gut of animal. [Originally described as plant, later interpreted as stuffed burrows of sediment-feeding animal (in German, *"Stopf-tunnel"*); for discussion of the interpretation, see Häntzschel (1938) and particularly Richter in Wilckens (1947, p. 44-45). Several "species" of *Muensteria* von Sternberg and *Caulerpites* von Sternberg have been placed by Schimper (in Schimper & Schenk, 1879, p. 46) in *Keckia; K. andina* Borrello, 1966 (U.Jur., Arg.) and *K. haentzscheli* Hundt, 1941 (L.Dev., Ger.) should probably not be assigned to *Keckia.*] *L.Cret.,* USA (Texas); *Cret.-Tert.,* Eu.(Ger.-Aus.-Czech.-Switz.-?USSR).——Fig. 47,*2.* **K. annulata,* U.Cret. (Cenoman.), Ger.; ×0.16 (Glocker, 1841).

Kingella Savage, 1971, p. 299 [**K. natalensis*; OD]. Impression subellipsoidal in outline 40 mm. long, 15 mm. wide; curved marks at "anterior" end indicating pair of antennae (about 10 mm. long) and perhaps one pair of antennules; at least 4 pairs of impressions of appendages. [Only one specimen known, defined as resting impression by Savage; undoubtedly impression of crustacean living in freshwater periglacial environment, possible producer of trails named *Umfolozia* Savage. It is questionable whether such an impression is still to be included in lebensspuren though interpreted as a resting trace.] *L.Perm. (Dwyka Gr.);* S.Afr.(N.Natal).——Fig. 47,*1. K. natalensis,* U.Carb. or L.Perm.; ×1.3 (Savage, 1971).

Kouphichnium Nopcsa, 1923, p. 146 [**Ichnites lithographicus* Oppel, 1862, p. 121; M] [=*Micrichnium* Abel, 1926, p. 150 (type, *M. scotti*); *Micrichnus* Abel, 1926, p. 35 *(nom. null.); Artiodactylus* Abel, 1926, p. 52 (type, *A. sinclairi); Hypornithes* Jaekel, 1929, p. 238 (type, *H. jurassica); Ornichnites* Jaekel, 1929,

p. 235 (type, *O. caudatus); Protornis* Jaekel, 1929, p. 216 *(non* Meyer, 1844) *(nom. nud.); Paramphibius* Willard, 1935, p. 47 (no type species designated); *Limuludichnulus* Linck, 1943, p. 10 (type, *L. nagoldensis); Limuludichnus* Linck, 1949, p. 46 (type, *L. variabilis);* for discussion of all these synonyms see Caster (1939, 1940, 1944), Nielsen (1949), Malz (1964)]. Heteropodous tracks of great variability; complete track consisting of 2 kinds of imprints, 1) 2 chevron-like series each of 4 oval or round holes or bifid V-shaped impressions or scratches, forwardly directed [made by anterior 4 pairs of feet], and 2) one pair of digitate or flabellar, toe-shaped or otherwise variable imprints [made by birdfoot-like "pushers" of 5th pair of feet, with their 4 or 5 leaflike movable blades]; track with or without median dragmark; occasionally preserved in the Upper Jurassic, Solnhofen Limestone, leading to carcass of producer *(Mesolimulus).* [These traces were originally misinterpreted as the work of fishlike amphibians, birds (even *Archaeopteryx!*), pterodactyls, or bipedal dinosaurs, or jumping mammals, later recognized as made by limulids, particularly by comparisons with tracks of Recent limulids (Caster, 1938). Some tracks are traceable for distances of 10 m. or more. Rarely, burrowing activity is recorded by lunate casts corresponding to the limulid prosoma (e.g., *K. rossendalensis,* U.Carb., Eng.; see Hardy, 1970). Incomplete patterns of well-preserved limulid tracks were recently interpreted as "undertracks" (duplicate imprints on lower surfaces as opposed to "surface tracks") (Goldring & Seilacher, 1971); composite types of these tracks apparently made by males and females during the mating season (Bandel, 1967b, p. 7); for interpretation of 2 sinuous grooves with different amplitude produced by telsons of a pair of limulids in nuptial embrace (U.Carb., Eng.), see King (1965).] *Dev.-Jur.,* Eu.-N.Am.-Greenl.——Fig. 47,*3a. K. didactylus* (Willard), U.Dev.(Chemung), USA(Pa.); ×1.4 (Caster, 1938).——Fig. 47,*3b. K. gracilis* (Linck), U.Trias.(Schilfsandstein), Ger.; ×0.6 (Linck, 1949).——Fig. 47,*3c. Limulus polyphemus* and its tracks (schem. drawing) (Malz, 1964).

Kulindrichnus Hallam, 1960, p. 64 [**K. langi*; M]. Stumpy, cylindrical or conical bodies with apex directed downward; oriented subvertically in bed; up to 13 cm. in length and 7.5 cm. in diameter; composed of shell aggregates, some aligned peripherally to margin; matrix may be phosphatic. [Interpreted as burrow (resting trail) produced by cerianthid sea anemone; somewhat similar "genera" are *Bergaueria* Prantl, *Conichnus* Myannil, and *Amphorichnus* Myannil.] *L.Jur.,* Eu.(Eng.-Ger.).——Fig. 48,*2.* **K. langi,* Blue Lias., Eng.; *2a,* long sec. with phosphatic sheath; *2b,* long sec. without phosphatic sheath;

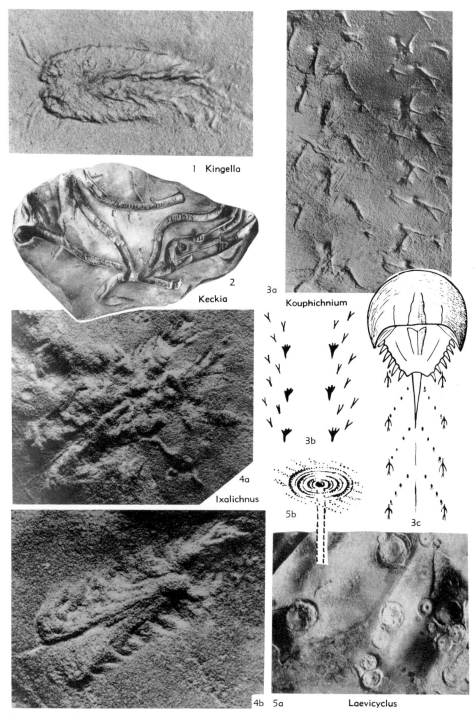

FIG. 47. Trace fossils (p. *W*75, 77-78).

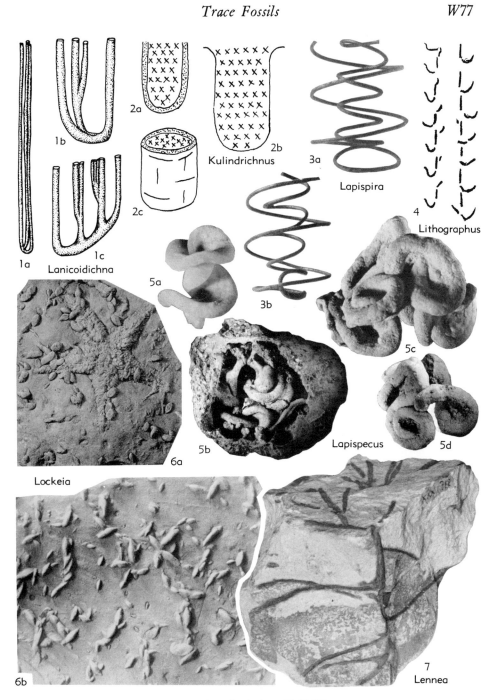

FIG. 48. Trace fossils (p. *W*75, 78-79).

2c, reconstr. burrow indicating calcite-filled cracks in phosphatic sheath, *ca.* ×0.3 (Hallam, 1960).

Laevicyclus QUENSTEDT, 1879, p. 577 [neither formal species name nor type species designated] [=*Cyclozoon* WURM, 1912, p. 127 *(partim)*]. Approximately cylindrical bodies standing at right angles to bedding plane; diameter variable in same specimen; perforated by central canal; visible on bedding planes as regular concentric circles with diameter of several cm. [Interpreted by QUENSTEDT (1879, p. 577) as coral, by PHILIPP (1904, p. 59) and WURM (1912, p. 128) as organism of unknown affinities, by SCHMIDT (1934, p. 18-27) as inorganic, made by gas-exhalations and water under pressure within sediment, and by SEILACHER (1953c, p. 270; 1955, p. 389) as trace fossil (feeding burrow) comparable with dwelling shaft and scraping circles of Recent annelid *Scolecolepis.* For comparison with *Palaeoscia* CASTER, 1942 (U.Ord., Ohio), see p. *W*147 and OSGOOD (1970, p. 396).] *L.Cam.,* Pak.; *?L.Carb.(Kulm),* Eu.(Ger.); *Trias.-Cret.,* Eu. (Eng.-Ger.-Spain-Italy)-N.Am.USA (Kans.). ——FIG. 47,5. *L.* sp., *5a,* U.Trias.(Campiler beds), Italy; ×0.22 (Schmidt, 1934); *5b,* reconstr., L.Cam., Pak.; ×2.4 (Seilacher, 1955).

Laminites GHENT & HENDERSON, 1966, p. 158 [*L. kaitiensis*; OD]. Large, long burrows; subcircular to nearly circular in cross section; slightly meandering or running straight for some distance; filled with fine parallel laminations convex in distal direction; maximum width up to 7.5 cm., length up to 0.5 m.; usually parallel to bedding plane. [Similar to *Keckia* GLOCKER, *Planolites* NICHOLSON, or other "*Stopftunnel,*" but of much larger size; *Beaconites* VYALOV, 1962, possibly is senior synonym of *Laminites* (see p. *W*45). Interpreted as periodic filling by separate packets of feces backwardly extruded into burrow; probably produced by holothurians; for discussion see GREGORY (1969, p. 6) and CHAMBERLAIN (1971a, p. 226).] *Penn.,* ?N.Am.(USA, Okla.); *Tert.(Mio.),* N.Z.

Lanicoidichna CHAMBERLAIN, 1971, p. 223 [*L. metulata*; M] [=*Lanicoidichnus* CHAMBERLAIN, 1971a, p. 216 *(nom. null.)*]. U-shaped burrows; vertical to bedding; 1 to 3 secondary galleries branching at base of U from main burrow and running parallel with it, yielding W-shaped structures; occasionally linked at bases by horizontal or oblique burrows; most additional burrows are smaller than primary gallery; individual burrows 2 to 7 mm. wide; interval of limbs of U- or W-shaped structure 2 to 3 cm.; length of entire system about 60 cm. or more. [Somewhat similar to occasional W-shaped tubes built by the Recent polychaete *Lanice* (SEILACHER, 1953a, p. 428, fig. 3).] *Penn.(Wapanucka Ls.),* N.Am. (USA, Okla).——FIG. 48,1. *Lanicoidichna* structure; *1a,* total structure, *ca.* ×0.1; *1b,c,* lower part of "U" structure, approx. ×0.3 (Chamberlain, 1971a).

Lapispecus VOIGT, 1970, p. 373 [*L. cuniculus*; OD]. Long cylindrical burrows, 1 to 4 mm. in diam., winding similar to tubes of serpulid *Glomerula*; preserved only as casts in cavities which are result of partial leaching of pebbles,

particularly small pebbles, in conglomerates. [Bladelike thin borders on concave or convex side of winding fillings of burrows not interpreted as spreite; borders are discontinuous along length of burrows; probably dwelling burrows of polychaetes.] *U.Cret.(Santon.)* [in pebbles of *U.Jur. (Kimmeridg.)* age], Eu.(Ger.).——FIG. 48,5. *L. cuniculus; 5a,* spindle-like spiral form, ×8; *5c,* ×4.5; *5b,d,* ×3 (Voigt, 1970).

Lapispira LANGE, 1932, p. 540 [*L. bispiralis*; M]. U-shaped tunnel with both legs spirally curved in same direction. *L.Jur.(low.Lias.),* Eu.(Ger.). ——FIG. 48,3. *L. bispiralis; 3a,b,* wire models of burrows, ×0.2 (Lange, 1932).

Lennea KRÄUSEL & WEYLAND, 1932, p. 189 [*L. schmidti*; M]. Vertical shaft about 1 cm. wide, with numerous narrower lateral tunnels branching off irregularly at right angles along whole length of vertical shaft; lateral branches at first approximately horizontal, then directed downward; branching dichotomously. [Originally interpreted as roots of plants; later recognized as trace fossil (feeding burrow) (KRÄUSEL & WEYLAND, 1934, p. 100); for detailed description and discussion see PAULUS (1957) and FISCHER & PAULUS (1969).] *Dev.,* Eu.(Ger.).——FIG. 48,7. *L. schmidti,* M.Dev., Ger.; ×0.3 (Paulus, 1957).

Lenticraterion KARASZEWSKI, 1971, p. 886 [*L. bohdanowiczi*; M]. Lenticular depressions (7 to 12 mm. long and 4 to 8 mm. wide) in epirelief, maximum depth 5 mm.; individual depressions commonly display 2 funnel-shaped hollows, of the same or different depths, at each end of long axis but do not possess peripheral collars characteristic of *Calycraterion* KARASZEWSKI, 1971a (see p. *W*49); convex structures on bottom of same rock slab correspond with depressions in epirelief. [Interpreted by KARASZEWSKI (1971b, p. 889) to have been produced by an unknown animal moving upward in the sediment.] *L.Jur. (low.Pliensbach.),* Eu.(Pol.). [Description supplied by W. G. HAKES.]

Lithographus HITCHCOCK, 1858, p. 156 [*L. hieroglyphicus*; SD LULL, 1953, p. 43]. Very similar or identical to *Copeza* HITCHCOCK but having oblique markings outside longitudinal ones. [?Insect trail.] *Trias.,* N.Am.(USA,Mass.). ——FIG. 48,4. *L. hieroglyphicus;* ×0.4 (Lull, 1953).

Lobichnus KEMPER, 1968, p. 72 [*L. variabilis*; M]. Very small or scooped hollows which form irregular main stem with unilateral pectinate branches comprised of very small leaf-shaped hollows also arranged unilaterally; systems are highly variable with many transitions between forms; limited to lobate configuration and thus resembling ammonite sutures; somewhat similar to *Lophoctenium*; endogenic and preserved exclusively in troughs of current ripple marks. [Interpreted as true grazing trails; KEMPER believed that *Lobichnus* was an indicator of shallow

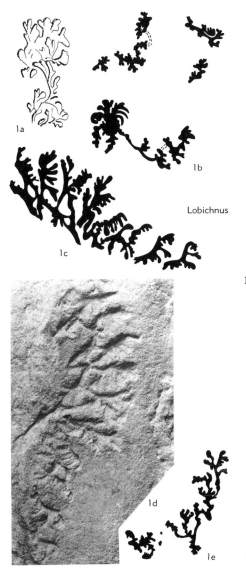

la

lb

Lobichnus

lc

ld

le

<figure>FIG. 49. Trace fossils (p. *W78-79*).</figure>

water in the Bentheimer Sandstein.] *L.Cret.(M. Valang., Bentheimer Sandstein), Eu.(Ger.).——* FIG. 49, *1.* **L. variabilis; 1a,c,* holotype, schem.; *1b,e,* schematic examples of the wide range of forms, all approx. ×0.5; *1d,* ×1 (Kemper, 1968).

Lockeia U. P. JAMES, 1879, p. 17 [**L. siliquaria;* M] [=*Pelecypodichnus* SEILACHER, 1953b, p. 105 (type, *P. amygdaloides*); for discussion see OSGOOD, 1970, p. 308-312]. Small almond-shaped oblong bodies preserved in convex hyporelief; tapering to sharp and obtuse points at both ends;

surface commonly smooth; mostly symmetrical; length varying from 2 to 12 mm. [Originally interpreted as algae; later regarded by J. F. JAMES (1885) as "ovarian capsules" of graptolites; now considered resting trails of small burrowing pelecypods, perhaps semi-sessile forms; for discussion on mode of formation and synonymy with *"Dawsonia"* NICHOLSON, 1873 (preoccupied by HARTT, in DAWSON, 1868), see OSGOOD, 1970, p. 208-212; regrettably, the most appropriate name *Pelecypodichnus* must be replaced by a very rarely used one published in an obscure journal!] *Ord.,* Eu. (Nor.-France)-USA (Ohio-Ky.)-Can., *?Ord.,* S.Am.(Arg.); *Penn.,* USA(Kans.-Okla.); *Trias.,* Eu.(Ger.-Swed.-Italy)-E.Greenl.; *Jur.,* Eu. (Eng.-France-Ger.-Swed.); *Cret.,* USA(Utah); *Tert.,* Eu.(Switz.)-Iraq.——FIG. 48,*6a. L. amygdaloides,* M.Jur.(Dogger β, Donzdorfer Ss.), Ger.; (shown with *Asteriacites quinquefolius* (QUEN-STEDT), ×0.5 (Seilacher, 1953b).——FIG. 48,*6b.* **L. siliquaria,* Ord. (up. Trenton. or low. Cincinnat.), Ludlow, Ky.; ×0.7 (Osgood, 1970).

Lophoctenium RICHTER, 1850, p. 199 (without formal species name) [**L. comosum* RICHTER, 1851, p. 563; SM] [=*Buthotrephis radiata* LUD-WIG, 1869, p. 114; *Criophycus* TOULA, 1906, p. 159 (type, *C. ramosus*)]. Bunches of closely spaced, inwardly bent "twigs" with comblike branches, joining to form main axis. [Formerly thought to have affinities with graptolites, sertularids, or algae; without doubt a feeding burrow; according to SEILACHER (1960, p. 49), *L. globulare* GÜMBEL (1879, p. 469) is identical to *"Schaderthalia"* HUNDT, 1953 (obscure nondescript "genus"); see also PFEIFFER, 1968, p. 671, who renounced establishment of a new name for this "species."] [Found in flysch deposits.] *Ord.-L.Carb.,* Eu.(Ger.-Port.)-N.Am.(USA,Okla.); *L.Tert.,* Eu.(Aus.-Switz.-Pol.).——FIG. 50,*1a.* **L. comosum,* M.Dev.(*Nereites* beds), Ger.; ×1.5 (Seilacher, 1954).——FIG. 50,*1b. L. ramosum* (TOULA), low. Eoc., Pol.; ×0.5 (from Książkiewicz, M., 1970, p. 284, in: *Trace Fossils* edited by T. P. Crimes & J. C. Harper, Geol. Jour. Spec. Issue 3, Seel House Press, Liverpool).

Macanopsis MACSOTAY, 1967, p. 32 [**M. pagueyi;* M]. Straight or somewhat bent burrows circular or oval in cross section, 1 to 3 cm. in diameter, not branched; burrows end with hemispherical hollow, 4 to 5 cm. in diameter; burrows perpendicular to bedding; usually slightly bent before enlarging to hemispherical hollow. *L.Tert. (Paleoc.-Eoc.),* S.Am.(Venez.).——FIG. 51,*1.* **M. pagueyi; 1a,b,* holotype and paratype, ×0.3 (Macsotay, 1967).

Mammillichnis CHAMBERLAIN, 1971, p. 238 [**M. aggeris;* M] [=*Mammillichris* CHAMBERLAIN, 1971a, p. 238 *(nom. null.); Mammillichnus* CHAMBERLAIN, 1971a, p. 217 *(nom. null.)*]. Subhemispherical teatlike protuberances 9 to 12 mm. wide, 7 mm. high, preserved in convex

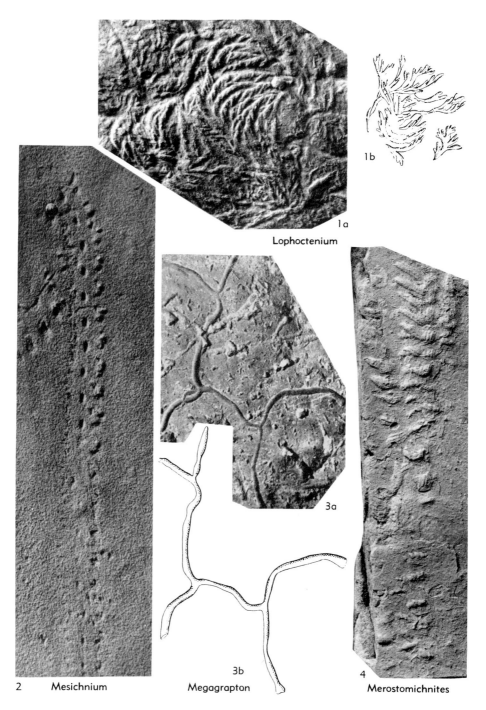

1a

1b

Lophoctenium

2   Mesichnium

3a

3b

Megagrapton

4   Merostomichnites

Fig. 50. Trace fossils (p. *W*79, 82).

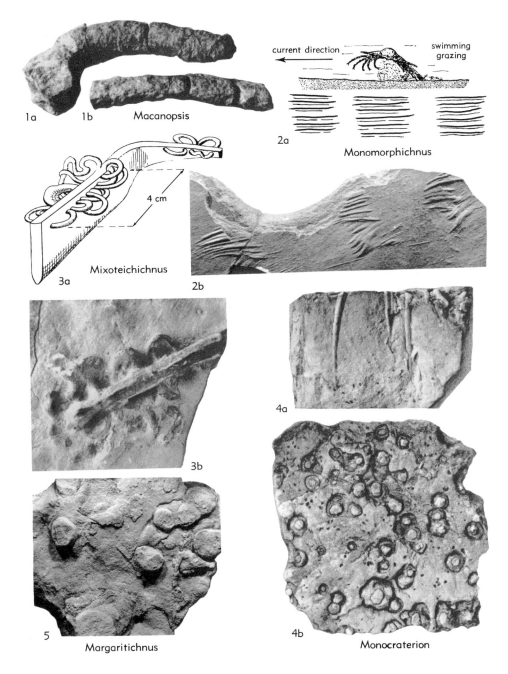

1a 1b Macanopsis

current direction   swimming grazing

2a

Monomorphichnus

4 cm

Mixoteichichnus

3a

2b

3b

4a

5

Margaritichnus

4b

Monocraterion

Fig. 51. Trace fossils (p. *W*79, 82, 84).

hyporelief, each mound consists of 3 to 5 mm. hemicircular apex and wide flange. [Origin unknown; the following three interpretations were discussed by CHAMBERLAIN (1971a): resting or hiding trace of an animal in the sediment; body fossil ("egg case" or juvenile deposited in sediment); excurrent end of burrow where animal worked sediment for food or formed fecal pellets;

impossible to make decision on origin of form from CHAMBERLAIN's (1971a) figures.] *Miss. (Jackfork Gr.)-Penn.(Atoka F.),* USA(Okla.).

Margaritichnus BANDEL, 1973, p. 1002 (*nom. subst. pro Cylindrichnus* BANDEL, 1967a, p. 6 (*non* TOOTS in HOWARD, 1966, p. 45)) [*Cylindrichnus reptilis* BANDEL, 1967a, p. 6; OD]. Vertically compressed ball structures 15 to 30 mm. in diameter; originally spherical; commonly arranged like string of pearls; rarely connected by ridges which show crescentic transverse grooves. ["Balls" interpreted as fecal pellets probably made by large wormlike sediment-eating animals (sipunculids?, priapulids?); "trail" possibly formed below the surface of the sediment.] *U.Penn. (Missouri.),* USA(Kans.); *Perm.,* W.Australia; *?Cret.,* USA.——FIG. 51,5. *M. reptilis* (BANDEL), U.Penn., Kans.; ×0.1 (Bandel, 1967a). [Also found in *U.Precam.,* S.Australia-USSR(Sib.Plat.).]

Megagrapton KSIĄŻKIEWICZ, 1968, p. 5, 14 [*M. irregulare;* OD] [=*Megagrapton* KSIĄŻKIEWICZ, 1961, p. 882, 888 (*nom. nud.*)]. Networks consisting of irregular polygons and rectangles which are never closed, formed by slightly curved or straight cylindrical strings, 1 to 5 mm. wide; rather regular intervals of branching at nearly right angles; possibly transitional to *Squamodictyon* VYALOV & GOLEV, 1960. [Evidently of postdepositional origin.] *L.Cret.,* Japan-Eu.(Pol.); *L.Tert.(low.Eoc.),* Eu.(Pol.).——FIG. 50,3. *M. irregulare; 3a,* L.Tert.(Eoc., flysch), Pol.; ×0.43 (Książkiewicz, 1961); *3b,* low.Eoc.(Beloveza Beds), Pol.; ×0.5 (Książkiewicz, M, 1970, p. 307, in: *Trace Fossils* edited by T. P. Crimes & J. C. Harper, Geol. Jour. Spec. Issue 3, Seel House Press, Liverpool).

Merostomichnites PACKARD, 1900, p. 67 [*M. beecheri;* SD HÄNTZSCHEL, 1962, p. *W205*] [=*Merostomchnites* OSGOOD, 1970, p. 355 (*nom. null.*)]. Two parallel rows of circular bow- or spindle-shaped feet impressions; transversely or slightly obliquely arranged, opposite to each other. [Paleozoic forms probably attributable to eurypterids, Triassic forms possibly to phyllopods; striding track *Merostomichnites* and burrowing trail *Isopodichnus* may have been produced by the same animal (SEILACHER, 1963, p. 88).] *Cam.-L.Trias.,* Eu.-Asia(AsiaM., Jordan)-N.Am.——FIG. 50,4. *M. strandi* STØRMER, Sil.-Dev. (Downton.), Nor.(Spitz.); ×1 (Størmer, 1934).

Mesichnium GILMORE, 1926, p. 34 [*M. benjamini;* M]. Two parallel lines of footprints with median row of suboval regularly spaced depressions; trackway about 20 mm. wide; stride (distance between depressions of median row) about 15 mm. long. [Crawling track; producer unknown.] *Perm.(Coconino Ss.),* N.Am.(USA, Ariz.).——FIG. 50,2. *M. benjamini,* Ariz.(Grand Canyon); *ca.* ×0.37 (Gilmore, 1926).

Mesonereis HATAI, 1968, p. 132 [*M. ragaensis;* M]. Name conditionally proposed. Burrow 0.5 to 1 cm. in diameter, lower part curved planispirally and upper part vertical. [Undoubtedly a trace fossil, considered by HATAI to have been made by "an undescribed kind of marine worm . . . . close in morphological feature to the living genus *Nereis.*"] *L.Cret.(Miyakoan),* Japan(NE.Honshu). [Most likely invalid.]

Micatuba CHAMBERLAIN, 1971, p. 238 [*M. verso;* M]. Rather irregularly arranged tubes radiating from a center (central gallery?), singly or multiply bunched, straight or more or less curved, sandcoated or filled, about 20 mm. long, 1 mm. wide. *Penn.(up.Atoka F.),* N.Am.(USA,Okla.).——FIG. 52,2. *M. verso; 2a,* plain view; *2b,* cross section and oblique view, schem. (Chamberlain, 1971a).

Minichnium PFEIFFER, 1968, p. 683 [*M. wurzbachense;* M]. Large systems of rather long feeding burrows which diverge clusterlike from a starting point and trend slightly downward, exhibiting distinct bioturbate structures. [Poorly figured.] *L.Carb.(Kulm),* Eu.(Ger.,Thuringia).

Mixoteichichnus MÜLLER, 1966, p. 720 [*M. coniungus;* OD]. Straight or slightly curved, retrusively formed, wall-like back-fill (*Versatzbauten*) burrows similar to *Teichichnus* SEILACHER, with simply curved and semicircular burrows that originate from their upper parts; these smaller burrows partly resemble *Rhizocorallium* and are constructed parallel to bedding. [Trail belonging to Fodinichnia.] *Low.M.Trias. (low.Muschelkalk),* Eu.(Ger.).——FIG. 51,3. *M. coniungus,* low. Muschelkalk, Ger.; *3a,* schem., ×0.5; *3b,* ×0.3 (Müller, 1966).

Monocraterion TORELL, 1870, p. 13 [*M. tentaculatum;* M] [=*Lepocraterion* STEHMANN, 1934, p. 17; no "species" name; *Monocraterium* VOLK, 1967, p. 98 (*nom. null.*)]. "Trumpet pipes"; funnel structure penetrated by central straight or slightly curved plugged tube, perpendicular to bedding plane, never branched; diameter commonly 5 mm., up to 8 cm. (max., 16) long; funnel simple or multiple (latter discernible in transverse section as a series of concentric rings); diameter of funnels usually 1 to 4 cm., greatest depth about 2 cm.; tubes commonly abundant but never crowded like *Skolithos.* Funnel obviously constructed by upward migration of animal inhabiting tube is reflected by downward warping of surrounding bedding planes toward central tube. [Dwelling burrow; probably belonging to gregarious, suspension-feeding wormlike organisms. *Lepocraterion* STEHMEN differs from *Monocraterion* only by the occurrence of a carbonaceous wall which is not considered to be sufficient taxonomic reason to establish a "genus." BOUČEK (1938, p. 249) and HÄNTZSCHEL (1962, p. *W218*), with reservation, regarded *Monocraterion* as synonym of the commonly annulated tubes "*Tigillites*"; HALLAM & SWETT (1966, p. 103) properly retained *Monocraterion* as a valid name for vertical funnel-shaped burrows; for dis-

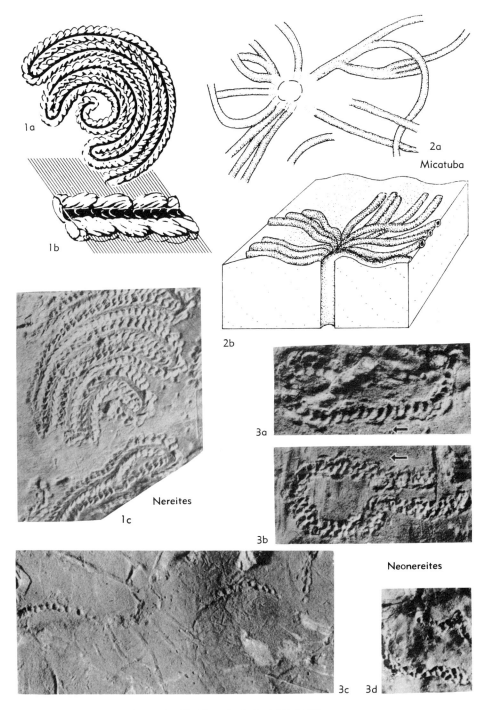

FIG. 52. Trace fossils (p. *W*82, 84-85).

cussion of relationship to *Skolithos, Histioderma,* and *Micrapium,* see WESTERGÅRD (1931, p. 12); also considered valid genus by FREY & CHOWNS (1972).] *Cam.* (Pleist. drift), Eu.(N.Ger.); *Cam.-Ord.,* Eu.(Swed.-Nor.-Eng.)-N.Am.-Asia (Jordan); *?U.Dev.,* Eu.(Eng.); *Carb.,* Eu.(Scot.); *?L.Trias.,* Eu.(Pol.); *L.Jur.,* Greenl.——FIG. 51,*4.* **M. tentaculatum;* L.Cam.(Lingulid ss.), Swed. (W.Gotl.); *4a,b,* ×0.5 (Westergård, 1931).

Monomorphichnus CRIMES, 1970, p. 57 [**M. bilinearis;* M]. Series of straight or slightly sigmoidal ridges associated in pairs; 1 ridge of each pair more prominent than the other; ridges 2 to 4 cm. long, sometimes repeated laterally; trail resembling *Dimorphichnus* SEILACHER, but without blunt markings and other markings to suggest sideways progression. [Produced by several clawed limbs of trilobites, perhaps members of the Olenidae; interpreted as swimming-grazing trail.] *U.Cam.(FjestiniogStage),* Eu.(Eng., N. Wales).——FIG. 51,*2a. M.* sp.; showing how trace is produced by swimming-grazing manner of trilobite locomotion (Crimes, 1970b).——FIG. 51, *2b.* **M. bilinearis;* ×0.4 (Crimes, 1970b).

Muensteria VON STERNBERG, 1833, p. 31 [*non* KROGERUS, 1931, *nec* DESLONGCHAMPS, 1835] [no type species designated]. Heterogeneous "genus," comprising bodily preserved fossils from the Jurassic Solnhofen limestone as well as trace fossils, particularly from European flysch deposits (e.g., *M. hoessii* HEER, 1877; *M. annulata* SCHAFHÄUTL, 1851; *M. involutissima* SACCO, 1888; *M. bicornis* HEER, 1877), these *"species"* are mostly stuffed burrows (German, *"Stopftunnel"*) with laminated structure ("segmentation") originated backfilling of the cylindrical burrows; *Muensteria* similar to *Taenidium* HEER but differs by being larger; it has been divided into "subgenera" by VON FISCHER-OOSTER, 1858 (*Eumuensteria, Keckia* GLOCKER, *Hydrancylus* VON FISCHER-OOSTER). *Jur.-Cret.,* Eu.(Ger.-Switz), N.Am.(Greenl.).

Myriapodites MATTHEW, 1903, p. 103 [published only as "*Myriapodites* sp."]. Two parallel rows of feet impressions, about 6 mm. apart, each row 2 mm. wide; linear prints closely set, arranged in double series of elongated scratches, mostly directed from outside to inside of row. [Tentatively interpreted as crawling track of myriapods.] *Carb.,* N.Am.(Can., Nova Scotia).

Neonereites SEILACHER, 1960, p. 48 [**N. biserialis;* OD] [=*Neonerites* CONYBEARE & CROOK, 1968, p. 276 *(nom. null.); Neoneretites* SEILACHER, 1969, p. 118 *(nom. null.);* former German names: *Punkt-Fährte* PUTZER, 1938, p. 418; *Perlspur* WEISS, 1940, p. 344; *Perlketten-Fährte* KUHN, 1952, p. 224]. Bimorphous, shape depending on its hypichnial or exichnial preservation; as negative epireliefs consisting of irregularly curved chains of deep, smooth-walled dimples; chain restricted in length, some bordered laterally by flabby structures caused by burrowing; correspond-

ing hypichnia form a median string, irregularly curving or straight or rarely meandering, consisting of single- or double-lined clay (fecal) pellets or small plates *(N. uniserialis, N. biserialis).* [Interpreted to be internal burrow, postdepositional; according to SEILACHER (1962, p. 233), *Neonereites* is possibly the irregular counterpart of *Helminthoida labyrinthica* HEER in sandy environment.] *Ord.,* Asia (Iraq); *?L.Carb.(Kulm),* Eu.(Ger.); *L.Jur.-M.Jur.,* Eu.(Eng.-Ger.); *L.Cret.,* Eu.(Ger.)-Asia(Japan); *L.Tert.(Eoc),* Eu.(Spain).——FIG. 52,*3a,c. N. uniserialis* SEILACHER, *3a,* L.Jur.(Lias a₂), Ger.; ×0.9 (direction of movement indicated by arrow in *3a*) (Seilacher, 1960); *3c,* low.Lias(Hettang.), Ger.(Helmstedt); ×0.9 (Häntzschel, 1968).——FIG. 52,*3b,d.* **N. biserialis,* M.Jur.(Dogger β), Ger.; *3b,d,* ×0.6 (direction of movement indicated by arrow in *3b*) (Seilacher, 1960).

Neoskolithos KEGEL, 1966, p. 22 [**N. picosensis;* M]. Similar to *Skolithos* but tubes not so crowded, shorter and more irregular; 4 to 5 cm. long, 0.5 to 1 cm. in diameter. *L.Dev.(Pimenteira F.),* S.Am.(Brazil,Piauí).

Nereites MACLEAY, 1839, p. 700 [*non* EMMONS, 1846] [**N. cambrensis;* SD HÄNTZSCHEL, 1962, p. W205] [=*Myrianites* MACLEAY, 1839, p. 700 (type, *M. macleaii); Nereograpsus* GEINITZ, 1852, p. 27 (name for the supposed "graptolite genera" *Nereites, Myrianites, Nemertites,* and *Nemapodia;* name "corrected" by HALL, 1865, p. 43, to *Nereograptus*); for synonymy of the type species and *N. tenuissimus* see PFEIFFER, 1968, p. 669, 670)]. Meandering trails, consisting of narrow median furrow, flanked on both sides by regularly spaced leaf-shaped, ovate, or pinnate lobes; closely spaced; commonly finely striated; meanders may be densely spaced (type of *"Geführte Mäander"* of RICHTER, 1924, p. 153); width of trail 1 to 2 cm.; meanders variable in form in width, shape, and size of the lateral lobe-like projections. [Formerly regarded as plants, bodily preserved worms, or graptolites or their impressions; lateral lobes explained as impressions of the setae of a worm; now interpreted as internal meandering grazing trails; according to SEILACHER (in SEILACHER & MEISCHNER, 1965, p. 615), *Nereites* occurs on top surface of thin turbidites, thus most probably produced in deep water environment ("*Nereites* facies" of SEILACHER). Various producers have been suggested: worms (e.g., RICHTER, 1928, p. 241), gastropods (e.g., RAYMOND, 1931a, p. 191; ABEL, 1935, p. 237), or crustaceans (FRAIPONT, 1915, p. 449); many "species" described, particularly by DELGADO (1910, p. 11-24), but not all of them definitely belonging to *Nereites;* for "psychologic analysis" of these meandering trails, see RICHTER (1928, p. 240) and SEILACHER (1967a, p. 297)]. [Found in flysch deposits.] *Ord.-Carb.,* Eu.-USA-S.Am.-N.Afr.; *Cret.,* Eu.(Spain-Italy); *Tert.(Eoc.),*

Japan.——Fig. 52,*1a,b. Nereites,* from Dev. and Carb.; schem. (Seilacher, 1967c).——Fig. 52,*1c. N. loomisi* Emmons, M.Dev., Ger.; ×0.3 (Richter, 1928).

Octopodichnus Gilmore, 1927. p. 30 [**O. didactylus;* OD]. Tracks of apparently 8-footed animal; feet impressions arranged in 4 groups: alternating; 2 anterior impressions of each group didactyle, 2 posterior, unidactyle. [Various interpretations have been advanced: made by crustaceans (Gilmore, 1927), arachnids (Abel, 1935, p. 265), large scorpionids (Brady, 1947, p. 469), producer unknown (Faul & Roberts, 1951, p. 272).] *L.Perm.(Coconino Ss.),* N.Am.(USA, Ariz.); *?Jur.(?Navajo Ss.),* N.Am.(USA,Colo.). ——Fig. 53,*2. *O. didactylus,* L.Perm.(Coconino Ss.), Ariz.; *2a,* ×0.8; *2b,* diagram of trackway, ×0.25 (Gilmore, 1927).

Oldhamia Forbes, 1849, p. 20 (publ. without formal species names; first description of species by Kinahan, 1858, p. 69) [**O. antiqua* Kinahan, 1858, p. 69; SD Häntzschel, herein] [=*Murchisonites* Goeppert, 1860, p. 441 (type, *M. forbesi* Goeppert)]. Bunches of fine rills, radiating from joints of sympodial axis; representing a grazing pattern. [Numerous explanations of origin: as remains of algae, hydrozoans, bryozoans or of inorganic origin; first(?) interpretation as trace fossil by Ruedemann (1942) (radiating feeding trails supposedly made by worms); prevalent in Lower Cambrian turbidite successions; sometimes regarded as index fossil of Lower Cambrian; the Ordovician *"O." pedemontana* of Mendoza (Rusconi, 1956) shown by Fritz (1965) to be bryozoan (*Hallopora* sp.); *"O." keithi* Ruedemann (1942) (Ord., Gaspé) according to Churkin & Brabb (1965, p. D123), different in appearance and thus probably not belonging to *Oldhamia.*] *L.Cam.-M.Cam.,* Eu.-N.Am.——Fig. 53,*3a. O. radiata* Forbes, Cam., Ire.; ×1.3 (Sollas, 1900).——Fig. 53,*3b. *O. antiqua,* Cam., Ire.; ×1.3 (Sollas, 1900).

Oniscoidichnus Brady, 1949, p. 573 [**Isopodichnus filiciformis* Brady, 1947, p. 470; M] [=*Isopodichnus* Brady, 1947, p. 470, obj. (*non* Bornemann, 1889, p. 25)]. Track with low, sinuous median ridge and forward-pointing bractlike footprints on each side at intervals of about 1 mm.; width of entire track about 1 cm. [Resembles tracks of Recent isopod *Oniscus.*] *L.Perm.(Coconino Ss.),* N.Am.(USA,Ariz.).——Fig. 53,*1a.* Trackway of *Oniscus* sp.; ×0.5 (Brady, 1947).——Fig. 53,*1b. *O. filiciformis* (Brady); ×0.5 (Brady, 1947).

Ophiomorpha Lundgren, 1891, p. 114 [*non* Szepliget, 1905] [**O. nodosa;* M] [=*?Ophiomorpha* Nilsson in Mantell, 1836, p. 25 (*nom. nud.*); Cylindrites spongioides Goeppert, 1842 (?1841), p. 115 (*partim*); Spongites saxonicus Geinitz, 1842, p. 96 (*partim*); ?Halymenites flexuosus von Fischer-Ooster, 1858, p. 55; Cylindrites tuberosus Eichwald, 1865, p. 8;

Phymatoderma dienvalii Watelet, 1866, p. 24; *Halymenites major* Lesquereux, 1873, p. 373; *?Broeckia bruxellensis* Carter, 1877, p. 382 (*nom. oblit.* if identical with *Ophiomorpha*); *Astrophora* Deecke, 1895, p. 167 (type, *A. baltica*); *Sabellastartites* Dudich, 1962, p. 108 (type, *S. arenaceus*); for discussion see Häntzschel, 1952, p. 144-149; for detailed list of synonyms of the type species see Kennedy & Macdougall, 1969, p. 460-461]. Three-dimensional burrow systems, vertical and horizontal; cylindrical tunnels (diam., 0.5-3 cm.) dichotomously branching, generally at acute angles; with local swellings close to or at points of branching; tunnels internally smooth, but outer surface of burrow lining characteristically mammillate due to presence of discoidal or ovoid pellets, which are several mm., rarely more, in diameter; tunnels may also be only partly lined by small pellets; longitudinal ridges occur on outer surface of some burrow fillings. Occasionally penetrating sediment for more than 1 m. in depth. [Doubtless to be ascribed to burrowing decapod crustaceans, particularly callianassids as proven by *Ophiomorpha*-like structures produced by Recent callianassids in modern sediments (Weimer & Hoyt, 1964); found associated in same rocks with *Callianassa* claws (Cret., Delaware) (Pickett, et al., 1971); swellings of the tunnels are "turn-arounds" of the animals; pellets cemented by the producer and put into the sides of the burrow; reticulate ridges on some burrows are scratches made by the inhabitant of the burrow, probably during initial burrowing; passing of warty exterior into smooth burrows observed (Kennedy & Sellwood, 1970, p. 108). *O. borneensis* Keij, 1965, has been observed rarely to exhibit vertical, spiral burrows while in close association with horizontal ones, similar forms occur in sandy Tertiary sediments of West Germany (Kilpper, 1962, p. 57); *Ophiomorpha* occasionally seen to pass into *Thalassinoides* (Ager & Wallace, 1970, p. 8) and rarely into wall-like structures similar to *Teichnichnus* (Hester & Pryor, 1972); cylindrical burrows with smooth walls (U.Cret., Saltholm Ls., Denm.) sometimes named *Ophiomorpha* in museum collections; generally regarded as indicator of marine environment, especially littoral, sublittoral, or upper neritic; for discussion of interpretation of occurrences in apparently brackish or freshwater environments see Kennedy & Macdougall (1969, p. 467); for a list of the very extensive literature on this trace fossil see Kennedy & Sellwood (1970, p. 101) and Müller (1969e, 1970b)]. *L.Jur.,* Greenl.; *M.Jur.-Pleist.,* cosmop.——Fig. 54,*1a. O. major* (Lesquereux), U.Cret., USA(N. Dak.); ×0.5 (Häntzschel, 1952).——Fig. 54,*1b. *O. nodosa,* ?U.Cret. or L.Tert., S.Swed. (Scania); ×0.4 (Häntzschel, 1952). [Chamberlain & Baer (1973, p. 80) have recently extended the

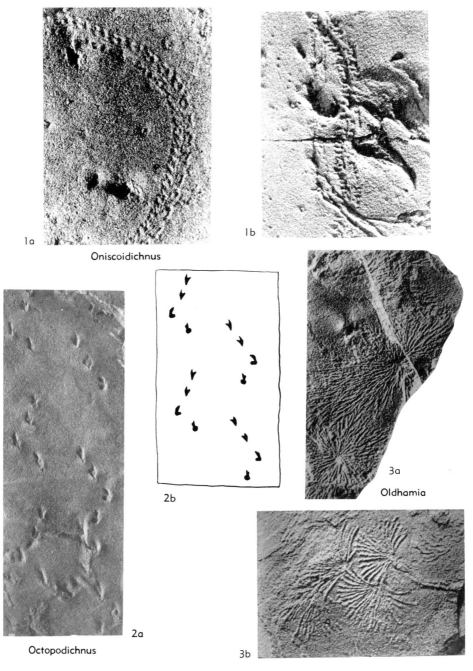

1a, 1b. Oniscoidichnus

2a. Octopodichnus

2b.

3a. Oldhamia

3b.

FIG. 53. Trace fossils (p. *W*85).

stratigraphic range of *Ophiomorpha*: *L.Perm.* (*Wolfcamp.*), N.Am.(USA,Utah); *low.U.Perm.* (*Zechstein*), Eu.——W. G. HAKES.]

Ormathichnus MILLER, 1880, p. 222 [*O. monili-formis*; M]. According to OSGOOD (1970, p. 372), "genus" comprising two different forms: 1) one

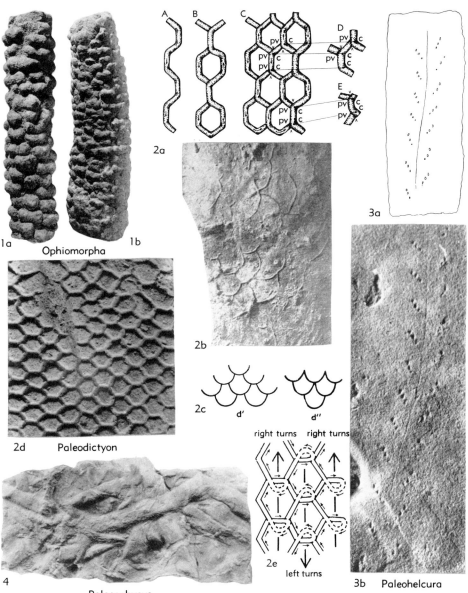

FIG. 54. Trace fossils (p. *W*85-86, 88-91).

syntype is a small trail, consisting of series of minute, disconnected *Rusophycus*-like bodies; 2) the other syntype was compared by MILLER with cast of column of *Heterocrinus*. [The first-mentioned specimen, according to OSGOOD (1970, p. 373) is trail of small arthropod (trilobite?), "a combination of *Cruziana* and *Rusophycus*"; the other syntype is certainly an inorganic tool mark (impression of a rolling crinoid stem) as supposed by JAMES (1886).] *U.Ord.(Cincinnat.),* USA (Ohio).

Palaeohelminthoida RUCHHOLZ, 1967, p. 512 [**P. hercynia*; OD]. Very regular "guided meanders" like *Helminthoida* SCHAFHÄUTL, differing from it by very narrow, median, cordlike ridge and by close contact of meanders; trail about 4 mm.

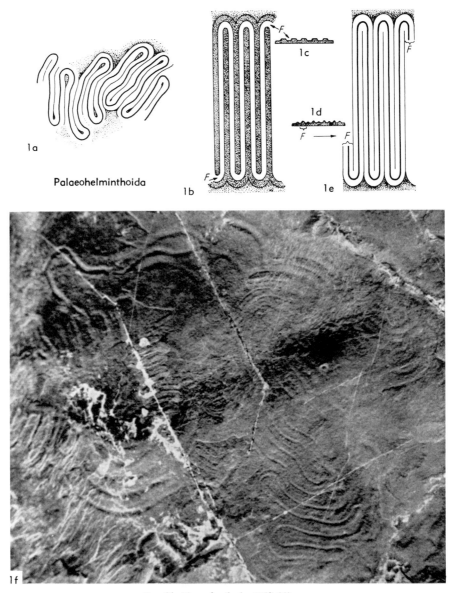

FIG. 55. Trace fossils (p. *W*87-88).

wide, length of meanders 3 to 6 cm.; rounded median ridge 1 mm. high, dividing trail into 2 smooth furrows. [Typical grazing trail.] *U.Dev., ?L.Carb.,* Eu.(Ger.,Hartz Mts.).——Fig. 55,*1a,f. *P. hercynia,* U.Dev., S.Harz Mts., Ger.; *1a,* schem., ×0.7; *1f,* holotype, ×0.9 (Ruchholz, 1967).——Fig. 55,*1b-e.* Schematic comparison of *Helminthoida* (*1b,* plan view; *1c,* cross sec. view) and *Palaeohelminthoida* (*1d,* cross sec. view; *1e,* plan view) [*F,* trail] (Ruchholz, 1967).

Palaeophycus HALL, 1847, p. 7 [**P. tubularis*; SD BASSLER, 1915, p. 939] [=?*Aulacophycos* MASSALONGO, 1859, p. 92 (as generic name only proposed for *Palaeophycus simplex* HALL, 1847, p. 63); *Palaeospongia prisca* BORNEMANN, 1886, p. 21]. Ichnogenus showing wide range of morphology; cylindrical or subcylindrical burrows, usually sinuous, oriented more or less obliquely to bedding; commonly unbranched, though may be branched occasionally; surface of walls smooth or rarely with faint longitudinal striae; up to about 20 cm. or more in length; 3 to 15 mm. in diameter; commonly intersecting one another. [Originally considered to be stems of "fucoids," interpreted by JAMES (1885) as trace fossil; belongs to repichnia of infaunal origin; pathways of various groups of errant animals; neither parts of constructed tubes as suggested by several authors nor stuffed burrows of sediment ingestors; "no one has studied the genus in detail" (OSGOOD, 1970, p. 375); very many "species" established from different environments; impossible to list all "species" erroneously placed in *Palaeophycus*, e.g., *P. kochi* LUDWIG, 1869, p. 110 (="Belorhaphe" kochi MICHELAU, 1955) and *P. flexuosus* JAMES, 1879 (inorganic, according to OSGOOD, 1970, p. 393); genus often compared with *Planolites* NICHOLSON, 1873, but in *Palaeophycus* there is no distinct difference in lithology of the burrows and the host rock as in *Planolites*; for discussion see OSGOOD, 1970, p. 375; *Spongillopsis* GEINITZ, 1862, p. 132, established for *Palaeophycus* in lacustrine sediments (but *S. recurva* FLICHE, 1906, p. 34, belongs to *Rhizocorallium*, perhaps also *S. triadica* FLICHE, 1906, p. 33).] Precam.-Rec., cosmop.——FIG. 54,4. *P. tubularis HALL, 1847, Ord.(Beekmantown beds), USA(Amsterdam, New York); ×0.25 (Osgood, 1970).

Paleodictyon MENEGHINI in MURCHISON, 1850, p. 484 [*P. strozzii; M] [=Paleodictyon, Palaeodictyum AUCTT., non HEER, 1865, p. 245 (=jun. homonym of *Phycosiphon* VON FISCHER-OOSTER, 1858, p. 59)] [=Reticulipora STOPPANI, 1857, p. 407 (no type species designated); *Glenodictyum* VAN DER MARCK, 1863, p. 6 (type, *G. hexagonum*); *Cephalites maximus* EICHWALD, 1865, p. 82; *Paretodictyum* MAYER, 1878, p. 80 (no type species designated); *Palaeodyction* DE STEFANI, 1879, p. 446 (*nom. null.*); *Retiofucus* KEEPING, 1882, p. 488 (type, *R. extensus*); *Retiphycus* ULRICH, 1904, p. 139 (type, *R. hexagonale*); *Palaeopiscovum* BANYAI, 1939, p. 83 (no type species designated); the following "genera" may also be regarded as synonyms of *Paleodictyon* or as subgenera: *Priodictyon* VYALOV & GOLEV, 1960, p. 176 (established for small *Paleodictyon* on upper surfaces of beds; no formal species name); *Squamodictyon* VYALOV & GOLEV, 1960, p. 178 (type, *S. squamosum*) (established for *Paleodictyon* with meshes in outline resembling fish scales); *Largodictyon* VYALOV & GOLEV, 1965, p. 111 ("subg." of *Squamodictyon*); "*Pleurodictyon*" FUCHS, 1895, p. 394, considered to be named erroneously for *Paleodictyon* (used only in heading and in explanation of figure; no description of differences between *Paleodictyon* and *Pleurodictyon*)]. Honeycomblike network of ridges in hyporelief, consisting of remarkably regular hexagonal polygons; may be also 4- to 8-sided; reticulate pattern of considerably varying size (in-

complete along margins) but diameter of meshes constant within individual net (from less than 1 mm. to about 50 mm.); walls of meshes 0.5 to 2 mm. wide and occasionally consisting of small circular or oval "pimples" closely arranged in rows which may cross one another regularly; networks may cover large areas up to about 1 sq. m.; polygons sometimes elongated due to current action. [One of the most famous Problematica, discussed for more than a century; interpreted as algae, sponges, corals, bryozoans, spawn of fishes or molluscs, and very often as inorganic in origin (infilled mud cracks, interference ripple marks, raindrop imprints); interpreted by FUCHS (1895) and ABEL (1935) as trace fossil; now thought to be grazing trails (SEILACHER, 1954, 1955); according to WOOD & SMITH (1959, p. 167), made by burrowing animals at interfaces of sandy and muddy sediment; manner of preservation in dispute, considered predepositional by SEILACHER (1962, p. 229) and WEBBY (1969a, p. 87) and postdepositional by SIMPSON (1967, p. 512) and KSIĄŻKIEWICZ (1970, p. 316) who believed that "evidence is conflicting" and "still open to argument"; preservation "pseudexogene" (SEILACHER, 1964, p. 292) = "preendogene" (WEBBY, 1969a, p. 91); for possible mode of origin discussed by WEBBY (1969a, p. 87) see Figure 54,2e (worms mining systematically in series through the sediment horizontally, regularly turning 120°, then overturning vertically in order to rejoin the tunnel at the last 120° section: this explanation assumes that producer is highly sensitive to thigmotaxis; for an explanation of polygons made by strictly planar feeding animal (simple meander pattern overlapping outside of each previous meander) see CHAMBERLAIN (1971a, p. 227); incomplete patterns and initial forms described as *Protopaleodictyon* KSIĄŻKIEWICZ (1970, p. 303) (see p. W97); more than 30 "species" have been named, many of them based on size of the meshes, their shape and thickness of walls; occurrences mainly in flysch deposits of all ages but also in facies intermediate between flysch and molasse and even (HÄNTZSCHEL, 1964) from epicontinental environment; representative for "*Nereites*" facies. The following papers discuss *Paleodictyon*, its origin, synonymy and history; ANTONIAZZI (1966); NOWAK (1959); OSGOOD (1970, p. 384-386); SACCO (1939); SILVESTRI (1911); VYALOV & GOLEV (1960, 1964, 1966a); WANNER (1949); WEBBY (1969a).] Ord.-Tert., Eu.-N. Am.-S. Am.-N. Afr.-Asia-Australia-N.Z.-Antarct.
——FIG. 54,2a,b. P. sp.; 2a, A-E, development of structure, schem. [c, curve, v or ʌ, angle of convergence or divergence; pv, pseudoangle of truncation] (after Chamberlain, 1971a); 2b, L. Cret. (Cieszyn limestones), Pol. (Coleszow), ×0.7 (Nowak, 1959).——FIG. 54,2c. *Paleodictyon* (="Squamodictyon" VYALOV & GOLEV),

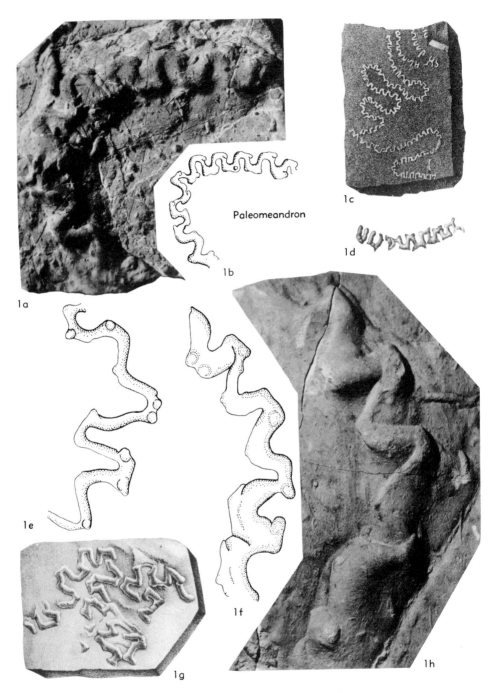

Paleomeandron

FIG. 56. Trace fossils (p. *W*91).

diagram. (Vyalov & Golev, 1960).——FIG. 54,2*d*. *P. regulare* SACCO, L.Tert.(flysch), Italy; ×0.4 (Seilacher in Häntzschel, 1962).——FIG. 54,2*e*.

P. sp., interpretation of possible origin, schem. (Webby, 1969a).
Paleohelcura GILMORE, 1926, p. 31 [**P. tridactyla*;

M] [=*Palaeohelcura* BRADY, 1939, p. 32; 1961, p. 201 *(nom. null.)*]. Long continuous tracks, consisting of 2 parallel rows of foot imprints arranged in groups of 3; between them undulating drag mark occasionally present; groups of foot impressions arranged in straight line, inclined at 60° to 80° angle to median line; average width of track 25 mm.; stride varying from 14 to 22 mm. [Cluster of foot impressions made by apparently tridactyl pointed extremities, clusters alternating on both sides; probably made by small scorpionids as concluded by BRADY (1939, 1947, 1961) from his experiments with living scorpionid *Centruroides.*] *L.Perm.(Coconino Ss.),* N.Am. (USA,Ariz.); *?L.Trias.,* Eu.(Ger.).——FIG. 54,*3*. **P. tridactyla,* Coconino Ss., Ariz.; *3a,* diagram of trackway; ×0.3; *3b,* ×0.3 (Gilmore, 1926).

Paleomeandron PERUZZI, 1881, p. 8 [**P. elegans*; SD HÄNTZSCHEL, herein] [=*Palaeomaeandron* FUCHS, 1895, p. 395 *(nom. null.)*]. Wide first order meanders consisting of small, mostly quadrangular, second order meanders with double-pointed corners; large meanders several cm. long, small meanders 1 to 5 mm. wide, rarely much larger (e.g., *P. robustum* KSIĄŻKIEWICZ, 1968). [Grazing trail.] [Found in flysch deposits.] *U. Cret.-L.Tert.,* Eu.(Aus.-Italy-Spain-Pol.)-S.Am. (Venez.).——FIG. 56,*1a,b,e*. *P. robustum* KSIĄŻKIEWICZ, low. and mid. Eoc. (Beloveza Beds), Pol.; *1a,* ×1 (Książkiewicz, 1968); *1b,e,* ×0.5 (Książkiewicz, M., 1970, p. 299, in: *Trace Fossils* edited by T. P. Crimes & J. C. Harper, Geol. Jour. Spec. Issue 3, Seel House Press, Liverpool).——FIG. 56,*1c,d*. **P. elegans,* Eoc., Italy; *ca.* ×0.67 (Fuchs, 1895, after Peruzzi, 1881).——FIG. 56,*1f,h*. *P.* sp. aff. *P. robustum* KSIĄŻKIEWICZ, mid. Eoc. (Łącko Beds), Pol.; *1f,h,* ×0.5, ×0.7 (*1f,* Książkiewicz, M., 1970, p. 299, in: *Trace Fossils* edited by T. P. Crimes & J. C. Harper, Geol. Jour. Spec. Issue 3, Seel House Press, Liverpool; *1h,* Książkiewicz, 1968).——FIG. 56,*1g*. *P. rude,* Eoc., Italy; *ca.* ×2 (Peruzzi, 1881).

Palmichnium RICHTER, 1954, p. 267 [**P. palmatum*; OD]. Large, plantlike track, about 11 cm. wide; opposed symmetrical rows of [leg] impressions; median keel, divided at regular intervals; bordered by longitudinally directed club-shaped impressions distinctly set off toward interior, but indistinctly toward exterior. [Crawling track; made by arthropod, probably eurypterid; a similar track from Devonian of Antarctica is *Beaconichnus antarcticus* GEVERS, 1971 (see GEVERS et al., 1971, p. 87, 93).] *L.Dev.,* Eu.(Ger.); *?L.Carb.(Kulm),* Eu.(Ger.).——FIG. 57,*5*. **P. palmatum,* L.Dev., Ger.; *ca.* ×0.2 (Richter, 1954).

Parahaentzscheliana CHAMBERLAIN, 1971, p. 236 [**P. ardelia*; M] [=*Parathaentzscheliana* CHAMBERLAIN, 1971a, p. 236 *(nom. null.)*]. Numerous small tubes, about 15 to 20 mm. long, 1 to 2 mm.

wide, radiating vertically and obliquely from common starting center within sediment upward to bedding plane, producing surface pattern (15 to 60 mm. in diameter) consisting of radially arranged openings which are mostly sediment-filled. *Penn.(AtokaF.),* USA(Okla.).——FIG. 57, *4*. **P. ardelia*; schem. drawing, *4a,* plan pattern, ×0.9; *4b,* complete perforation of sediment, ×0.9 (Chamberlain, 1971a).

Paratisoa GAILLARD, 1972, p. 150 [**P. contorta*; OD]. Series of branching, straight to curved galleries (up to 40 mm. in diam.) with a small, characteristic axial tube (about 4 mm. in diam.); size of entire burrow system as much as 55 cm.; branchings can be either T- or Y-shaped; galleries may also possess distinct swellings and are commonly calcareous; axial tube commonly filled with ferruginous material. [Considered to have been produced by a marine burrowing annelid; similar to *Tisoa* (p. *W*117) but *Tisoa* does not branch and axial tube is U-shaped.] *Jur.(Oxford.),* Eu.(France). [Description supplied by W. G. HAKES.]

Pennatulites DE STEFANI, 1885, p. 99 [*non* COCCHI, 1870, p. 116 *(nom. nud.)*] [**P. longespicata*; M] [=*Paleosceptron* DE STEFANI, 1885, p. 100 (type, *P. meneghinii*) (*non* COCCHI, 1870, p. 116, *nom. nud.*); *Virgularia presbytes* BAYER, 1955 (Tert. forms only)]. Thick cylindrical stalk (diameter about 4 cm.) followed by club- or ear-shaped part, with deep median furrow, consisting of biserially arranged overlapping rows of leaves; surface of ear-shaped part nodose, nodes arranged in parallel rows. [Regarded as alcyonarian by DE STEFANI (1885); certainly branching *Spreiten-bau*; interpreted as feeding burrow by SEILACHER (1955, fig. 5). [Found in flysch deposits.] *Cret.-L.Tert.,* Eu. (Italy-Greece)-W.Indies (Trinidad).——FIG. 57,*3a,b*. **P. longespicata,* U.Cret., Italy; *3a,* model, ×0.17 (Seilacher, 1955); *3b,* ×0.4 (de Stefani, 1885).

Permichnium GUTHÖRL, 1934, p. 174 [**P. völckeri*; OD]. Two parallel, equal, and equidistant rows of V-shaped foot impressions, open to exterior; indicative of equal walking feet with 2 claws each; somewhat similar to *Bifurculapes* HITCHCOCK, 1858. [Running track of insect (?blattoid).] *L.Perm.(Rotl.),* Eu.(Ger.); *L.Trias.,* Eu.(Eng.).——FIG. 57,*1*. **P. voelckeri,* L.Perm. (Rotl.), Ger.; holotype, ×1.2 (Guthörl, 1934).

Petalichnus MILLER, 1880, p. 221 [**P. multipartitus*; M]. Simple or complex tracks of varied morphology, consisting of numerous transversely elongated unifid or bifid imprints, in complete series varying from 10 to 12; about 1 to 2 cm. wide. [Tracks indicating straight-ahead or slightly oblique movement of producer; tentatively regarded by MILLER (1880) to have been made by cephalopods (see TEICHERT, 1964b, p. *K*487); most probably tracks of moderately sized trilo-

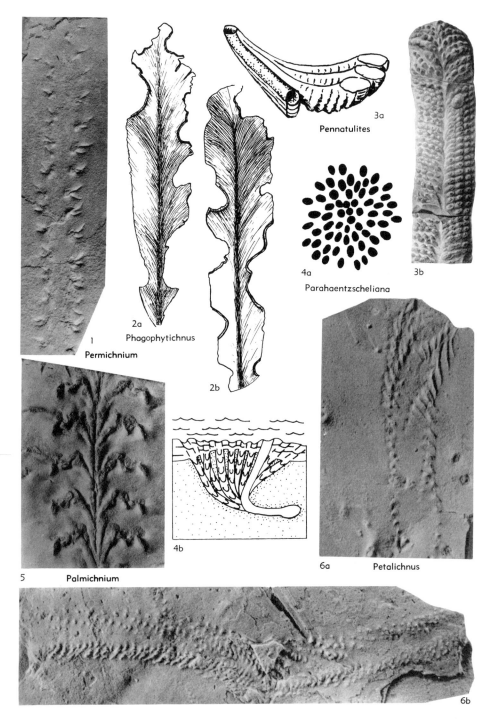

FIG. 57. Trace fossils (p. W91, 93).

bites, perhaps particularly by *Flexicalymene,* also by other trilobite genera, or even by other arthropods; Osgood (1970, p. 362) regarded the following as synonyms of type species: *Trachomatichnus permultus* MILLER, 1880, *T. cincinnatiensis* MILLER, 1880, and "*Merostomichnites* sp." (CASTER, 1938, p. 34).] *U.Ord.(Cincinnat.),* USA(Ohio).——FIG. 57,6. **P. multipartitus; 6a,* Eden beds, Ohio; ×0.9; *6b,* Southgate beds, Ohio (Hamilton Co.); *ca.* ×0.9 (Osgood, 1970).

Petaloglyphus VYALOV, GORBACH & DOBROVOLSKA, 1964, p. 94 [**P. krimensis;* OD]. Starlike trace fossil, insufficiently figured and described only in Ukrainian language. [Grazing trace with dwelling burrow.] *L.Cret.,* USSR(Crimea).

Phagophytichnus VAN AMERON, 1966, p. 182 [**P. ekowskii;* M]. Malformations of leaves of *Neuropteris praedentata* and *Glossopteris* sp., consisting of damaged margins nibbled by insects, leaving hemispherical or oval scallops, sometimes also broad concave or small uniform convex ones, rarely reaching midrib; ridge of scallops clearly marked and mostly thickened. *U.Carb.(Westphal. C),* Eu.(N.France), *U.Carb.(Stephan. B),* Eu. (N.Spain, Prov.Léon); *Permo-Carb.,* S.Afr.-Eu. (Spain).——FIG. 57,2. **P. ekowskii,* Perm.-Carb., Spain; *2a,b,* at a leaf of *Glossopteris,* ×0.7 (van Ameron, 1966).

Phoebichnus BROMLEY & ASGAARD, 1972, p. 29 [**P. trochoides;* OD]. Central shaft 6 to 8 cm. in diam., nearly vertical to bedding, with numerous, long, straight radial burrows oriented more or less parallel to bedding; radial burrows about 1.5 cm. in diam. including distinct, annulated wall lining about 5 mm. thick; mica flakes infilling of radial burrows oriented in discrete concavo-convex planes, concave toward central shaft; total length of shaft and tunnels unknown. [Central shaft interpreted as dominichnia, radial burrows as fodinichnia of same unknown animal; radial burrows actively filled.] *L.Jur.-M.Jur. (Bajoc.-Callov.),* Greenl.(Jameson Land).——FIG. 58,1. **P. trochoides,* low.Callov.; *1a,* schem. reconstr., ×0.15; *1b,* portion of holotype, ×0.25 (Bromley & Asgaard, 1972). [Description supplied by R. W. FREY.]

Pholeus FIEGE, 1944, p. 415 [**P. abomasoformis;* OD]. Large compactly cylindrical burrow with longitudinal axis parallel to bedding; anterior and posterior ends closed and rounded with 2 or more rounded tubes, oriented obliquely or vertically to bedding, leading to surface; walls lined with flakes. [Dwelling burrow, probably made by decapod crustaceans.] *M.Trias.(Muschelkalk); Eu.(Ger.).——FIG. 59,1. **P. abomasoformis,* L. Muschelkalk, Ger.; ×0.4 (Fiege, 1944).

Phycodes RICHTER, 1850, p. 205 [*non* GUENEE, 1852; *nec* MILNE-EDWARDS, 1869] [**P. circinnatum* RICHTER, 1853, p. 20 (?=*Fucoides circinnatus* BRONGNIART, 1828); SM (see MÄGDEFRAU, 1934, p. 260)] [=*Licrophycus* BILLINGS,

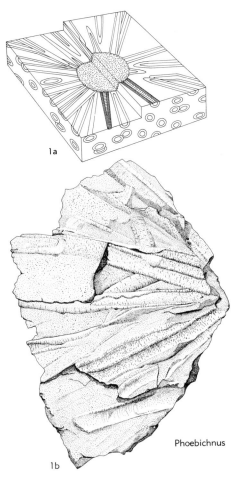

Phoebichnus

FIG. 58. Trace fossils (p. *W93*).

1862, p. 99 (type, *L. ottawaensis*); *Vexillum rouvillei* DE SAPORTA, 1884, p. 43; *Lycrophycus* TWENHOFEL, 1928, p. 83, 99 (*nom. null.*); for discussion see MÄGDEFRAU (1934, p. 270) and OSGOOD (1970, p. 342)]. Bundled structures of flabellate or broomlike pattern, consisting of horizontal tunnels; proximal part of main tunnels unbranched, distal tunnels divide at acute angles into several free cylindrical tunnels showing delicate annulation beneath thin smooth "bark"; main branches may show structure similar to retrusive spreiten (absent in *P. flabellum* from Cincinnatian of USA); other "species" (e.g., *P. pedum* and *P. flabellum*) vary considerably in morphology from type species which is also variable (e.g., falcate or featherstitch-like pattern of feeding tunnels); about 15 cm. long in entirety; generally preserved as convex hyporeliefs in quartzites. [Originally interpreted as "fucoids"

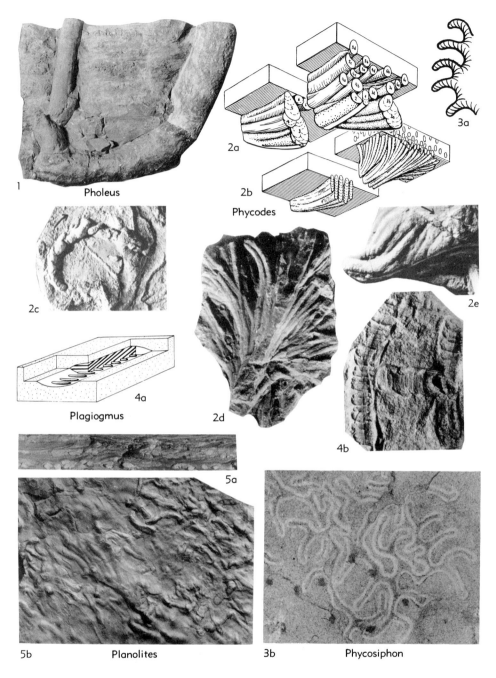

FIG. 59. Trace fossils (p. *W*93, 95-97).

or even as inorganic structures; certainly feeding structure of typical flabellate pattern; probably produced by sediment-feeding wormlike animal;

relations of *Phycodes* to *Teichichnus* were discussed by HÄNTZSCHEL & REINECK (1968, p. 26); *Arthrophycus* HALL, 1852, regarded by

SEILACHER (1955, p. 386) as junior synonym; *P. pedum* SEILACHER, according to OSGOOD (1970, p. 342), should be assigned to a separate genus; for detailed discussion see MÄGDEFRAU (1934), SEILACHER (1955, p. 383-388), and OSGOOD (1970, p. 341-343).] *?U.Precam.,* Australia; *Cam.,* Eu. (Eng.-Swed.-Nor.-Spain)-USA (Ariz.)-Asia (Pak.)-Australia; *Ord.,* Eu.(Eng.-France-Ger.)-N.Am.-S. Am.-Asia(Iraq)-N.Afr.(Libya); *L.Carb.,*Eu.(Scot.); *Jur.,* Eu.(Ger.-France-Swed.); *?Tert.(Mio.),* N.Z.
———FIG. 59,2a. *P.* cf. *P. palmatum* (HALL), L. Cam., Pak.; model, ×0.7 (Seilacher, 1955).———FIG. 59,2b,d. *P. circinnatum; 2b,* L.Ord., Ger.; model, ×0.3 (Seilacher, 1955); *2d,* Ord.(Galena F.), USA(Minn.); ×0.7 (Mägdefrau, 1934).———FIG. 59,2c. *P. pedum* SEILACHER, Cam., Pak. (Salt Range); ×0.7 (Seilacher, 1955).———FIG. 59,2e. *P. palmatum* (HALL), Cam., Pak.(Salt Range); (arrow indicates direction of movement of producer), ×0.7 (Seilacher, 1955).

Phycosiphon VON FISCHER-OOSTER, 1858, p. 59 [*P. incertum;* M] [=*Palaeodictyon* HEER, 1865, p. 245 (type, *P. singulare*) (*non Paleodictyon* MENEGHINI in MURCHISON, 1850, often erroneously spelled *Palaeodictyon*); *Reticulum* DE STEFANI, 1879, p. 446 (type, *R. textum*) (*nom. nov. pro Palaeodictyon* HEER, 1865); *Eterodictyon* PERUZZI, 1881, p. 8 (type, *E. textum*); *Lophoctenium richteri* DELGADO, 1910, p. 51; "*Polydora?*" GÓMEZ DE LLARENA, 1946, p. 153]. Small U-shaped loops; frequently branched; in large numbers forming antler-shaped systems; similar to asymmetrical very small *Rhizocorallium;* parallel or oblique to bedding planes. [Feeding burrows; regarded by FISCHER & PAULUS (1969, p. 90) as true spreiten burrows, protrusively built; various forms (L.Carb., Kulm; Ger.) have been placed in *Phycosiphon* by PFEIFFER (1968, p. 676).] *Ord.-Carb., Jur.-Tert.,* Eu.-USA(Okla.-Alaska).
———FIG. 59,3. *P. incertum; 3a,* U.Cret., Aus.; ×1 (Seilacher, 1955); *3b,* Eoc., Italy; ×2 (Seilacher in Häntzschel, 1962).

Phyllodocites GEINITZ, 1867, p. 1 [*Crossopodia thuringiaca* GEINITZ, 1864a, p. 3; SD HÄNTZ-SCHEL, 1962, p. W210]. Curved or meandering trails, similar to *Nereites,* up to several cm. wide; consisting of narrow median furrow (about 5 mm. wide); smooth or articulated, flanked on either side by oval lateral markings, mostly overlapping one another, somewhat irregularly but closely placed, resembling "foliaceous outgrowths." [Formerly regarded as parapodia of polychaetes, now interpreted as originating by turbation of sediment along sides of median string (the latter perhaps of fecal origin). Originally considered by GEINITZ (1867) to be impressions of the bodies of polychaetes related to *Phyllodoce;* interpreted by RAYMOND (1931a, p. 188) as feeding trails of branchiopods or phyllocarids; according to ABEL (1935, p. 241), made by gastropods. For discussion of *Phyllodocites* interpreted as endogene feed-

ing burrows see also PFEIFFER, 1968, p. 686-687.] *Paleoz.,* Eu.-N.Am.

Phytopsis HALL, 1847, p. 38 [*P. tubulosum* HALL, 1847; SD HÄNTZSCHEL, 1966, p. 72]. Vertical inosculating tubes, straight or flexuous, nearly circular in section (5 to 10 mm. in diam.); variously branching, lined with dark material. [Originally described as probably a marine plant; according to RAYMOND (1931b, p. 195), probably burrows of polychaetes; another "species," *P. cellulosum* HALL, 1847, has been transferred to the tabulate corals as *Tetradium cellulosum* (HALL) (RAYMOND, 1931b, p. 197).] *Ord.,* USA (Ky.-Tenn.-N.Y.).

Pilichnia CHAMBERLAIN, 1971, p. 223 [*P. elliptica;* M] [=*Pilichna* CHAMBERLAIN, 1971a, p. 215, 224 *(nom. null.)*]. Large vertical or horizontal burrows, about 60 mm. wide; oval or elliptical in cross section. [Ill-defined form; name unnecessary.] *Penn.(WapanuckaLs.),* USA(Okla.).

Plagiogmus ROEDEL, 1929, p. 51 [*P. arcuatus;* SD HÄNTZSCHEL, 1962, p. W210]. Smooth, flat, concave ribbon (1 to 2 cm. wide), straight or slightly curved; with pronounced single transverse ridges, mostly straight, usually not extending to sides, at regular or irregular intervals, also occasionally closely spaced, passing into obliquely textured band (backfill of trail) consisting of sandy laminae; rarely faint longitudinal furrows. [Formerly regarded as epichnial trail; according to GLAESSNER (1969, p. 387), endichnial burrow parallel to bedding; perhaps made by ancestral mollusk with foot and mantle feeding in sediment and backfilling its trail with rejected sediment; smooth surface of burrow cemented by mucus.] *?L.Cam.-?M.Cam.,* N.Am.(USA, Wyo.); *L. Cam.,* Eu.(Swed.-Nor.)-Greenl.-Australia; *L.Cam.* (Pleist. drift), Eu.(N.Ger.).———FIG. 59,4. *P. arcuatus,* Cam.(Pleist. drift), Ger.; *4a,* block diagram explaining endichnial burrow interpretation, filling shown by cross hatching (Glaessner, 1969); *4b,* ×0.4 (Roedel, 1929). [Also found in U.Precam., USSR(Russ. Plat.).]

Plangtichnus MILLER, 1889, p. 580 [*P. erraticus;* OD]. Simple narrow trail, smooth, irregularly zigzagging in every direction. [Made by larva or pupa of palaeodictyopterous insect?] *L.Carb. (Kaskaskia Gr.),* USA(Ind.).

Planolites NICHOLSON, 1873, p. 289 [*P. vulgaris* NICHOLSON & HINDE, 1875, p. 139 (=*P. vulgaris* NICHOLSON, 1873, p. 290, *nom. nud.*); SD HOWELL, 1943, p. 17] [=*?Scolecites* SALTER, 1873, p. 2, 10 (without formal species name)]. Cylindrical or subcylindrical infilled burrows (diam. up to 15 mm.), straight to gently curved, nonbranching; usually more or less horizontal or oblique to bedding planes, penetrating sediment in irregular course and direction, may cross one another. [Interpreted as infilled endichnial burrows (German, "*Stopftunnel*"); the name *Plano-*

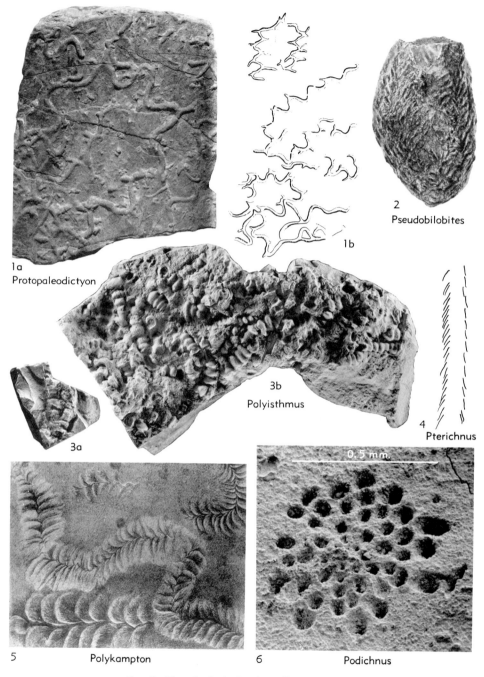

1a
Protopaleodictyon

1b

2
Pseudobilobites

3a

3b
Polyisthmus

4
Pterichnus

0.5 mm.

5   Polykampton

6   Podichnus

Fig. 60. Trace fossils, boring (p. *W*97, 99, 131).

lites explicitly established by Nicholson (1873, p. 288) for "burrows filled up with the sand or mud which worm has passed through its alimentary canal"; simple burrows showing transverse annulation ("packing structure," "backfilling") have been placed in *Planolites* by several authors (e.g., by CHISHOLM (1970, p. 24) for trace fossils from Carboniferous of Scotland). *Planolites* is often difficult to distinguish from morphologically similar *Palaeophycus* HALL; for discussion see OSGOOD (1970, p. 376) (fillings of *Palaeophycus* are generally regarded as apparently not having been passed through gut of animals); several "species" assigned to *Palaeophycus, Chondrites,* and even *Arthrophycus* more correctly referable to *Planolites; P. rugulosus* REINECK, 1955, type species of *Scoyenia* WHITE; *P. ophthalmoides* JESSEN, 1950, type species of *Opthalmidium* PFEIFFER, 1968 (superfluous name), for discussion of that "species" see SEILACHER (1963, p. 84), "guide fossil" for Upper Carboniferous *"Augenschiefer"* of Westphalia; Precambrian "species" described by WALCOTT (1899, 1914) were recently interpreted by CLOUD (1968, p. 55) as "algal?".] *Precam.-Rec.,* cosmop.——FIG. 59,5. *P. montanus,* U.Carb., Ger.; *5a,* transv. sec., ×0.7; *5b,* ×1 (Richter, 1937).

Polyisthmus BARTHEL, 1969, p. 128 [*P. enigma*; OD]. Fragmentary burrows, long-cylindrical to tapering; regularly widening and narrowing (3 to 9 mm. wide); intervals of constrictions sometimes variable; cross section circular to oval; wall smooth; greatest length of fragments observed about 3 cm.; whole burrow consists of 2 parallel pieces which converge V-shaped downward; mostly only washed-out fragments of whole broken burrows preserved. [Interpretation of construction of burrow difficult.] *U.Jur.(U.Tithon., U.Neuburg F.),* Eu.(Ger.), Bavaria).——FIG. 60,3. *P. enigma; 3a,* holotype; *3b,* paratype, both ×0.7 (Barthel, 1969).

Polykampton OOSTER, 1869, p. 23 [*P. Alpinum;* M] [=*Polycampton* FUCHS, 1895, p. 433, *nom. null.*]. Central zigzag-shaped stalk, at angles of stalk featherlike bunches grow out at both sides with backwardly directed curvature; externally similar to *Sertularia.* [Interpreted originally as hydrozoan; explained by FUCHS (1895, p. 433) as spawn ribbons of gastropods (see also EHRENBERG, 1941, p. 303); according to SEILACHER (1959, p. 1070), feedng burrow with alternating fanlike feeding fields.] [Found in flysch deposits.] *Trias., Cret.-L.Tert.,* Eu.(Switz.-Aus.-Spain).——FIG. 60,5. *P. alpinum,* Trias., Switz.; *ca.* ×0.3 (Ooster, 1869).

Protichnites OWEN, 1852, p. 214 [*P. septemnotatus;* SD HÄNTZSCHEL, 1962, p. *W210*] [=*Protichnides* CHAPMAN, 1878, p. 490 *(nom. null.)*]. Two rows of bifid or trifid imprints and a commonly narrow, intermittent double drag trail in the middle; tracks irregularly and closely set; trackway in places connected with the resting trace *Rusophycus.* [Interpreted as tracks of limulids, crablike crustaceans or most probably trilobites moving straight forward (WALCOTT, 1912b, p. 275; 1918, p. 174), but also of gastropods; for discussion of the "genus" see BURLING (1917,

p. 387) and OSGOOD (1970, p. 352); several "species," according to STØRMER (1934, p. 22), belong to *Merostomichnites;* see also ÖPIK (1959, p. 8).] *L.Cam.,* Asia(Pak.)-Can.; *U.Cam.,* N.Am.-Asia M.; *?Cam., ?Ord.,* W.Australia.——FIG. 61,*1b. P.* sp., L.Cam.(Jutana Dol.), Pak. (Salt Range); trail starting from *Rusophycus* impression, ×1.5 (Seilacher, 1955).——FIG. 61,*1a.* *P. septemnotatus,* U.Cam.(Potsdam Ss.), Can.(Que.); track, ×0.5 (Walcott, 1912b).——FIG. 61,*1c. P. logananus* MARSH, U.Cam., USA(N.Y.); track, ×0.1 (Walcott, 1912b).

Protopaleodictyon KSIĄŻKIEWICZ, 1970, p. 303 [*P. incompositum;* OD] [=*Protopaleodictyon* KSIĄŻKIEWICZ, 1958, expl. pl. 2 *(nom. nud.); Protopalaeodictyum* NOWAK, 1959, p. 119, 125; *(nom. nud.); Protopalaeodictyon* KSIĄŻKIEWICZ, 1960, p. 737, 745 *(nom. nud.); ?Unarites* MACSOTAY, 1967, p. 38 (type, *U. suleki); ?Spinorhaphe* PFEIFFER, 1968, p. 681 (type, *Palaeophycus spinatus* GEINITZ, 1867a, p. 16); *?Pseudopaleodictyon* PFEIFFER, 1968, p. 674 (type, *Palaeophycus hartungi* GEINITZ, 1867a, p. 16)]. Initial, irregular forms of *Paleodictyon,* quite variable, less regular, not strictly polygonal pattern; mostly meanders with ramifications on their apices; sometimes representing transitional forms from *Cosmorhaphe* or *Belorhaphe* to *Protopaleodictyon,* therefore is a combination of features of these ichnogenera. [Found in flysch deposits.] *?L.Carb.(Kulm),* Eu.(Ger.); *L.Cret.,* Japan; *Cret.-L.Tert.,* Eu.(Pol.-Aus.-Spain); *?Cret.-L.Tert.,* S. Am.(Venez.).——FIG. 60,*1a. P.* sp., low.Eoc. (Beloveza Beds), Pol. (Carpath.); ×0.7 (Książkiewicz, 1960).——FIG. 60,*1b.* *P. incompositum,* mid. Eoc. (Hieroglyphic beds), Pol.(Przykrzec); ×0.3 (Książkiewicz, M., 1970, p. 302, in: *Trace Fossils* edited by T. P. Crimes & J. C. Harper, Geol. Jour. Spec. Issue 3, Seel House Press, Liverpool).

Protovirgularia M'COY, 1850, p. 272 [*P. dichotoma* (=*?Cladograpsus nereitarum* RICHTER, 1853, p. 450; *Triplograpsus nereitarum* RICHTER, 1871, p. 251); M] [=*Provirgularia* GÜMBEL, 1879, p. 469 *(nom. null.)*]. Small keel-like trail, a few mm. wide, mostly straight or slightly curved, may branch dichotomously; consisting of an elevated median line and lateral wedge-shaped appendages, alternating on both sides. [Regarded by M'COY as octocoral because of its similarity to the Recent octocoral *Virgularia;* as late as 1952 considered an octocoral *incertae sedis* by ALLOITEAU (1952, p. 415); ascribed to graptolites by RICHTER (1853a, 1871); undoubtedly a trail as recognized by NATHORST (1881a, p. 85), belonging to "group" *Ichnia spicea* RUDOLF RICHTER (1941, p. 229); for detailed discussion see HÄNTZSCHEL, (1958, p. 84) and VOLK (1961); producing animal unknown.] *Ord.,* Eng.; *L.Dev.-M.Dev.,* Eu.(Ger.); *L.Carb.(Kulm),* Eu.(Ger.).——FIG. 61,2. *P. nereitarum* (RICHTER), M.Dev.

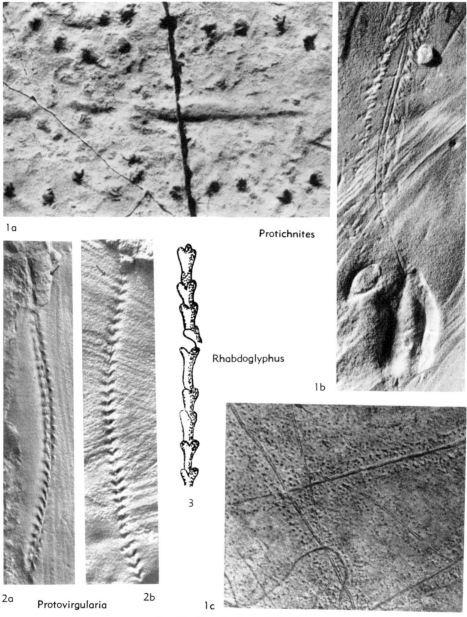

1a

Protichnites

Rhabdoglyphus

1b

3

2a   Protovirgularia   2b   1c

FIG. 61. Trace fossils (p W97-99).

("Nereiten-Schiefer"), Ger.(Thuringia); 2a,b, ×2.25 (Volk, 1961).

Psammichnites TORELL, 1870, p. 9 [*Arenicolites gigas* TORELL, 1868, p. 34; SD FISCHER & PAULUS, 1969, p. 91] [=?*Cymaderma* DUNS, 1877, p. 352 (no formal species named)]. Large ribbonlike trails with narrow longitudinal median ridge; convex upper surface; mostly very flexuous, about 2 to 5 cm. wide; with very fine transverse ridges closely spaced. [Usually interpreted as trails made by burrowing gastropods; according to GLAESSNER (1969, p. 389) by mollusks without shells; belonging to the "group" *Scolicia* DE QUATREFAGES, 1849; interpreted by HÖGBOM (1926) as sandy excrements of worms or content of their intestines; regarded by HADDING (1929, p. 58) as worm trails.] *L.Cam.,* Eu.(Ire.-S.Swed.)-Can.; *L.Cam.*(Pleist.drift), Eu.(Ger.-Denm.); *?M.Dev.,* Eu.(Ger.); *?Penn.,* USA(Okla.).——FIG. 62,2*a,b. P.* sp., L.Cam.; *2a,* Swed., ×0.25; *2b,* loc. unknown, ×0.7 (Häntzschel, 1964b).——FIG. 62,2*c. *P. gigas* (TORELL), L.Cam., S.Swed.; ×0.7 (Torell, 1868).

Pseudobilobites KENNEDY, 1967, p. 153 [*P. jefferiesi* KENNEDY, 1967, p. 154; OD] [=Pseudobilobite BARROIS, 1882, p. 175; *Pseudobilobites* BARROIS (in LESSERTISSEUR, 1955, p. 45 *(nom. nud.)*)]. Ovoid or rounded masses of shell fragments and sand-size microfossils cemented by calcite; 3 to 7 cm. long; upper surface mostly flat or slightly concave, smooth or somewhat granular; lower surface convex, covered by groups of short more or less parallel ridges inclined to long axis of structure. [Apparently surface trace made by crustaceans; for discussion see KENNEDY, 1967, p. 155.] [Author of this ichnogenus is neither BARROIS, 1882 (intended by him as a vernacular name) nor LESSERTISSEUR, 1955, p. 45, as attributed by KENNEDY (1967, p. 154); LESSERTISSEUR neither published a diagnosis nor designated a type species; conditions for the establishment of a valid generic name were fulfilled only by KENNEDY (1967).] *U.Cret.,* Eu.(Eng.-France).——FIG. 60,2. *P. jefferiesi,* mid.Cenoman.(Lower Chalk), S.Eng.(Buckinghamsh.); holotype, ×0.7 (Kennedy, 1967).

Pterichnus HITCHCOCK, 1865, p. 14 [*Acanthichnus tardigradus* HITCHCOCK, 1858, p. 151; OD]. Two rows of numerous [foot] imprints, turned outward from median line at angle of 15 to 20 degrees; width of track about 12 mm., foot imprints 3 mm. long. [?Myriapod track.] *Trias.,* USA(Mass.).——FIG. 60,4. *P. tardigradus* (HITCHCOCK); ×0.7 (Hitchcock, 1858).

Pteridichnites CLARKE & SWARTZ, 1913, p. 545 [*P. biseriatus; M]. Two rows of small pits bordered by narrow elevated margin; about 4 mm. wide; median ridge crenulated; pits nearly equidimensional, alternating in position; somewhat similar to *Nereites.* [Interpreted as crawling trail of arthropod or annelid.] *U.Dev.,* USA (Md.).——FIG. 62,3. *P. biseriatus,* Jennings F.; ×1 (Clarke & Swartz, 1913).

Quebecichnus HOFMANN, 1972, p. 196 [*Q. lauzonensis; OD]. Large, uniformly branching burrow systems along bedding planes; containing cylindroidal to ellipsoidal fecal pellets. Branches generally nearly rectilinear, fairly uniform in length (10-30 cm.) and width (1.5 cm.), developed by repeated equal, distally directed, lateral forking from opposite points along distal half of individual segments. Burrows show multiple laminations indicative of upward displacement of burrows during successive stages of

occupation, similar to *Teichnichnus.* [Possible interpretation as being produced by one or several worms systematically traversing sediment.] *L. Ord.,* Can.——FIG. 62,4. *Q. lauzonensis,* Quebec Gr.; ×0.16 (Hofmann, 1972b). [Description supplied by W. G. HAKES.]

Radiichnus KARASZEWSKI, 1973, p. 159 [*R. staszic; M]. Starlike trace fossil preserved in convex hyporelief; 6 to 7 cm. in diameter, maximum thickness (11 cm.) in central region (diam. 6-10 mm.) from which radiate approximately 30 ridges (1.5 mm. wide, 2-4 mm. in relief) grouped in "bundles"; ridges commonly bifurcate toward margins and occasionally reach margins. [Mold of a structure produced in sand by the movement of the antenna of a worm living buried in the sediment or the accumulation of undigested mud arranged by worm.] *Jur. (Bathon.),* Eu.(Pol.). [Description supplied by W. G. HAKES.]

Radionereites GREGORY, 1969, p. 10 [*R. ballancei; OD]. Featherlike structures of uniform size, arranged in radiating clusters consisting of sand-filled tubes; single burrows with narrow central rounded axis 2 to 4 mm. wide and about 10 cm. long; flanked bilaterally by closely set, opposed, leaf-shaped, lobate extensions, each up to 1 cm. long, arranged regularly at equal intervals and diverging at acute angles. [In first description by BARTRUM (1948, p. 489) "fucoid" or sponge affinities were suggested; later interpreted as feeding burrows by BALLANCE (1964, p. 492) and GREGORY (1969, p. 10).] *U.Tert.(low.Mio., Waitemata Gr.),* N.Z.——FIG. 62,1. *R.* sp., Auckland; ×0.3 (Bartrum, 1948).

Radomorpha VYALOV, 1966, p. 72 [*R. ferganensis; OD]. Straight, curved, or branching burrows, either single or forming complex patterns, characterized by longitudinal furrows. *Tert.(Oligo.),* USSR(Ferghana). [Description supplied by CURT TEICHERT.]

Rauffella ULRICH, 1889, p. 235 [*R. filosa; OD] [=*Raufela* SARDESON, 1896, p. 78; *nom. null.*] Only *R. palmipes* ULRICH a true trace fossil similar to *Arthrophycus* HALL; other species sponges or *incertae sedis* (see DE LAUBENFELS, 1955, p. E107). *U.Ord.,* USA.

Rhabdoglyphus VASSOEVICH, 1951, p. 61 [*R. grossheimi; M]. Cylindrical tubes consisting of short, closely spaced, invaginated "calyces," some with short branches; preserved in convex hyporelief. [Trail of uncertain origin; considered post-depositional by KSIĄŻKIEWICZ (1970, p. 315-316). FUCHS (1895, p. 391) described "Rhabdoglyphen" from Austrian flysch deposits, several of his forms similar to paper bags packed one inside another.] *Cret.(Cenoman.),* USSR(Azerbaidj.).——FIG. 61, 3. *R. grossheimi,* U.Cret., Caucasus; ×1 (Vyalov, 1971).

[This trace fossil has a somewhat confused nomenclatural history. HÄNTZSCHEL (1965, p. 78) felt that an adequate

1
Radionereites

2a
Psammichnites

2b

3
Pteridichnites

2c

4
Quebecichnus

Fig. 62. Trace fossils (p. *W*98-99).

description of *Rhabdoglyphus* had not been provided by
Vassoevich in either 1951 or 1953. He therefore claimed
that the conditions of availability for the name had subsequently been met by Bouček & Eliáš (1962, p. 146)
nearly a decade later. However, after inspection of a
rare copy of Vassoevich (1951), kindly lent to us by

M. Książkiewicz, it was found that an adequate description appears in the explanations of Plate V, figure 3 and
Plate VI, figures 3 & 4, both on p. 219. Bouček & Eliáš
seem to have only expanded the description of *Rhabdoglyphus* Vassoevich and complicated matters by figuring
specimens much different from the material described by

VASSOEVICH originally. This has been pointed out by KSIĄŻKIEWICZ (1970, p. 286-287). As a result, the figured specimens of BOUČEK & ELIÁŠ have been mistakenly considered *Rhabdoglyphus* by HÄNTZSCHEL (1965, p. 75; 1966, p. 15) and OSGOOD (1970, p. 369).

VYALOV (1971, p. 90) finally clarified matters by introducing the new name *Fustiglyphus* for the material figured by BOUČEK & ELIÁŠ and restricting the name *Rhabdoglyphus* to the material described by VASSOEVICH (see *Fustiglyphus*, p. *W*64).—CURT TEICHERT, W. G. HAKES.]

Rhizocorallium ZENKER, 1836, p. 219 [**R. jenense* (=*Spongia rhizocorallium* GEINITZ, 1846, p. 695); M] [=*?Lithochela* GÜMBEL, 1861, p. 411 (type, *L. problematica*); *Glossifungites* LOMNICKI, 1886, p. 99 (type, *G. saxicava*); *?Myelophycus* ULRICH, 1904, p. 145 (type, *M. curvatum*); *Spongillopsis recurva* FLICHE, 1906, p. 34; *?Spongillopsis triadica* FLICHE, 1906, p. 33; *Lissonites* DOUVILLÉ, 1908, p. 367 (provided for *Taonurus saportai* DEWALQUE, 1882, to be ascribed to *Rhizocorallium*); *Glossofungites* FRITEL, 1925, p. 35 *(nom. null.); Cavernaecola* BENTZ, 1929, p. 1181 (type, *C. baertlingi*); *Upsiloides* BYRNE & BRANSON, 1941, p. 261 (type, *U. permiana*); *Rhizocorallum* SULLIVAN & ÖPIK, 1951, p. 13 *(nom. null.); Rhyzocorallium* HARY, 1969, p. 120 *(nom. null.);* for discussion of *Cavernaecola* see HAMM, 1929, p. 105, and KEMPER, 1968, p. 64-67)]. Simple U-tubes with spreite, generally protrusive, or somewhat oblique to bedding; "arms" more or less parallel, several cm. apart; very rarely branched, occasionally with lateral flaps; tubes relatively thick (1 cm. or more), commonly initially vertical for several cm. downward, then sharply bending at right angle; outer side of many tubes often marked by numerous striae interpreted as scratch markings indicative of crustaceans (see WEIGELT, 1929); pills of ellipsoidal excrements may be incorporated in walls or within tube; median line of U often curved; horizontal forms on bedding planes characteristically winding. [Tentatively interpreted originally as sponges or corals; now regarded as burrows of deposit-feeding animals, or perhaps as dwelling burrows of plankton-feeding animals (VEEVERS, 1962, p. 10: "protective nest") for discussion of mode of life of *Rhizocorallium* animal see SELLWOOD, 1970, p. 494; parallel orientation of *Rhizocorallium* tubes observed in Jurassic of England (AGER & WALLACE, 1970, p. 14); interpreted by FARROW (1966, p. 132, 146) as orientation in response to tidal currents, oblique or horizontal position possibly depending on water depth (see AGER & WALLACE, 1970, p. 15); horizontal tubes of 70 cm. and more long have been observed (Jur., Eng.; see FARROW, 1966, fig. 7-9); very large screwlike form (30 cm. in diam.) described by FIRTION as *R. uliarensis* (FIRTION, 1958; U.Jur., France); other specimens (M.Trias., Ger.) consist of one vertical limb surrounded spirally by the other (MÜLLER, 1956b, p. 405); sometimes also vertically retrusive forms have been assigned to *Rhizocorallium* (e.g., RIOUET,

1960, p. 8; SELLWOOD, 1970, p. 492); for transitions to *Teichichnus* see SELLWOOD, 1970, p. 494; reworked burrows rarely observed (SCHLOZ, 1968, p. 697; L.Jur., S.Ger.).] Cam.-Tert., cosmop.——FIG. 63,*1. R.* sp.; *1a,* U.Cret., France, ✕0.8 (Abel, 1935); *1b,* L.Cam., Pak., model, ✕0.6 (Seilacher, 1955).

Rosselia DAHMER, 1937, p. 532 [**R. socialis*; M]. Cylindrical pencil-thick burrows, commonly oblique (30° or more) to bedding; lower end not observed; opening expanded and filled with concentric layers of matrix which as a rule are strongly weathered. [According to DAHMER (1937, p. 533), dwelling burrow; interpreted by SEILACHER (1955, p. 389) as feeding burrow, recently regarded by him (1969, p. 122) as a junior synonym of *Asterosoma* VON OTTO, 1854 (see Fig. 25,1).] L.Cam., Asia(Pak.), L.Dev., Eu.(Ger.); *?Penn.,* USA(Okla.); *?Jur.,* Eu.(Ger.); *U.Cret.,* N.Am.(USA, Utah).——FIG. 63,2. **R. socialis,* L.Dev.(low. Taunus Quartzite), Ger.; *2a,* opening, ✕0.5; *2b,* upper end of dwelling burrow with opening, ✕0.5 (Dahmer, 1937).

Rusophycus HALL, 1852, p. 23 [**Fucoides biloba* VANUXEM, 1842, p. 79; "OD"] [=*Rhyssophycus* EICHWALD, 1860, p. 54 *(nom. null.); Rusichnites* DAWSON, 1864, p. 367 *(nom. van.); Rysophycus* DE TROMELIN & LEBESCONTE, 1876, p. 627 *(nom. null.); Rhysophycus* SCHIMPER, in SCHIMPER & SCHENK, 1879, p. 54 *(nom. null.); Rhizophycus* BUREAU, 1900, p. 148 *(nom. van.)*]. Short bilobate bucklelike forms, resembling shape of coffee beans; lobes transversely wrinkled by anterolaterally directed coarse of fine striae; with deep median furrow; outline mostly elliptical; generally width equal to one-half to two-thirds length; bilobate pits deeply excavated or only shallowly dug; quite variable in size and shape (size of Cincinnatian specimens from 1-25 cm.); morphology variable and dependent on mechanics of burrow excavation, and therefore difficult to render an unobjectionable "diagnosis." ["The most famous of all the 'fucoids'" (OSGOOD, 1970, p. 301); originally interpreted as of plant origin; undoubtedly resting excavations made by trilobites digging in sediment to rest there temporarily, interpretation given by DAWSON (1864, p. 365, 366: "for shelter or repose" or "places of incubation"); other less probable interpretations: feeding structures or egg depositories; well-preserved specimens may show imprints of segments, pygidia, pleural spines, and other parts of the trilobite; in several cases (U.Ord., USA, Ohio) the producer of the burrow has been found preserved *in situ* (see *Rusophycus pudicum* HALL with *Flexicalymene meeki* (OSGOOD, 1970, pl. 57, fig. 6)). CRIMES (1970c, p. 114) has shown that several "forms" of *Rusophycus* have restricted time range (U.Cam. or L.Ord.) and thus are usable as guide fossils. Many "species" were

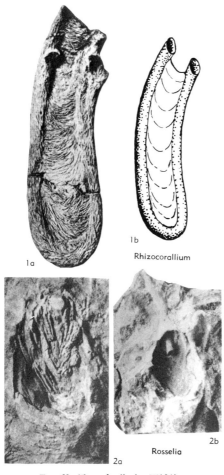

FIG. 63. Trace fossils (p. *W*101).

established only on small differences in shape; for discussion of nomenclatural status of *Rusophycus* see OSGOOD (1970, p. 303); with regard to intermediate specimens between *Rusophycus* and *Cruziana, Rusophycus* was often regarded as synonym of *Cruziana,* but LESSERTISSEUR (1955, p. 45), SEILACHER (1955, p. 366), and OSGOOD (1970, p. 303) recommended *Rusophycus* for the short bilobate resting trails of trilobite origin, and this is approved by the present author. However, SEILACHER (1970, p. 455) recently proposed combining all presumable "resting tracks," "resting nests," and "resting burrows" of trilobites in the one ichnogenus *Cruziana;* for detailed discussions see SEILACHER (1955, p. 358-364) and OSGOOD (1970, p. 301-305).] *U.Precam.-Dev.,* cosmop.——FIG. 63A,*1a.* *R. bilobatus* (VANUXEM), L. Cam., Pak.; ×0.5 (Seilacher, 1955).——FIG. 63A,*1b,c. R. didymus* (SALTER), L.Cam., Pak.; *1b,* ×0.5; *1c,* ×1 (Seilacher,

1955).——FIG. 63A,*1d-f. R. pudicum* HALL, 1852, U.Ord.(Corryville beds), Ohio (Cincinnati area); *1d,* convex hyporelief, ×1; *1e,* the originator of the trace, *Flexicalymene meeki, in situ,* ×0.9 (Osgood, 1970); *1f,* Mt. Hope beds, convex hyporelief associated with 6 specimens of *Lockeia siliquaria,* ×0.9 (Osgood, 1970).——FIG. 63A,*1g. R. carleyi* (J. F. JAMES), loc. unknown; convex hyporelief, ×0.6 (Osgood, 1970).

Sabellarifex RICHTER, 1921, p. 50 [*S. eifliensis;* M] [=*Sabellarites* RICHTER, 1920, p. 226 (*non Sabellarites* DAWSON, 1890, p. 605)]. Similar to *Skolithos,* but individual tubes less straight and not as crowded; never branched. [Regarded by RICHTER (1920, p. 226; 1921) as constructed tubes comparable to those of the Recent annelid *Sabellaria alveolata* LAMARCK; according to WESTERGÅRD (1931, p. 14-15), forms intermediate between *Sabellarifex* and *Skolithos* have been observed in Lower Cambrian of Sweden; "*Sabellarifex dufrenoyi* (ROUAULT)" (=*Tigillites dufrenoyi* ROUAULT) described from Lower Paleozoic of Jordan (HUCKRIEDE in BENDER, 1963, p. 253-254) differs from *Sabellarifex* by its distinct annulation; these forms should be named *Tigillites.*] *L.Cam.,* Eu.(Swed.)-?N.Am.; *L.Dev.,* Eu.(Ger.).——FIG. 64,5. *S. eifliensis,* L.Dev., Ger.; *5a,* ×0.65; *5b,* ×0.6 (Richter, 1921).

Sabellarites DAWSON, 1890, p. 605 [*non* RICHTER, 1920, p. 226] [*S. trentonensis* DAWSON, 1890, p. 608; SD HÄNTZSCHEL, 1962, p. *W*215]. Somewhat tortuous tubes, 1 to 3 mm. in diameter, about 3 cm. long; walls thick, composed of grains of sand and minute calcareous organic fragments cemented by organic substance; some in groups of 2 or more attached together. [Similar to Recent genus *Terebella.*] *U.Precam.,* Eu.(Eng.); *U.Ord.(Trenton.),* Can.; *?M.Dev.,* Eu.(Ger.).

Saerichnites BILLINGS, 1866, p. 73 [*S. abruptus;* M]. Track consisting of 2 parallel rows of semicircular or subquadrate pits, about 15 mm. in diameter; alternating with each other uniformly; somewhat curved in outline on outer margin; anterior and posterior margin nearly straight. [Very tentatively interpreted by BILLINGS as made by mollusks, perhaps cephalopods (see TEICHERT, 1964b, p. *K*487); according to TWENHOFEL (1928, p. 100), also comparable to impressions of fucoids (giant kelp of the North Atlantic).] *Ord.,* Can. (Anticosti).——FIG. 64,7. *S. abruptus,* English Head F.; ×0.1 (Twenhofel, 1928).

Sagittichnus SEILACHER, 1953, p. 115 [*S. lincki;* M]. Trails suggestive of arrowheads with median keel; up to 5 mm. long. [Resting trail; producer unknown, belonging to epipsammons; occurring in masses equally oriented rheotactically.] *U. Trias.(mid.Keuper, Schilfsandstein),* Eu.(S.Ger.); *?Tert.(Oligo.),* Eu.(Pol.).——FIG. 65,5. *S. lincki;* ×2 (Seilacher, 1953b).

Saportia SQUINABOL, 1891, p. XX [*Zonarides striatus* SQUINABOL, 1888, p. 554; M] [=*Saportaia*

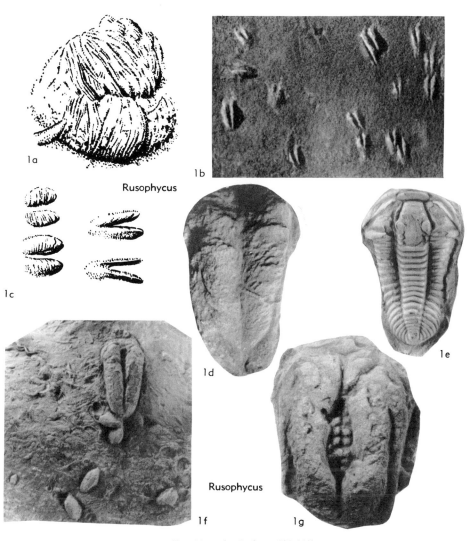

1a

Rusophycus

1b

1c

1d

1e

Rusophycus

1f 1g

FIG. 63A. Trace fossils (p. *W101-102*).

WILCKENS, 1947, p. 47 *(nom. van.); Palaeo-saportia* BORRELLO, 1966, p. 20 (type, *P. loedeli*)]. Long large cylindrical burrows, 1 to 2 cm. in diameter, commonly in dendriform arrangement, branching dichotomously; surface with rhombic pattern produced by delicate arched parallel striations in 2 systems. [Interpreted by RICHTER (in WILCKENS, 1947) as fillings of burrows made by animals and deposited posteriorly after passing through alimentary canal; in German, *"Stopftunnel mit Kotfüllung."*] [Found in flysch deposits.] *L.Tert.(Eoc.),* Eu.(N.Italy).——FIG. 64, 4. **S. striata* (SQUINABOL); ×0.3 (Squinabol, 1891).

[BORRELLO (1966, p. 20) observed that there are only small differences in shape between his *Palaeosaportia* from the Ordovician of South America(Arg.) and *Saportia* from the Tertiary of Italy. In my opinion, a new generic name is not required for such burrows which vary considerably in shape.]

Scalarituba WELLER, 1899, p. 12 [**S. missouriensis*; M]. Subcylindrical burrows, 2 to 10 mm. (max.) in diameter; sinuous; parallel, oblique or nearly vertical to bedding; marked by transverse "scalariform" ridges situated at average distances of 2 to 3 mm., which may be only poorly preserved or lacking in argillaceous rocks. [In "unbelievable abundance" in silty sequences (e.g., the Hannibal F. of Missouri), to be interpreted as internal trail; according to HENBEST (1960, p. B383) and CONKIN & CONKIN (1968, p. 5),

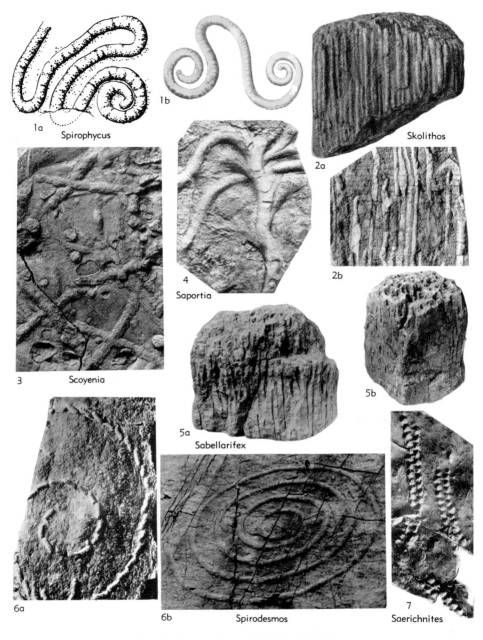

FIG. 64. Trace fossils (p. *W*102-103; 106-108).

made by sediment-eating worm or wormlike organism living in shallow marine, possibly estuarine, or even (CONKIN & CONKIN, 1968) tidal-flat environment; SEILACHER (1964c, p. 309) listed occurrences of this ichnogenus from all of his three trace fossil communities (*Cruziana,* *Zoophycos, Nereites* facies) related to depth, which would mean occurring from epicontinental to geosynclinal environments; SEILACHER & MEISCHNER (1965, p. 615) compared *Scalarituba* with *Nereites* and *Neonereites,* referring to the similar general structure.] *Ord.,* Eu.(Nor.)-USA

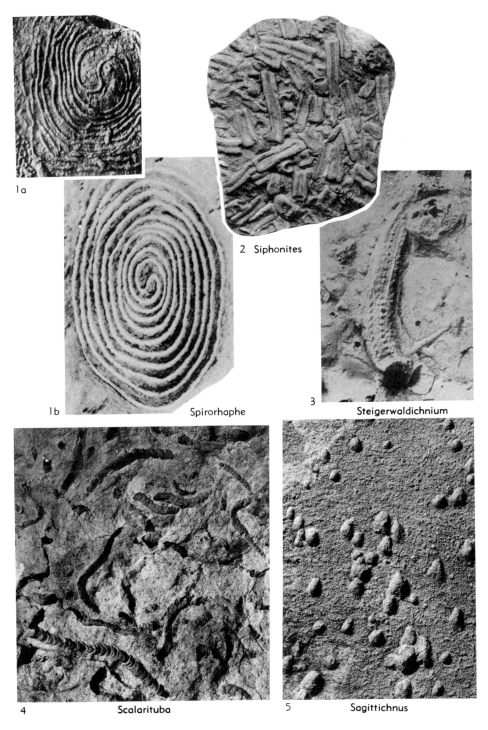

1a

1b   Spirorhaphe

2   Siphonites

3   Steigerwaldichnium

4   Scalarituba

5   Sagittichnus

Fig. 65. Trace fossils (p. *W*102-104, 106, 108-109, 111).

(Ill.); *M.Dev.*, Eu.(Aus.)-USA; *Miss.-Penn.*, USA (Ala.-Ky.-Ind.-Mont.-N.Mexico-Ill.-Ohio-Mo.-Ark.-Okla.-Utah); *Perm.*, Mexico.——Fig. 65,4. **S. missouriensis*, L.Miss.(Kinderhook), USA(Mo.); ×0.8 (Häntzschel, 1962).

Scolicia DE QUATREFAGES, 1849, p. 265 [**S. prisca*; M] [The following ichnogenera belong to the "*Scolicia* group" but are not classifiable as true synonyms: *Nemertilites* SAVI & MENEGHINI, 1850, p. 421 (type, *N. strozzii*); *Nereiserpula* STOPPANI, 1857, p. 334 (no type species designated); *Psammichnites* TORELL, 1870, p. 9 (see p. W98); *Cymaderma* DUNS, 1877, p. 352 (no formal species named; ?jun. synonym of *Psammichnites*); *?Phyllochorda* SCHIMPER in SCHIMPER & SCHENK, 1879, p. 50 (no type species designated); *?Bolonia* MEUNIER, 1886, p. 567 (type, *B. lata*); *Scolithia* KINDELAN, 1919, p. 187 *(nom. null.);* *Palaeobullia* GÖTZINGER & BECKER, 1932, p. 379 (no formal species named); *Subphyllochorda* GÖTZINGER & BECKER, 1932, p. 380 (no formal species named); *Olivellites* FENTON & FENTON, 1937b, p. 452 (type, *O. plummeri*); *Paleobulla* CLINE, 1960, p. 92]. Horizontal bilaterally symmetrical gastropod trails of great variability, long, band-like; morphology depending on their origin as surface trails or internal trails; varied sculpture caused by different methods of burrowing, creeping, and removing sediment; up to about 4 cm. wide; two main types: 1) type species (*Scolicia s. str.* = "group" *Palaeobullia* GÖTZINGER & BECKER, 1932) representing a "true trail" as surface trail of negative epirelief, consisting of variably shaped median axis, ribbonlike or ridge-like, ribbed; lateral parts transversely striated (the striae slanting backward from the midline) or of "gill-like" structure, width larger than or equal to median axis; type species briefly described by DE QUATREFAGES represents this type; 2) internal trails as sole trails, varied full relief burrows ("group" *Subphyllochorda* GÖTZINGER & BECKER, 1932); bandlike, trifid, with varied longitudinal markings; on both sides of the median ribbon characteristic narrow carinate ridges common; both of these types occasionally traceable over great distances. [Originally interpreted by DE QUATREFAGES as long annelid about 2 m. long; now regarded as creeping or feeding trail (or both) of burrowing gastropods; of wide facies range (*Nereites* and *Cruziana* facies); large bedding planes of European flysch deposits furrowed by countless trails of *Palaeobullia* type; nomenclatural treatment of these variable trails difficult (e.g., KSIĄŻKIEWICZ, 1970, p. 289, used *Scolicia* only for the *Palaeobullia* type, and retained *Subphyllochorda* GÖTZINGER & BECKER); for detailed discussion of *Scolicia* see GÖTZINGER & BECKER (1932, p. 377-384; 1934, p. 82-84); AZPEITIA MOROS (1933, p. 9-17); ABEL (1935, p. 219-237); SEILACHER (1955, p. 373-376)]. *Cam.-Tert.*, cosmop.——Fig. 66,1a-i.

Palaeobullia; schem. drawing of different forms (Götzinger & Becker, 1934).——Fig. 66,2a-d. *Subphyllochorda*, schem. drawing of different forms (Götzinger & Becker, 1934).——Fig. 66,2e. *Subphyllochorda* striata, low.Eoc.(Beloveza Beds), Pol. (Lipnica Wielka); ×0.4 (from Książkiewicz, M., 1970, p. 291, in: *Trace Fossils* edited by T. P. Crimes & J. C. Harper, Geol. Jour. Spec. Issue 3, Seel House Press, Liverpool).——Fig. 66,3. **Olivellites plummeri* FENTON & FENTON, Penn. (Cisco F.), Texas; ×0.48 (Fenton & Fenton, 1937b).——Fig. 66,4. **S. prisca, 4a,* low.Eoc. (Beloveza Beds), Pol.(Zubrzyca Gorna); ×0.4 (from Książkiewicz, M, 1970, p. 291, in: *Trace Fossils* edited by T. P. Crimes & J. C. Harper, Geol. Jour. Spec. Issue 3, Seel House Press, Liverpool); *4b,* Eoc.(flysch), Aus., Italy; ×0.4 (Seilacher, 1955).

Scoyenia WHITE, 1929, p. 115 [**S. gracilis*; M] [= *cf. Spongillopsis dyadica* GEINITZ, 1862 (*non* POTONIÉ, 1893, p. 18); *Planolites rugulosus* REINECK, 1955, p. 79; for discussion see REINECK, 1955, p. 81-82]. Slender burrows with ropelike sculpture; 2 to 10 mm. (max. 20) in diameter; in half or full relief or flattened; linear and commonly curved, not branched, often crossing each other; parallel or vertical or oblique to bedding; sometimes showing slight "peristaltic" thickenings; outside covered by fine clustered wrinkles densely arranged; inner structure as on stuffed burrows with backfilling, visible if preserved in full relief. [According to MÜLLER (1969c, p. 926, 927), probably made by same animal (polychaete worm?) or one similar to that making ichnogenus *Tambia* MÜLLER; *Scoyenia*, index trace fossil for "*Scoyenia* facies" (SEILACHER, 1967, p. 415), representing nonmarine sand and shales, commonly red beds.] *Perm.*, Eu.(France-Ger.)-USA(Ariz.).——Fig. 64,3. **S. gracilis*, Hermit Sh., Ariz.; *ca.* ×0.7 (White, 1929).

Siphonites DE SAPORTA, 1872, p. 110 [**S. heberti*; M]. Tubes, several cm. long and about 1 cm. in diameter, with sandy lining, mostly washed out and collapsed on bedding planes. [For detailed description and discussion see GARDET, LAUGIER, & LESSERTISSEUR (1957); regarded erroneously by some authors as synonym of *Palaeophycus* HALL.] *U.Trias.(Rhaet.),* Eu.(France). ——Fig. 65,2. **S. heberti*; ×0.35 (Laugier in Häntzschel, 1962).

Skolithos HALDEMANN, 1840, p. 3 [**Fucoides ?linearis* HALDEMANN, 1840, p. 3; M] [=*Scolithus* HALL, 1847, p. 2 (and all later authors dealing with this "genus" till HOWELL (1943, p. 6) who detected HALDEMANN's spelling *Skolithos*) (type, *S. linearis*); *Scolecolithus* F. ROEMER, 1848, p. 171 *(nom. van.); Scolites* SALTER, 1857, p. 204 (no species name) *(nom. null.); ?Haughtonia* KINAHAN, 1859, p. 119 (type, *H. poecila*)]. "Ordinary pipes"; straight tubes or pipes perpendicular to bedding and parallel to each other,

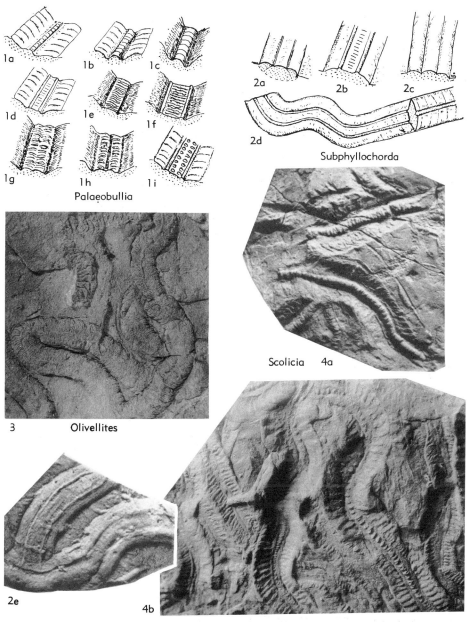

FIG. 66. Trace fossils (p. *W*106).

subcylindrical, unbranched; 1 to 15 mm. in diameter, constant for each tube; few cm. up to 30 cm. (max. 100) long; inner walls may be finely annulated; tubes commonly closely crowded (particularly the *Skolithos* in the Cambrian of Sweden), but also may show widely spaced gradations; frequent in arenaceous sediments; forming Cambrian "pipe rocks" of Scotland. [For a detailed discussion and treatment of ichnogenus see JAMES (1892b), RICHTER (1920, 1921), FENTON & FENTON (1934d), WESTERGÅRD (1931), HOWELL (1943); originally interpreted as marine plants *in situ,* but Scottish occurrences referred to "*Sabella* or other marine worm" by McCULLOCH

(1814, p. 461); interpreted as made by annelids (e.g., NICHOLSON, 1873, p. 288) or by brachiopods (PERRY, 1872, p. 139), phoronids (FENTON & FENTON, 1934d, p. 348), or even as of inorganic origin (e.g., HÖGBOM, 1915). HALLAM & SWETT (1966, p. 104) rejected RICHTER's interpretation as "reefs" built by colonial worms, proposing that *Skolithos* tubes were made during periods of negligible sedimentation by the same animal that produces *Monocraterion* tubes by upward movement due to influx of sand. *Sabellarifex* RICHTER, 1921, regarded by FENTON & FENTON (1934d, p. 344) as synonym of *Skolithos*; other authors (PÉNEAU, 1946, p. 78; SEILACHER, 1969b, p. 118) considered *Tigillites* ROUAULT a synonym of *Skolithos*; this question not yet cleared up, particularly due to cursory descriptions of *Tigillites* and *Skolithos* itself, along with missing type species and figures in the first descriptions. Therefore, in agreement with OSGOOD (1970, p. 326): "at present the genus remains in a state of confusion . . . it is badly in need of a monographic study"; this concerns the "species" of *Skolithos* as well as its synonyms.] *U.Precam., N.Australia; Cam.-Ord., Dev.,* cosmop.; *U.Penn.,* USA(Texas); *U.Carb.,* S.Afr.; *L.Cret.,* USA(Colo.); *L.Jur.,* Greenl.——FIG. 64,2. *S. linearis,* L.Cam.; *2a,* Swed. (Öland), ×0.6; *2b,* Swed., ×0.5 (Westergård, 1931).

Spirodesmos ANDRÉE, 1920, p. 85 [*S. interruptus*; M]. Large spiral-shaped form, diameter up to about 30 cm.; consisting of individual parts 2 to 3 cm. long and up to 10 mm. wide; in outer coils parts are displaced toward interior with respect to each other; *S. archimedeus* HUCKRIEDE (1952) differs from type species by uninterrupted spiral band; type species possibly part of large double spiral such as *Spirophycus*. [Interpreted by ANDRÉE (1920) and HUCKRIEDE (1952) as strings of spawn of gastropods; more likely trace fossil (see PFEIFFER, 1968, p. 674); HÜLSEMANN (1966, p. 455) discussed similarity of *Spirodesmos* to some large Recent trails in the form of coiled or spiral pattern observed on abyssal sea floor of the Pacific and other oceans (BOURNE & HEEZEN, 1965).] *L.Carb.(Kulm),* Eu.(Ger.).——FIG. 64, *6a. *S. interruptus;* ×0.17 (Andrée, 1920).——FIG. 64,*6b.* *S. archimedeus* HUCKRIEDE; ×0.2 (Huckriede, 1952).

Spirophycus HÄNTZSCHEL, 1962, p. W215 [*Muensteria bicornis* HEER, 1877, p. 165; SM HÄNTZSCHEL, 1962, p. W215 (=*Muensteria caprina* HEER, 1877, p. 163; *M. involutissima* SACCO, 1888, p. 168)] [=*Ceratophycos* SCHIMPER in SCHIMPER & SCHENK, 1879, non FISCHER DE WALDHEIM, 1824]. Cylindrical bulges, about 5 to 20 mm. thick, transversely folded or rugose; curved like horns or bent spirally at ends; similar to *Taphrhelminthopsis* SACCO, 1888. [Grazing trail, according to SEILACHER (1962, fig. 1), pre-depositional.] [Found in flysch deposits.] *Miss.-Penn.,* USA(Okla., Ouachita Mts.); *Cret.-L.Tert.,* Eu.(Aus.-Switz.-Spain-Italy-Pol.)-S.Am. (Venez.). ——FIG. 64,*1.* *S. bicornis; 1a,* Eoc., Aus., ca. ×0.4 (Seilacher, 1955); *1b,* Eoc., Switz., ca. ×0.3 (Heer, 1877).

Spirophyton HALL, 1863, p. 78 [*S. typum* HALL, 1863, p. 80; OD (SCHIMPER in SCHIMPER & SCHENK, 1879, p. 55, incorrectly designated *S. cauda-galli* (HALL, 1863, p. 79) as type species; see Art. 68(b) ICZN)] [=*Zoophycos* MASSALONGO, 1855, p. 48 (*partim,* see p. W120); list of true synonyms only possible after thorough monographic treatment of closely related cosmopolitan ichnogenus *Zoophycos*]. Similar to spirally coiled forms of *Zoophycos* but differing by smaller size and by circular outline of laminae (spreite) which are also composed of lamellae; laminae (=whorls) not tending to lobate forms, 1 to 4 mm. thick, sloping outward from axis, then flattening and bent upward to margin in dextrogyrate or sinistral spirals, curving ridges on laminae convex in the sense of the rotation; diameter of last whorl up to about 10 cm., central axis J-shaped. [For older and newer interpretations (plants, inorganic, feeding burrows) see *Zoophycos*; it is difficult to decide which forms described as *Spirophyton* belong to *Zoophycos* and vice versa (*S. cauda-galli* HALL, 1863, to be ascribed to *Zoophycos*); SIMPSON (1970) is correct in regarding *Spirophyton* as a separate ichnogenus, with the name *Spirophyton s. str.* maintained for forms such as *S. eifeliense* KAYSER, 1872 (Dev., Ger.); this "species" placed in the Recent polychaete genus *Spirographic* VIVIANI, 1805, by PLIČKA (1968, p. 843); for discussion of *Spirophyton* see ANTUN (1950) and SIMPSON (1970).] *?Sil.,* N.Afr.; *Dev.,Carb.,* Eu.-Afr.-N.Am.——FIG. 67,*1a-c.* *S. eifeliense* KAYSER; *1a,* schem. (Antun, 1950); *1b,* Dev. (Ems.), Luxembourg; tang. sec., ×0.7 (Antun, 1950); *1c,* L.Dev.(Eifel.), Ger.(Prüm); a sinistral specimen viewed from above, ×0.47 (PliČka, 1968, after Kayser, 1872, pl. 28, fig. 1c).——FIG. 67,*1d.* *Zoophycos crassus* (HALL) ["*Spirophyton crassum*" HALL], U.Dev., USA; ×2.7 (Hall, 1863).

Spirorhaphe FUCHS, 1895, p. 395 [*Helminthopsis involuta* DE STEFANI, 1895, p. 16; SD HÄNTZSCHEL, herein] [=*Gilbertina* ULRICH, 1904, p. 140 (non MORLET, 1888; nec JORDAN & STARKS, 1895) (nom. nud.); *Helminthopsis? concentrica* AZPEITIA MOROS, 1933, p. 46 (see SEILACHER, 1959, p. 1068); *Spiroraphe* of many authors (nom. null.); "*Spirodictyon* ABEL" OSGOOD, 1970, p. 386 (nom. null.)]. Spirally coiled threads, 0.5-3 mm. thick, running from outside inward, with diameter of spiral 5 to about 30 cm., turning at center and looping backward between primary whorls; simple closely coiled spirals not reversing direction at the center have been assigned to

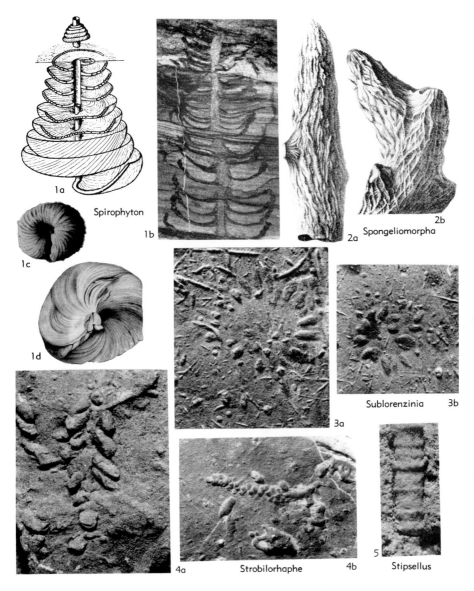

FIG. 67. Trace fossils (p. *W*108-109, 111-112).

Spirorhaphe (e.g., *S. minuta* KSIĄŻKIEWICZ, 1970, p. 305). [Grazing trail; for a new interpretation as a multifloored three-dimensional tunnel, see SEILACHER (1967c, p. 75-76).] [Found in flysch deposits.] *Cret.-L.Tert.,* Eu.(Aus.-Spain-Italy-Pol.-Greece)-N.Am. (Alaska)-?Asia (Japan).——FIG. 65,*1. S.* sp.; *1a,* Tert.(Greifensteiner Ss.), Aus.; *ca.* ×0.5 (Abel, 1935); *1b,* *("Gilbertina"),* U. Cret.(Yakutat F.), Alaska; ×1 (Ulrich, 1904).

Spongeliomorpha DE SAPORTA, 1887, p. 298 [**S. iberica;* M] [=*Spongiliomorpha* DARDER, 1945,

p. 405 *(nom. null.)*]. Thick, elongate burrows, cylindrical; suggestive of antlers; with ramifications and lateral tapering offshoots; surface with network of ?scratching traces crossing each other at acute angles. [Originally regarded as sponges (DE LAUBENFELS, 1955, p. *E*36); according to REIS (1922, p. 231), burrows similar to *Rhizocorallium;* "a rather unsatisfactory ichnogenus" (KENNEDY, 1970, p. 272); most probably arthropod dwelling burrows; for synonymy and discussion of the relations to *Thalassinoides* see

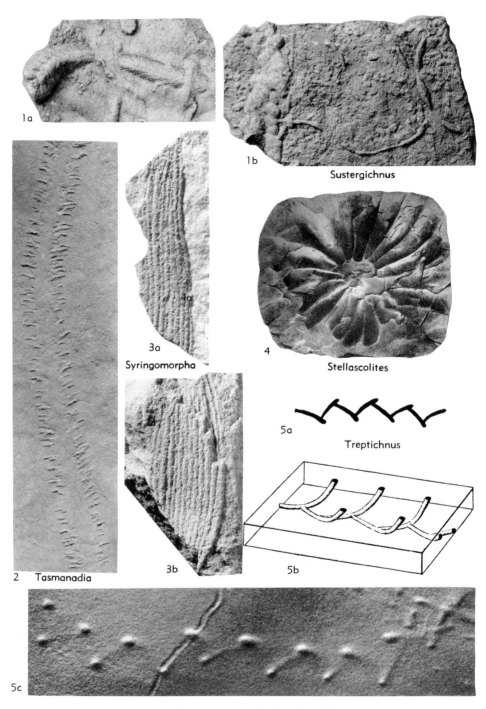

1a

1b

Sustergichnus

3a

Syringomorpha

4a

4

Stellascolites

5a

Treptichnus

2   Tasmanadia

3b

5b

5c

FIG. 68. Trace fossils (p. *W*111-112, 114, 117).

1a

1b

Stelloglyphus

1c

FIG. 69. Trace fossils (p. *W*111-112).

KENNEDY, 1967, p. 150-151.] *Trias.-Tert.,* Eu.; *?Trias.-Tert., L.Cret.,* USA.——FIG. 67,2. **S. iberica,* ?Tert., Spain; *2a,b,* ×0.7 (de Saporta, 1887).

[FÜRSICH (1973, p. 728) considered *Ophiomorpha* and *Thalassinoides* to be subjective synonyms of *Spongeliomorpha.*—CURT TEICHERT.]

Steigerwaldichnium KUHN, 1937, p. 366 [**S. heimi;* M] [=*Steigerwaldichnites* KUHN, 1937, p. 368 *(nom. null.)*]. Straight, rarely curved, tunnel traces parallel to bedding with distinct longitudinal rows of tiny projections and impressions from doubtful parapodia. [Probably made by polychaetes; holotype lost, no other specimens preserved.] *U.Trias.,* Eu.(S.Ger., Bavaria).—— FIG. 65,3. **S. heimi,* M. Keuper; *ca.* ×1.5 (Kuhn, 1937).

Stellascolites ETHERIDGE, 1876, p. 109 [**S. radi-atus;* M]. Radiate or stellate disclike impression with 16 rays of nearly equal length radiating from central round space, becoming broader at their extremities which are not clearly defined; diameter 20 to 25 cm. [Name only very rarely used.] *L.Ord.,* Eu.(Eng.); *?Miss.,* N.Am.(USA, Mont.).——FIG. 68,4. **S. radiatus,* Ord., Eng.; ×0.17 (Etheridge, 1876).

Stelloglyphus VYALOV, 1964, p. 112 [**S. turkomanicus;* OD] [=*Stelleglyphus* VYALOV, 1968, expl. pl. 2 *(nom. null.)*]. Large rosette-like trace fossils, consisting of about 25 very closely spaced "rays," without central smooth field; diameter about 7 cm. *?Penn.,* N.Am.(USA,Okla.); *Perm.,* S.Afr.; *U.Cret. (Santon.-Turon.),* USSR (Turkmenistan-Crimea).——FIG. 69,1a. **S. turkomanicus,* U.Cret.(Turon.), W.Kopet Dag, Turkmen.;

SCOTT W. STARRATT

×1 (Vyalov, 1968).——Fig. 69,*1b,c. S. giganteus* Vyalov, U.Cret.(Turon.), Kopet Dag; ×0.25 (Vyalov, 1968).

Stipsellus Howell, 1957, p. 18 [**S. annulatus*; OD] [=*Stripsellus* Howell, 1957b (correct spelling only in the title of Howell's paper; B. F. Howell, pers. commun., 1957) *(nom. null.)*]. Perpendicular, cylindrical burrows, spaced about 2 cm. apart in sediment, diameter about 1 cm.; differing from *Skolithos* by distinct ringlike expanded belts regularly distributed throughout length of tube; perhaps identical with *Trachyderma serrata* Salter, 1864. *Cam.(Tapeats Ss.),* N.Am.(USA, Ariz.); *?Penn.,* USA(Md.)-?Arabia. ——Fig. 67,*5. *S. annulatus,* Tapeats Ss., Ariz.; ×1 (Howell, 1957b).

Strobilorhaphe Książkiewicz, 1968, p. 8 (Pol.) and 15 (Engl.), [**S. clavata*; OD]. Short narrow string, with 3 to 4 ranges of small pearl-like knobs about 7 mm. long, laterally protruding from string; entire trail loose-coiled, usually 3 to 4 cm. long, 1 to 1.5 cm. wide. [Found in flysch deposits.] *Tert.(low.Eoc.-mid.Eoc.),* Eu. (Pol.).——Fig. 67,*4a. *S. clavata,* low.Eoc. (Beloveza Beds), Pol.; ×1.1 (Książkiewicz, 1968).——Fig. 67,*4b. S. pusilla* Książkiewicz, low. Eoc.(Beloveza Beds), Pol.; ×2.2 (Książkiewicz, 1968).

Subglockeria Książkiewicz, 1974, *nom. subst.,* herein, *pro Asterichnus* Książkiewicz, 1970, p. 310 (*non* Bandel, 1967a) [**Asterichnus nowaki* Książkiewicz, 1970, p. 310; OD] [=*Asterichnus* Nowak, 1961, p. 227, *nom. nud.*]. Rosetted trace, up to 16 cm. in diameter, fairly structureless central area (4 to 6 cm. in diameter surrounded by an aureole of ribs, variable in length, which always point outward; central area may possess a central knob. *U.Jur.(Tithon.)-L.Cret. (Hauteriv.),* Eu.(Pol.). [Description supplied by W. G. Hakes.]

Sublorenzinia Książkiewicz, 1968, p. 10 (Pol.) and p. 15 (Engl.) [**S. plana*; OD]. Similar to *Lorenzinia* da Gabelli; midfield large and flat, encircled by ring of 12 to 20 knobs; diameter 3 to 6 cm.; differing from *Lorenzinia* by irregular (not circular) form of ring and by different shape of knobs, which vary from round to elongate. [Found in flysch deposits.] *U.Cret.(Cenoman.-Turon.),* Eu.(Pol., W.Carpathians).——Fig. 67,*3. *S. plana; 3a,b,* ×0.7 (Książkiewicz, 1968).

Sustergichnus Chamberlain, 1971, p. 231 [**S. lenadumbratus*; M]. Carinate burrows, irregularly sinuous, 1 to 10 mm. wide, 1 to 7 mm. high; numerous fine striae crossing exterior surface obliquely and converging near lower apex; this outer structure not always present; inner structure consisting of sand rod with smooth external surface, almond-shaped in cross section; preserved as hyporelief and full relief. [According to Chamberlain, made by animal pulling itself through the sediment following the sand/mud interface, disturbing it by feeding and pulling, then forming the smooth-walled tunnel when drawing its body forward; inner sand-packing perhaps fecal; fossil named "Arkansas Razorback" by petroleum geologists.] *Miss.-Penn.,* USA (Okla.).——Fig. 68,*1. *S. lenadumbratus,* Ouachita Mts.; *1a,* Penn.(?Johns Valley Sh.), ×1; *1b,* L.Penn.(Atoka F.), ×1 (Chamberlain, 1971a).

Syringomorpha Nathorst, 1886, p. 47 [**Cordaites ?nilssoni* Torell, 1868, p. 36; OD]. Cylindrical sticks several cm. long and 1 to 2 mm. wide lying close together; slightly arched; touching each other along whole length and forming complete slab; occurring in large numbers independent of bedding. [Interpretation difficult; according to Richter (1927b, p. 267), perhaps work of gregarious worms on flat substratum.] *L.Cam.,* Eu. (Swed.-Nor.-N.Ger., Pleist. drift).——Fig. 68,*3. *S. nilssoni* (Torell), L.Cam., drift boulder, Berlin; *3a,b,* ×1 (Richter, 1927b).

Taenidium Heer, 1877, p. 117 [**T. serpentinum* Heer, 1877, p. 117; SD Häntzschel, 1962, p. W218] [The following ichnogenera are not strictly considered as synonyms of *Taenidium* but all are stuffed burrows (German, *Stopftunnel*) that exhibit transverse annulations (some names are invalid and others are less frequently used than *Taenidium*): *Muensteria* von Sternberg, 1833 *(partim); ?Eione* Tate, 1859 (*non* Rafinesque, 1814) *(nom. nud.); Volubilites* Lorenz von Liburnau, 1900; *Pseudocrinus* Anelli, 1935 (*non* Pearce, 1843, *nec* Geinitz, 1846, *nom. nud.); Notaculites* Kobayashi, 1945 (=*Notakulites* Kobayashi, 1945, *nom. null.); Scolecocoprus* Brady, 1947 (=*Scolecoprus* Häntzschel, 1965, *nom. null.); Tebagacolites* Mathieu, 1949; *?Rhizocorallites* Müller, 1955]. Cylindrical burrows with distinct stuffed structure, mostly branched, typical *Taenidium* (*T. fischeri* Heer, 1877) umbellated, rootlike system of burrows radiating downward; burrows with transverse segmentation reminiscent of "*Orthoceras*"; segmentation may also be observed on outside as annular constrictions; similar to *Keckia* Glocker and *(partim) Muensteria* von Sternberg (see p. W75, W84) but commonly smaller. [*Taenidium* was originally interpreted as alga (see Lorenz von Liburnau, 1900, p. 528-567), but originates in feeding burrows by periodic filling of tunnel in backward direction; it occurs in wide range of environments. Stuffed burrows have been discussed by Richter in Wilckens (1947, p. 44-45) and by Toots (1967).] *?Carb., Perm.-Tert.,* Eu.-N.Am.-(Japan)-?N.Z.-Antarctic.——Fig. 70,*1. T.* sp., U.Cret., Aus.; *1a,* ×0.7 (Papp, in Häntzschel, 1962); *1b,* ×0.27 (Seilacher, 1955).

Tambia Müller, 1969, p. 924 [**T. spiralis* (=gen. inc. *spiralis* Müller, 1956a, p. 149); OD]. Spirally coiled structures with circular outline; diameter 2 to 3 cm.; surface covered by

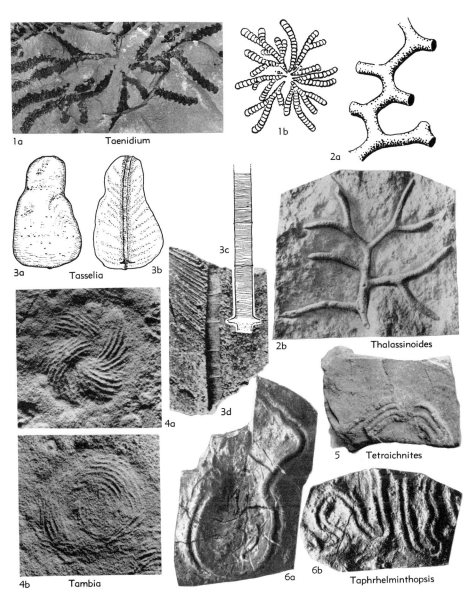

FIG. 70. Trace fossils (p. *W*112-115, 117).

streaks either running parallel to periphery of circular outline or arranged in subspiral fanlike manner; streaks sometimes transversely annulated. [Probably part of feeding burrow dipping into sediment at low angle.] *L.Perm.(up. Rotliegendes),* Eu.(Ger., Thuringia).——FIG. 70,4. **T. spiralis; 4a,b,* ×0.9, ×1.4 (Müller, 1969c).

Taphrhelminthopsis SACCO, 1888, p. 170 [**T. auricularis;* SD HÄNTZSCHEL, 1962, p. *W*218].

Bilobate trails, 1 to 3 cm. wide; mostly very long; morphology varying: more or less straight (*T. recta* SACCO), freely winding (*T. auricularis* SACCO), or even meandering with distinct rather large median furrow 3 to 10 mm. wide, flat; lateral ridges may be transversely striated; trails varying in size and relief. [Most probably gastropod grazing trail, description of *Taphrhelminthopsis* as having tightly coiled spirals and meanders

(Häntzschel, 1962, p. W218) was based on drawing given by Seilacher (1955, fig. 5, no. 76), which does not represent true *T. auricularis* as described and figured by Sacco, 1888, p. 172, pl. 2, fig. 3; see also Książkiewicz (1970, p. 290); coiled spirals or meanders named *Taphrhelminthopsis* have been figured only by Książkiewicz (1968, pl. 6, fig. 3) ("*T. sp. ind.*") and by Müller (1962, p. 16, fig. 12) ("*T. auricularis*").] [Found in flysch deposits.] *Cret.-Tert.*, Eu.——Fig. 70,6*a*. **T. auricularis* (Sacco), low. Eoc.(Beloveza Beds), Pol.(Lipnica Mala); ×0.08 (Książkiewicz, 1970).——Fig. 70,6*b*. *T. convoluta* (Heer), low.Eoc.(Beloveza Beds), Pol. (Sidzina); ×0.08 (Książkiewicz, 1970) (*6a,b,* from Książkiewicz, M., 1970, p. 293, 297, in: *Trace Fossils* edited by T. P. Crimes & J. C. Harper, Geol. Jour. Spec. Issue 3, Seel House Press, Liverpool).

Tasmanadia Chapman, 1929, p. 5 [**T. twelvetreesi*; M]. Double row of very sharp transverse imprints, commonly single but some joined internally or rarely externally to form bifid impressions. [Originally interpreted by Chapman (1929) as bodily preserved worm with its bristles preserved as sets of imprints. Glaessner (1957, p. 103) conclusively proved it to be arthropod track. The age of the Australian representatives of this ichnogenus, originally reported as Cambrian by Chapman, is now known to be Late Carboniferous (Gulline, 1967; Glaessner, 1973b).] *Precam.-Cam.*, India; *U.Carb., ?L.Perm.*, Australia(Tasm.).——Fig. 68,2. **T. twelvetreesi*, U.Carb., Tasmania; part of holotype, ×0.8 (Glaessner, 1957).

Tasselia de Heinzelin, 1965, p. 505 [**T. ordam* (=*T. ordamensis; nom. correct.,* herein); M]. Sideritic and phosphoritic concretions of cylindrical, pyriform or subspherical shape with an axial straight unbranched tube; concretions 3 to 30 cm. long, 2 to 15 cm. wide, primarily found with vertical orientation in fine marine sands and usually occurring in groups; tube 1 to mostly 3 mm. in diameter, with segmentation intervals of 2.5 to 6 mm., each segment exhibits very fine transverse annulation with striae at intervals of about 0.2 mm., tube ending in small flat chamber near bottom of concretion, but lower part of tube often continues more indistinctly downward several decimeters into underlying sediment. [Tubes tentatively interpreted as made by *Pogonophora*.] *L.Pleist.(Merxem.),* Eu.(Belg.).——Fig. 70,3. **T. ordamensis; 3a,b,* concretion and long. sec., ×0.8; *3c,* fine transverse annulation in the tube, diagram. (all van Tassel, 1965); *3d,* holotype, ×1.4 (de Heinzelin, 1965).

Teichichnus Seilacher, 1955, p. 378 [**T. rectus*; M]. *Spreiten-bauten* formed by series of long horizontal burrows stacked vertical to bedding, resembling stacked flat U-shaped roof gutters with pipe at top; wall-shaped laminar body straight

or slightly sinuous; generally not branching; commonly retrusive built but can also be protrusive; up to about 50 cm. long (in M.Cam. of Öland up to 135 cm.), about 10 cm. or more in height. [Endogenic burrows, belonging to fodinichnia; producer unknown, but, due to the very long time range of this ichnogenus, certainly made by different groups of animals; comparable modern structures made by the Recent polychaete *Nereis diversicolor* (see Seilacher, 1957, p. 203); Martinsson (1965, p. 216) explained Cambrian specimens as combinations of retrusive and protrusive digging activity; transitional forms to *Rhizocorallium* (L.Carb., Scotland) were described by Chisholm (1970b); as shown by Sellwood (1970, p. 494), a limb of a vertically retrusive *Rhizocorallium* may be mistaken for *Teichichnus*; tunnels of *Ophiomorpha* have been observed to grade into *Teichichnus*-like structures (Eoc., N.Am., Miss.) (Hester & Pryor, 1972, p. 686); the relationships of *Teichichnus* to *Phycodes* were discussed by Häntzschel & Reineck (1968, p. 26).] *Cam.*, Eu.(Nor.-Swed.-Spain)-N.Am.(USA,Ariz.)-Asia(Pak.); *Ord.*, Eu.(Ger.-Eng.)-N.Am.(Can.)-Asia(Iraq); *U.Dev.*, Eu. (Eng); *L.Carb.*, Eu.(Scot.-USSR); *M.Trias.*, Eu. (Ger.); *Jur.*, Eu.(France-Ger.-Swed.)-Greenl.; *U. Cret.*, USA(Kans.-Utah); *Tert.*, Eu.(Eng.-Belg.).——Fig. 71,4*a,c*. **T. rectus*, L.Cam. (Kusak F.), Pak.(Salt Range); *4a,* model, ×0.4; *4c,* ×0.7 (Seilacher, 1955).——Fig. 71,4*b*. *T. sp.*, Cam. (Tapeats Ss.), Ariz. (Grand Canyon); ×0.7 (Seilacher, 1956).——Fig. 71,4*d*. Large teichichnian burrow, M.Cam., Swed. (Äleklinta, Öland); ×0.3 (Martinsson, 1965).

Teratichnus Miller, 1880, p. 221 [**T. confertus*; M] [=*Tetraichnus* Flower, 1955, p. 857 *(nom. null.)*]. Complex track, sickle-shaped; 17 mm. wide; consisting of numerous bifid imprints, 9 per set arranged in elliptical pattern; in part very confused, probably resulting from rotation of body of animal; 3 sharply defined median grooves visible between the 2 series, indicating medial posterior terminal spine or appendage. [Only type specimen is known; originally interpreted by Miller (1880) as made by cephalopod (see Teichert, 1964b, p. K487); detailed interpretation as crawling track of an unknown small arthropod with bifid dactyls (?trilobite, aglaspid such as *Neostrabops* Caster & Macke, 1952?) given by Osgood (1970, p. 368-369).] *U.Ord. (Cincinnat.),* USA(Ohio).

Tetraichnites de Stefani in de Stefani, *et al.*, 1895, p. 15 [**T. majorianus*; M]. Flexuous trail, 2 cm. wide; consisting of 4 parallel ridges, smooth, 3 mm. wide; 1 to 3 mm. wide furrows between ridges. [Regarded by de Stefani as probably made by crustaceans; placed by Seilacher (1955, p. 374) in group *Scolicia* de Quatrefages, interpreted as creeping trails of burrowing gastropods.] *L.Tert.*, Eu.(Medit., Isle

FIG. 71. Trace fossils (p. *W*114, 117).

of Kárpathos).——FIG. 70,5. *T. majorianus*;
×0.47 (de Stefani in de Stefani *et al.,* 1895).
Thalassinoides EHRENBERG, 1944, p. 358 (*emend.*
KENNEDY, 1967, p. 132) [**T. callianassae*; OD]
[For detailed synonymy of the "species" *T. sax-*

onicus, cf. T. suevicus, and *T. paradoxicus* see
KENNEDY, 1967, and MÜLLER, 1970]. Cylindrical
burrows forming 3-dimensional branching systems consisting of horizontal networks connected
to surface by more or less vertical shafts; burrows

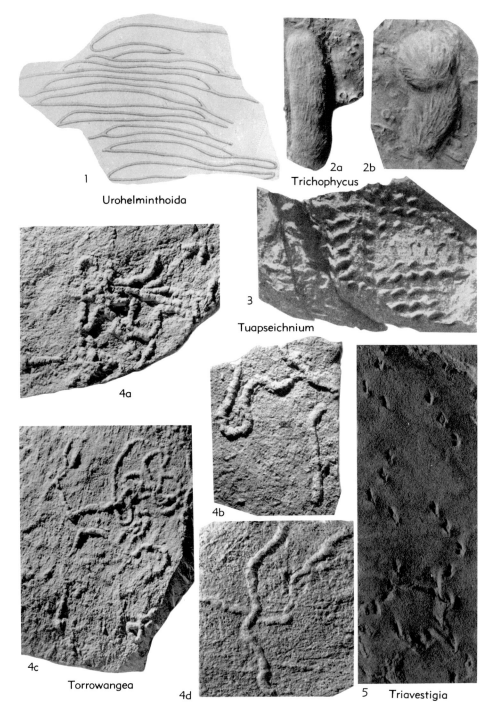

1
Urohelminthoida

2a   2b
Trichophycus

3
Tuapseichnium

4a

4b

4c
Torrowangea

4d

5   Triavestigia

Fig. 72. Trace fossils (p. W117-118, 120).

1 to about 20 cm. (typically 10-15 mm.) in diameter; regularly branching, Y-shaped bifurcations; in horizontal systems forming polygons; typical swellings at points of branching or elsewhere; rare transitional forms with tuberculate structure of *Ophiomorpha* have been described (MÜLLER, 1970b). [Formerly interpreted as algae or horny sponges (Ceratospongidae); undoubtedly feeding and dwelling burrows of crustaceans; sometimes associated with actual remains of callianassids (EHRENBERG, 1938; MERTIN, 1941, GLAESSNER, 1947, MÜLLER, 1970b), described with *Glyphea* crustacean inside burrow (SELLWOOD, 1971); Recent lebensspuren comparable to *T. saxonicus* described from modern burrows of callianassids; producers most likely living in sublittoral environment; burrows and burrow systems in hardgrounds more irregular (lacking widenings, branching) than those in soft chalk; for discussion of filling mechanism of such crustacean burrows (fill channels on their crests) see SEILACHER, 1968, p. 200.] *Trias.-Tert.*, Eu.-Asia(Iraq)-Taiwan-USA (Kans.-Utah)-Australia (Vic.); *L.Jur.*, Greenl.-G.Brit.; *L.Cret.*, USA(N.Mex.-Texas); *U.Cret.(Cenoman.)*, Eu.(Pol.).——FIG. 70,*2a*. *T. sp.*, Mio.(marine molasse), Switz.; *ca.* ×0.07 (Seilacher, 1955).——FIG. 70,*2b*. *T. saxonicus* (GEINITZ), U.Cret.(up. Cenoman.), Ger.(Sax.); ×0.08 (Müller, 1970b).

Tigillites ROUAULT, 1850, p. 740 [**T. dufrenoyi*; SD HÄNTZSCHEL, 1962, p. W218] [=*Foralites* ROUAULT, 1850, p. 742 (type, *F. pomeli*; SD HÄNTZSCHEL, 1965, p. 36)]. Simple vertical burrows without special lining; smooth or regularly annulated; openings may be funnel-shaped; not crowded. [Dwelling burrow; e.g., *Tigillites habichi* LISSON (1904, p. 41) (Jur. or Cret.; S. Am., Peru) is U-shaped burrow with spreite and type species of *Polyupsilon* HOWELL, 1957, p. 151 (according to GOLDRING, 1962, p. 238, junior synonym of *Diplocraterion* TORELL); whether *Tigillites* is to be regarded as a synonym of *Skolithos* HALDEMAN, 1840, or *Monocraterion* TORELL, 1870, has been under discussion for more than a century—see SALTER (1864b, p. 289), BOUČEK (1938, p. 249), PÉNEAU (1946, p. 78), HALLAM & SWETT (1966, p. 103), SEILACHER (1969a, p. 118, 122); thorough studies of many specimens of these three ichnogenera are required before questions of synonymy will be resolved.] *Cam.-Jur.*, Eu. (G.Brit.)-N. Am.-?S. Am. (Arg.)-Antarct.-Arabia; *?L.Cret.*, Eu.(Ger.); *U.Cret.*, USA(Kans.); *?Tert.*, N.Z.——FIG. 71,*2*. *T. sp.*, Ord., France (Normandy); ×0.7 (Haug, 1911).

Tisoa DE SERRES, 1840, p. 6 (*emend.* FREY & COWLES, 1969, p. 21) [**T. siphonalis*; M] [=*?Tissoa* REYNES, 1868, p. 65 *(nom. null.)*]. Vertical U-shaped cylindrical tubes with closely appressed limbs; individual tubes 2 to 3 mm. in diameter, lying 1 to 15 mm. apart, rarely branched; principally form axis of elongated conical concretions

1 m. or more long; basal part of "U" commonly not preserved; burrow walls usually lined, occasionally striated; transitional forms difficult to distinguish from *Arenicolites*. [Dwelling burrow; according to FREY & COWLES (1969, p. 20; 1972) probably made by a shrimp or amphipod-like arthropod rather than by worm; for history of various interpretations (e.g., siphons of pelecypod having extremely degenerate valves, worm burrow, or algal origin), see FREY & COWLES (1969, p. 20); for bibliography see GOTTIS (1954, p. 190).] *Jur.*, Eu.(France-Ger.)-Madagascar; *L.Cret.*, USSR; *Tert.*, N.Afr.(Tunisia)-USA(Wash.-Ore.).——FIG. 71, *3a,b*. **T. siphonalis*, L.Jur. (Lias), France; *3a,b*, *ca.* ×0.7 (de Serres, 1840).——FIG. 71,*3c,d*. *T. sp.*, Eoc.(Numidian), Tunisia; *3c*, individual, *ca.* ×0.7; *3d*, colony, *ca.* ×0.7 (Gottis, 1954).

Torrowangea WEBBY, 1970, p. 99 [**T. rosei*; OD]. Trails, sinuous to meandering, 1 to 2 mm. wide; characterized by crudely transverse annulation produced by irregularly spaced constrictions, mainly at 1 to 4 mm. intervals; trails tending to form random meshwork. *U.Precam.(up.Torrowangeegr.)*, Australia(NewS.Wales).——FIG. 72, *4*. **T. rosei*; *4a,b,d*, paratypes, ×1; *4c*, holotype, ×0.7 (Webby, 1970b).

Trachomatichnus MILLER, 1880, p. 219 [**T. numerosus*; SD MILLER, 1889, p. 454]. Trackway consisting of 2 rows of crowded, poorly defined polydactylous imprints, ?9 to 11 per set; width about 5 to 15 mm.; track straight ahead; no dimorphism in the 2 rows of imprints; morphology of trackway varying along its length resulting from different types of movement. [Tentatively interpreted by MILLER (1880) as made by cephalopod (see TEICHERT, 1964b, p. K487); for detailed discussion and interpretation as trilobite tracks, probably made by *Cryptolithus* GREEN, see OSGOOD (1970, p. 367-368); *T. permultum* MILLER and *T. cincinnatiensis* MILLER lacking sufficient features; according to OSGOOD (1970, p. 362), possibly tracks of *Flexicalymene* SHIRLEY and belonging to *Petalichnus multipartitus* MILLER.] *U.Ord.(Cincinnat.)*, USA(Ohio).——FIG. 71,*1*. **T. numerosus*, Eden Gr.; convex hyporelief, movement from bottom to top, ×0.7 (Osgood, 1970).

Treptichnus MILLER, 1889, p. 581 [**T. bifurcus*; OD] [="Feather-stitch trail" WILSON, 1948, p. 57]. Straight or curved row of short individual burrows of equal length, arranged alternating to right and left, tending upward, resulting in a zigzag featherstitch pattern, comparable to sympodial ramification of plants. [Feeding burrow (SEILACHER & HEMLEBEN, 1966, p. 49).] *L.Cam.* E.Greenl.(*Bastion F.*)-Eu.(N.Nor.)(*Breivik F.*); *Cam.*, USA(Ariz.); *Ord.(Trenton.)*, Can.; *L.Dev.* (*Hunsrück Sh.*), Eu.(Ger.); *L.Carb.*, USA(Ind.); *L.Jur.*, Eu.(Ger.); *L.Cret.(Valang., Bentheim Ss.)*, Eu.(N.Ger.).——FIG. 68,*5*. "Feather-stitch trail" WILSON; *5a,* schem. drawing (Wilson, 1948); *5b,* schem. reconstr.; *5c,* L.Dev., Hunsrück Sh., Ger.; ×1.7 (*5b,c,* Seilacher & Hemleben, 1966).

Triavestigia GILMORE, 1927, p. 32 [**T. ninigeri;* M]. Trackway consisting of 3 rows of [foot] impressions between 2 of which faintly impressed "tail" drag; longer axes of foot markings slightly diagonal to direction of movement, alternating; feet ?unidactyl. [Origin of third row with most distinct imprints dubious; arthropod (?insect) trackway.] *L.Perm.(Coconino Ss.),* N.Am.(USA, Ariz.).——FIG. 72,5. **T. ninigeri;* ×0.6 (Gilmore, 1927).

Trichichnus FREY, 1970, p. 20 [**T. linearis;* OD]. Threadlike, cylindrical burrows, 10 mm. to 35 mm. long; diameter less than 1 mm.; straight or very slightly curved; branched or unbranched; typically vertical but also inclined to bedding plane or horizontal; with distinct walls, commonly lined with diagenetic minerals such as pyrite or rarely calcite. [Possibly combined feeding-dwelling burrow of very small deposit-feeding animal.] *U.Cret.(Niobrara Chalk),* USA(Kans.).——FIG. 73,1. *T.* sp.; ×2.6 (from Frey, R. W., & Howard, J. D., 1970, p. 147, in: *Trace Fossils* edited by T. P. Crimes & J. C. Harper, Geol. Jour. Spec. Issue 3, Seel House Press, Liverpool).

Trichophycus MILLER, & DYER, 1878, p. 1 [**T. lanosus;* M]. Large cylindrical burrows showing slight constrictions, 15 to 25 cm. long, diameter 1 to 3 cm.; floor of burrow ornamented by fine striae radiating from midline; some forms (e.g., *T. venosus* MILLER, 1879) with a few vertically directed secondary branches; backfill structure of burrows similar to *Teichichnus* or *Pennatulites;* type species *T. lanosus* consists of sinuous trails ending (?anteriorly) in buttonlike depression from which radiate fine striae; the ichnogenus better typified by more common ichnospecies *T. venosus* (=*Cyathophycus siluriana* JAMES, 1891). [Interpreted by SEILACHER & CRIMES (1969, p. 148) as feeding burrows probably made by small trilobites (trinucleids?), striation of burrows (=scratches) indicate lateral movement of animals in burrows. For history and interpretation of trace fossil (originally described as alga, later as inorganic in origin), including a very detailed discussion of synonymy, see OSGOOD (1970, p. 346-350). Entire morphology of the two ichnospecies, however, still requires some study; particularly of *T. lanosus,* now regarded by OSGOOD (1970, p. 347) as perhaps "a behavioral variant of the same organism that produced *T. venosum.*"] *Ord.,* Eu. (Ger.-Nor.)-N.Am. (Ohio-Newf.)-Asia (Iraq).——FIG. 72,2a. *T. venosus* MILLER, loc. unknown; ×0.4 (Osgood, 1970).——FIG. 72,2b. **T. lanosus,* U.Ord.(Eden Gr.), Ohio; ×0.2 (Osgood, 1970).

Trisulcus HITCHCOCK, 1865, p. 18 [**T. laqueatus;* M]. Sinuous trail, about 1 cm. wide; consisting

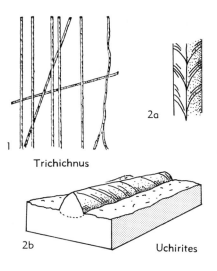

FIG. 73. Trace fossils (p. *W*118).

of 3 continuous grooves with intermediate ridges. [Originally interpreted by HITCHCOCK (1865) as made by annelids; according to LULL (1915, p. 69) perhaps mollusk trail.] *Trias.,* USA(Mass.).

Tuapseichnium VYALOV, 1971, p. 86 [**T. ramosum,* p. 86; OD]. Paired traces occurring as 2 rows of short cylinders that do not touch and give off long free branches. *U.Cret.,* Eu.(?Aus.)-USSR(Caucasus).——FIG. 72,3. **T. ramosum,* Caucasus; ×0.8 (Vyalov, 1971). [Description supplied by CURT TEICHERT.]

Tylichnus OSGOOD, 1970, p. 371 [**Rusophycus asper* MILLER & DYER, 1878a, p. 25; OD]. Weakly bilobate burrow, preserved in convex hyporelief subquadrate in cross section; showing an unusual pustulose ornamentation consisting of 3 to 9 parallel rows of transversely elongated nodes in form of zipper-like pattern; nodes may in addition be distributed randomly over surface. [Uncommon crawling trail.] *U.Ord.(Cincinnat.),* USA (Ohio).——FIG. 74,2. **T. asper* (MILLER & DYER), Eden Gr., Ohio(Cincinnati); *2a,* enl., diagram., ×5; *2b,* several superimposed trails, loc. unknown; ×1.8; *2c,* ×1 (Osgood, 1970).

Uchirites MACSOTAY, 1967, p. 37 [**U. triangularis;* M]. Elevated ribs of triangular cross section; with sharp edge projecting over the bedding plane; about 3 mm. high; both sides very finely striped; both ends gradually tapering. *L.Tert.(Paleoc.),* S.Am.(Venez.).——FIG. 73,2. **U. triangularis; 2a,b,* dorsal view, three-dimensional diag. (after Macsotay, 1967).

Umfolozia SAVAGE, 1971, p. 221 [**U. sinuosa;* OD]. Biserial trackway, 20 to 25 mm. wide, consisting of paddle-shaped impressions, indicating repetition every 4 pairs; cross-interval of first pair smaller than that of fourth pair; between

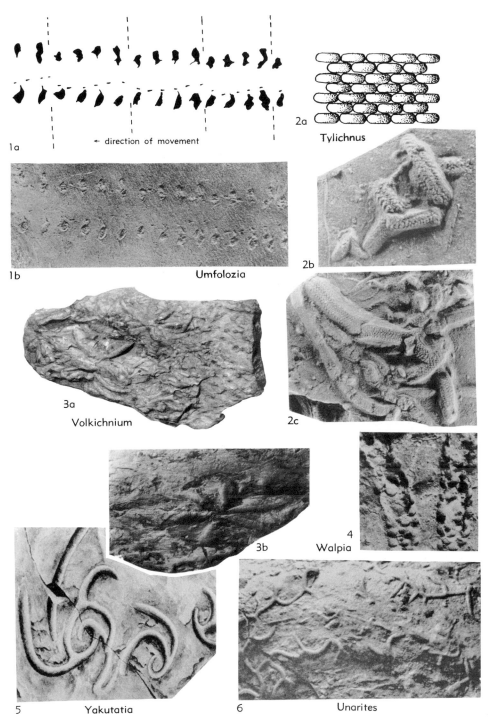

1a

← direction of movement

2a

Tylichnus

1b Umfolozia

2b

3a

Volkichnium

2c

3b Walpia 4

5 Yakutatia

6 Unarites

FIG. 74. Trace fossils (p. *W*118, 120).

the 2 rows of imprints a series of small oval marks [according to SAVAGE, telson marks], closer to one side than the other, arranged in regularly sinuous pattern with 6 marks to each curve. [Belonging to *"Diplichnites* group"; suggested to have been made by crustaceans, perhaps syncarids or peracarids, probably half swimming, half walking, living in freshwater periglacial environment.] *L.Perm.(Dwyka Gr.)*, S.Afr. (N. Natal).——FIG. 74,*1*. **U. sinuosa; 1a,* tracing of trail, ×0.78; *1b,* holotype, ×0.8 (Savage, 1971).

Unarites MACSOTAY, 1967, p. 38 [**U. suleki*; M] [=*Cylindrites submontanus* AZPEITIA MOROS, 1933, p. 44; placed in *Palaeochorda* McCOY by KSIĄŻKIEWICZ (1970, p. 302)]. Very irregular winding and branching trail, may be straight or broadly curving; strings 1 to 3 mm. wide, circular in cross section; commonly with rather short thornlike ramifications. [Grazing trail similar to, perhaps even identical with, *Protopaleodictyon* KSIĄŻKIEWICZ, 1970, and *Acanthorhaphe* KSIĄŻKIEWICZ, 1970; GÓMEZ DE LLARENA (1946, p. 141) regarded *Cylindrites submontanus* AZPEITIA MOROS as an irregular net of *Paleodictyon*.] [Found in flysch deposits.] *L.Tert.(Paleoc.)*, S. Am.(Venez.).——FIG. 74,*6*. **U. suleki;* ×0.6 (Macsotay, 1967).

Urohelminthoida SACCO, 1888, p. 183 [**Helminthoida appendiculata* HEER, 1877, p. 168; SD HÄNTZSCHEL, 1962, p. W219] [=*Hercorhaphe* FUCHS, 1895, p. 395 (no type species designated, no formal species name established)]. Threadlike reliefs forming meanders with tail-like appendage at each turn; forking strings 1 mm. to about 2 mm. thick. [Grazing trail.] [Found in flysch deposits.] *Cret.-L.Tert.*, Eu.(Aus.-Switz.-Italy-Spain-Pol.)-S.Am.(Venez.).——FIG. 72,*1*. **U. appendiculata* (HEER), Eoc., Switz.; ×0.3 (Heer, 1877).

Volkichnium PFEIFFER, 1965, p. 1266 [**V. volki*; M]. Starlike trace fossil, about 5 cm. in diameter; consisting of 6 to 8 tunnel-shaped "rays"; vertical shaft not observed. [Feeding burrow; very similar to *Bifasciculus* VOLK; ?made inside sediment]. *?L.Cam.*, Eu.(N.Nor.); *L.Ord.(Phycodes beds)*, Eu.(Ger.); *?L.Carb.(Kulm)*, Eu.(Ger.).——FIG. 74,*3*. **V. volki, Phycodes* beds, Thuringia; *3a,* holotype, ×0.8; *3b,* ×0.7 (Pfeiffer, 1965).

Walpia WHITE, 1929, p. 117 [**W. hermitensis*; M]. Tunnels lined with flattened, lenticular, smooth pellicles of rather leathery texture; irregularly crowded or imbricated; probably representing excrement packed against walls of burrows. [?Made by worms or crustaceans.] *Perm. (Hermit Sh.)*, USA(Ariz.).——FIG. 74,*4*. **W. hermitensis;* ×0.9 (White, 1929).

Yakutatia HÄNTZSCHEL, 1962, p. W220 [**Gyrodendron emersoni* ULRICH, 1904, p. 140; M] [=*Gyrodendron* ULRICH, 1904, p. 140, obj.(*non*

QUENSTEDT, 1880, p. 797)]. Cylindrical burrows, varying in thickness from 2 to 6 mm.; bifurcating 1 to 3 times, forming 1.7 volutions about acuminate inner extremity; outer end obtuse. [Originally interpreted as of plant origin, undoubtedly trace fossil.] *U.Cret.(YakutatF.)*, N.Am.(Alaska).——FIG. 74,*5*. **Y. emersoni* (ULRICH); ×0.5 (Ulrich, 1904).

Zoophycos MASSALONGO, 1855, p. 48 [Type species questionable; "genus" first published by MASSALONGO, in 1851, p. 39 (without description), and founded on *Zonarites? caputmedusae* = *Zoophycos caputmedusae* MASSALONGO, 1855, p. 48; PLIČKA (1968, p. 840) regards *Fucoides circinnatus* BRONGNIART, 1828, p. 83, as type species; TAYLOR (1967, p. 4), "*Zoophycus laminatus* SIMPSON" (*nom. nud.*); other authors, *Fucoides brianteus* VILLA, 1844, p. 22] [Due to lack of a thorough monographic treatment of the "genus," its confused nomenclature, and the many discussions of it still in flux, it is impossible to establish a list of valid synonyms; several of the following genera and species are certainly synonyms but some of them will probably be retained as separate ichnogenera if the "genus" is subsequently subdivided (see SIMPSON, 1970, p. 506): *?Umbellularia longimana* FISCHER DE WALDHEIM, 1811, p. 31; *Zoophycos* MASSALONGO, 1851, p. 39 (*nom. nud.*); *Chondrites scoparius* THIOLLIÈRE, 1858, p. 718; *Taonurus* VON FISCHER-OOSTER, 1858, p. 41 (*partim*) (type, *Fucoides brianteus* VILLA, 1844, p. 22); *Spirophyton* HALL, 1863, p. 78 (*partim*, for discussion see SIMPSON, 1970, p. 506, and herein, p. W108); *Sagminaria* TRAUTSCHOLD, 1867, p. 46 (=*Umbellularia longimana* FISCHER DE WALDHEIM, 1811, p. 31); *Alectorurus* SCHIMPER, 1869, p. 203 (type, *Fucoides circinnatus* BRONGNIART, 1828, p. 83); *?Physophycus* SCHIMPER, 1869 (*partim*) (type, *Caulerpites marginatus* LESQUEREUX, 1869, p. 314); *Zoophycus* SCHIMPER, 1869, p. 210 (and several subsequent authors) (*nom. null.*); *Cancellophycus* DE SAPORTA, 1872, p. 126 (type, *Chondrites scoparius* THIOLLIÈRE, 1858, p. 718); *?Glossophycus* DE SAPORTA & MARION, 1883, p. 103 (type, *G. camaillae*); *?Flabellophycus* SQUINABOL, 1890, p. 198 (type, *F. ligusticus* SQUINABOL); *Zoophicos* VASSOEVICH, 1953, p. 41 (*nom. null.*); *Palaeospira* PLIČKA, 1965, p. 1 (type, *P. ensigera*); *Spirographis carpatica* PLIČKA, 1968, p. 843 (*Spirographis*= Recent genus!); *Palaeospirographis* PLIČKA, 1962, p. 359 (type, *P. hrabei*) (regarded by PLIČKA (1968, p. 840) as synonym of *Zoophycos*)]. Complex spreiten structures with numerous morphological variations; divided into 2 basic forms: 1) helicoidal, and 2) flat or planar. Shallowly-conical, spiral form, consisting of 3 main parts: spirally coiled spreite (=lamina, plate), major and minor lamellae contained within the lamina, and a cylindrical tunnel (marginal and axial);

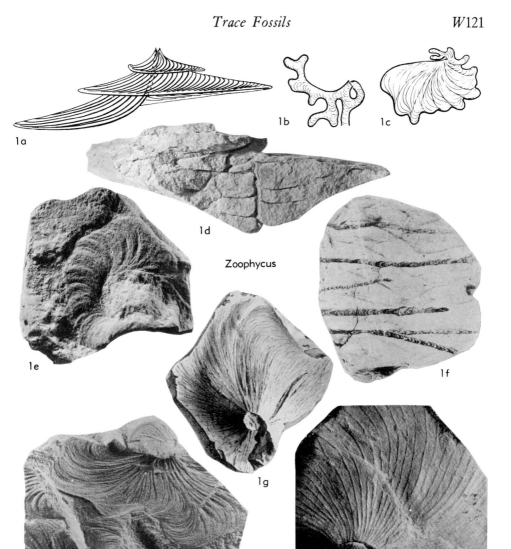

1a   1b   1c

Zoophycus

1d

1e   1f

1g

1h   1i

Fig. 75. Trace fossils (p. *W*120-122).

axis of spiral vertical to bedding; height small; single volutions conelike, sloping outward; diameter of successive whorls generally increasing downward; occasional inverse direction of coiling; basal diameter of structure (particularly in flysch deposits) up to 60 cm. or more (max., 1.45 m., Tert., N.Z.); whorls comprising lamina variable in outline: circular, arcuate, or lobate (broadly based or tonguelike); occasionally first volutions lobate and larger and deeper ones nearly circular in outline; laminae exhibit major and minor lamellae (ridges), appear lunate in cross section, and curve radially from axis of spiral; major lamellae branch at acute angle toward axis forming minor lamellae; cylindrical tunnel with axial and marginal part forms the axis of spreite, has same thickness as spreite, may continue for a part or for whole length of lamina and then may be open to sediment at both ends. Planar forms of *Zoophycos* similar to closed spiral spreite, may also be antler-like; thickness 1 to 7 mm. [One of the most discussed problematic fossils; originally interpreted as imprints of marine algae, later as body fossils (sponges, corals), as of inorganic origin (produced by eddy currents); as trace fossils (by ABEL, 1935; SEILACHER, 1954; LESSERTISSEUR, 1955, and others), tentatively regarded as feeding burrows made by soft-bodied wormlike animals, produced by systematic helicoid mining and foraging through sediment which shifted lobes of burrow (SEILACHER, 1967c, p. 80); other interpretation as imprints of discarded prostomial parts of sedimentary polychaetes (Sabellidae) (PLIČKA, 1962, and especially 1968, 1969) accepted by only few authors; for new interpretation as of plant origin see PLUMSTEAD (1967); "no single interpretation has yet found general acceptance" (TAYLOR, 1967, p. 11) and "much remains to be discovered" (SIMPSON, 1970, p. 505) as may be seen from many controversial discussions during recent years; for Recent "*Zoophycos* burrows" from great depth of the Pacific see SEILACHER, 1967b (cross section with lunate lamellae as in fossil *Zoophycos,* pl. 1, fig. E); complex spiral forms mostly in deep *Nereites* facies (SEILACHER, 1967b, p. 421), flat forms typical of *Zoophycos* facies, but also occasionally in neritic or even shallower marine environment (OSGOOD, 1970, p. 403); nomenclature very confused, the "cauda galli" (*Spirophyton cauda-galli* HALL, 1863) according to SIMPSON (1970, p. 506) undistinguishable from *Zoophycos*; BISCHOF (1968) proposed to restrict name

to *Z. brianteus* (VILLA), for other proposals see TAYLOR (1967, p. 19); for the older history see BARSANTI (1902); (see also BISCHOF, 1968; SIMPSON, 1970; LEWIS, 1970).] Ord.-Tert., cosmop. ——FIG. 75,*1a. Z. crassus* (HALL) ["*Spirophyton crassum*" HALL], U.Dev., USA; schem. drawing (Sarle, 1906b).——FIG. 75,*1b,c,f. Zoophycos;* *1b,* schem. drawing, antler-shaped form, ×0.08 (Seilacher, 1959); *1c,* schem. drawing, regular spiral form, ×0.05 (Seilacher, 1959); *1f,* Tert. (probably Mungaroa Ls. = Kaiwhata Ls.), N.Z.; ×0.5 (Webby, 1969b).——FIG. 75,*1d,h, Z. circinnatus* (BRONGNIART), Czech. (Carpath. flysch); *1d,* Paleoc., long. sec., imprint of prostomial lobe with gill rays, ×0.3; *1h,* Eoc., spiral imprint of gill organs, ×0.25 (Plička, 1968).——FIG. 75,*1e. Z.* sp., Cret., Czech.(Carpath. flysch); planar imprint of an uncoiled spiral of the gill rays; ×0.27 (Plička, 1968). FIG. 75,*1g. Z. brianteus* (VILLA), Eoc., Italy; ×0.4 (Massalongo, 1855). ——FIG. 75,*1i.* "*Zoophycos,*" prob. up. Mio.(up. Tongaporutan beds), N.Z.(Gower R.); dextral specimen, ×0.08 (Stevens, 1968) (from Plička, M., 1970, p. 367, in: *Trace Fossils* edited by T. P. Crimes & J. C. Harper, Geol. Jour. Spec. Issue 3, Seel House Press, Liverpool).

[BRADLEY (1973, p. 118-122) has proposed that *Zoophycos* could have been produced by the feeding activities of a sea pen or similarly related animal. Such an animal would be positioned so that its calyx remained in a relatively stationary position near the sediment water interface and its tubular rhachis protruded into the sediment, free to move. The volution and accompanying lateral movement of the rhachis would account for the characteristic spiral structure of *Zoophycos.*—W. G. HAKES.]

BORINGS

Borings in shells, bones or other hard parts of invertebrates or vertebrates, in sedimentary rocks or in wood, occupy a special position among trace fossils, which entitles them to a chapter of their own. Borings are known as far back as the early Paleozoic, and may be produced by plants or by animals. Those of plant origin are made by algae, fungi, or lichens. Within the animal kingdom, boring organisms are known from the following groups: Porifera, Bryozoa, Phoronidea, Sipunculidea, Polychaeta, Turbellaria, Brachiopoda, Gastropoda, Amphineura, Bivalvia, Cephalopoda, Arthropoda (Isopoda, Amphipoda, Insecta), Cirripedia, Echinoidea, and perhaps also Foraminiferida.

In the fast few years, Recent and, especially, fossil borings have attracted much

interest among paleontologists. New and important publications are by BROMLEY (1970, with many bibliographical references), BOEKSCHOTEN (1966, 1967), and CAMERON (1969b). For additional papers, one may refer to CARRIKER, SMITH, & WILCE (1969). These papers were presented at the International Symposium *Penetration of calcium carbonate substrates by lower plants and invertebrates,* which was held in Dallas in 1968. However, as the title suggests this symposium was restricted to borings and their producers only in calcareous substrates. In this symposium, CARRIKER & SMITH (1969, p. 1012) introduced the following concepts:

Calcibiocavitology. The study dealing with the hollowing out of spaces in hard, calcareous substrata by organisms.

Calcibiocavite. An organism that hollows out a space (burrow, borehole, caries) in hard, calcareous substrata (*calciphytocavite,* plant; *calcizoocavite,* animal).

Calcibiocavicole. An organism inhabiting a self-excavated space in a hard, calcareous substratum.

Calcicavicole. An organism inhabiting a space excavated by another organism or by nonbiogenic forces in a hard, calcareous substratum.

Communities of boring organisms in lithified sediments have been named *lithophocoenoses* by Radwański (1964).

According to Martinsson's (1970, p. 326) suggested toponomic (stratinomic) classification of trace fossils, most borings in lithified sediments have to be placed in his Endichnia, although some can be classified as Exichnia. However, these concepts are not applicable to borings in pebbles or hard parts of organisms such as shells. If some general terms should be needed, Martinsson (1970, p. 328) proposed the terms Ichnidia or Endichnidia for these types of borings.

For the paleontologist, it is generally difficult, if not impossible, to find the creator of the boring and often it cannot be determined if a boring is of plant or animal origin. Gatrall and Golubic (1970), with the help of stereoscan pictures of Jurassic and Recent material, have been the first to find characteristics that make it possible to distinguish between algal and fungal borings. In some cases, it is not even clear whether an organism found in a boring is the actual borer (e.g., borings containing shells). Even for many Recent borings, it is still unknown if they are due to chemical or mechanical processes. Likewise, it is often not certain what purpose the borings serve. Most were made as dwelling chambers, but in other cases, such as borings made by predatory snails, they were made in the search for food.

Some fossil borings have been given the names of their supposed makers (e.g., *Cliona cretacea*). I agree with Bromley (1970, 1972) who has emphasized that borings should not be given the name of the actual or supposed boring organisms. Especially names of Recent borers should be used only for the organisms themselves. For their borings special ichnological names are necessary, as Bromley (1970) showed on the example of the application of the name *Entobia* Bronn to all borings produced by sponges. Also, his new proposal (Bromley, 1972) is aiming in the same direction, e.g., that all pocket-shaped borings with a single opening should be named *Trypanites* Mägdefrau (Fig. 76).

Recently Hölder (1972) published an excellent study of endozoans and epizoans on belemnite rostra. Hillmer & Schulz (1973) have described Upper Cretaceous polychaete borings, some of which possess secondary excavated cavities. These secondary cavities have been interpreted as brood chambers. The authors believed that the presence and absence of these cavities with relation to overall boring size is an expression of sexual dimorphism of the borings' producers (see *Ramosulichnus,* p. *W*131).

Evolution of the boring habit in Recent gastropod taxa can be traced only as far back as the Upper Cretaceous. Borings attributable to predation in pre-Upper Cretaceous and Paleozoic brachiopod and some mollusc shells do not exhibit the tapering sides and countersunk features characteristic of gastropod borings. Their origin is considered unknown. This situation has been given particular attention by Carriker & Yochelson (1968) with respect to cylindrical borings in Middle Ordovician brachiopod shells. If these and other Paleozoic borings are considered to be the work of predatory gastropods, then a boring habit for gastropods must have evolved independently long before the ancestors of the present day groups appeared in the geologic record (Sohl, 1969, p. 733).

Abeliella Häntzschel, 1962, p. *W*228 [**A. riccioides* Mägdefrau, 1937, p. 60; OD] [=*Abeliella* Mägdefrau, 1937, p. 60, *nom. nud.,* estab-

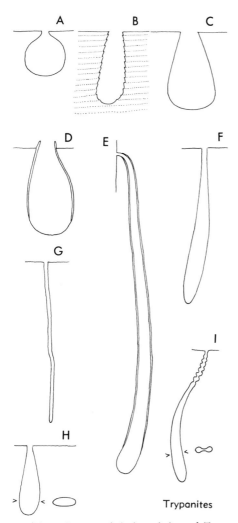

FIG. 76. Various morphologic variations of *Trypanites (A to I)*. *B* shows boring in wood; *D* and *E* show calcareous lining (mod. from Bromley, 1972).

Trypanites

lished without designation of type species]. Dichotomously branching starlike borings in fish scales; width of individual borings 4 to 8 microns, of whole system 0.25 to 0.5 mm. [?Produced by algae or fungi.] *U.Jur.-Oligo.,* Eu. (Ger.-Eng.).——FIG. 77,6. **A. riccioides* MÄGDEFRAU, Oligo., Ger.; (in fish scale), ×110 (Mägdefrau, 1937).

Anobichnium LINCK, 1949, p. 185 [**A. simile*; OD]. Smooth cylindrical perforations in fossil wood, 1 to 1.5 mm. in diameter, with numerous openings to each gallery; very similar to borings of Recent beetles of genus *Anobium*. *U.Trias.,*

Eu.(Ger.).——FIG. 78,5. **A. simile,* Keuper, Ger.; (in wood), ×0.7 (Linck, 1949a).

Bascomella MORNINGSTAR, 1922, p. 156 (*emend.* CONDRA & ELIAS, 1944, p. 538; *emend.* ELIAS, 1957, p. 390) [**B. gigantea*; OD]. First described as ctenostome parasitic, boring bryozoan characterized by large egg-shaped "vesicles" connected by narrow tubular "stolons." [CONDRA & ELIAS (1944, p. 538) doubted that interpretation, classed genus *incertae sedis* and pointed out great similarity of the "vesicles" to borings, particularly to the immature development of *Caulostrepsis* CLARKE; ELIAS (1957, p. 390) restricted diagnosis of *Bascomella* to oval to pear-shaped "vesicles" which he regarded as excavations and compared with borings made by Recent cirriped *Alcippe*; according to ELIAS (1957), "stolon"-like part of *Bascomella* should be placed in *Condranema* BASSLER, 1952; for discussion of combination of two borings of different origin see BROMLEY (1970, p. 58); see BASSLER (1953, p. G36, fig. 9,5).] *Penn.-Perm.,* USA(Pa.-Ohio).

Brachyzapfes CODEZ & DE SAINT-SEINE, 1958, p. 706 [**B. elliptica*; OD]. Short and broad borings; longitudinal cross section elliptical; depth half length; observed in belemnoids and pelecypods. [Borings of barnacles.] *L.Cret.,* Eu.(France)-Antarct.——FIG. 77,3. **B. elliptica*; schem. drawings, *3a-d,* long. sec., opening, tang. sec. (max.), chamber (Codez & de Saint-Seine, 1958).

Calcideletrix HÄNTZSCHEL, 1962, p. W222 [**C. flexuosa* MÄGDEFRAU, 1937, p. 57; OD] [=*Calcideletrix* MÄGDEFRAU, 1937, p. 57, *nom. nud.,* established without designation of type species]. Cavity systems in belemnoids; one or more openings, shrublike, ramified; sometimes dendritic networks of tunnels; diameter of branches 0.02 to 0.1 mm. [Probably made by algae. MARCINOWSKI (1972) interpreted this form as result of hard substrate borers in abandoned belemnite rostra.] *Jur.-U.Cret.,* Eu.(Eng.-Ger.-Pol.).——FIG. 77,4a. *C. breviramosa* MÄGDEFRAU; (in *Actinocamax*), ×8 (Mägdefrau, 1937).——FIG. 77,4b. **C. flexuosa* MÄGDEFRAU; (in *Belemnitella*), ×8 (Mägdefrau, 1937).

Calciroda MAYER, 1952, p. 455 [**C. kraichgoviae*; M]. Cylindrical boring tunnels up to 1 mm. wide; usually constructed parallel to outer surface in shells of mollusks or in stalk members of *Encrinus*; may be ramified, cutting through or crossing each other. [According to MÜLLER (1956b, p. 410) and present author, probably identical with *Trypanites* MÄGDEFRAU (p. W136).] *M.Trias.(Trochiten-Kalk),* Eu.(Ger.).

Caulostrepsis CLARKE, 1908, p. 169 [**C. taeniola*; M] [=*Polydorites* DOUVILLÉ, 1908, p. 365 ("genus" without species name; according to BATHER (1910), not intended as an independent generic name)]. U-shaped tunnels with spreite, corresponding to tiny *Rhizocorallium*, sometimes radiating inward from commissure of brachiopods;

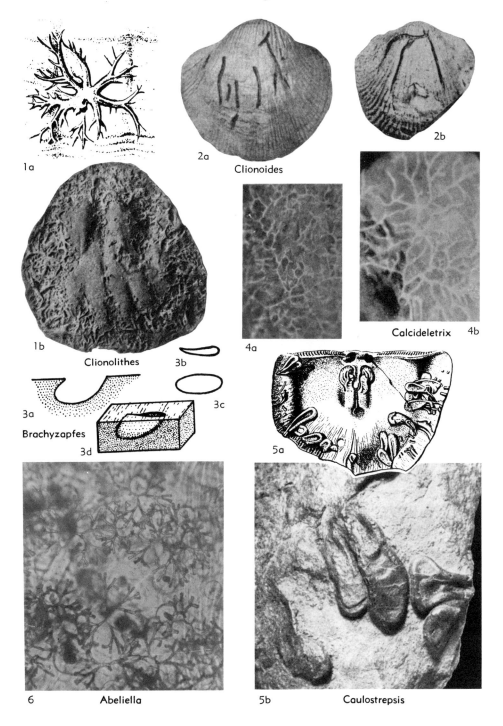

1a

2a
Clionoides

2b

1b
Clionolithes   3b

4a   Calcideletrix   4b

3a
Brachyzapfes
3c
3d   5a

6   Abeliella   5b   Caulostrepsis

Fig. 77. Borings (p. *W*124, 126-127).

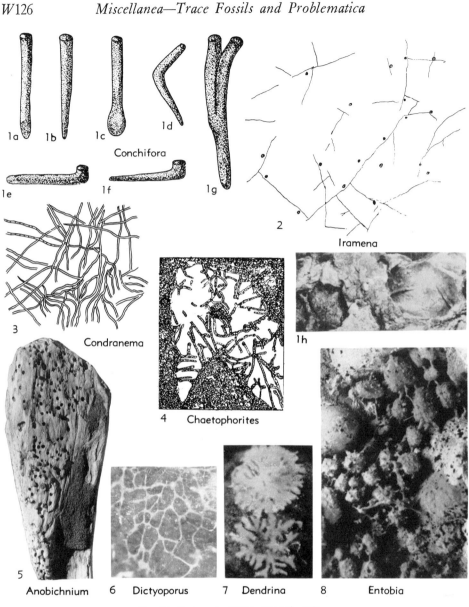

FIG. 78. Borings (p. *W*124, 126-127, 129).

up to 2 cm. long and 5 mm. wide; commonly found in shells of brachiopods, mollusks, and echinoids. [Interpreted tentatively as borings of worms (Spionidae) (CAMERON, 1969b); according to BROMLEY (1970, p. 50), possibly not true borings but embedment cavities; named "pseudoborings" by CONDRA & ELIAS (1944, p. 549) and thus placed in "Problematica."] *L.Dev.,* Eu. (Ger.); *Penn.-Perm.,* USA; *U.Trias., ?L.Jur.,* Eu. (Eng.), *Tert.,* Eu.(Port.)-Australia.——FIG. 77,5. **C. taeniola,* L.Dev., Ger.; *5a,* in shell of

Stropheodonta, ×0.75 (Clarke, 1908); *5b,* up. Ems., Ger.(Taunus), *ca.* ×1.9 (Häusel, 1965).

Chaetophorites PRATJE, 1922, p. 301 [**C. gomontoides;* M]. Ramifying tunnels in rostra of belemnoids and shells of brachiopods and mollusks; usually straight; diameter less than 0.02 mm.; located close to surface of shell. [Probably made by algae or (as supposed by BROMLEY, 1970, p. 55) fungi; according to E. VOIGT (pers. commun., 1971), *C. cruciatus* MÄGDEFRAU, 1937, belongs to boring bryozoans.] *Jur.-Tert.(Plio.),*

Eu.——Fig. 78,4. **C. gomontoides,* L.Jur.(Lias δ), Ger.; (in pelecypod shell), ×106 (Pratje, 1922).

Clionoides Fenton & Fenton, 1932, p. 47 [**C. thomasi;* OD]. Tubular borings, widely spaced, somewhat flexuous or straight, irregularly branched; 0.5 to 1.5 mm. in diameter; round perforations extending throughout length of tubes. Generally excavated in brachial valves of thick-shelled specimens of *Atrypa.* [Origin by sponges related to Recent *Cliona* has been suggested; regarded by Jux (1964) as produced presumably by polychaetes living in commensalism with brachiopods; according to Elias (1957, p. 381), *Clionoides* is possibly junior synonym of the bryozoan genus *Vinella* Ulrich, 1890 (Bassler, 1953, p. G35).] *U.Dev.,* Eu.(Ger.)-USA(Iowa).——Fig. 77,2. **C. thomasi,* Dev., Iowa; upon the brachial valve of *Atrypa waterlooensis* Webster; *2a,* tubes, *2b,* tubes and perforations, ?×1 (Fenton & Fenton, 1932).

Clionolithes Clarke, 1908, p. 168 [**C. radicans;* SD Fenton & Fenton, 1932, p. 43] [=*Pyritonema?* *gigas* Fritsch, 1908, p. 10 (non M'Coy, 1850); *Olkenbachia* Solle, 1938, p. 156 (type, *O. hirsuta*); for discussion see Teichert, 1945, p. 202]. Bent or cracked borings, generally radiating in one plane to all sides from very small central cavity; commonly branching dichotomously; diameter several mm.; always etched into shell or some host animal. [Made by sponges (e.g., *C. querens* Ruedemann, 1925, p. 38), algae, or worms; according to Jordan (1969), certain astrorhizae of Stromatoporoidea (M.Dev.,Ger.) are morphologically identical to *C. radicans* and might be made by parasitic boring organisms; nomenclature of the "species" not yet resolved; *C. reptans* Clarke, 1908, and similar forms may be placed in the "genus" *Filuroda* Solle, 1938; see also de Laubenfels, 1955, p. E40.] *Ord.,* Eu.(Czech.); *Dev.-Carb.,* Eu.(Ger.)-USA-China. ——Fig. 77,1. **C. radicans,* U.Dev.(Chemung Ss.), USA; *1a,* in *Atrypa* shell, ×6; *1b,* in shell of *Dalmanella superstes,* ×0.5 (Clarke, 1921).

Conchifora Gisela Müller, 1968, p. 68 [**C. zylindriformis zylindriformis;* OD]. Variously shaped straight or slightly sinuous tunnels in shells of brachiopods, seldom in pelecypods or gastropods; not branched; walls smooth; commonly with 1 or rarely 2 openings; sometimes with enlarged ends or conical; ends rounded or somewhat acute; 1 to 30 mm. long, diameter 0.1 to 1.4 mm., diameter of openings 0.2 to 0.5 mm.; seven "varieties" named, but these names are unavailable (*Code,* Art. 15). [?Made by polychaetes.] *L.Dev.(mid.Siegen.-low.Ems.),* Eu.(W. Ger.).——Fig. 78,1. **C. zylindriformis; 1a-h,* infillings of seven "varieties"; *1a-c,* ×7; *1d,* ×5; *1e,f,* ×4.7; *1g,* ×9.5; *1h,* ×1.3 (Gisela Müller, 1968).

Conchotrema Teichert, 1945, p. 203 [**C. tubulosa;* OD]. Narrow tubular borings in shells (diameter about 0.2 mm), communicating with surface, but buried completely in shell of host; straight or gently curved; branching. [Probably made by worms; according to Teichert (1945), *Clionolithes canna* Price, 1916, may also be placed in this "genus."] *U.Dev.,* USA(N.Y.); *L.Carb.,* Eu.(Scot.); *Miss.-Penn.,* USA(Ark.-W. Va.); *Perm.,* W.Australia; *?U.Cret.,* Eu.(Eng.). ——Fig. 79,7. **C. tubulosa,* Perm. (Wandagee F.), W.Australia; in *Taeniothaerus* valve, ×2 (Teichert, 1945).

Condranema Bassler, 1952, p. 381 [**Heteronema capillare* Ulrich & Bassler, 1904, p. 278; OD] [=*Heteronema* Ulrich & Bassler, 1904, p. 278 (type, *H. capillare*) (non Dujardin, 1841)]. Straight or somewhat curved cylindrical borings in shells, immersed tunnels very close to surface of shell; zooecial scars present. [Interpreted as creeping stolons of ctenostome bryozoan; see Bassler, 1953, p. G35.] *Ord.-Perm.,* Eu.(Swed.)-USA.——Fig. 78,3. **C. capillare* (Ulrich & Bassler), Sil., Gotl.; ×10 (Ulrich & Bassler, 1904).

Dendrina Quenstedt, 1848, p. 470 (published without species name) [**Talpina dendrina* Morris, 1851, p. 87 (=*Dendrina belemniticola* Mägdefrau, 1937, p. 55); SD Häntzschel, 1965, p. 30]. Borings just below surface in brachiopods and in rostra of belemnoids; without aperture; forming rosettes 1.5 to 6 mm. in diameter; ramifying intensely and irregularly; diameter of borings about 0.05 mm. [?Made by algae. Radwański (1972) interpreted *Dendrina* as result of hard substrate borers on abandoned belemnite rostra.] *Ord.,* (Pleist. drift), Eu.(Ger.); *M.Trias. (low. Muschelkalk),* Eu.(Ger.); *U.Cret.,* Eu.(Eng.-France-Ger.-Pol.).——Fig. 78,7. *D. belemniticola* Mägdefrau, U.Cret., Ger.; in *Belemnitella,* ×5 (Mägdefrau, 1937).

Dictyoporus Mägdefrau, 1937, p. 55 [**D. nodosus;* M]. Borings in rostra of belemnoids; without exterior aperture; distinctly netlike; canals about 0.07 mm. wide. [Producer unknown.] *?L.Jur.,* Eu.(S.Ger.); *M.Jur.,* Eu.(Pol.); *U.Cret.,* Eu.(Eng.-Ger.-Pol.).——Fig. 78,6. **D. nodosus,* U.Cret., Ger.; in *Belemnitella,* ×5 (Mägdefrau, 1937).

Electra Lamouroux, 1816 (see Bassler, 1953, p. G157) [Borings of this Recent cheilostome bryozoan have been observed in bivalve and gastropod shells from Pliocene of Belgium (Boekschoten, 1966, p. 366; 1967, p. 322); Recent species *Electra monostachys* (Busk) lives in brackish environments and tidal flats.]

Entobia Bronn, 1838, p. 691 [**E. cretacea* Portlock, 1843, p. 360; SD Häntzschel, 1962, p. *W*230]. Borings consisting of globular chambers (max. diam. about 1 cm.), mostly crowded, connected by very short slender canals (diam. 0.1-1.0 mm.); walls of chambers with few small surface

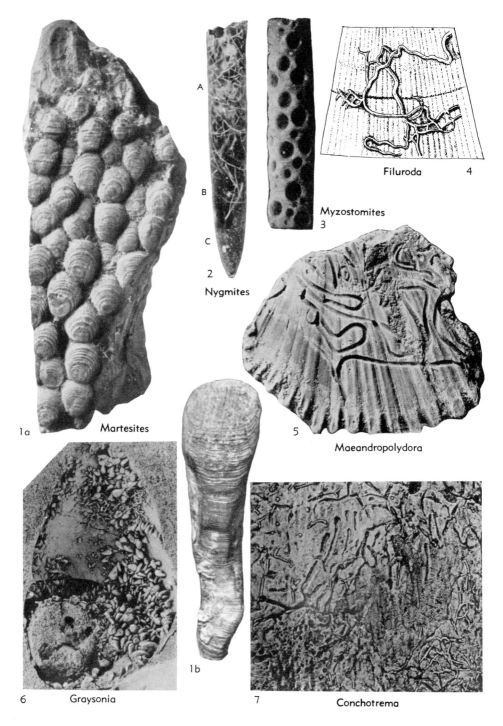

Fig. 79. Borings (p. *W*127, 129-131).

pores and penetrated by canals and by more slender holes, in steinkern preservation appearing as spines radiating from chambers; occurring in brachiopods, pelecypods (particularly *Inoceramus*), ammonites, and belemnites. [Made by sponges of family Clionidae; for selected synonymy of the type species (known as *Cliona cretacea* since 1854), see BROMLEY, 1970, p. 78; borings in trilobites of Silurian age also placed in this genus by PORTLOCK, 1843 *(E. antiqua).*] *?Sil.*, Eu. (Ire.); *U.Jur.(low.Tithon.)-Tert.*, Eu.(Ire.-Eng.- France-Ger.-Pol.).——FIG. 78,8. **E. cretacea* PORTLOCK, Cret.(Chalk Rock, Turon.), Eng. (Herts.), ×2.7 (Bromley, R. G., 1970, p. 81, in Trace Fossils edited by T. P. Crimes & J. C. Harper, Geol. Jour. Spec. Issue 3, Seel House Press, Liverpool).

Filuroda SOLLE, 1938, p. 158 [**Clinolithes reptans* CLARKE, 1908; OD] [=*Clionolithes* CLARKE, 1908, p. 168 *(partim)* (type, *C. radicans*); for discussion see TEICHERT, 1945]. Threadlike, strongly curved borings in shells, running closely below surface of shell. [Possibly made by boring sponges; see also DE LAUBENFELS, 1955, p. E40.] *L.Dev.-M.Dev.*, N.Am.(USA)-Eu.(Ger.)——FIG. 79,4. **F. reptans* (CLARKE), L.Dev.(Oriskany Ss.), USA; in *Leptostrophia, ca.* ×2 (Clarke, 1908).

Graysonia STEPHENSON, 1952, p. 52 [**G. berg- quisti*; OD]. Borings in shells of pelecypods and gastropods, preferentially in thicker shells; "zoarium" consisting of a compound system of "tubular stolons" and "vesicles (internodes)"; "stolons" rather irregularly distributed, forming connected series of little arches which may form complicated meshworks, "vesicles" irregularly subovate, ranging in size from microscopic to 4.5 mm., often crowded together but also widely scattered, intermingled with the "stolons." [Interpreted as a boring bryozoan of the family Vinellidae, perhaps living commensally rather than parasitically; regarded by BROMLEY (1970, p. 58) as a compound genus ("mixture of acrothoracican borings and thread borings") appearing to include worm borings or embedment traces; according to E. VOIGT (pers. commun. 1971), certainly no boring bryozoan; *Graysonia anglica* CASEY (1961, p. 573) from the Aptian of England represents only the "stolon"-like part of the fossil.] *L.Cret.(Apt.)*, Eu.(Eng.); *U.Cret.(Cenoman., Woodbine F.)*, USA(Texas).——FIG. 79,6. **G. bergquisti*, Cenoman.(Woodbine F.), Texas; holotype, in *Gymnentome valida* (gastropod), ×1.5 (Stephenson, 1952).

Iramena BOEKSCHOTEN, 1970, p. 45 [**I. danica*; OD] [=*?Terebripora antillarum* FISCHER, 1866, p. 300; for discussion see BOEKSCHOTEN, 1970, p. 45]. Diminutive borings of *"Penetrantia"*-type in oyster shells, gastropods *(Buccinum)*, and coral branches producing irregular network of long stolon tunnels 3 microns wide; reniform or circular apertures with diameters of 0.03 to 1.0 mm.,

situated in alternating positions laterally and at distance of 0.01 to 0.1 mm. from stolon tunnels; apertures spaced 0.5 to 2.5 mm. from each other. [Probably made by ctenostome bryozoans.] *L. Tert.(mid.Dan.)*, Eu.(Denm.); *U.Tert.(Plio.)*, Eu. (Belg.); *Rec.*(Neth.-France-Ire.).——FIG. 78,2. **I. danica*, Dan., Denm.; camera lucida tracings of the type zoarium borings of *Iramena*, ×11 (Boekschoten, G. J., 1970, p. 46, in: *Trace Fossils* edited by T. P. Crimes & J. C. Harper, Geol. Jour. Spec. Issue 3, Seel House Press, Liverpool).

Maeandropolydora VOIGT, 1965, p. 204 [**M. decipiens*; OD]. Long, meandering furrows sunk into outer or inner side of Cretaceous oysters and pectinids; width 0.5 to 1.2 mm.; resembling U-shaped tubes of *Polydora* but without spreite. [Probably made post-mortem by polychaete worms of family Spionidae.] *U.Cret.*, Eu.(Ger.-Neth.- Swed.).——FIG. 79,5. *M. sulcans* VOIGT, U.Cret. (L.Santon.), W.Ger(Gross Bülten); in *Neithea quinquecostata* (SOWERBY), ×1.8 (Voigt, 1965).

Martesites VITÁLIS, 1961, p 6, 16 [**M. vadaszi*; M]. Very closely crowded borings of pelecypods (probably *Martesia* sp.) in driftwood, lying approx. 45° oblique to the annual rings; clayey fillings of borings with circular rills produced by animal's boring activity; 5 to 7 cm. in length, diameter of opening of boring 1.0 to 1.5 cm. *L.Tert.(low.Mio.-mid.Mio., low.Helvet.)*, Eu.(N. Hung.).——FIG. 79,1. *M.* sp., low.Mio., Hung.; *1a*, in wood (concentric stripes on steinkerns of boreholes correspond to given rings of the wood), ×0.4; *1b*, in wood, steinkern of a borehole, ×1 (Abel, 1935).

Mycelites ROUX, 1887, p. 246 [**M. ossifragus*; M]. General ecologic name for various irregularly branching tunnels about 2 to 6 microns wide in hard parts (shells, bones, teeth, scales) of invertebrates and vertebrates. [According to BERNHAUSER, 1953, 1962, made by green algae; interpreted by W. J. SCHMIDT, 1954, as borings of fungi; for detailed discussion, see PEYER, 1945; for similar borings in Paleozoic fossils from freshwater sediments, BYSTROW, 1956, used name *Paleomycelites*.] *?Sil.*, *Carb.-Rec.*, cosmop.—— FIG. 80,3. *M. conchifragus* SCHINDEWOLF; *3a*, U.Jur.(up.Volg.), USSR (Moscow); hyphae of fungi in the dissolved shell of *Craspedites* sp. cf. *C. okensis* (D'ORBIGNY), ×30 (Schindewolf, 1963); *3b*, L.Jur., Ger.; in *Coroniceras rotiforme* (SOWERBY), schem. reconstr. [*a*, horizontal borings parallel to overlying layer; *b*, layer containing cavities almost perpendicular to overlying layer; *c*, horizontal borings parallel to underlying layer] (Schindewolf, 1962).

Myzostomites CLARKE, 1921, p. 58 (published without species) [**M. clarkei*; SD HOWELL, 1962, p. W167]. Gall-like protuberances on crinoid stems, with a central perforation. [Compared with modern worm *Myzostomum* causing similar

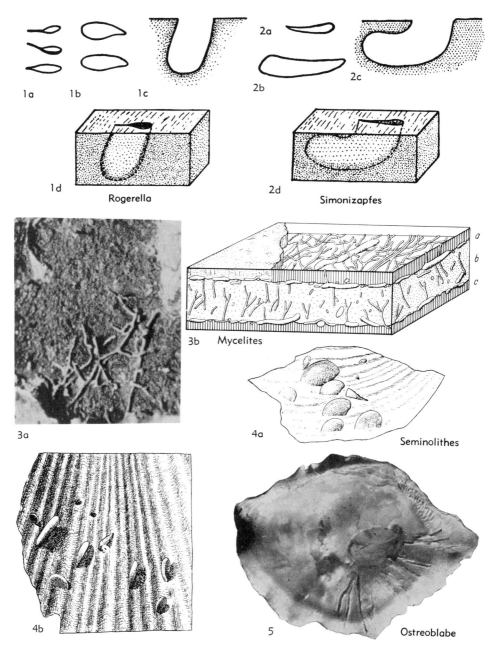

1a   1b   1c   2a   2b   2c

1d   Rogerella   2d   Simonizapfes

3b Mycelites

3a   4a   Seminolithes

4b   5   Ostreoblabe

FIG. 80. Borings (p. *W*129, 131, 133).

cysts or swelling on its host; see HOWELL (1962), p. *W*167.] *Ord.-Perm.,* USA. [According to HOWELL (1962, p. *W*167), also found in Trias.-Jur. and cosmop.]——FIG. 79,3. *M.* sp. CLARKE, Carb., loc. unknown; x1 (Clarke, 1921).

Nygmites HÄNTZSCHEL, 1962, p. *W*230 [**Talpina solitaria* VON HAGENOW, 1840; OD] [=*Talpina* VON HAGENOW, 1840 *(partim)* (type, *T. ramosa*; SD HÄNTZSCHEL, 1962, p. *W*231); *Nygmites* MÄGDEFRAU, 1937, p. 56, *nom. nud.,* established without designation of type species]. Simple, unbranched tunnels in rostra of belemnoids; oblique to surface, open to exterior, leading from outside inward. [Type species perhaps made by algae or fungi; according to E. VOIGT (pers. commun., 1971), *Nygmites pungens* (QUEN-STEDT) (=*Talpina pungens* QUENSTEDT) probably identical with boring bryozoan *Spathipora prima* VOIGT, 1962.] *L.Jur.*, Eu.(Ger.); *U.Cret.*, Eu.(France-Ger.-Pol.-USSR).——FIG. 79,2. *N. solitarius* (VON HAGENOW), U.Cret., Ger.; in *Belemnitella mucronata*; *A, Talpina cf. T. ramosa* VON HAGENOW; *B, N. solitarius*; *C, Terebripora pungens* (QUENSTEDT), ×0.87 (Voigt, 1972b).

Ostreoblabe VOIGT, 1965, p. 200 [*O. perforans*; OD]. Tubes in shells of Cretaceous oysters, sunk into shell material; straight or slightly curved; directed centripetally toward muscle scar, proceeding from round external opening perforating shell; resembling mud blisters of Recent oysters; [Obviously made by parasitic polychaete worms and representing *intra vitam* deformation of shell.] *U.Cret.(Turon.-Santon.)*, Eu.(W.Ger.). ——FIG. 80,5. *O. perforans*, mid.Turon., W. Ger.; in *Lopha semiplana* SOWERBY, ×1.4 (Voigt, 1965).

Palaeachlya DUNCAN, 1876, p. 210 [*P. perforans*; M]. Small tubes, average diameter 0.2 mm., usually straight, rarely flexuous; running inward in all directions to surface or parallel to it; sometimes branching. [Interpreted as made by parasitic algae; observed particularly in corals.] *Sil.-Dev.; Tert.*; Eu.-N.Am.(Can.)-Australia.

Palaeosabella CLARKE, 1921. p. 91 [*Vioa prisca* McCOY, 1855, p. 260; M] [=*Paleosabella* CLARKE, 1921, p. 91 *(nom. null.)*; *Paläosabella* SOLLE, 1938, p. 157 *(nom. null.)*]. Possible synonym of *Topsentopsis* DE LAUBENFELS, 1955, p. E4.

Palaeopede ETHERIDGE, 1899, p. 127 [*P. white-leggei*; M]. Borings in *Favosites*; consisting of chains of moniliform cells(?); longest chain 0.5 mm. [Referred to an endophytic alga similar to *Nostoc*.] *Dev.*, Australia(NewS.Wales).

Palaeoperone ETHERIDGE, 1891, p. 97 [*P. endophytica*; M]. Pinshaped, straight, tubular, tapering to ?distal end, ?proximal end inflated into globular chamber; observed in *Stenopora crinita* LONS-DALE, occurring in matted clusters and irregularly arranged bundles. [Tentatively interpreted as fungi.] *Perm.*, Australia(NewS.Wales).

Paleobuprestis HÄNTZSCHEL, 1962, p. *W*230 [*P. maxima* WALKER, 1938, p. 138; OD] [=*Paleobuprestis* WALKER, 1938, p. 138, *nom. nud.*, established without designation of type species]. Channels under bark of *Araucarioxylon arizonicum*; diameter 2 to 10 mm.; recognizable all around tree; channels resembling work of Recent buprestids. *Trias.*, USA(Ariz.).——FIG. 81,*1a*. *P. maxima* WALKER, Chinle F., Petrified Forest

Natl. Mon.; ×0.54 (Walker, 1938).——FIG. 81,*1b*. *P. minima* WALKER, Chinle F., Petrified Forest Natl. Mon., ×0.5 (Walker, 1938).

Paleoipidus HÄNTZSCHEL, 1962, p. W230 [*P. perforatus* WALKER, 1938, p. 140; OD] [=*Paleoipidus* WALKER, 1938, p. 140, *nom. nud.*, established without designation of type species]. Tunnels and burrows penetrating heartwood of *Araucarioxylon arizonicum* (see also *Paleobuprestis* and *Paleoscolytus*); diameter 2 to 5 mm.; boring near bark or through wood. *Trias.*, USA (Ariz.).

Paleoscolytus WALKER, 1938, p. 139 [*P. divergus*; M]. Channels under bark of *Araucarioxylon arizonicum*; diameter 5 mm.; running in all directions; not filled with castings; resembling channels of Recent bark beetles of family Scolytidae. *Trias.*, USA(Ariz.).——FIG. 81,*3*. *P. divergus*, Chinle F., Petrified Forest Natl. Mon.; ca. ×0.7 (Walker, 1938).

Penetrantia SILÉN, 1946 (see BASSLER, 1953, p. G37). [Recent boring bryozoan; the genus *Penetrantia* has been based on the anatomy of the producer, and not the morphology of the boring; VOIGT & SOULE (1973) described first fossil Cretaceous species *P. gosaviensis* from Upper Cretaceous, Austria(Gosau); according to BOEKSCHOTEN (1970) name should not be applied to fossil borings of *Penetrantia* type, thus BOEKSCHOTEN (1966, 1967) described such borings from the Pliocene, Netherlands, and Pleistocene, Europe (England), as "*Penetrantia*."]

Podichnus BROMLEY & SURLYK, 1973, p. 363 [*P. centrifugalis*; OD]. More or less compact groups of pits or cylindrical holes in hard, calcareous substrates; pits at center of group more or less perpendicular to surface, more peripheral pits typically deeper and larger, entering substrate obliquely, centrifugally; size of pits up to ca. 200 mμ. [Recent examples are produced by brachiopod pedicles; see BROMLEY (1970, p. 61).] *L.Cret.-U.Cret.*, Eu.(Eng.-Swed.-Ger.); *Rec.*, Eu. (Nor., N.Sea).——FIG. 60,6. *P. centrifugalis*, U. Cret., Eng.; ×80 (Bromley & Surlyk, 1973). [Description supplied by R. G. BROMLEY.]

Pseudopolydorites GŁAZEK, MARCINOWSKI, & WIERZBOWSKI, 1971, p. 441 [*P. radwanskii*; OD]. U-shaped burrows without spreite; limbs rather closely spaced, circular in cross section, highest parts near openings somewhat curved; 3 to 5 cm. long; 6 to 8 mm. in diameter. [Borings in hardgrounds.] *U.Cret.(low.Cenoman.)*, Eu.(Pol.).—— FIG. 81,*5*. *P. radwanskii*, Sudót; *5a*, ×0.7; *5b,c*, showing two sides of the same section of "*Potamilla*" type B containing opening of boring, ×0.7 (Głazek, et al., 1971).

Ramosulichnus HILLMER & SCHULZ, 1973, p. 9 [*Polydora biforans* GRIPP, 1967; OD]. Long, unbranched, commonly weakly curved borings, gradually expanding distally and ending blindly;

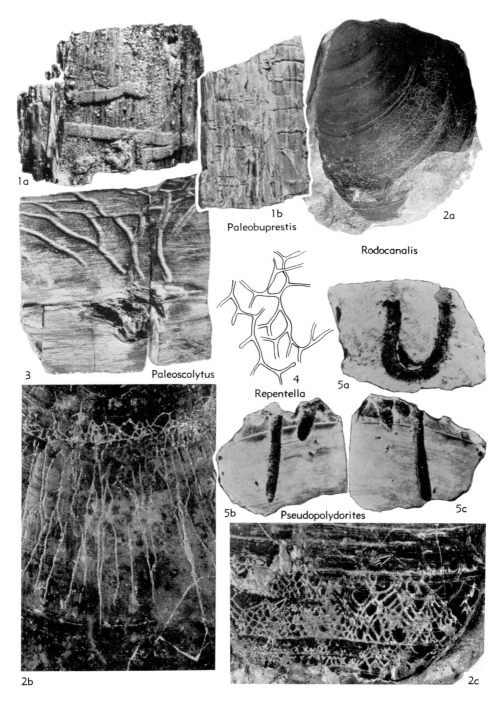

FIG. 81. Borings (p. *W*131, 133).

about 1 to 4 mm. in diameter and 1 to 5 cm. long; small aperture commonly connected to surface by numerous, diverging grooves or furrows; transverse section is nearly circular near aperture becoming oval to weakly dumbbell in outline toward the distal end; specimens representing the larger members of the genus have 4 rows of dimple-like chambers while the smaller ones do not. [Type species divided into two morphological groups which were thought by the authors to express the sexual dimorphism of the producers. The cavities near the apertural end of the larger borings were interpreted as brood chambers produced by female polychaetes while the smaller borings without these secondary cavities were interpreted to have been produced by males (see p. *W*123); produced in belemnite rostra.] *U.Cret.* (*Santon.-low. Maastricht.*), Eu. (Denm.-Eng.-N. Ger.-USSR). [Description supplied by W. G. HAKES.]

Repentella GISELA MÜLLER, 1968, p. 86[**R. maior*; OD]. Netlike arranged tunnels in shells of brachiopods (e.g., *Spirifer, Stropheodonta*), forming irregular polygons; tunnels straight or slightly sinuous, branched; walls smooth; 0.2 to 0.5 mm. in diameter, enlarging to 2.5 mm. on ramifications. *L.Dev.(mid.Siegen.)*, Eu.(W.Ger.).——FIG. 81, 4. *R. fragilis*; schem. drawing, ×2.7 (Gisela Müller, 1968).

Rodocanalis SCHLOZ, 1972, p. 164 [**R. reticulatus*; OD]. Netlike pattern of grooves on outer surface of pelecypod shells; grooves nearly as deep as wide, about 0.1 to 0.3 mm., which do not seem to connect with other borings. [Interpreted as being produced by etching.] *L.Jur.(Hettang.-Sinemur.)*, Eu.(Ger.).——FIG. 81,2. **R. reticulatus* on *Plagiostoma giganteum* SOWERBY, Hettang.-Sinemur. boundary, S.Ger.; *2a*, ×0.6; *2b*, upper part, *R. reticulatus*; lower, *R. sp.*, ×2.2 (Schloz, n, I.G.P. Stuttgart, cat. no. S.1121); *2c*, arrows point to *Talpina ramosa* HAGENOW, ×3 (Schloz, 1972). [Description supplied by W. G. HAKES.]

Rogerella DE SAINT-SEINE, 1951, p. 1053 [**R. lecontrei*; OD] [=*Rodgerella* NEWMAN, ZULLO & WITHERS, 1969, p. R252, R272 (*nom. null.*)]. Very deep borings of barnacles; cross section short and broad; observed in shells of corals, brachiopods, bivalves, gastropods, and echinoids. *Perm.*, USA(Texas). [According to BROMLEY, 1970, p. 69, *Clionites mantelli* WETHERELL, 1852, is identical with *Rogerella mathieui* DE SAINT-SEINE, 1956.] *M.Jur.-U.Cret.*, Eu.(Eng.-France-Ger.-Pol.)-USA; *Tert.(Mio.)*, Eu.(France); *Tert. (Plio.)*, Afr.(Morocco).——FIG. 80,1. *R. mathieui* DE SAINT-SEINE, Cret., France; *1a,b*, schem., various kinds of openings and tang. secs.; *1c*, long. sec.; *1d*, chamber (Codez & de Saint-Seine, 1958).

Seminolithes HYDE, 1953, p. 215 [**S. linii*; M]. Thin lenticular cavities in shells of brachiopods; usually perpendicular or inclined to surface, rarely

almost parallel to it; shape and size variable, similar to flax seed; 2 mm. long, 0.3 to 0.5 mm. wide; somewhat resembling *Caulostrepsis* and *Bascomella*. [Producer unknown.] *Miss.*, USA (Ohio-Okla.-?Mo.).——FIG. 80,4. **S. linii*, Logan F., Ohio; *4a,b*, sediment-filled borings from shell of *Spirifer striatiformis*, both ×1 (Hyde, 1953).

Simonizapfes CODEZ & DE SAINT-SEINE, 1958, p. 704 [**S. elongata*; OD]. Long, narrow borings of barnacles; length (max.) 4.5 mm., width (max.) 1.1 mm.; shallow; observed in hard parts of corals, oysters, gastropods, belemnoids and other fossils. *Jur.*, Eu.(Eng.-France-Ger.-Pol.).——FIG. 80,2. **S. elongata*, France; *2a-d*, schem., opening, tang. sec. (max.), long. sec., chamber (Codez & de Saint-Seine, 1958).

Spathipora FISCHER, 1866, p. 986 (see BASSLER, 1953, p. G37). [Borings of bryozoans; according to BOEKSCHOTEN (1970, p. 44) and BROMLEY (1970, p. 57), to be regarded as ichnogenus. The taxonomy is confused; a few Recent species have been erected on strength of anatomical criteria only, others are based on pattern of their borings system.]

Specus STEPHENSON, 1952, p. 51[**S. fimbriatus*; OD]. Small club-shaped borings in shells of gastropods and pelecypods *(Breviarca, Ursirivus)*, straight, curved, or irregular in trend, circular in cross section, diameter increasing to rounded end (from 0.2 to 0.75 mm.); maximum length about 8 mm. [Possibly made by sponges, questionably referred to Clionidae, commensal rather than parasite?; interpreted by VOIGT (1970, p. 377) as made by worms or wormlike organisms; perhaps not borings, according to BROMLEY (1970), who noted that distribution and orientation resemble embedment structures.] *U.Cret.(Cenoman., Woodbine F.)*, USA(Texas).——FIG. 82,3. **S. fimbriatus*; ferruginous casts of sponge borings in shells of bivalve mollusks (shell substance removed in solution), ×3 (Stephenson, 1952).

Stichus ETHERIDGE, 1904, p. 257 [**S. mermisoides*; M]. Borings in pelecypod *Fissilunula clarkei* (MOORE), related to *Palaeopede* ETHERIDGE, 1899. [Made by ?algae or fungi.] *U.Cret.*, Australia (New S.Wales).

Talpina VON HAGENOW, 1840, p. 671 [**T. ramosa*; SD HÄNTZSCHEL, 1962, p. *W*231]. Straight tunnel systems in rostra of belemnoids, commonly branched, diameter *ca.* 0.2 mm.; numerous oval or circular openings toward exterior. [According to MORRIS (1851) and LESSERTISSEUR (1955, p. 81), produced by boring bryozoans; VOIGT considered only *Talpina pungens* probably identical with boring bryozoan *Terebripora prima* (VOIGT, 1962); type species *T. ramosa* interpreted as phoronid, not bryozoan, boring (VOIGT, 1972); *T. solitaria* VON HAGENOW, 1840 = type species of *Nygmites* HÄNTZSCHEL, 1962, p. *W*230; *T. dendrina* MORRIS, 1851 (=*Dendrina belemniticola* MÄGDEFRAU, 1937, p. 55) = type species of

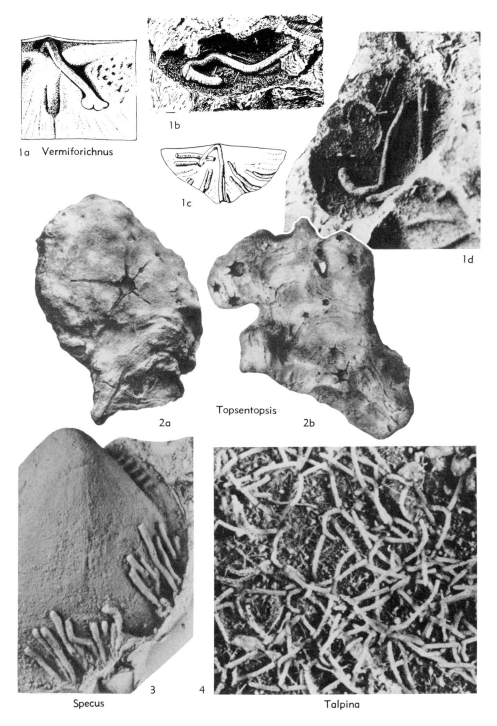

1a   Vermiforichnus

1b

1c

1d

Topsentopsis

2a   2b

Specus   Talpina

FIG. 82. Borings (p. *W*133, 135-136).

Dendrina QUENSTEDT, 1849 (SD HÄNTZSCHEL, 1965, p. 30).] *?Jur.,U.Cret.,* Eu.(Eng.-France-Ger.-Pol.-USSR); *U.Cret.(Campan.),* Eu.(Ire.).——FIG. 82,4. **T. ramosa,* Danian-Montian(?), France(Vigny, near Paris); steinkerns of tunnel systems, shell dissolved, ×5.7 (Voigt, 1972).

Tarrichnium WANNER, 1938, p. 398 [**T. balanocrini*; M]. Irregularly branched, ribbonlike, sharply entrenched traces on stalks of *Balanocrinus*; surface of ribbons slightly convex, some divided by 1 or 2 very thin longitudinal furrows; with fine bowl-shaped impressions. [Made by ?hydrozoan; see HILL & WELLS (1956, p. *F*88).] *U.Tert. (Mio.),* E.Indies.——FIG. 83,*1.* **T. balanocrini; 1a,* ×2.5; *1b,* ×1.4 (Wanner, 1938).

Terebripora D'ORBIGNY, 1842, p. 22 (see BASSLER, 1953, p. *G*37). [Borings of bryozoans; according to BOEKSCHOTEN (1970, p. 44) and BROMLEY (1970, p. 57), to be regarded as ichnogenus; taxonomy of *Terebripora* and that of *Spathipora*, p. *W*133, similarly confused.] *Jur.-Rec.,* Eu.-Asia-Atl.-Pac.

Teredolites LEYMERIE, 1842, p. 2 [**T. clavatus*; OD]. Clusters of clublike tubes, about 2 cm. long. [Apparently made by bivalves; name based on tubes only. See TURNER, 1969, p. *N*740 and Fig. E214.] *Cret.,* Eu.(France).

Thalamophaga RHUMBLER, 1911, p. 229 [**T. ramosa*; SD LOEBLICH & TAPPAN, 1964, p. *C*183] [For synonymy with *Orbithophage* SCHLUMBERGER, 1903; *Marsupophaga, Tubophaga, Nummophaga* RHUMBLER, 1911; *Arthalamophagum* RHUMBLER, 1913, see LOEBLICH & TAPPAN (1964, p. *C*183)]. Very small borings in tests of foraminifers, consisting of irregular "chambers" 2 to 8 microns in diameter and connected by stolonlike tubes. [Regarded by RHUMBLER (1911, p. 228) as of doubtful systematic position, possibly representing boring foraminifer morphologically modified by its parasitic mode of life; LOEBLICH & TAPPAN (1964, p. *C*183) placed it in the Foraminiferida (family Allogromiidae RHUMBLER), mentioning only Recent occurrences (Atlantic), whereas RHUMBLER (1911, p. 229-230) cited descriptions of similar borings in fossil shells and tests which he suggested as made by *Thalamophaga* or other (synonymous) genera; according to BOEKSCHOTEN (1966, p. 344), *Thalamophaga* probably belongs to group of algal or fungal borings.] *?Jur.,* Eu.(Ger.), *Rec.,* Atl.-Medit.

Topsentopsis DE LAUBENFELS, 1955, p. E41 [**Topsentia devonica* CLARKE, 1921, p. 88; M] [=*Topsentia* CLARKE, 1921, p. 88 (obj.) (*non* BERG, 1899); *Palaeosabella* CLARKE, 1921, p. 91 (*partim*) (type, *Vioa prisca* McCoy, 1855, p. 260); for discussion of the confused nomenclature and synonymy, see TEICHERT, 1945 (p. 200) and CAMERON, 1969a (p. 189)]. Borings of quite variable size, consisting of cavities and tubes or channels; cavity central, irregularly spheroidal or ovoid; tubes radiating from it, simple or branching, sometimes enlarging distally; diameter of central cavity 1 to 10 mm., of tubes 0.5 to 3 mm. [Interpreted as sponge borings; according to DE LAUBENFELS (1955, p. *E*41), sponge affinities doubtful; observed in many stromatoporoids.] *?Sil.,* USA; *Dev.,* USA.——FIG. 82,2. **T. devonica*

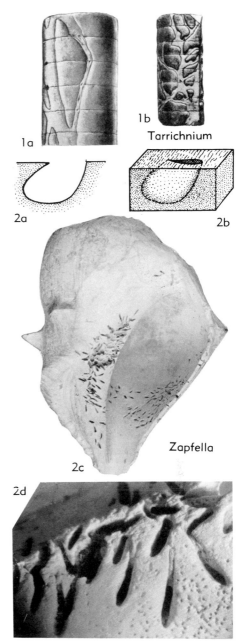

1a 1b

Tarrichnium

2a 2b

2c

2d

Zapfella

FIG. 83. Borings (p. *W*135-136).

CLARKE, Dev., Iowa; on lower surface of a stromatoporoid, *2a,* from Shellrock stage, reduced; *2b,* Cedar Valley beds, ×0.2 (Fenton & Fenton, 1932).

Trypanites MÄGDEFRAU, 1932, p. 151 [**T. weisei*; M] [=*?Calciroda* MAYER, 1952, p. 455 (type, *C. kraichgoviae*); see A. H. MÜLLER, 1956b, p. 410]. More or less straight tunnels, usually vertical, 1 to 2 mm. wide, without ramification; closely spaced; occasionally contain excrement of producer (see Fig. 76). [Made by rock and hardground borers; apparently polychaetes.] *L. Ord., USSR; Sil.* (Pleist. drift), *Eu.*(Ger.); *U. Dev., USSR; M.Trias.(Muschelkalk),* Eu.(Ger.); *U.Jur.,* Eu.(Pol.).

Vermiforichnus CAMERON, 1969, p. 190 [**V. clarkei*; M] [=*Gitonia* CLARKE, 1908, p. 154 (*?partim*) (type, *G. corallophila*); for discussion of this and the additional synonyms *Clionolithus priscus* (McCOY) and *Palaeosabella prisca* (Mc-COY), see CAMERON, 1969b, p. 692]. Borings, straight to slightly curved; rarely irregular, hooked or coiled; unbranched, smooth; nonintersecting; sometimes with subclavate termination; diameter 0.05 to 3 mm., commonly 0.2 to 2 mm. [Interpreted as worm borings, perhaps of Spionidae, possibly by the worm *Vermiforafacta rollinsi* CAMERON, 1969, which has been found in such a boring (see CAMERON, 1969b, p. 694); observed in calcareous algae, corals, bryozoans, brachiopods, mollusks; best known from Devonian strata.] *Ord.-Perm.,* cosmop.——FIG. 82,*1a-c.* **V. clarkei* (=*Palaeosabella prisca* (McCOY)), USA(N.Y.); *1a,* M.Dev.(Hamilton Gr.), portion of thickened substance of shell of brachiopod (*Leptostrophia*), flattened form in thinner part, ×?; *1b,* L.Dev. (Oriskany Ss.), hook-shaped boring in cast of brachiopod, ×?; *1c,* M.Dev.(Hamilton Gr.), sketch to show bend in tube where shell is thickest, ×? (Clarke, 1921).——FIG. 82,*1d. Vermiforichnus* tubes in *Meristella,* L.Dev.(Oriskany Ss.), USA; ×2 (Clarke, 1921).

Zapfella DE SAINT-SEINE, 1956, p. 449 [**Z. pattei*; M]. Saclike bore holes, 1 to 4 mm. long, 0.5 to 1 mm. wide, and up to 5 mm. deep; slitlike opening. [Made by barnacles (Acrothoracica); found in corals, brachiopods, mollusks, echinoids, and solid rock.] *Jur.-Tert.,* Eu.(Eng.-France-Aus.-Hung.-Italy-Pol.)-N.Afr.(Alg.)-N.Z.——FIG. 83,*2. Zapfella* borings in *Galeodes (Volema) cornuta* AGASSIZ, Mio., Hung.; *2a,* long. sec., schem.; *2b,* chamber, schem.; *2c,* ×0.9; *2d,* ×4.5 (Codez & de Saint-Seine, 1958).

GENERIC NAMES OF RECENT BORING ORGANISMS USED FOR FOSSIL BORINGS

PORIFERA

Cliona GRANT, 1826, p. 78. The name of the widely distributed marine boring sponge *Cliona* has often been used for similar borings observed in Mesozoic and Tertiary, and even Paleozoic fossils. Names of Recent species of *Cliona* or new species names have been applied to fossil borings apparently made by clionids (e.g., from the Upper Paleozoic of USA (ELIAS, 1957), from the Cretaceous of USA (STEPHENSON, 1941, 1952) and Europe (NESTLER, 1960; SCHREMMER, 1954), and from the Tertiary of Europe (BOEKSCHOTEN, 1966; RADWAŃSKI, 1964)). However, as shown recently by BROMLEY (1970, p. 77), the suitable ichnogeneric name for such Mesozoic and Tertiary borings is *Entobia* BRONN, 1838 (*emend.* BROMLEY, 1970). This name alone should be used (for Paleozoic sponge borings other ichnogeneric names are available). A detailed description of Recent clionid borings, their morphology and ecology, and their significance as agents of erosion and sedimentation has been given by BROMLEY (1970, p. 70-77).

BRYOZOA

The question of whether borings of bryozoans are to be regarded as ichnofossils is still a matter of dispute. BOEKSCHOTEN (1970) and BROMLEY (1970) are undoubtedly correct for considering them to be true ichnofossils. As BROMLEY (1970, p. 57) has stated, *Terebripora* D'ORBIGNY and *Spathipora* FISCHER "are in reality ichnogenera, since they were erected for empty borings, and they therefore rightly belong to the ichnologist rather than to the zoologist." The criteria employed in establishing these two ichnogenera were based entirely on the morphology of the borings and not the soft-body morphology of the producers. BOEKSCHOTEN and BROMLEY further emphasized that the names of Recent boring Bryozoa such as *Immergentia* SILÉN and *Penetrantia* SILÉN should be used for only the Recent animals and not their borings. Ichnogeneric names should be used for all fossil bryozoan borings even if they are morphologically congruent with these Recent Bryozoa.

The opposing viewpoint has been most recently expressed by VOIGT & SOULE (1973). They have pointed out that the tunnel systems made by bryozoans correspond well to the soft-body morphology of the animals, particularly to the structure of the zooids. In their opinion, bryozoan borings can not be compared with other borings generally regarded as trace fossils with ichnogeneric names. (See also POHOWSKY, 1974.)

Bryozoan borings may have an exceptional position among all the other borings. However, they are also cavities for which generic names based on the anatomy of the producer should not be applied.

Because of these two differing opinions, there now exists a regrettable state of taxonomic confusion with respect to the establishment of names for Recent and fossil bryozoan borings. Several genera are composed of both "biospecies" (zoological species) and ichnospecies (species determined solely on the morphologic pattern of the tunnel system). This situation has been recently emphasized by BOEKSCHOTEN (1970, p. 43-44). It is hoped that future studies employing modern techniques (casting-embedding procedure, scanning electron microscopy) will help to clear up this unfortunate circumstance.

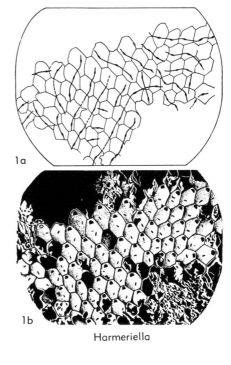

1a

1b

Harmeriella

FIG. 84. Borings (p. *W*137).

Harmeriella? cretacea VOIGT, 1957, p. 348. Very small borings in the cheilostome bryozoan *Strichomicropora membranacea* (VON HAGENOW, 1839), oblong, about 0.05 mm. long, *ca.* 0.03 mm. in diameter; away from apertures of bored zooecia; longitudinal axes of borings often arranged parallel to one another or even aligned. [Probably made by a colony of sessile organisms, presumably by a ?parasitic ctenostome bryozoan; according to VOIGT, 1957, p. 354, has doubtful affinities with Recent boring bryozoan *Harmeriella* BORG, 1940.] U.Cret.(low.Maastricht.), Eu.(Ger., Isle of Rügen).——FIG. 84,1. *Harmeriella? cretacea* borings in *Stichomicropora membranacea* (VON HAGENOW); *1a*, schem. drawing, borings connected by lines; *1b*, ×10 (Voigt, 1957).

Immergentia SILÉN, 1946, p. 6 (see BASSLER, 1953, p. G37). *Rec.* [Single *Immergentia*-like fossil described (*Immergentia? lissajousi* WALTER, 1965, from Oxfordian of France) does not belong to this genus (BOEKSCHOTEN, 1970, p. 44, and E. VOIGT, pers. commun., 1971).]

BIVALVIA

Recent bivalve borings in rocks are rather well known, thus their identification is relatively simple. Such borings or their steinkerns have commonly been named after their producer. This concerns mainly the following genera known from the Upper Cretaceous or Lower Tertiary to the Recent which are distributed all over the world: *Aspidopholas* FISCHER, 1887 (TURNER, 1969, p. *N*712); *Gastrochaena* SPENGLER, 1783 (KEEN, 1969b, p. *N*699); *Jouannetia* DES MOULINS, 1828 (TURNER, 1969, p. *N*718); *Lithophaga* RÖDING, 1798 (SOOT-RYEN, 1969, p. *N*276); *Petricola* LAMARCK, 1801 (KEEN, 1969a, p. *N*689).

RADWAŃSKI (1964, 1965, 1969) has studied the "lithophocoenoses" of Miocene littoral sediments in Poland. His publications serve as excellent examples of the description and environmental interpretation of bivalve borings produced by the genera listed above. (See also KAUFFMAN, 1969, p. *N*168-*N*170, and especially BROMLEY, 1970, p. 64.)

POLYCHAETA

Dodecaceria OERSTED, 1843, p. 44. VOIGT (1970, p. 375) described small borings in boulders of Upper Cretaceous (Santonian) limestones in North Germany (Lower Saxony). Similar borings have been found in the Upper Maastrichtian "Tuffkreide" of Holland (VOIGT, 1971). They are slightly clubshaped, straight, or somewhat curved, with a faint median depression, oblong or oval in cross section, 5 to 15 mm. long, 2 to 4 mm. in diameter, resembling the tubes of the Recent genus *Dodecaceria* OERSTED. The Santonian specimens were named *Dodecaceria(?)* sp., and later were united with the Dutch borings in a new species *D. cretacea* (VOIGT, 1971).

The borings of the Recent species *Dodecaceria concharum* from the North Sea occur in soft sandstones and limestones, in shells of mollusks or in calcareous algae. Another modern species, *D. fistulicola* EHLERS, described from Upper Tertiary and Pleistocene rocks of the United States (Ore., Calif.) lives in colonies only and represents a nonboring species of this genus (HOWELL, 1962, p. *W*163; REISH, 1952).

Very small meandering borings (0.5 mm. in diameter) have been observed in shells of *Cypricardia* from the Pliocene in Italy. ROVERETO (1901, p. 228) named them *"Dodekaceria"(?)* sp. According to VOIGT (1970, p. 373), these borings should be placed in the genus *Maeandropolydora* VOIGT.

Polydora BOSC, 1802, p. 150. Small U-shaped borings with spreite (diameter of tubes *ca.* 0.5 mm.) as made by the Recent spionid *Polydora ciliata* (JOHNSTON) have sometimes been placed in the ichnogenus *Polydora* (e.g., *P. biforans* in Upper Cretaceous belemnites found on the beach of the Baltic Sea) (GRIPP, 1967). Borings of the *Polydora* type have commonly been observed particularly in shells of Tertiary mollusks and in rocks (DOUVILLÉ, 1908; GEKKER & USHAKOV, 1962; PAPP, 1949; RADWAŃSKI, 1964; TAUBER, 1944, and others); see BOEKSCHOTEN, 1966, p. 357.

Some Recent species (e.g., *Polydora hoplura*) live in "pseudo-borings" or blisterlike cavities. Differences between true borings and pseudoborings originated by embedment have been discussed by BROMLEY (1970, p. 50); see also BLAKE & EVANS (1973).

The paleontologic history of *Polydora* up to 1908 was reviewed by BATHER (1909); for a later discussion see VOIGT (1965, p. 206).

Potamilla MALMGREN, 1865, p. 401. Isolated cylindrical borings, several cm. long, 1 to 6 mm. in diameter, straight or somewhat curved, not branched, with rounded end "like a glove finger," vertical or oblique to the bedding plane, have been compared with borings of the Recent polychaete genus *Potamilla* and sometimes given

the same name. They have been observed in Carboniferous and Jurassic hardgrounds (e.g., Eng., Switz.; see HÖLDER & HOLLMANN, 1969), Rhaetic-Liassic dolomite pebbles (Poland, Tatra Mts.; RADWAŃSKI, 1959), Upper Cretaceous hardgrounds (France; ELLENBERGER, 1947), and Miocene littoral sediments (Poland; RADWAŃSKI, 1964-69).

SIPUNCULIDEA

A boring (narrow entrance tunnel and expanded inner chamber) in the test of an *Echinocorys* (U. Cret., Isle of Wight) has been tentatively interpreted by JOYSEY (1959, p. 398) as made by an echiuroid and compared with burrows of *Thalassema neptuni* VON GAERTNER.

Sabella LINNÉ, 1767, p. 1268. Branched cylindrical borings (3-5 cm. long, 3-5 mm. diam.) in shells of *Cypricardia* (Plio., Italy) have been named *Sabella?* sp. by ROVERETO, 1901, p. 228. [*Sabella* is more favorably compared with tubes of *Potamilla*. Recent species of *Sabella* (in the present sense of the genus) do not bore.]

PHORONIDEA

Phoronis WRIGHT, 1856, p. 167. A branched burrow system in the test of an Upper Cretaceous echinoid (*Echinocorys*) from the Isle of Wight has been compared by JOYSEY (1959, p. 398) with borings of *Phoronis ovalis* WRIGHT which are known to occur in shells of Recent mollusks. The Cretaceous burrow system has not been named.

Phoronopsis GILCHRIST, 1908, p. 153. Straight or slightly curved vertical borings (length up to 3 cm., diam. 0.1-1 mm.) in Maastrichtian and, rarely, Tertiary limestones of Israel have been interpreted by AVNIMELECH (1955) as made by phoronids, particularly the genus *Phoronopsis*.

Small vertical borings (diam. 0.5-2 mm.) observed in hardgrounds of the Upper Maastrichtian "Tuffkreide" of Maastricht (Netherlands) have been compared by VOIGT (1970) with tunnels of Recent phoronids. They have been named *(?) Phoronopsis* sp. Their upper ends are frequently surrounded by agglutinating foraminifers (e.g., *Bdelloidina vincetownensis* HOFKER).

CIRRIPEDIA

Trypetesa NORMAN, 1903, p. 369. TOMLINSON (1963, p. 164) described fan-shaped burrows with elongate slitlike apertures which he observed in shells of myalinids from Pennsylvanian and lowermost Permian rocks of USA (Kans., Texas, Okla.). He recorded them as the largest fossil acrothoracican burrows (length about 1 cm., width *ca.* 5 mm.) and compared them with those made by the Recent cirriped *Trypetesa* NORMAN (1903) and thus named them *T. caveata.* However, SEI-LACHER (1969b, p. 709) interpreted them as borings of the *Polydora* type.

Ulophysema oeresundense BRATTSTRÖM, 1936, p. 1. Conical borings (outer diam. 0.8-0.9 mm., inner diam. 4-5 mm.) in the Upper Cretaceous echinoid *Echinocorys* from North Jutland (Denmark) have been ascribed by MADSEN & WOLFF (1965) to the Recent cirriped *Ulophysema oeresundense* BRATT-STRÖM. These borings are morphologically identical with those made by *Ulophysema* living parasitically in the echinoids *Echinocardium* and *Brissopsis* of the North Sea.

COPROLITES

The term coprolite has been defined in different ways (AMSTUTZ, 1958, p. 498). The shortest definition, "fossilized excrements of animals," seems to be best because it is independent of size and chemical composition of the "fossils" in question. It includes larger excrements, small fecal pellets (composing "coprogene sediments"), and microcoprolites.

In regard to their "systematic position" within paleozoology, ABEL (1935) classified them as lebensspuren, together with tracks, trails, borings, and other structures. However, coprolites do not correspond entirely to the widely accepted definition of trace fossils as structures left by living organisms in the sediment or on hard substrates. Thus, the special position of the coprolites requires them to be considered separately.

General questions about coprolites (size, shape, composition, occurrences, preservation, fossilization) are briefly discussed in the introduction to the annotated bibliography of coprolites (HÄNTZSCHEL, EL-BAZ, & AMSTUTZ, 1968). This work lists nearly 400 publications dealing exclusively or in part with coprolites.

This section describes only those coprolites that are identified by generic and specific names, and which are of undoubted invertebrate origin. For all other forms that have not been named, the reader is referred to the above-mentioned bibliography.

Names have been given mostly to microcoprolites observed in thin sections of sedimentary rocks. Especially, crustaceans of the order Anomura produce readily distinguishable excrements of which Recent and fossil examples are known (BRÖNNI-MANN, 1972). According to BRÖNNIMANN, transverse sections of different anomuran coprolites such as *Favreina, Helicerina, Palaxius, Parafavreina,* and *Thoronetia* reveal internal canals with different morphologies and arrangements. A single name *Tibikoia* (HATAI, KOTAKA, & NODA, 1970) has been introduced for larger fecal particles up to 5 mm. in length.

The term Coprolichnia, proposed by MACSOTAY (1967) for all coprolites is etymologically incorrect, because the ending *-ichnia* means tracks, and should not be used for coprolites. Besides, the term is superfluous.

VYALOV (1972a) coined the term Coprolithidii for "coprolites proper" which he further subdivided into genetic groups according to their makers.

Aggregatella ELLIOTT, 1962, p. 40 [**A. pseudo-hieroglyphicus;* OD]. Microcoprolites forming clusters or tangles of pellets, 0.5 to 1.0 mm. long, similar to but smaller than those of Recent ophiuroids or brittle stars. *U.Jur.,* SW.Asia(Iraq). ——FIG. 85,2. **A. pseudohieroglyphica,* Najmah F., Duliam Liwa; thin section, ×13 (Elliott, 1962).

Bactryllium HEER, 1853, p. 117 [**B. canalicula-tum;* SD HÄNTZSCHEL, herein] [=*Bactryllum* AZPEITIA MOROS, 1933, p. 52 *(nom. null.); Bactrydium* EMBERGER, 1968 *(nom. null.)*]. Small rounded or flat bacilliform bodies, few mm. to 1 cm. long, about 0.6 mm. wide; smooth or mostly with delicate transverse striations and 1 or 2 longitudinal furrows; ends rounded; material siliceous. [Interpretation as ?diatoms (HEER, 1853, 1877; FLICHE, 1906) very improbable; STEINMANN's interpretation (1907) as small dorsal plates of predatory worms has escaped notice; most probably are fecal pellets (excrements of gastropods, ROTHPLETZ, 1913; ALLASINAZ, 1968).] *Trias.-Jur.,* Eu.——FIG. 86,*1a.* **B. canaliculatum,*

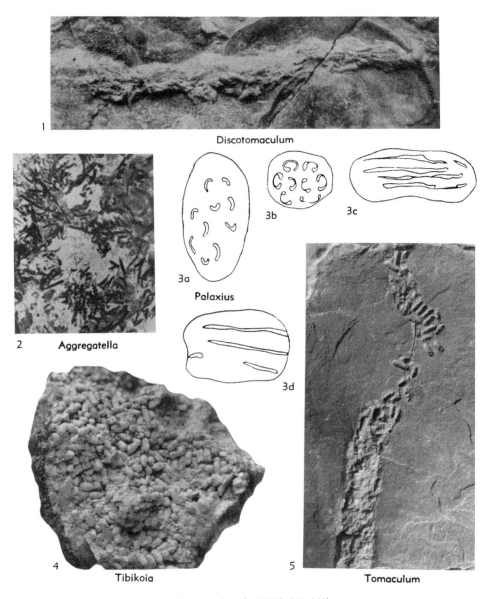

FIG. 85. Coprolites (p. *W*139, 141, 143).

L.Trias.(Carn.), Italy(Lago del Predil); ×6 (Allasinaz, 1968).——FIG. 86,*1b. B. heeri* ALLASINAZ, L.Trias.(Carn.), Italy(Lago del Predil); ×6 (Allasinaz, 1968).——FIG. 86, *1c. B. striolatum* HEER, U.Trias.(Rhaet.), Italy(Vedeseta); ×12 (Allasinaz, 1968).——FIG. 86,*1d. B. deplanatum* HEER, U.Trias.(Rhaet.), Italy(Gerosa); ×12 (Allasinaz, 1968).

Coprolithus PARÉJAS, 1948, p. 512. Name used for coprolites of crustaceans, proposed as informal term, not a "genus"; nevertheless, three "species" (*C. salevensis, C. prusensis,* and *C. decemlunulatus*) have been erected and described by PARÉJAS, 1948; see also *Favreina* BRÖNNIMANN, 1955.

Coprulus RICHTER & RICHTER, 1939, p. 163. Mechanical-ecological subsidiary name, proposed as neutral and informal name for excrement in form of isolated, loose pills, but designated as *Coprulus* "n.g." without a species name; used as "genus" by MAYER (1952) with "species" *C. oblongus* and *C. sphaeroideus* from Middle Triassic (up. Muschelkalk) of southern Germany.

Discotomaculum CHIPLONKAR & BADWE, 1972, p. 2 [**D. variabilis*; OD]. Zigzag burrow, 0.4 to 1.0 cm. wide, preserved in convex epirelief, filled with tiny discoid flakes of variable orientation (crisscross, parallel to subparallel, or transverse) to length of burrow. No evidence of burrow lining. Similar to *Tomaculum* but flakes may not be fecal in origin. [Interpreted as domichnia.] *L.Cret.*, India.——FIG. 85,*1*. **D. variabilis,* Bagh Beds; ×1 (Chiplonkar & Badwe, 1972). [Description supplied by W. G. HAKES.]

Favreina BRÖNNIMANN, 1955, p. 40 [**F. joukowskyi* (="Organisme B" JOUKOWSKY & FAVRE, 1913, p. 315; *"Coprolithus" salevensis* PARÉJAS, 1948, p. 512); OD]. Subtriangular and rounded dark organic remains of apparently homogeneous texture; 0.5 to 1.5 mm. long, 0.2 to 0.4 mm. wide; longitudinal section showing long, thin, straight and parallel canals distributed in regular but intermittent pattern; transverse section showing minute pores either arranged in 2 or more flattened, oblong rings or distributed irregularly; diameter of pores 12 to 40 microns. [Interpreted by PARÉJAS (1948) as coprolites of crustaceans, by CUVILLIER (in CUVILLIER & SACAL, 1951) as primitive Charophyta, by BRÖNNIMANN (1955) as microfossils *incertae sedis,* and by BRÖNNIMANN & NORTON (1960) as coprolites of crustaceans (Anomura). KENNEDY, JAKOBSON, & JOHNSON (1969) described an association of *Thalassinoides* with the microcoprolite *Favreina* from the Great Oolite Series of England.] *M.Trias.,* SW.Asia (N.Iraq); *L.Jur.-U.Tert.(mid.Mio.),* Eu.(France-Eng.-Switz.-Yugosl.-Hung.-Romania-S. Italy-Turkey-?USSR)-N.Afr.(Morocco-?N.Alg.-Libya)-Asia (Arabia-Israel-N.Iraq-Qatar-Iran)-USA (Texas)-C. Am. (Guatemala)-Gulf Mexico-W. Indies (Cuba-Trinidad).——FIG. 87,*2a,b*. **F. joukowskyi,* U. Jur.(mid.Portland.), W.Indies(Cuba); *2a,b,* transv. sec., long. sec.; ×26 (Brönnimann, 1955).——FIG. 87, *2c. F. asmarica* ELLIOTT, mid.Mio.(up. Asmari F.), Iran(AsmariMt., Masjid-i-Sulaiman); ×20 (Elliott, 1962).——FIG. 87,*2d. F. martellensis* BRÖNNIMANN & ZANINETTI, M.Trias,, S. France; diag. transv. sec. (Brönnimann, 1972).

Helicerina BRÖNNIMANN & MASSE, 1968, p. 154 [**H. spinosa*; OD]. Rod-shaped coprolites, up to 0.5 mm. long, oval in cross section and provided with groovelike depression; with 1(?), 2, 3, or 5 longitudinal canals showing bilaterally symmetric pattern, "upper" canals interconnected by fissural spaces; median canal of forms with 3 or 5 canals with spinelike extension breaking through to exterior; much shorter extensions developed on lateral canals of coprolites with 5 canals; similar to *Favreina* and *Palaxius* but differing from both by angular shape of canals and by spinelike extensions sometimes connected with exterior. [Produced by Anomura.] *L.Cret.(uppermost Barrem. or lower-*

Bactryllium

FIG. 86. Coprolites (p. *W*139-140).

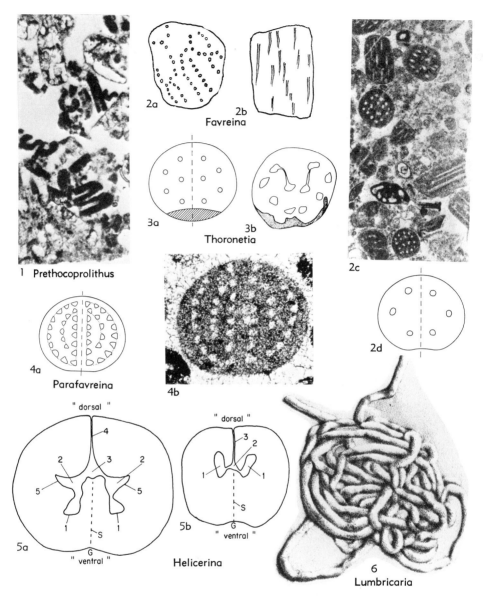

FIG. 87. Coprolites (p. *W*141-143).

most Apt.), Eu.(S.France).——FIG. 87,5a. **H. spinosa,* up.Barrem. or low. Apt.; [*1,* basal canals; *2,* lat. canals; *3,* median canal; *4,* spinelike "dorsal" extension of median canal; *5,* spinelike extension of lateral canals; *S,* plane of bilateral symmetry; *G,* "ventral" median groove], *ca.* ×330 (Brönnimann & Masse, 1968).——FIG. 87,5b. *H. alata* BRÖNNIMANN & MASSE, up.Barrem. or low. Apt.; [*1,* lat. canals; *2,* median canal; *3,* spinelike "dorsal" extension of median canal; *S* and *G* as

in *5a*], *ca.* ×330 (Brönnimann & Masse, 1968). **Lumbricaria** MÜNSTER (in GOLDFUSS), 1831, p. 222 [**L. intestinum*; SD HÄNTZSCHEL, 1962, p. *W*202] [=vermiculites PARKINSON, 1811, p. 93 (name not intended for genus); *Medusites* GERMAR, 1827, p. 108 (long unused name seemingly intended for rare, very thin tangles); Cololithen AGASSIZ, 1833, p. 676 (clearly not intended as generic name); *Lumbricites* (GOLDFUSS: "auctt.")]. Entangled intertwined strings, cross section somewhat round, diameter 1 to 4 mm.; length (max.) up to 170 cm., sometimes narrowed at irregular intervals; calcitic, rarely (e.g., *L. recta*) phosphatic; surface rough; strings consisting of very small fragments of planktonic crinoid *Saccocoma pectinata* (GOLDFUSS). [Interpreted as disgorged guts of fish and other animals (AGASSIZ, 1833; FRISCHMANN, 1853; O. KUHN, 1966) or ejected entrails of holothurians (GIEBEL, 1857; FENTON & FENTON, 1934c), as worms or wormlike animals (MÜNSTER [in GOLDFUSS], 1831; DE QUATREFAGES, 1846), as coprolites (EHLERS, 1868), as ?excrements of annelids (BROILI, 1924), partly of fish (MAYR, 1967; MÜLLER, 1969); intestinal fillings of worms (MAYR, 1967); according to JANICKE, (1970) certainly coprolites of cephalopods (Ammonoidea or more probably Teuthoidea) as earlier suggested by GOLDFUSS (1862). In the opinion of JANICKE (1970) the rare species *Lumbricaria coniugata* and *L. filaria* (=*Medusites* GERMAR, 1827) are not coprolites but perhaps conglobated filaments of algae; *L. antiqua* and *L.? gregaria* PORTLOCK (1843, p. 361) from the Silurian of Ireland identified as trails or ?burrows; *L. flexuosa* and *L. spiralis* SAVI & MENEGHINI (in MURCHISON, 1850, p. 491) from the Tertiary (Macigno) of Italy are unrecognizable.] *L.Jur.(up.Lias.)*, Eu.(S.Ger.); *U.Jur.(low.Tithon., Solnhofen Limestone)*, Eu.(S.Ger., Bavaria).——FIG. 87,6. *L. intestinum*, U.Jur.; ×0.8 (Goldfuss, 1831).

Palaxius BRÖNNIMANN & NORTON, 1960, p. 838 [*P. habanensis*; OD]. Coprolites of oval to subpentagonal or subrectangular shape; width 0.5 to 2 mm.; breadth about 0.5 mm.; pierced by crescent or hookshaped longitudinal canals (max. length 45-140 microns, width 15-35 microns), arranged in 2 symmetrical groups. [Structurally closely related to coprolites of Recent thalassinid *Axius stirhynchus*.] *L.Cret.*, Eu.(Hung.); *L.Tert.(Eoc.)*, C.Am.(Guatemala); *U.Tert.(Mio.)*, W.Indies (Cuba)-N.Afr.(Libya).——FIG. 85,*3a,d*. *P. habanensis*, Mio., Cuba; ×26 (Brönnimann & Norton, 1960).——FIG. 85,*3b,c*. *P. petenensis* BRÖNNIMANN & NORTON, Eoc., Guatemala; ×26 (Brönnimann & Norton, 1960).

Parafavreina BRÖNNIMANN, CARON, & ZANINETTI, in BRÖNNIMANN, 1972, p. 100 [*P. thoronetensis*; OD]. Rod-shaped, about 250 microns in diameter, "ventral" side slightly compressed; perforated by two bilaterally symmetric groups of longitudinal canals, which in transverse section resemble isosceles triangles. [Interpreted as anomuran coprolites.] *Trias.(Nor.-Rhaet.)-L.Jur.(M.Lias)*, Eu.(Aus.-France-Spain)-N.Afr.(Alg.).——FIG. 87,

4. *P. thoronetensis*, Trias.(Rhaet.), S. France; *4a*, diagram. transv. cross sec. (BRÖNNIMANN, 1972); *4b*, transv. cross sec., ×116 (BRÖNNIMANN et al., 1972b). [Description supplied by W. G. HAKES.]

Prethocoprolithus ELLIOTT, 1962, p. 38 [*P. centripetalus*; OD]. Rodlike, elongate cylindrical bodies, hollow, tapering to rounded ends, circular in cross section, straight or curved, with central tubular cavity; 0.75 to 1 mm. long, 0.25 to 0.5 mm. in diameter; resembling the coprolites of Recent gastropod genera *Patina*, *Trochus*, and *Gibbula*. *U.Jur.*, Asia(Iraq).——FIG. 87,*1*. *P. centripetalus*, Najmah F., Iraq(Dulaim Liwa); thin section, ×24 (Elliott, 1962).

Thoronetia BRÖNNIMANN, CARON, & ZANINETTI, in BRÖNNIMANN, 1972, p. 100 [*T. quinaria*; OD]. Rod-shaped, about 300 microns in diameter, possessing "ventral cap" of denser material than rest of coprolite; in transverse section, internal canals appear subcircular to tear-shaped in outline. [Interpreted as galatheid anomuran coprolite.] *Trias.(Rhaet.)*, Eu.(France).——FIG. 87,*3*. *T. quinaria*; *3a*, diagram cross sec. (BRÖNNIMANN, 1972); *3b*, holotype transv. sec., ×90 (BRÖNNIMANN et al., 1972a). [Description supplied by W. G. HAKES.]

Tibikoia HATAI, KOTAKA, & NODA, 1970, p. 8 [*T. judoensis*; M]. Oblong fecal pellets, cylindrical, sometimes ovoid or of short rodlike shape; circular in cross section; both ends bluntly and flatly rounded; surface smooth; about 1 mm. long, diameter 0.5 mm. [Regarded as excrements of worms.] *Cenoz.*, Asia(Japan).——FIG. 85,*4*. *T. sp.*, Kogata F.; upper view of fecal pellets, ×4 (Hatai, Kotaka & Noda, 1970).

Tomaculum GROOM, 1902, p. 127 [*T. problematicum*; M] [=*Syncoprulus* RICHTER & RICHTER, 1939a, p. 164 (type, *S. pharmaceus*)]. Strands of elliptical fecal pellets (="*Coprulus*" RICHTER & RICHTER, 1939a, p. 163) up to 10 cm. long and 1 to 2 cm. broad; lying on bedding planes; within strands pellets commonly lumped together in clusters; length of pellets 1 to 5 mm., diameter 0.5 to 1.5 mm. [Interpreted by BARRANDE (1872) as "*oeufs d'origine indéterminée*," by GROOM (1902) as eggs, possibly of trilobites, and by RICHTER & RICHTER (1939a) as coprolites. Similar structures have been described by CHAMBERLAIN & CLARK (1973, p. 677) from the Oquirrh Formation (Pennsylvanian-L. Permian of Utah).] *Ord.*, Eu.(Eng.-Ire.-France-Spain-Ger.-Czech.).——FIG. 85,*5*. *T. problematicum*, Herscheid slates, Ger.; ×2.5 (Richter & Richter, 1939a).

TRACE FOSSILS OR MEDUSAE INCERTAE SEDIS

Starlike fossils reminiscent of medusae are known since the early Paleozoic. Their affinities are uncertain, and most of them have been described as "medusoid." Some have been placed in the Trachylinidae *incertae sedis* (HARRINGTON & MOORE, 1956c, p. *F73*), some in Medusae *incertae sedis,* and others were regarded as unrecognizable (HARRINGTON & MOORE, 1956e, p. *F153*).

Probably some of the controversial forms are body fossils that may be related to medusae or may even represent genuine medusae. However, the suspicion exists that in other cases we are dealing with trace fossils. This is true especially for some starlike fossils found in Mesozoic to Cenozoic European flysch deposits. For example, many authors have regarded the genera *Atollites* MAAS, *Bassaenia* RENZ, and *Lorenzinia* DA GABELLI as trace fossils. NOWAK (1957) has looked upon these genera as starlike feeding burrows of crustaceans (?brachyuran). The large unnamed "star," from 30 to 50 cm. in diameter, found in Polish flysch has been compared by NOWAK (1957) and HÄNTZSCHEL (1970) with grazing trails of worms. However, no extensive investigations regarding this abundant fossil material have as yet been made.

On the other hand, the interpretation of some of these forms as medusae is uncertain and very controversial. This is indicated by the example of the previously mentioned genera *Atollites* and *Lorenzinia*. KIESLINGER (1939) suggested that both are perhaps medusae and that *Atollites* may be a synonym of *Lorenzinia*. Contrary to this, HARRINGTON & MOORE (1956b, p. *F43*; 1956c, *F73*) considered *Lorenzinia* as belonging to the Scyphomedusae and *Atollites* to the Hydrozoa.

Considering the scarcity of body fossils in flysch deposits, which for the most part have been interpreted as turbidites, it is unlikely that such delicate animals as me-

dusae, even if abundant in this type of environment, would be preserved. The interpretation of these forms as trace fossils seems more acceptable, since these problematical fossils occur together with many proven trace fossils on the same bedding planes.

All authors interpreting Jurassic starlike fossils such as *Palaeosemaeostoma* have had to offer interpretations which were either improbable or unproven. Thus, it was supposed that the animals died from desiccation after a rapid, tectonically controlled retreat of the sea. An interpretation as trace fossils presents no difficulties although we do not yet have well-documented Recent counterparts of such stellate imprints. It is to be hoped that a better knowledge of the biology of sessile medusae will help to solve these problems.

Under these circumstances, it seems best, in the author's opinion, to treat all these problematical fossils in an individual section, as is done below.

Atollites MAAS, 1902, p. 320 [**A. zitteli*; SD KIESLINGER, 1939, p. A88] [=*Attolites* LUCAS & RECH-FROLLO, 1965, p. 167 *(nom. null.)*]. Starlike but of varying morphology; central area small, circular, surrounded by 12 to 14 narrow, radial bands changing into an external zone of pyriform lobes, thicker and wider at periphery. [Originally described as medusa; according to HARRINGTON & MOORE (1956c, p. *F73*), belonging to hydrozoan medusae (?Trachylinida *incertae sedis*); however, interpreted by NOWAK (1957) as trace fossil, explained tentatively as feeding burrows made by crustaceans; likewise, SEILACHER (1959, p. 1070) and VYALOV (1968a, p. 332) mentioned the genera *Atollites* and *Lorenzinia* among trace fossils, which seems logical. Some "species" of *Lorenzinia* have been placed in *Atollites* and vice versa; KIESLINGER (1939, p. A88) considered *Atollites* a junior synonym of *Lorenzinia* or at least as subgenus, but most authors distinguish between the two genera; GRUBIĆ (1970, p. 185) regarded *Atollites* as "undoubtedly true fossil medusae" but (GRUBIĆ, 1961; 1970, p. 187) interpreted *Lorenzinia* as a true trace fossil (see Fig. 88, *4a*).] [Found in German flysch deposits.] *Cam.,* USSR(Sib. Plat.), *M.Jur.,* USSR; *L.Cret.-Tert.,* Eu.(Ger.).——FIG.

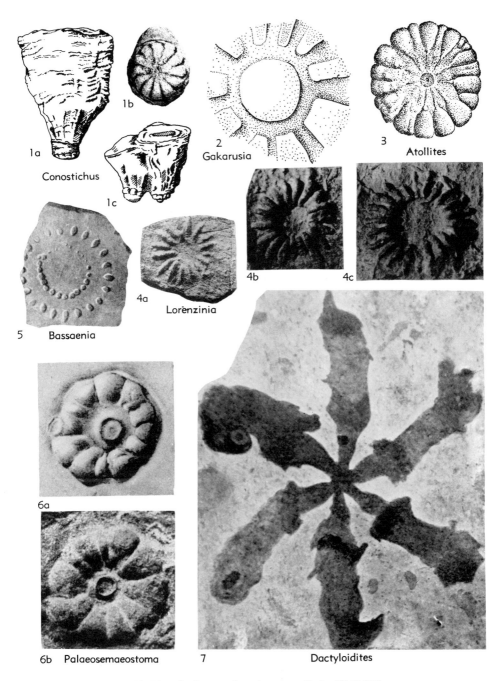

1a 1b 1c Conostichus

2 Gakarusia

3 Atollites

4a Lorenzinia

4b 4c

5 Bassaenia

6a

6b Palaeosemaeostoma

7 Dactyloidites

Fig. 88. Trace fossils or medusae incertae sedis (p. *W*144-148).

88,3. **A. zitteli,* L.Cret., Ger.; ×1 (Kieslinger, 1939).

Bassaenia Renz, 1925, p. 222 [**Lorenzinia (Bas-

saenia) moreae*; M]. Very similar to *Lorenzinia* DA Gabelli, differing from it by a second circle of 22 separated small knobs. [Originally described as problematic imprints of medusae; regarded by RENZ (1925) and KIESLINGER (1939, p. A88) as "subgenus" of *Lorenzinia*; HARRINGTON & MOORE (1956b, p. F43) hesitatingly assigned *Bassaenia* to Scyphomedusae (?family Callaspididae); interpreted as rosetted trail by HÄNTZSCHEL (1962, p. W185), VYALOV (1964a, p. 113), GRUBIĆ (1970, p. 187), and KSIĄŻKIEWICZ (1970, p. 313).] [Found in flysch deposits.] *U.Cret.-L.Tert.,* Eu. (Greece-Pol.).——FIG. 88,5. **B. moreae* RENZ, U.Cret., Greece; ×0.5 (Renz, 1925).

Brooksella canyonensis BASSLER, 1941, p. 522 [for genus *Brooksella* WALCOTT, 1896 (p. 611) and its synonym *Laotira* WALCOTT, 1896 (p. 613), see HARRINGTON & MOORE, 1956a, p. F23]. Stellate disclike structure, 7 cm. in its major diameter; consisting of 8 to 10 radiating lobes of fairly equal size, rather uniformly arranged, terminating with a distinct edge; most lobes with a few radial grooves. [Various interpretations have been 1) body fossil: supposed jellyfish (BASSLER, 1941); *Protomedusa* (HARRINGTON & MOORE, 1956a); 2) inorganic: reverse imprint of a subradial fracture system of unknown origin (CLOUD, 1960); resembling structures produced by gas evasion from sediments or by compaction around compressible or soluble objects such as gas domes or crystals (CLOUD, 1968, 1973); 3) trace fossils: perhaps starlike feeding burrow (SEILACHER, 1956a); result of metazoan life process, probably sediment-feeder, perhaps an annelid, better named *Asterosoma? canyonensis* (GLAESSNER, 1969).] *U.Precam.(Nankoweap Gr., Grand Canyon Ser.),* N.Am.(USA); *M.Cam.-U.Cam.,* USSR(Sib.Plat.)-USA(Ala.-Wyo.).——FIG. 89,1. *B. canyonensis,* Precam., Ariz.; ×0.48 (van Gundy, 1951).

Conostichus LESQUEREUX, 1876, p. 142 [**C. ornatus*; M] [=*Conostychus* LESQUEREUX, 1880, p. 14 *(nom. null.); ?Duodecimedusina* KING (in HARRINGTON & MOORE), 1955, p. F154 (type, *D. typica); Conostiches* POGUE & PARKS, 1958, p. 1629 *(nom. null.);* for discussion of several species of *Conostichus* as synonyms of *Duodecimedusina* see BRANSON, 1960, p. 195 (as *Duodecimedusa,* erroneously); *Consotichus* CHAMBERLAIN, 1971a, p. 242 *(nom. null.)].* Biogenic sandstone structures of variable shape, 4 to 9 cm. high, 3 to 14 cm. wide; mostly conical or subconical but also high forms with flat or rounded twelve-lobed basal discs and nearly parallel sides; bodies commonly fluted by transverse constrictions and longitudinal furrows and ridges; internally concentric conical sand laminae. [Regarded as of plant origin by STOUT (1956); compared with sponges by LESQUEREUX (1880); according to FUCHS (1895, p. 411), probably a member of the "group" Alectoruridae representing strobilation stage of medusa (for this interpretation see also BRANSON, 1960, p. 195; 1961, p. 134); according to BRANSON (1959, 1960, 1961), scyphomedusa or at least of scyphomedusan affinity (order

Brooksella

Rotamedusa

FIG. 89. Trace fossils or medusae incertae sedis (p. *W146, 148).

Coronatida, fam. Conostichidae); HENBEST (1960, p. B384) considered *Conostichus* trace fossil (with apex down, sand-filled trace of a sedentary burrowing animal); interpreted as trace fossil by CASTER (oral commun. in MARPLE, 1956, p. 29); CHAMBERLAIN (1971a, p. 220) regarded *Conostichus* as dwelling burrow of animal having greater affinities with Actinaria than with Scyphomedusae; *Conostichus*-like structures from Devonian of Bolivia have been interpreted as the feeding cones of an *Arenicola*-like worm (BARTHEL & BARTH, 1972, p. 579); for detailed discussions see BRANSON (1959, 1960, 1961, 1962) and CHAMBERLAIN (1971a, p. 220).] *Dev.,* S.Am. (Bol.); *?Carb.-L. Perm. (Singa F.),* NW. Malay; *Penn.,* USA(Okla.-Mo.-Ohio-Ill.); *?L.Perm.,* USA (Texas).——FIG. 88,1a. **C. ornatus,* Penn.(Potts-

ville ser.), Ohio; ×0.3 (Marple & Stout, 1956).
——Fig. 88,*1b*. *C. pulcher* Branson, Penn.
(HoldenvilleSh.), Okla.; ×0.7 (Branson, 1961).
——Fig. 88, *1c*. *C.* sp., Penn., Ohio; ×0.3
(Marple & Stout, 1956).

Dactyloidiscus Ślączka, 1965, p. 470 [**D. beskidensis*; M]. Discoid starlike impressions, 2 to 5
cm. in diameter; convex, consisting of 14 to 18
radiating transversely wrinkled lobes of mostly
unequal length. [Regarded as medusa; convex
upper surface interpreted as exumbrella; similar
to the "medusa" described by Zahálka (1957)
(U.Cret., Czech.) and compared by him with
Palaeosemaeostoma Rüger & Rüger-Haas; description of *Dactyloidiscus* up to 1971 only in
Polish language, not yet figured.] [Found in
flysch deposits.] *U.Cret.(Istebna Ss.),* Eu.(Pol.).

Dactyloidites Hall, 1886, p. 160 [**D. bulbosus*
(=*Buthotrephis? asterioides* Fitch, 1850, p. 862);
M]. Starlike impressions of varying sizes and
shapes, with 4 to 7 (commonly 6 or 7) "rays."
[Interpreted as algae or sponges (Hall, 1886;
Ruedemann, 1934; Resser & Howell, 1938), as
imprints of medusae (Walcott, 1890), as bodily
preserved medusae (Walcott, 1898), and as
worms or starlike worm trails (Ruedemann,
1934). Distinctly rosette-like "species" *D. edsoni*
(Ruedemann, 1934) in all probability is a starlike
trace fossil (very similar to unnamed starlike trace
fossils from Paleozoic of North America and
Bohemia).] *L.Cam.,* N.Am., USA(N.Y.); *?M.
Cam.,* N.Am.(USA,Vt.).——Fig. 88,*7*. **D. asterioides* (Fitch), L.Cam., N.Y.; ×1 (Walcott,
1898).

Gakarusia Haughton, 1964, p. 258 [**G. addisoni*;
M]. Central disc, 2 cm. in diameter, somewhat
elevated, with 10 or 11 short "rays" of different
width and trapezoidal cross section, beginning
some distance in from margin of disc. [Interpreted
by Haughton as "medusoid."] *U.Precam.(Transvaal Syst., Pretoria Ser.),* S.Afr.——Fig. 88,*2*. **G.
addisoni*; ×0.73 (Haughton, 1964).[1]

Kirklandia Caster, 1945, p. 175 [**K. texana*;
OD]. Problematic starlike fossil described in detail by Harrington & Moore (1956c, p. F70,
fig. 54). [Interpreted by Caster (1945) as belonging to Hydromedusae; according to Harrington & Moore (1956c, p. F69) "unquestionable
trachylinid medusa"; specimen of "*?Kirklandia*
sp." from the M.Jur. of Germany (Harrington
& Moore, 1956c, p. F72, fig. 55) was at first
described by Lörcher (1931) as *Medusina* sp.
and assigned "rather certainly" by Rüger (1933,
p. 39) to *Palaeosemaeostoma* Rüger & Rüger-
Haas (1925); Caster (1945, p. 198) called
"genus" "a perplexing fossil"; *Kirklandia multiloba* Ślączka (1964, p. 482) (Paleoc., Pol.) similar to *Atollites zitteli* Maas; possible interpretation of these various starlike Problematica as trace

---

[1] Considered a concretion by Cloud (1968). [W. G.
Hakes.]

fossils (feeding and dwelling burrows) was recently discussed by Häntzschel (1970, p. 206-
208); more thorough investigations of these problematic medusoid "genera" are required to clarify
their true nature.] *M.Jur.,* Eu.(Ger.); *L.Cret.
(Comanch. Ser.),* USA(Texas); *L.Tert.(Paleoc.),*
Eu.(Pol.).——Fig. 90,*2*. **K. texana,* L.Cret.
(PawpawF.), Texas(Denton Co.); ×0.7 (Caster,
1945).

Lorenzinia da Gabelli, 1900, p. 77 [**L. apenninica*; M]. Starlike; circular or elliptical rings
consisting of 16 to 26 (20 on an average) cylindrical or spindle-shaped ribs of equal length or
small roundish knobs encircling smooth flat central area; ribs or knobs rather regularly spaced
or arranged; diameter of star 2 to 5 cm.
[Originally, and by many authors today, regarded
as medusa (listed by Grubić, 1970, p. 187; see
also Harrington & Moore (1956b, p. F43),
where described as ?scyphomedusa). Divergent
opinions for interpretations as feeding burrows,
see Seilacher (1955, fig. 5, no. 88, without discussion; 1962, p. 229), Nowak (1957), Grubić
(1961), Ślączka (1964), Vyalov (1968a), and
Książkiewicz (1970); supposed to have been
made by crustaceans; according to Seilacher
(1962, p. 229), predepositional, not surface trail;
questionable whether *Atollites* Maas, 1902, should
be regarded as synonym of *Lorenzinia* as suggested (e.g., by Kieslinger, 1939, p. A88); see
also Häntzschel (1970, p. 208, 210, and p.
W144).] [Found in flysch deposits.] *?Ord.,* Eu.
(Ire.); *?L.Carb.(Kulm),* Eu.(Ger.); *Cret.-Tert.*
Eu.(Pol.)-Japan.——Fig. 88,*4a*. **L. apenninica,*
?Cret.-Eoc., Italy; holotype, ×0.7 (Gortani,
1920).——Fig. 88,*4b*. *L. gabellii* Vyalov, U.
Cret., Carpath.; ×1 (Vyalov, 1968a).——Fig.
88,*4c*. *L.* sp. aff. *L. kulcynskii* (Kuzniar), U.
Cret., Carpath.; ×0.84 (Vyalov, 1968a).

"Medusina" tergestina Malaroda, 1947, p. 57.
Feeding burrow, according to Seilacher (1959,
p. 1070). [Discussion on confused situation and
nomenclatorial status of *Medusina* Walcott, 1898,
has been offered by Caster (1945, p. 196, footnote 7) and Harrington & Moore (1956e, p.
F153).] [Found in flysch deposits.] *U.Cam.,*
Sib.; *L.Tert.(Eoc.),* Eu.(Aus.-Spain-Italy).

Nimbus Bogachev, 1930, p. 103 [jr. hom.; *non*
Mulsant & Rey, 1870] [**N. helianthoides*; M].
Large starlike fossil with 32 rays; central elliptical
field, 6 and 9 cm. in diameter; somewhat re-
sembling *Atollites* and similar forms. [Explained
as belonging to Trachymedusae or Narcomedusae.]
[Found in flysch deposits.] *L.Tert.(low.Eoc.),*
USSR.

Palaeoscia Caster, 1942, p. 26 [**P. floweri*; OD].
Disclike impressions, circular in outline, composed
of series of regular or irregular circles; several cm.
in diameter; small porelike depression in center;
about 16 slightly impressed grooves (*ca.* 1 cm.
long) may radiate from center of depression.

1a
Palaeoscia

1b

2
Kirklandia

3
Staurophyton

Fig. 90. Trace fossils or medusae incertae sedis (p. *W*147-148).

[Originally interpreted as belonging to order Siphonophorida, family Porpitidae (Harrington & Moore, 1956d, p. F150); according to Osgood, 1970, p. 395-397, perhaps partly feeding traces similar to those of Recent *Scolecolepis*, somewhat resembling sweep marks comparable to *"Dystactophycus"* Miller & Dyer, 1878b.] *U.Ord.*, USA (Ohio).——Fig. 90,*1*. *P. floweri*, Corryville beds, Ohio (Stonelick Creek, Clermont Co.); *1a,b*, concave epireliefs, ×0.5, ×0.53 (Osgood, 1970).

Palaeosemaeostoma Rüger & Rüger-Haas, 1925, p. 17 [*Medusina geryonides* von Huene, 1901, p. 1 (=*Medusa gorgonoides* Wagner, 1932, p. 163, nom. null.); M]. Starlike, about 5 cm. in diameter; rosette of 10 to 12 pillowy sectors sharply defined by grooves (for description see Harrington & Moore (1956c, p. F76)). [Regarded by most authors as body fossils belonging to medusae; according to Rüger & Rüger-Haas (1925), sessile scyphomedusa; assigned by Harrington & Moore (1956c, p. F76) with some doubt to order Trachylinida (*incertae sedis*) of the Hydrozoa. Fuchs (1901) did not consider *Medusina geryonides* a medusa but rather related to *Gyrophyllites* Glocker. Seilacher (1955, fig. 5) interpreted it as feeding burrow and also referred it to the ichnogenus *Gyrophyllites*; for discussion of interpretation as medusa or trace fossil see Häntzschel (1970, p. 206-208).] *M.Jur.*, Eu.(Ger.); *U.Cret.*, Eu.(Czech.).——Fig. 88,*6*. *P. geryonides* (von Huene). M.Jur., Ger.; *6a*, holotype, ×1 (von Huene, 1901); *6b*, another specimen, ×0.5 (Kieslinger, 1939).

Rotamedusa Simpson, 1969, p. 698, 700 [*R. roztocensis*; OD]. Subcircular imprints (max. diam. 1-2.5 cm.); consisting of a central circular depression, featureless, surrounded by 2 low concentric ridges, innermost ridge flat, symmetrical in cross section and covered by up to 24 very narrow radial ribs mostly terminating abruptly at margins; outer ridge intermittent. [?Starlike trace fossil interpreted by Simpson as the counterpart of a medusa [outer wall = velum, narrow ribs = counterparts of radial canals, surface = exumbrellar, central depression = central orifice]; provisionally placed *incertae sedis* in hydrozoan order Trachylinida (see Harrington & Moore, 1956c, p. F68-76); probably deposited by a suspension current, together with silt-size sediment fraction.] *L.Tert.* (*mid.Eoc., Hieroglyphic Beds, Magura Ser.*), Eu. (Pol.).——Fig. 89,*2*. *R. roztocensis*, mid.Eoc., Pol.(Stryszawa-Roztoki); *2a,b*, ×1.5, ×1.8 (Simpson, 1969).

Staurophyton Meunier, 1891, p. 134 [*S. bagnolensis*; M]. Similar to *Radiophyton* Meunier (1887, p. 59). [Originally described as of plant origin; ?trace fossil; see Harrington & Moore (1956c, p. F23), also Häntzschel, 1965, p. 18.] *Ord.*, Eu.(France).——Fig. 90,*3*. *S. bagnolensis*, L.Ord.; ×1 (Meunier, 1891).

BODY FOSSILS

This chapter contains descriptions of "genera" of doubtful or completely uncertain classificatory status. Many of them were described only once and never discussed again. Additional "genera" of this type may be found in the section on "unrecognizable genera." HOFMANN (1972a, p. 28) suggested that the term dubiofossil be used for fossils whose taxonomic origin is uncertain or unknown. With this usage, dubiofossils occupy a place intermediate between body fossils with an assigned taxonomic position and pseudofossils (see p. W168).

Some genera, which were considered to be body fossils in the first edition of Part W of the *Treatise*, have now been included in the new chapters Microproblematica and Coprolites.

Precambrian Metazoa, most of which belong to the Ediacara fauna, are not being considered here. Their position in the zoological system is, for the most part, no longer problematical because it has now been demonstrated that they are coelenterates, annelids, and arthropods. These forms will be fully discussed in Part A of the *Treatise* by M. F. GLAESSNER (Adelaide).

Anthonema WALTHER, 1904, p. 142 [**A. problematica*; M]. Small oblong bodies, finely serrated, 5 to 7 mm. long; tapering to one end, broad end 1.5 mm. (max.) wide. [Interpretation left undecided by WALTHER; according to JANICKE (1967, p. 82, R. FÖRSTER, pers. commun.), very probably larvae of crustaceans.] *U.Jur.(Solnhofen Limestone),* Eu.(S.Ger.).

Anzalia TERMIER & TERMIER, 1947, p. 65 [**A. cerebriformis*; M]. Reef-forming organisms of brainlike aspect, with large central cavity and very numerous small apertures resembling oscula of sponges. [For new discussion of systematic position see TERMIER & TERMIER (1964).] *Cam.,* Afr.(Morocco).——FIG. 91,7. **A. cerebriformis*; ×0.4 (Termier & Termier, 1947).

Ceramites LIEBMANN in FORCHHAMMER, 1845, p. 162 [*non* MASSALONGO, 1859, p. 11] [**C. hisingeri*; M]. Described from Alum Shale (U.Cam.) of southern Sweden and Bornholm as a "fucoid," probably represents species of *Dictyonema* HALL,

1851, perhaps *D. flabelliforme* EICHWALD (Dr. CHRISTIAN POULSEN, pers. commun., 1956).

Cestites CASTER & BROOKS, 1956, p. 183 [**C. mirabilis*; M]. Fringed ribbon reduced to carbonaceous film, with longitudinal lines. Regarded as lobe of fossil cestid ctenophoran, but identification questionable. *Ord.,* USA(Tenn.).——FIG. 91,3. **C. mirabilis*; ×2 (Caster & Brooks, 1956).

Charniodiscus FORD, 1958, p. 213 [**C. concentricus*; OD]. Disclike structure, 5 to 30 cm. in diameter; central area rough-surfaced; smooth flange with or without concentric corrugations. [Possibly associated with frondlike fossil *Charnia* FORD; interpreted by FORD (1958) as basal part of *Charnia*, and by GLAESSNER (1959) as medusalike base of coelenterate related to the Pennatulacea.] *Precam.,* Eu.(Eng.); *U.Proteroz.(low. Vend.),* USSR(Russ. Plat.).

Chuaria WALCOTT, 1899, p. 234 [**C. circularis*; M]. Disclike bodies resembling conical shells of discinoid or patelloid shape, 2 to 5 mm. in diameter; concentrically wrinkled; dark bituminous matter covering surface. [Originally interpreted as brachiopod-like fossils (remains of a compressed conical discinoid shell); according to SCHINDEWOLF (1956) possibly small, wrinkled clay galls or concretions; CLOUD (1968) regarded the type species and *Chuaria wimani* BROTZEN as algae; EISENACK (1966) considered *C. wimani* (Precam., Swed.) unrecognizable, neither gastropod nor brachiopod, nor eggs of trilobites, nor hystrichosphaerid, nor megaspore, but perhaps ?chitinous foraminifer; HOFMANN (1971, p. 24) considered the genus to be compressed globular bodies of biologic or nonbiologic origin. GUSSOW (1973, p. 1111) considered *Chuaria* to be of definite organic origin, either a large planktonic organism or a cyst or spore sac. FORD & BREED (1973a, p. 1257; 1973b, p. 547) regarded *Chuaria* to be algal in origin and classified it as a sphaeromorphid acritarch.] *Precam.,* USA-Can.-Eu. (Swed.)-USSR.——FIG. 91,6. **C. circularis*; U. Precam., USA(Ariz.); *6a,b,* ×12 (Walcott, 1899); *6c,* ×7 (Gussow, 1973); *6d,* ×7 (Ford & Breed, 1973b).

Curculidium HANDLIRSCH, 1907, p. 665 [**"Curculionites senonicus"* KOLBE, 1888, p. 136; M] [=*Curculionites* KOLBE, 1888, p. 136 (*non* HEER, 1847; *nec* GIEBEL, 1856); *nom. nud.*]. Name proposed for burrow of curculionid, presumably in wood; recognized by QUENSTEDT (1932b, p. 182) as belonging to *Doratoteuthis syriaca* WOODWARD. *U.Cret.(Senon.),* SW.Asia (Lebanon).

Diorygma BIERNAT, 1961, p. 20 [**D. atrypophilia*; OD]. Protuberances growing upward from floor of pedicle valves of *Desquamatia subzonata* (BIERNAT) on either or both lateral margins of

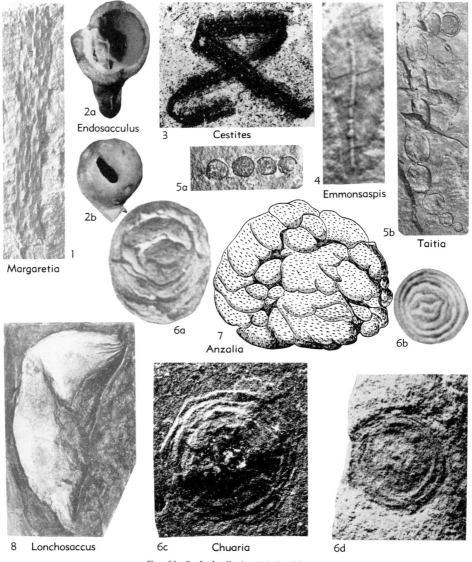

FIG. 91. Body fossils (p. *W*149-152).

ventral diductor muscle field, enclosing 2 contiguous tubes united along their entire length, with round or somewhat elliptical apertures; tubes straight or slightly sinuous, opening into mantle cavity region of brachiopod; ventral tube larger than the dorsal; tubes probably developed by simultaneous growth of their inhabitants and brachiopod. [First interpreted as a boring made by annelid-like parasite; MACKINNON & BIERNAT, 1970, regarded *Diorygma* cohabitant of *Desquamatia subzonata* living within shell, probably to be placed within the Phoronidea.] *M.Dev.,* Eu.

(Pol.).——FIG. 92,2. **D. atrypophilia,* in *Desquamatia subzonata* (BIERNAT); ×2 (Biernat, 1961).

Emmonsaspis RESSER & HOWELL, 1938, p. 233 [**Phyllograptus? cambrensis* WALCOTT, 1890, p. 604; OD]. Oval-shaped impression, more blunt at one end than other, with rod beginning about one-third of way back and extending almost to posterior end, mostly with ribbing beginning at about center line and extending to outer margins. [Possibly a chordate.] *L.Cam.,* USA(Vt.).—— FIG. 91,4. **E. cambrensis* (WALCOTT), RomeF.

(*Olenellus* Z.); mag. unknown (Resser & Howell, 1938).

Endosacculus Voigt, 1959, p. 219 [**E. moltkiae*; OD]. Globular, gall-like deformations in internodes of octocoral *Moltkia minuta* Nielsen; diameter about 5 mm.; with narrow ventral slitlike opening, about 2.5 mm. long; interior of "cyst" smooth. [Interpreted as made by barnacles (Ascothoracida); according to Bromley (1970, p. 67), not borings but more probably result of embedding.] *U.Cret.(Campan.-U.Maastricht.),* Eu. (Neth.-Swed.-?USSR).——Fig. 91,2. **E. moltkiae,* Maastricht., Neth.; *2a,* cyst with somewhat damaged opening, ×3; *2b,* cyst opened, showing the thin walls, ×3 (Voigt, 1959).

Escumasia Nitecki & Solem, 1973, p. 903 [**E. roryi;* OD]. Bilaterally symmetrical, flattened body (75 to 205 mm. in length) consisting of 2 arms, a trunk, stalk, and attachment disc; arms protrude from trunk, are long, slender, rounded, and commonly equal in length, and are located at each end of a slitlike opening; trunk (longer than wide) possesses "anal opening" on one side and tapers rapidly to stalk at basal end; attachment disk rounded and expanded basally. [Not assigned to any phylum by authors who very tentatively suggested that *Escumasia* may have been derived from the Coelenterata as an unsuccessful lineage; unequal length in arms considered to be the result of predation.] *M.Penn.,* USA(Ill.).——Fig. 92A,*1.* **E. roryi,* Carbondale F. (Francis Cr. Sh.); reconstr., ×1 (Nitecki & Solem, 1973). [Description supplied by W. G. Hakes.]

Halysium Swidzinski, 1934, p. 146 [**H. problematicum;* M] [=*Halimeda saportae* Fuchs, 1894, p. 204; *Arthrodendron* Ulrich, 1904, p. 138 (*non* Seward, 1898; *nec* Scott, 1900) (*nom. nud.*); perhaps =*Hormosira moniliformis* Heer, 1877, p. 161]. Ovate capsules, commonly flattened, smooth or minutely granulated, consistency differing from matrix; some specimens with carbonaceous lining; capsules forming branching rows. [?Alga.] [Found in flysch deposits.] *U. Cret.-L.Tert.,* Eu.-N.Am.(Alaska).——Fig. 92,*3.* **H. problematicum,* U.Cret., Italy; ×0.6 (Seilacher, 1962).

Leckwyckia Termier & Termier, 1951, p. 187 [**L. aenigmatica;* M]. Smooth, sharply pointed, acutely conical tube; upper end widening regularly and showing transverse units separated by constrictions. [Originally regarded as problematic in origin; interpreted by Destombes (1964) as rachis of trilobite, perhaps of *Dalmanitina.*] *Ord.,* Afr. (Morocco).

Lonchosaccus Ruedemann, 1925, p. 84 [**L. uticanus;* M]. Formed like bent bag, length more than twice width, with thick, substantial wall, now carbonized; 2 "extremities" drawn into apertures. [Systematic position unknown, ?anne-

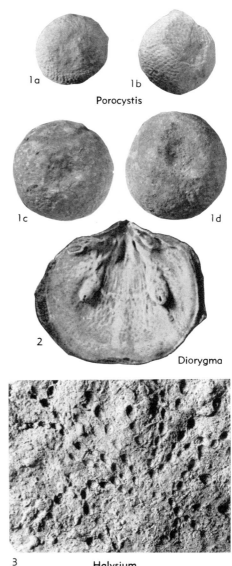

1a, 1b **Porocystis**

1c 1d

2 **Diorygma**

3 **Halysium**

Fig. 92. Body fossils (p. *W*149-152).

lid.] *Ord.,* USA(N.Y.).——Fig. 91,*8.* **L. uticanus,* Utica Sh.; holotype, ×? (Ruedemann, 1925).

Margaretia Walcott, 1931, p. 2 [**M. dorus;* OD]. Thin membranous sheet with elongate oval perforations arranged on longitudinal and obliquely transverse lines; tegument presumably leathery. [Compared with algae and alcyonarians.] *M. Cam.,* N.Am.(Can., B.C.)-USA(Idaho).——Fig. 91, *1.* **M. dorus,* Burgess Sh., B.C.; holotype, ×0.7 (Walcott, 1931).

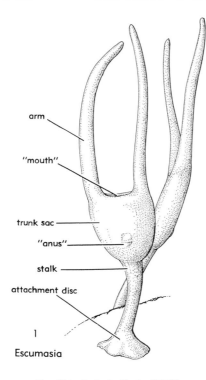

arm

"mouth"

trunk sac

"anus"

stalk

attachment disc

1

Escumasia

FIG. 92A. Body fossils (p. *W*151).

Palaeobalanus schmidi "VON SEEBACH, 1876." Name sometimes used erroneously for borings in Lower Muschelkalk of Thuringia; for details see MÄGDEFRAU, 1932, p. 150-151.

Porocystis CRAGIN, 1893, p. 165 [**P. pruniformis*; (=*Siphonia globularis* GIEBEL, 1853; *Araucarites? wardi* HILL, 1893); M]. Spheroids, generally prolate, with flattened, slightly protuberant area; whole surface covered with ridges and oval or circular depressions; commonly arranged rather irregularly in rows; diameter about 2 cm. [Interpreted by GIEBEL (1853) as alga, by HILL (1889-93) as fruit of *Goniolina, Parkeria,* or *Araucarites,* by CRAGIN (1893) as cheilostomatous bryozoan, by RAUFF (1895) as calcareous alga, by JARVIS (1905) as gigantic monothalamian foraminifer. For bibliography see ADKINS, 1928, p. 57-58.] *L.Cret.,* USA(Texas).——FIG. 92,*1.* **P. pruniformis,* L.Alb. (large specimens)-M.Alb. (small specimens); *1a-d,* ×1 (Häntzschel, 1962).

Taitia CROOKALL, 1931, p. 175 [**T. catena*; M]. Small chains commonly composed of 6 to 7 (max., 11) circular or oval bodies; adjacent bodies united by thin isthmus 1 mm. long and 1 mm. wide; bodies generally constant in size (diam., 1 cm.), some progressive diminution in size toward extremity; characteristic but problematical fossil of Scottish Downtonian rocks. *U.Sil.,* Eu.(Scot.).

——FIG. 91,*5.* **T. catena; 5a,b,* ×0.7, ×1 (Crookall, 1931).

Tullimonstrum RICHARDSON. 1966, p. 76 [**T. gregarium*; M]. Bilaterally symmetrical soft-bodied animal with head region, trunk, and tail; complete specimens very rare; total length of longest and smallest individuals known, estimated from fragmentary material, ranges from 8 to 34 cm.; head tapering to long proboscis bearing at its distal end jaw-like apparatus; jaws bearing minute stylets; entire proboscis constitutes one-third of animal's length but rather rarely preserved; head region poorly defined; transverse bar delimits head and trunk, consisting of medial plate and thin rod terminating in small ovoid bodies; trunk mostly segmented, narrowing to spatulate to nearly circular tail which shows 8 to 12 segments; tail lobe laterally expanding into flexible, triangular ribs. [Probably marine organism; impossible to assign *Tullimonstrum* to any known phylum al-

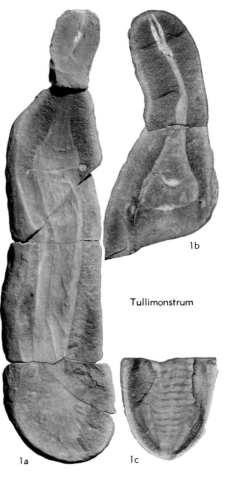

1b

Tullimonstrum

1a   1c

FIG. 93. Body fossils (p. *W*152-153).

though several thousand specimens with a documented geographical range of 200 miles have been investigated; ? a relic in the Pennsylvanian of a more ancient group.] *M.Penn.*, N.Am.(USA,

Ill.).——Fig. 93,*1.* **T. gregarium*, Francis Creek Sh., Ill.; *1a,* concretion, ×0.6 (Johnson & Richardson, 1970); *1b,c,* proboscis, and spadelike tail, both ×0.53 (Johnson & Richardson, 1969).

MICROPROBLEMATICA

In the first edition of *Treatise,* Part W (Häntzschel, 1962), and in the supplement to that volume (Rhodes, *et al.,* 1966), only very few microfossils of uncertain taxonomic position were included in the section on Body Fossils. However, in this revised and expanded edition, a separate section is devoted to Microproblematica, excluding microcoprolites, which are covered in the section on coprolites. A complete record of all the Microproblematica, such as originally proposed by Elliott (1958), is neither intended nor practical. For such a goal to be attained, the entire micropaleontological literature of the world must be reviewed. The author wishes to thank Dr. G. Deflandre (Paris) and Dr. A. Eisenack (Reutlingen) for help in supplying references.

Many problematical microorganisms are known only from thin sections of sedimentary rocks as, for example, the many genera which Flower (1961) described from the Ordovician of the United States. Photomicrographs do not completely and accurately reflect the three-dimensional shape of these fossils, and definite identification with some plant or animal group is very difficult. The same problem is encountered with forms having less characteristic shapes, such as small pellets of calcareous, siliceous, or pyritic composition. Furthermore, the state of preservation can make the interpretation of such microorganisms very difficult, e.g., that of the chloritic pellets from the Ordovician of France described as a *Papinochium* by Deflandre & Ters (1966).

Numerous forms are in dispute which may be of either plant or animal origin or which may be inorganic. For example,

Distichoplax Pia (1934) was regarded by Pia and by Elliott (1962) as an alga, but Lemoine (1960) interpreted it as belonging to the Rhabdopleuridae or a closely related family. In spite of the uncertainty of their position in the system, these microproblematica have proven to be, in some cases, stratigraphically useful fossils. In France, Deflandre & Ters (1966) determined the age of previously undifferentiated Lower Paleozoic rocks as Ordovician by studying certain microorganisms (very probably Acritarcha). In Lower Cretaceous limestones of Cuba, microproblematica have been used in stratigraphic correlation (Brönnimann, 1955).

It is hoped that in the future the systematic classification of many microproblematica will be clarified through electron microscope and stereoscan investigations.

Aeolisaccus Elliott, 1958, p. 422 [**A. dunningtoni*; OD]. Small thin-walled tubes, slightly irregular or somewhat curved, hollow, gently tapering, probably open at both ends; maximum length 1.7 mm., diameter commonly 1.0 mm.; walls consisting of crystalline calcite, with septate or camerate structure(?) [Probably pelagic organism; doubtfully interpreted as shells of small extinct pteropods; compared with the calcareous alga *Tubulites* Bein, 1932 (U.Perm., Ger.), by Hecht, 1960.] *Perm.*, SW. Asia(S.Turkey-Arabia); *U. Trias.-M.Jur.,* SW.Asia(N.Iraq); *U.Jur.(Kimmeridg.-Portland.),* L.Cret. *(Alb.)-U.Cret. (Maastricht.),* Eu.(Yugosl.).——Fig. 94,*5.* **A. dunningtoni.* U.Perm., Arabia; *5a,* approx. long. sec. of irregular elongate tube; *5b,* sec. showing numerous individuals; both ×50 (Elliott, 1958). [Also occurs in M.Trias.-U.Trias.(Anis.-Nor.), Eu.(Czech.).]

Ampelitocystis Deunff, 1957, p. 1 [**A. feuguerollensis*; OD]. Chitinous shell, oviform or of wide-bellied or bulgy shape, with one opening; 4 to 7 spinelike processes (50-200μ long) attached to its thickened margin in rather symmetrical arrangement; shell (max.) 50 to 90 microns in diameter,

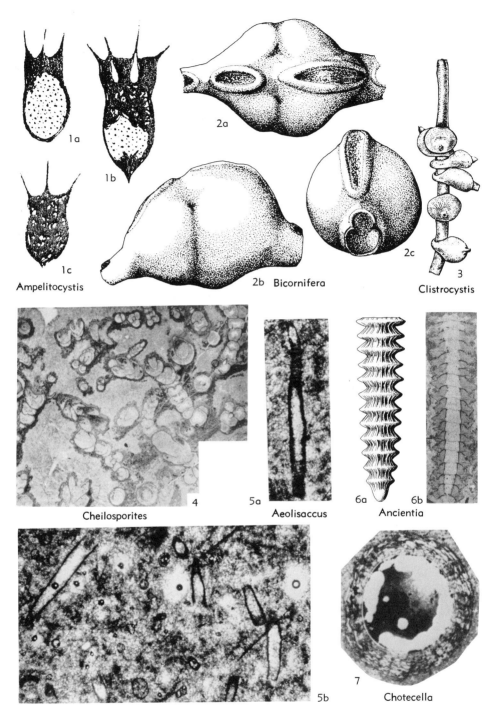

FIG. 94. Microproblematica (p. *W*153, 155-156).

120 to 170 microns high (without spines). [Systematic position unknown; morphologically somewhat similar to certain Ciliata.] *M.Sil. (Wenlock.),* Eu.(France).——Fig. 94,*1*. **A. feuguerollensis,* Calvados; *1a-c,* ×195 (Deunff, 1957).

Ampulites FLOWER, 1961, p. 115 [**A. vasiformis*; OD]. Short, simple, vase-shaped tubes, 1 mm. long, circular in section, basally broad, contracting to a neck; wall thin, calcitic; attached to corals; observed only in thin sections. [Systematic position unknown.] *Ord.,* USA(N.Mex.).

Ancestrulites FLOWER, 1961, p. 115 [**A. tubiformis*; OD]. Cylindrical, thick-walled tubes, 1 mm. long, about 0.5 mm. wide, calcitic; forming small colonies attached to corals; known only from thin sections. [Systematic position unknown.] *Ord.,* USA(N.Mex.).

Ancientia Ross, 1967, p. 39 [**A. ohioensis*; OD]. Small, hollow, calcitic structures, about 0.4 mm. long, 0.3 to 0.5 mm. high, diameter about 2 mm.; tubes having well-defined longitudinal series of imbricate rings that extend distally and partially overlap; externally prominent longitudinal costae or striae; proximal region smoothly rounded; longitudinal section displaying 2 longitudinal series of dentate imbricate segments whose microstructure consists of inclined fine laminae; tubes occurring only as fragments (greatest length observed 6.5 mm.); external features resembling *Cornulites sterlingensis* MEEK & WORTHEN, differing from it by much smaller size, cylindrical tube, and more strongly developed striation. [Phylum, class, order uncertain.] *U.Ord.(Cincinnat., Richmond Gr., Waynesville F.),* USA(Ohio). ——Fig. 94,*6a*. *A.* sp.; reconstr. of external aspect, ×7.5 (Ross, 1967).——Fig. 94, *6b*. **A. ohioensis*; long. sec., ×7.5 (Ross, 1967).

Bacinella RADOIČIĆ, 1959, p. 89 [**B. irregularis*; OD]. Aggregate (3 by 10 mm. in size) of irregularly shaped or polygonal chambers (0.1-0.4 mm. wide); walls of chambers composed of micritic calcite. [Systematic position unknown.] *M. Jur.(Anis.)-L.Cret.(Barrem./Apt.),* Eu.(Czech.); *Cret.,* Eu.(Yugosl.). [Description obtained from JABLONSKY (1973, p. 418) by W. G. HAKES.]

Bicornifera LINDENBERG, 1965, p. 22, emend. KEIJ, 1969, p. 243 [**B. alpina*; OD]. Calcareous shells consisting of 2 chambers of different size separated by double wall, with small round tube of unknown original length at both ends or at one end only; walls of shell hyaline, externally smooth; length 0.5 mm., height 0.3 mm. [Systematic position unknown.] *L.Tert.(Oligo.),* Eu. (S.Ger.-SW.France-W. Aus.-NW.Yugosl.-Turkey)- USA(Ala.).——Fig. 94,*2*. **B. alpina,* Oligo., Aus.; *2a-c,* different views, ×75 (Lindenberg, 1965).

Birrimarnoldia HOVASSE & COUTURE, 1961, p. 1054 [*Arnoldia antiqua HOVASSE, 1956; OD] [=*Arnoldia* HOVASSE, 1956, p. 2584, obj. (*non* MAYER, 1887, *nec* KIEFER, 1895, *nec* VLASENKO, 1931)]. Siliceous or iron oxide globules, 35 to 800 microns in size; wall 12 to 20 microns thick, apparently consisting of arenaceous matter; probably without openings; obviously consisting of several chambers arranged in row. [Interpretation as foraminifer is questionable, according to DEFLANDRE, 1957; LOEBLICH & TAPPAN (1964, p. *C786*) believe it to be inorganic.] *L.Proteroz.(Birrim.),* Afr.(Côte-d' Ivoire).

Calcisphaera WILLIAMSON, 1881, p. 521 [**C. robusta*; SD S. A. MILLER, 1889, p. 155] [=*Granulosphaera* DERVILLE, 1931, p. 132 (type, *Calcisphaera laevis* WILLIAMSON, 1879, p. 521); *Calcisphaerula* BONET, 1956, p. 44 (type, *C. innominata,* p. 56; OD)]. Hollow calcitic spheres, ranging in size from less than 0.1 to 0.5 mm.; thickness of shells varies from 3 to more than 200 microns. [Technically, *Granulosphaera* may be valid name; similar objects have been described as *Cytosphaera, Diplosphaerina* (=*Diplosphaera*), *Palaeocancellus* (=*Cancellus*), and *Polyderma* by DERVILLE (1931, 1950); as *Asterosphaera, Radiina,* and *Radiosphaera* by REITLINGER (1957); as *Fibrosphaera* by DE LAPPARENT (1924); as *Cadosina* and *Stomiosphaera* by WANNER (1940). Of uncertain affinities, variously interpreted as foraminifers, acritarchs, and algae (?charophytes, ?dasycladaceans); not all objects necessarily of the same affinities; for discussion and literature references see KONISHI (1958), TEICHERT (1965), RUPP (1966), STANTON (1966), WRAY (1967), FLÜGEL & HÖTZL (1971).] *Cam.-Rec.,* cosmop. [Description supplied by CURT TEICHERT.]

Cayeuxipora GRAINDOR, 1957, p. 2075. Established without designation of type species for small siliceous bodies with a reticulate surface ornamentation; about 10 microns in diameter. [Regarded as foraminifer by GRAINDOR (1957), and regarded by DEFLANDRE (1957) as resulting from bacterial activity.] *Proteroz.(Briovér.),* Eu. (France). [Description supplied by CURT TEICHERT.]

Cayeuxistylus GRAINDOR, 1957, p. 2077. Proposed without species for form similar to *Cayeuxipora* but having a long spine. *Proteroz.(Briovér.),* Eu.(France). [Description supplied by CURT TEICHERT.]

Cheilosporites WÄHNER, 1903, p. 100 [**C. tirolensis*; M]. Large arborescent colonies, only fragmentarily preserved; height of single shrubby colonies about 5 cm., "branches" composed of uniserial but branching rows of chambers (diameter 0.6 to 4.0 mm. max.), penetrated by axial siphon; chambers enlarging and altering their shape distally from casklike to bowl-like and finally in uppermost parts of "branches" vaselike; wall 0.01 to 0.04 mm. thick, consisting of calcite grains 0.01 to 0.02 mm. in diameter; probably guide fossil for Rhaetian. [WÄHNER (1903), LEUCHS (1928), SIEBER (1937), and LOEBLICH &

TAPPAN (1964) regarded *Cheilosporites* as alga; PIA (1939) compared it with Sphinctozoa resembling *Amblysiphonella*; tentatively referred by FISCHER (1962) to Foraminiferida, representative of new family Cheilosporitidae; see also LOEBLICH & TAPPAN, 1964, p. C786.] *U.Trias.(Rhaet.)*, Eu.(Ger.-Aus.-N.Alps-Yugosl., S.Alps).——FIG. 94,4. **C. tirolensis*, Trias. (Rhaet.), Aus.; ×2.5 (Fischer, 1962).

Cheneyella FLOWER, 1961, p. 113 [**C. clausa*; OD]. Low-arched tiny body, 0.7 mm. long, 0.2 mm. high, covered with a rather thick plate; broadly attached to *Catenipora*; observed only in thin sections. [Systematic position doubtful.] *Ord.*, USA(N.Mex.).

Chisibyllites DEFLANDRE, 1961, p. 126 [**C. kerguelenensis*; M]. Lenticular bodies, calcareous, nearly always with ellipsoidal inclusions of unknown nature and origin; observed only in limestone and only in thin sections. [Systematic position questionable, somewhat similar to radiolarians or foraminifers.] In limestone of unknown age; Ind.O.(Kerguelen I.).

Chotecella OBRHEL, 1964, p. 217 [**C. leiotheca*; M]. Small hollow globules with smooth surface, diameter 500 to 800 microns; wall 85 to 170 microns thick; formed by several very thin irregularly adjacent layers; globules showing organic structure consisting of carbonaceous matter; somewhat similar to *Leiosphaeridia* EISENACK, 1958, and *Tasmanites* NEWTON, 1875. [Origin unknown, plant or animal.] *Uppermost Sil.-lowermost Dev.*, Eu.(Czech.).——FIG. 94,7. **C. leiotheca*, Dev., Czech. (Choteč, near Prague); holotype, thin sec., ×50 (Obrhel, 1964).

Claviradix FERGUSON, 1961, p. 140 [**C. ashi*; OD]. Small cone-shaped bodies with small central elevations on upper surface; size about 2 mm.; 8 to 10 tapering radii growing from edge; stem projecting from underside of body and terminating in root which may be hollow; whole finely striated and pitted; similar to *Palaeocoryne* DUNCAN & JENKINS, 1869 (*emend.* FERGUSON, 1961). [Neither hydrozoan nor algal nor bryozoan in origin.] *U.Carb.(low.Namur.)*, Eng.—— FIG. 95,2. **C. ashi; 2a*, lower surface; *2b*, upper surface; *2c*, body, stem, and roots, all ×17 (Ferguson, 1961).

Clistrocystis KOZŁOWSKI, 1959, p. 273 [**C. graptolithophilius*; M]. Padlock-like chitinous forms bearing very small cone about 0.5 mm. long; individuals side by side on stipes of *Mastigograptus* sp. and embracing them; longitudinal axis perpendicular to graptolite stipes. [Systematic position unknown; possibly cysts of aquatic invertebrates; compared by KOZŁOWSKI (1965) with egg capsules of *Sepia* and explained as those of cephalopods.] *M.Ord.* (Pleist. drift), Eu.(Pol.).——FIG. 94,3. **C. graptolithophilius*; on a stipe of *Mastigograptus* sp., ×25 (Kozłowski, 1959).

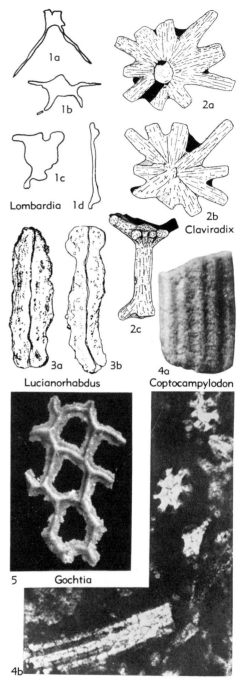

1a   1b   1c   Lombardia   1d

2a

2b   Claviradix

2c

3a   3b   4a
Lucianorhabdus   Coptocampylodon

5   Gochtia

4b

FIG. 95. Microproblematica (p. *W156-157, 161*).

Coelenteratella KORDE, 1959, p. 627 [**C. antiqua*; M] [=*Coelenterella, nom. null.* in translation of KORDE's paper, no. 2233, by the Bureau des Recherches Géologiques et Minières, Paris]. Small cuplike bodies; height about 7 mm., wall thickness *ca.* 0.15 mm.; fixed by foot about 8 mm. long. [Questionable coelenterate.] *M.Cam.,* USSR (Sib.).

Coptocampylodon ELLIOTT, 1963, p. 297 [**C. lineolatus*; OD]. Calcareous cylindrical bodies, solid, slightly curved or irregular, mostly up to 3.0 mm. long (incomplete), 0.25 to 1.0 mm. in diameter, ends irregularly rounded; outer surface commonly smooth but with 5 to 8 deep equidistant longitudinal grooves; transverse section resembling stellate structure with truncated rays. [Often regarded as remains of dasyclad alga *Acicularia* but certainly not alga or spicular elements of calcareous sponges; great similarity to calcareous joints of octocoral *Moltkia*; probably dissociated calcareous skeletal remains of small octocoral.] *L.Cret.,* Asia(Iraq-Borneo).——FIG. 95,4. **C. lineolatus; 4a,* NE.Iraq (Sulaimania Liwa); lat. view, ×37 (Elliott, 1963); *4b,* L. Cret., Iraq(Dulaim Liwa); thin section, ×37 (Elliott, 1963).

Cucurbita JABLONSKY, 1973, p. 420 [**C. infundibuliforme*; OD]. Club-shaped structure (0.1-0.3 mm. long; 0.05-0.1 mm. max. diam.) with a curved, convex funnel-like collar (0.2-0.3 mm. max. diam.) projecting from narrow end; funnel-like collar creates an opening to a central "cavity"; walls of structure composed of dark, micritic calcite. [Systematic position unknown; may be related to the tintinnids.] *?M.Trias., U. Trias.,* Eu.(Czech.). [Description supplied by W. G. HAKES.]

Cystosphaera FLOWER, 1961, p. 113 [**C. rotunda*; OD]. Round body, 1 mm. in size, covered by numerous small thin plates; thin walls enclosing round calcitic bodies with dark carbonaceous centers; broadly attached to *Catenipora*; known only from thin sections. [Systematic position unknown.] *Ord.,* USA(N.Mex.).

Dicasignetella KEIJ, 1969, p. 21 [**D. eocaenica*; OD]. Calcareous test, ovate or globular in front view, consisting of thick solid shield, enclosing 2 unequal chambers connected by large pore; height about 0.45 mm., width 0.3 mm., frontal shield 0.15 mm. wide; frontal shields of chambers ornamented by 2 rows of costae; on shields a frontal circular orifice flanked by paired blunt perforated spines; distal chamber smaller than proximal one. [Systematic position unknown; several features (e.g., separate chambers) in common with cheilostomatous Bryozoa, but other ones (e.g., the thick shield) uncharacteristic of Bryozoa.] *L.Tert.(up.Eoc., Barton.),* Eu.(Belg.).——FIG. 96,1. **D. eocaenica; 1a,b,* holotype, dorsal side, frontal shield of proximal chamber, ×120 (Keij, 1969b).

Draffania CUMMINGS, 1957, p. 407 [**D. biloba*;

OD]. Flask-shaped or pyriform rounded, "hollow" bodies, consisting of two unequal hemispherical globes with an elongate neck 0.25 mm. long originating between them; aperture of neck circular in outline, internal diameter 0.02 mm.; walls calcareous, relatively thick (0.06 to 0.12 mm.), layered, commonly perforate; greatest dimension, 0.85 mm.; maximum transverse cross section 0.75 by 0.65 mm. [Origin unknown. Originally compared with Foraminifera or other protozoans, pedicellaria of Echinodermata, and several plants, notably the Charophyta, with no systematic assignment; see BELFORD (1967).] *L.Carb.,* Eu.(G.Brit.-Belg.)-W.Australia.——FIG. 96A,1. **D. biloba,* Broadstone Ls., Scot.; *1a-c,* lat. side, and apertural views, all ×45; *1d,* vert. sec., diagram., ×55; *1e,* horiz. transv. sec., ×70 (Cummings, 1957). [Description supplied by W. G. HAKES.]

Eliasites FLOWER, 1961, p. 112 [**E. pedunculatus*; OD]. Spherical body, 0.8 to 0.9 mm. in size; wall thick, fibrous, composed of few plates; with central cavity; attached to *Catenipora*; observed only in thin sections. [Systematic position doubtful.] *Ord.,* USA(Texas).

Eotaeniopsis EISENACK, 1955, p. 184 [**E. articulata*; OD]. Rectangular chitinous integuments, flattened, black or brown; joined to short "chains" of 2 or 3 links; 90 to 470 microns long, 50 to 215 microns wide; single links of unequal size; corners rounded; surface smooth, bright. *U.Sil.* (*Beyrichia limestone*, Pleist. drift), Eu.(N.Ger.). ——FIG. 96,4. **E. articulata; 4a-f, ca.* ×45 (Eisenack, 1955).

Fentonites FLOWER, 1961, p. 117 [**F. irregularis*; OD]. Small planispiral shells or tests, 1 mm. in diameter, calcitic, thick-walled; internal cavity small and greatly reduced; surfaces irregular; attached to corals; observed only in thin sections. [Systematic position unknown.] *Ord.,* USA(N. Mex.).

Gochtia EISENACK, 1968, p. 305 [**G. rete*; M]. Finely meshed network of irregular 4- to 7-sided polygons formed by very thin rounded ribs 20 to 50 microns thick, with an axial channel; polygons 180 to 500 microns long; attached to thin flat basal plate; only fragments up to 2 mm. in size found, consisting of dahllite. [Systematic position unknown.] *U.Sil.(up.Ludlov.)* (*Beyrichia limestone*, Pleist. drift), Eu.(N.Ger., Isle of Hiddensee).——FIG. 95,5. **G. rete;* ×30 (Eisenack, 1968).

Goldringella FLOWER, 1961, p. 117 [**G. plana*; OD]. Tiny planispiral shells, thin-walled; whorl cavity rounded; 1.2 mm. in diameter, 0.6 mm. high; attached to corals by the broad flat side; observed only in thin sections. [Systematic position unknown.] *Ord.,* USA(N.Mex.).

"Guttulae" HILTERMANN & SCHMITZ, 1968, p. 301. Informal name for problematical bodies of microscopic size from freshwater sediments; re-

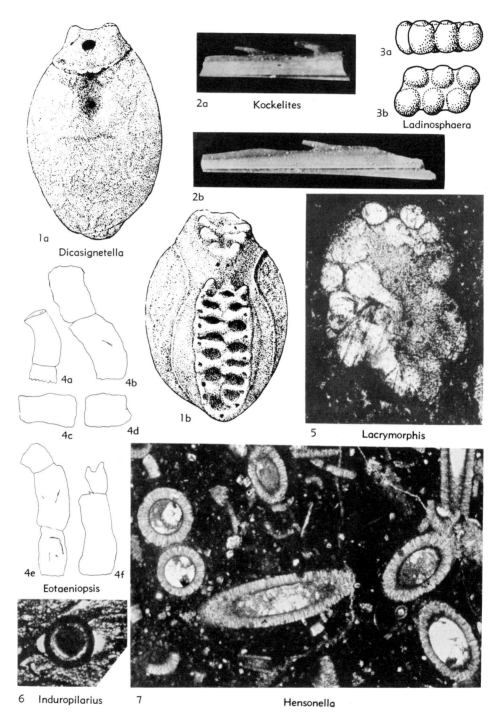

Dicasignetella

Kockelites

Ladinosphaera

Eotaeniopsis

Lacrymorphis

Induropilarius

Hensonella

FIG. 96. Microproblematica (p. *W*157, 159-160).

sembling in shape drops of liquid; white or gray, bright, surface smooth, 0.2 mm. long, 0.12 mm. wide; composed of apatite; in thin sections marginal layer (microcrystalline and somewhat translucent) and large darker inner part distinguishable. [May perhaps be isolated parts of some larger organisms, but inorganic origin is not to be excluded. The generic name *Guttula* and the species name *randeckensis* were proposed by HILTERMANN & SCHMITZ (1968) as conditional names and have no standing under the *Code*, Art. 15.] *U.Tert.(Mio.)*, Eu.(S.Ger.).——FIG. 97,5. "Guttulae"; *5a*, numerous specimens, ✕49; *5b*, cross section showing internal structure, ✕1,300 (Hiltermann & Schmitz, 1968).

Harjesia FLOWER, 1961, p. 111 [**H. anomala*; OD]. Tiny flask- or vase-shaped bodies, 1.4 mm. long, 1.2 mm. wide, with long neck above and solid calcitic body in central cavity; thick-walled; calcitic, with rodlike inclusions; only one specimen known from thin section; attached to coral. [Systematic position unknown.] *Ord.*, USA(N. Mex.).

Hensonella ELLIOTT, 1960, p. 229 [**H. cylindrica*; OD]. Calcareous tubes, hollow, cylindrical, straight, slightly tapering, up to 2.5 mm. long (?incomplete); diameter 0.1 to 0.5 mm.; walls consisting of very thin dark impervious inner layer and thick outer layer of aragonite with radiate structure. [Affinities doubtful; according to ELLIOTT (1960, 1962), not calcareous alga; perhaps small scaphopod.] *?L.Cret.*, Eu.(Yugosl.); *L.Cret.*, SW.Asia(Arabia-Israel-Iran-Iraq)-N.Afr. (Algeria)-E.Indies(Borneo).——FIG. 96,7. **H. cylindrica*, NE. Iraq; ✕30 (Elliott, 1960).

Hikorocodium ENDO, 1951, p. 126 [**I. Elegantae*; M] [*=Hicorocodium* KOCHANSKY & HERAK, 1960, p. 90 *(nom. null.)*]. Cylindrical bodies, rather straight, with rounded end, occasionally branched or with rounded protuberances; "thallus" composed of 1 or 2 central "stems" (diam. 0.1-2 mm.), made up of mass of very fine threadlike filaments, and dichotomously branched tubular pores (diam. 0.1-0.2 mm.) in peripheral part; outer part of "thallus" calcified. [Systematic position uncertain; originally interpreted by ENDO (1951a) and others as codiacean alga allied to *Gymnocodium* PIA; compared with hydrozoans by KOCHANSKY-DEVIDÉ & RAMOVS (1955), KOCHANSKY-DEVIDÉ (1958), E. FLÜGEL (1959), and KOCHANSKY & HERAK (1960); H. FLÜGEL (1963) recommended research on possible sponge nature.] *L.Carb.-Perm.*, Asia (Turkey-Iran-Japan); *Perm.*, Eu.(Yugosl.-S.Anatol.).

Hydrocytium(?) **silicula** MATTHEW, 1890, p. 146. Minute oval bodies; 0.5 mm. long, 0.25 mm. wide; with strong cuticle and pedicle-like knob at one end. *Cam.*, Can.(N.Scotia).

Induropilarius DEFLANDRE & TERS in TERS & DEFLANDRE, 1966, p. 342 [**I. aenigmaticus*; OD]. Silico-organic globules 0.1 mm. in diameter, with

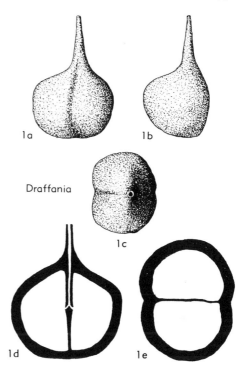

Draffania

FIG. 96A. Microproblematica (p. *W157*).

thick membrane; globule consisting entirely, or only peripherally, of fine radial-fibers. *Ord.*, Eu. (W.France).——FIG. 96,6. **I. aenigmaticus* Vendée, France; ✕160 (Ters & Deflandre, 1966).

Ivesella FLOWER, 1961, p. 108 [**I. adnata*; OD]. Capsule-like bodies, 0.8 mm. wide, 0.6 mm. high, high-arched; wall thin, consisting probably of a single piece; origin material perhaps chitinous; attached to colonies of *Palaeophyllum*; observed only in thin sections. [Systematic position unknown.] *Ord.*, USA(Texas).

Kockelites ALBERTI, 1968, p. 129 [**K. longus*; OD]. Conical bodies, tapering, 1.5 to 3.5 mm. long, cross section flat, oval; on upper side 1 to 5(?) tiny teeth, with forward inclination; on posterior part of lower side 2 furrows converging at acute angle; in anterior of body a hollow with ramifications leading into the teeth; only fragments known, rendering complete diagnosis impossible. [Systematic position unknown.] *Sil.-M.Dev.(Eifel.)*, Eu.(W.Ger.). [Unpublished occurrences (H. ALBERTI, pers. commun., 1971): *L.Dev.-M.Dev.*, Eu. (W.France-Aus.)-Afr. (Morocco), *Sil.-Dev.*, Eu.(Swed.,Gotl.-Aus.).]——FIG. 96,2. **K. longus*, Dev.(up.Ems.-low Eifel.), Harz Mts.; *2a,b*, ✕30 (Alberti, 1968).

Kruschevia FLOWER, 1961, p. 111 [**K. verruca*; OD]. Small fibrous bodies, narrowly elevated,

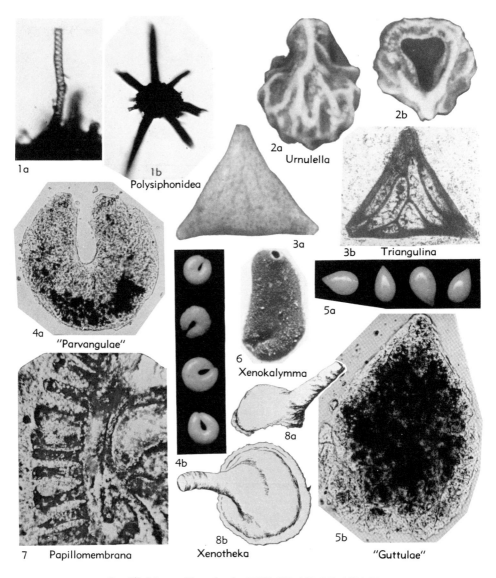

FIG. 97. Microproblematica (p. *W*157, 159, 163, 165, 167-168).

tip rounded, height and width 0.2 to 0.3 mm.; attached to *Catenipora*; observed only in thin sections. [Systematic position unknown.] *Ord.*, USA(Texas-N.Mex.).

Lacrymorphus ELLIOTT, 1958, p. 424 [**L. perplexus*; OD]. Small hollow bodies, 65 to 80 microns in diameter, tiny spherical, pear-, acorn-, or retort-shaped, not sections of tubes; very thin-walled; occurring in clusters, often nearly touching each other, but never with polygonal outline; ?aggregations fortuitous. [?Clusters of unicellular green algae.] *U.Trias.*, SW.Asia(N.Iraq); *L.Cret.-*

U.Cret.(Cenoman.-Turon.), Eu.(Yugosl.).——FIG. 96,5. **L. perplexus,* U.Trias., N.Iraq; ×109 (Elliott, 1958).

Ladinella OTT in KRAUS & OTT, 1968, p. 273 [**L. porata*; M]. Tubes (0.025-0.04 mm. in diam.) grouped together in node- or tongue-shaped "colonies," about a mm. in size; in cross section, walls surrounding cavities display "pseudosepta" but tubes themselves are not similarly partitioned. [Systematic position unknown, but producers were considered by author to have lived in "communities" probably with a commensal relationship to a variety of organisms.] *M.Trias.*, Eu.(S.Aus.); *M.Trias.-U.Trias.*, Eu.(Czech.). [Description supplied by W. G. HAKES.]

Ladinosphaera OBERHAUSER, 1940, p. 44 [**L. geometrica*; M]. Small globules (6 or 9) linked in one plane forming regular geometric figures; diameter about 0.5 mm., surface of globules obviously perforated or retiform. [?Plant, ?animal, ?inorganic; see LOEBLICH & TAPPAN (1964, p. C786).] *M. Trias.(Ladin.),* Eu.(Aus.).——FIG. 96,3. **L. geometrica; 3a,b,* ✕50 (Oberhauser, 1960).

Lamellitubus OTT *in* KRAUS & OTT, 1968, p. 274 [**L. cauticus*; OD]. Double-walled tube (approx. 1 mm. external diam. and approx. 0.5 mm. internal diam.); partially branched; whole specimens create surfaces of branched tubes; internal surface smooth and lined with micritic calcite; external surface uneven and undulating; internal wall thicker than external wall and connected to it by thin lamellae (0.05-0.10 mm. apart). [Systematic position not assigned by author.] *M.Trias.,* Eu.(S.Aus.); *M.Trias.-U.Trias.,* Eu.(Czech.). [Description supplied by W. G. HAKES.]

Lenaella KORDE, 1959, p. 626 [**L. reticulata*; OD]. Cylindrical calcareous organism, about 1 mm. long and 0.5 mm. wide; wall perforated by very fine holes. [Systematic position unknown, ?hydrozoan.] *L.Cam.*, USSR(Sib.).

Linotolypa EISENACK, 1962, p. 136 [**L. arcuata*; OD]. Thin ?chitinous threads arranged as links or meshes which form minute hollow globules. [?Planktonic organism.] *?Ord.*(Pleist. drift), Eu. (Ger.), *M.Ord.,* USA(Va.); *M.Sil.(Wenlock.),* Eu.(Swed., Gotl.); *M.Trias.(Muschelkalk),* Eu. (N.Ger., Holstein); *L.Cret.,* Australia; *L.Cret. (up.Apt.),* Eu.(N.Ger.)——FIG. 98,2. **L. arcuata,* M.Trias., Ger.; ✕500 (Eisenack, 1962).

Lithraphidites DEFLANDRE, 1963, p. 3486 [**L. carniolensis*; OD]. Rodlike, calcareous; cross section cruciform; 26 microns long, 2 microns wide; apparently pierced by thin canal. *L.Cret.(up. Apt.)-U.Cret.,* Eu.(France)-Australia, reworked in *Oligo.,* Eu.(France).——FIG. 98,5. **L. carniolensis,* Cret., France; *5a,* fragment, ✕5,000; *5b,* holotype, ✕3,200 (Deflandre, 1963).

Lombardia BRÖNNIMANN, 1955, p. 44 [**L. arachnoidea*; OD] [=*Formes découpées* LOMBARD, 1938; "*Sections de thalles*" LOMBARD, 1945]. Free, calcareous, transparent microfossils; spined, broad-branching or angularly bone-shaped; symmetrical; central body of variable size and shape and granular in aspect; extensions with dark median line; diameter up to 1.5 mm. [Interpreted by LOMBARD (1945) as algae, by PARÉJAS *in* LOMBARD (1938) as remains of sponge skeletons, and by BRÖNNIMANN (1955) as sections of microscopic symmetrical holothurian remains or microscopic planktonic crinoids or ophiuroids.] *U.Jur.,* Eu.(France-Switz.)-W.Indies(Cuba).—— FIG. 95,*1a,b.* **L. arachnoidea,* Portland., Cuba;

1a,b, ✕62 (Brönnimann, 1955).——FIG. 95,*1c. L. perplexa* BRÖNNIMANN, Portland., Cuba; ✕62 (Brönnimann, 1955).——FIG. 95,*1d. L. angulata* BRÖNNIMANN, Portland., Cuba; ✕62 (Brönnimann, 1955).

Lucianorhabdus DEFLANDRE, 1959, p. 142 [**L. cayeuxi*; OD]. Very small rodlike microorganisms, shape varying from cylindrical or subcylindrical to slightly curved or seldom even fungiform, 8 to 30 microns long, 7 to 8 microns wide; one end conical or spherical; rods consisting of 4 parallel elements closely connected, each element rhomboidal in cross section; surface wrinkled or granular. [First described by CAYEUX (1897) as "*bâtonnets de nature indéterminée*" and suggested to be calcareous algae; systematic position uncertain, ?related to coccolithophorids. Stratigraphic use questionable, as these forms have been found redeposited in Tertiary sediments.] *U.Cret.,* Eu.(France-Eng.-Pol.)-W.Australia.—— FIG. 95,3. **L. cayeuxi,* U.Cret. (Maastricht.), France(Vanves, Seine); *3a,b,* ✕1,300 (Deflandre, 1959).

Microrhabdulinus DEFLANDRE, 1963, p. 3486 [**M. ambiguus*; OD]. Rodlike, cylindrical, straight or slightly curved; cross section polygonal or roundish; 55 microns long, 3 to 4 microns wide; microstructure homogenous but very unique. *U. Cret.,* Eu.(France).——FIG. 98,3. **M. ambiguus,* Senon., France(Saint-Denis de Maronval); *3a,b,* ca. ✕3,200, ✕2,000 (Deflandre, 1963).

Microrhabduloides DEFLANDRE, 1963, p. 3486 [**Microrhabdulus rugosus* BOUCHÉ, 1962, p. 92; OD]. Rodlike, calcareous, 7 to 35 microns long; cross section roundish or angular; with or without thin canal; microstructure homogeneous or irregularly heterogeneous. *L.Tert.(Eoc., Lutet.),* Eu.(France).

Microrhabdulus DEFLANDRE, 1959, p. 140, *emend.* DEFLANDRE, 1963, p. 3486 [**M. decoratus;* OD]. Very small calcareous rods, cylindrical or spindle-shaped, straight or slightly curved, with narrow rather distinct axial canal; both ends blunt; 16 to 33 microns long, 1.5 to 2 microns wide. [Probably entire organisms, not fragments; ?related to coccolithophorids.] *U.Cret.,* Eu.(France-Eng.-Pol.)-USA(Texas)-W.Australia; reworked in *M.Eoc., Oligo.,* Eu.(France).——FIG. 98,1. **M. decoratus,* U.Cret.(Maastricht.), France(Vanves, Seine); ✕3,200 (Deflandre, 1959).

Microtubus E. FLÜGEL, 1964, p. 75 [**M. communis*; OD]. Very small cylindrical tubes, mostly curved, seldom straight, probably articulated transversely; length 0.2 to 2.0 mm. (commonly 0.3-0.5 mm.), diameter 0.05 to 0.2 mm.; walls smooth, without distinct structure, obviously not agglutinated, 0.02 to 0.04 mm. thick. [Probably sessile organism, belonging to worms (?Serpulidae).] *U.Trias.(Rhaet.,* reef ls.), Eu.(S.Ger.-N. Alps-S.Alps-NW.Yugosl.-AegeanSea-?C.Italy).——

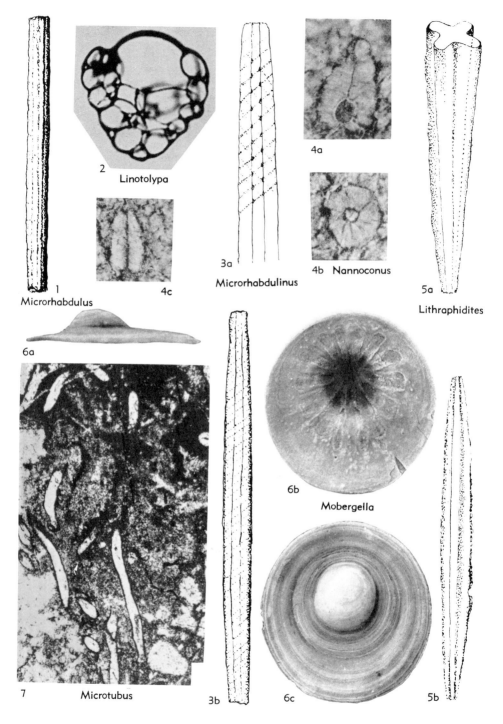

2
Linotolypa

4a

4b Nannoconus

3a
Microrhabdulinus

1
Microrhabdulus

4c

5a

Lithraphidites

6a

7   Microtubus

3b

6b

Mobergella

6c

5b

Fig. 98. Microproblematica (p. *W*161, 163).

FIG. 98,7. **M. communis*, Dachstein-Riffkalk, Aus.(Donnerkogel); ✕30 (Flügel, 1964).

Mobergella HEDSTRÖM, 1923, p. 5 [**Discinella holsti* MOBERG, 1892, p. 5; M]. Phosphatic shell with excentrical apex, circular or ovate, convex, flattened, diameter 0.6 to 6.5 mm.; 7 pairs of muscle imprints, bilaterally arranged, radiating from apical region on inner side; great morphological variability in shell types, e.g., in degree of convexity (three species distinguishable). [MOBERG (1892) described *Mobergella* as species of brachiopod *Discinella* HALL, 1872; HEDSTRÖM (1923) referred it to patellacean gastropods; POULSEN (1963) suggested affinities to Monoplacophora; according to FISHER (1962, p. *W*132), "undoubtedly a hyolithelminth operculum"; ÅHMAN & MARTINSSON (1965) regarded it as probably a sedentary hyolithellid, and BENGTSON (1968) interpreted this form as operculum of a ?sedentary tube-dwelling organism.] *L.Cam.* (chiefly in glacial drift boulders), Eu.(Swed.-Nor.)-NE.Asia(Sib.).——FIG. 98,6. **M. holsti* (MOBERG), Swed.(Venenäs); *6a-c*, lat., concave, and convex views, ✕15 (Bengtson, 1968). [Concerning the validity of the name *Mobergella* given by HEDSTRÖM (1923) as a junior objective synonym of *Discinella* HALL, 1872, and his use of the name homonymously for a genus separated from *Discinella* and based on *Discinella holsti* MORBERG, 1892, see BENGTSON, 1968, p. 330.]

Mooreopsis FLOWER, 1961, p. 112 [**M. rotundus*; OD]. Solid round bodies, 1 mm. in diameter, consisting of finely granular calcite; surface rounded, ?hemispherical; without central cavity; broadly attached to *Catenipora*; observed only in thin sections. [Systematic position doubtful.] *Ord.,* USA(Texas).

Moundia FLOWER, 1961, p. 108 [**M. fibrosa*; OD]. Low-arched bodies, 2 mm. long, 1.5 mm. wide, 1 mm. high, consisting of few thick calcareous plates with vertical fibrous appearance; small central cavity with aperture in ?anterior end; attached to coral *Catenipora*; observed only in thin sections. [Systematic position unknown.] *Ord.,* USA(Texas).

Nannoconus KAMPTNER, 1931, p. 288 [**N. steinmanni*; OD]. Microscopically small peg-shaped microorganisms with axial canal or large central cavity, 5 to > 50 microns (commonly 15 to 20μ) long, 5 to 10 microns wide; outline conical, spherical, pear-shaped, barrel-shaped, cylindrical, or U-shaped; composed of numerous wedge-shaped individual elements; wedges arranged in mounting spiral or spirals, oblique to axis; 2 terminal apertures. [Systematic position "still obscure" (BRÖNNIMANN, 1955); ?skeletal remains of planktonic Protozoa; regarded as embryonic stage of *Lagena* or *Lagena* proper (DE LAPPARENT, 1931), as alga (CADISCH, *fide* COLOM, 1945) and as belonging to *Fibrosphaera* DE LAPPARENT (COLOM, 1945); ?relationship to oogonia of algae (BRÖNNIMANN, 1955); see also CAMPBELL, 1954, p. *D*170-*D*171.] *U.Jur.-U.Cret.,* Eu.-

N.Afr.-C.Am.(W.Indies(Cuba)-?Mexico).——FIG. 98,4a,b. **N. colomi* (DE LAPPARENT), L.Cret., Cuba; *4a,b*, long., transv. secs., *ca.* ✕1,575 (Brönnimann, 1955).——FIG. 98, 4c. *N. steinmanni* KAMPTNER, L.Cret., Cuba; (slightly retouched), *ca.* ✕1,575 (Brönnimann, 1955).

Niccumites FLOWER, 1961, p. 113 [**N. oculatus*; OD]. Spherical or somewhat flattened body, fine-grained calcitic, seemingly finely granular texture; attached to *Catenipora* by irregular masses of coarsely crystalline calcite; observed only in thin sections. [Systematic position unknown.] *Ord.,* USA(N.Mex.).

Palambages WETZEL, 1961, p. 338 [**P. morulosa*; OD]. Spheroidal bodies, morphologically similar to mulberries; about 45 to 120 microns in diameter; composed of 8 to (?) 18 oval membranous cells, 12 to 50 microns in diameter. [?Egg-balls of copepods or coenobia of algae or hystrichosphaerids.] *U.Cret.(Senon.)-Paleoc.(Dan.),* Eu. (Denm.-Pol.).

Palamphimorphium DEFLANDRE & TERS in TERS & DEFLANDRE, 1966, p. 342 [**P. speciosum*; OD]. Microorganism consisting of 2 different parts, somewhat round central body 0.13 mm. long and 0.06 mm. wide, silico-organic, enveloped by translucent quartzose material of similar oblong shape; central body with blackish membrane containing globules of delicate structure. *Ord.,* Eu. (W.France).

Paleocryptidium DEFLANDRE, 1955, p. 184 [**P. cayeuxi*; OD]. Tiny spherical or ellipsoidal hollow bodies with entirely smooth shell of organic material; diameter about 9 microns. [Regarded as ?acritarch by DEFLANDRE (1955), as possible sponge by PACLTOVÁ (1972) whose material, however, may not be congeneric.] *Proteroz.(Briovér.),* Eu.(France-?Czech.). [Description supplied by CURT TEICHERT.]

Papillomembrana SPJELDNAES, 1963, p. 63 [**P. compta*; M]. Compressed bodies of primarily cylindrical or spherical shape, 0.4 to 0.5 mm. in diameter; having thin outer (?carbonaceous) membrane with dense papilla-like protuberances, thin-walled and hollow. [Systematic position unknown, somewhat resembling dasycladacean algae.] *U.Precam.(Esmark.),* Eu.(Nor.)——FIG. 97,7. **P. compta*; ✕355 (Spjeldnaes, 1963).

Papinochium DEFLANDRE & TERS in TERS & DEFLANDRE, 1966, p. 342 [**P. dubium*; OD]. Apparently globular, sometimes deformed or crushed, with hollow blunt appendices; diameter of globule without appendices 0.15 mm.; often fossilized in fibrous chlorite. *Ord.,* Eu.(W. France).

"Parvangulae" HILTERMANN & SCHMITZ, 1968, p. 301. Informal name for white problematical bodies of microscopic size from freshwater sediments; in shape resembling horseshoes; surface smooth, about 0.15 mm. in diameter; ends of horseshoe somewhat tapering and inclined toward

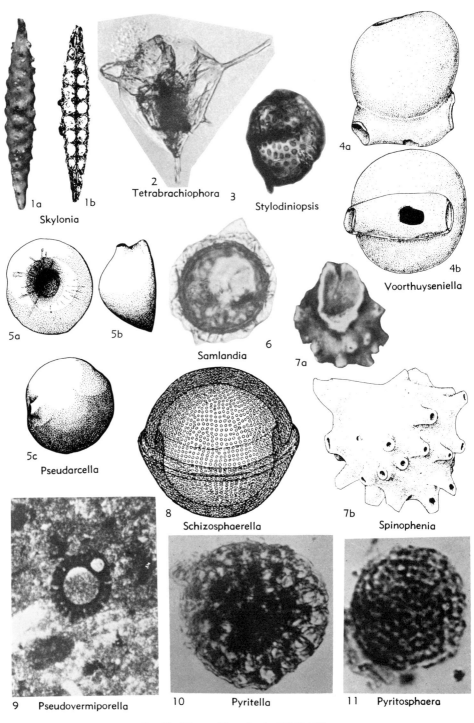

1a 1b
Skylonia

2
Tetrabrachiophora 3
Stylodiniopsis

4a

4b
Voorthuyseniella

5a 5b

Samlandia
6

7a

5c
Pseudarcella

8
Schizosphaerella

7b
Spinophenia

9   Pseudovermiporella

10   Pyritella

11   Pyritosphaera

FIG. 99. Microproblematica (p. *W165-167*).

each other; thin, marginal, rather translucent layer and less translucent inner part, both of microcrystalline apatite, distinguishable in thin sections. [Parvangulae may be interpreted as isolated parts of larger organisms, but inorganic origin is not to be excluded. The generic name *Parvangula* and the species name *randeckensis* were proposed by HILTERMANN & SCHMITZ (1968) as conditional names and have no standing under the *Code*, Art. 15.] *U.Tert.(Mio.),* Eu.(S.Ger.).──FIG. 97,4. "Parvangulae"; *4a,* cross section showing internal structure, ×1,870; *4b,* numerous specimens, ×70 (Hiltermann & Schmitz, 1968).

Pedicillaria FLOWER, 1961, p. 114 [*P. bifurcata;* OD]. Resembling echinoderm pedicillaria; stalk narrowing from broad attachment, terminating in bifurcated tip; attached to corals; observed only in thin sections. [Systematic position unknown.] *Ord.,* USA(N.Mex.).

Pictonicopila DEFLANDRE & TERS, 1966, p. 240 [*P. polymorpha;* OD]. Vesicular globules with thin membrane, 3 to 60 microns in diameter; mostly united to irregular loose "colonies," 0.3 to 0.4 mm. in size, composed of individuals of equal or various size; globules mostly hollow, some with surface layer consisting of very small polygonal bodies (1-3 microns in size), closely spaced. *Ord.,* Eu.(W.France).

Plutoneptunites DEFLANDRE, 1961, p. 127 [*P. antarcticus;* M]. Interlacing and occasionally coalescing filaments of varied shape, straight or curved; one end rounded, without septa; often with concentric integument, 2 to 3 microns thick; sometimes acute-angled or rectangularly branched; similar to colonies of blue algae. [Systematic position unknown; *"au sein des Protocaryotes"* (DEFLANDRE, 1961), ?Cyanophyceae, ?fungal hyphae.] In limestones of unknown age; Ind.O. (Kerguelen I.).

Polysiphonidia EISENACK, 1971, p. 458 [*P. enigmatica;* OD]. Oval central body, somewhat irregular in outline, 72 to 180 microns in diameter; with very small tubes radiating irregularly; 2 kinds of tubes to be distinguished; 7 to 20 larger ones, structureless, 10 to 15 microns in diameter, 420 microns long (max.); 2 to 5 smaller ones, short, annulated, 52 microns long (max.); central body and tubes chitinous. [Systematic position unknown.] *U.Sil.(Beyrichia limestone,* Pleist. drift); Eu.(Pol., Pomerania).──FIG. 97,1. *P. enigmatica;* holotype, *1a,* ×467; *1b,* ×130 (Eisenack, 1971).

Pseudarcella SPANDEL, 1909, p. 199 [*emend.* LE CALVEZ, 1959; *emend.* LINDENBERG, 1965] [*P. rhumbleri;* M] [=*Pseudoarcella* SZCZECHURA, 1969 *(nom. null.)*]. For description as foraminifer, see LOEBLICH & TAPPAN (1964, p. C522). [According to LINDENBERG (1965, p. 28), not a protozoan; LOEBLICH & TAPPAN (1968) assigned this genus to the tintinnids.] *L.Tert.(low.Eoc.),*

Eu. (?France-Belg.); *L. Tert. (up. Eoc.),* Eu. (?France-SE.Pol.-Belg.); *L.Tert.(Oligo.),* Eu. (Ger.-Aus.),──FIG. 99,5. *P. rhumbleri* SPANDEL, L.Tert.(Oligo.), Aus.; *5a-c,* different views, ×73 (Lindenberg, 1965).

Pseudovermiporella ELLIOTT, 1958, p. 419 [*P. sodalica;* OD] [=?*Vermiporella* STOLLEY, 1893, p. 140 (type, *V. fragilis*); for discussion see KOCHANSKY & HERAK, 1960, p. 72; ELLIOTT, 1962, p. 40]. Small calcitic tubes, diameter up to 1.4 mm., meandriform, forming tangled coils or loops; tubes consisting of an innermost, thin, dark compact-walled tube and an outer tubular layer pierced by numerous radial pores 0.03 to 0.04 mm. in diameter; pores approximately at right angles to surface and forming distinct regular mesh, which has a dark calcareous layer on its inner surface; tubes showing creeping habit, occurring free or attached to other ones or to shell fragments. [Tentatively considered as unusual primitive dasyclad alga.] *Sil., Perm.,* Eu. (Yugosl.-USSR)-Asia (Arabia-N.Iraq-?Japan)-?N. Afr.(?Tunisia).──FIG. 99,9. *P. sodalica,* Arabia; transv. sec. of small individual, ×50 (Elliott, 1958).

Pyritella LOVE, 1958, p. 433 [*P. polygonalis;* OD]. Oval to round microorganisms, diameter 20 to 55 microns, composed of closely packed translucent cells, appearing roughly polygonal at surface, separated by walls up to 1 micron thick; found in pyrite aggregates of the *"Kies-Kügelchen"* type (NEUHAUS, 1940). [Systematic position unknown, probably of plant origin (?fungi).] *L. Carb.,* Eu.(Scot.).──FIG. 99,10. *P. polygonalis,* holotype; *ca.* ×1,000 (Love, 1958).

Pyritosphaera LOVE, 1958, p. 433 [*P. barbaria;* OD]. Spherical or subspherical microorganisms, ranging in diameter from 2 to 35 microns, with uniform and closely packed radial spines covering outer surface; spiny processes rapidly tapering, 1 to 2 microns long; surface of organisms closely associated with iron pyrite and always coated by this mineral; organisms obtained from framboidal granules of the *"Kies-Kügelchen"* type (NEUHAUS, 1940). [Systematic position unknown, perhaps of plant origin, but probably neither bacteria nor actinomycetes.] Described under name *Pyritosphaera* only from Cambrian and Lower Carboniferous of Europe, but occurring worldwide in marly sediments of every age. *Cam.(up. Revin.),* Eu.(Belg.); *L.Carb.,* Eu.(Scot.).──FIG. 99,11. *P. barbaria,* L.Carb., Scot.; *ca.* ×2,000 (Love, 1958).

Samlandia EISENACK, 1954, p. 76 [*S. chlamydophora;* OD]. Globular or slightly ellipsoidal body, 70 to 90 microns in size; with inner thick-walled integument, enveloped by second very thin integument, both connected by numerous small pillars; apical slip-hole. [Perhaps an acritarch.] *Tert.(low.Oligo.),* Eu.(Samland, formerly NE.

Ger., now USSR).——Fig. 99,6. *S. chlamydophora;* ×420 (Eisenack, 1954).

Schizosphaerella DEFLANDRE & DANGEARD, 1938, p. 1116 [*S. punctulata;* M] [=*Nannopatina* STRADNER, 1961, p. 78 (DEFLANDRE, 1971, pers. commun.)]. Calcareous globules, 12 to 30 microns in size, consisting of 2 valves of dissimilar shape, commonly occurring separated from one another; one with marginal circular furrow into which other dome-shaped valve is fitted; both valves with sometimes irregular punctation; this ornamentation similar to that of diatom *Pyxidicula*. [Planktonic organism of unknown systematic position.] *M.Jur.(Bajoc.)-U.Jur.(low.Oxford.),* Eu.(W.France, Normandy).——Fig. 99,8. *S. punctulata,* U.Jur.; oblique view, ×1,875 (Deflandre & Dangeard, 1938).

Skylonia THOMAS, 1961, p. 359 [*S. mirabilis;* OD]. Small spindle-shaped fossil, calcareous, slender, fusiform, tapering symmetrically, 2 mm. long, chambered; chambers in close contact, arranged quadriserially, hexagonal externally, *ca.* 15 chambers in each longitudinal row, length and width of median chamber 0.17 mm., maximum width 0.35 mm. [Systematic position unknown.] *Tert.(low.Mio.),* Afr.(Kenya).——Fig. 99,1. *S. mirabilis; 1a,* surface view, ×25; *1b,* long. thin section, ×23 (Thomas, 1961).

Slocomia FLOWER, 1961, p. 111 [*S. quadrata;* OD]. Calcareous bodies, subquadrate in cross section, 2 mm. long, 1 mm. high; surface extensively perforate; attached to corals; observed only in thin sections. [Systematic position unknown.] *Ord.,* USA(N.Mex.).

Spinophenia SZCZECHURIA, 1969, p. 89 [*S. multituba;* OD]. Calcareous test, hyaline, spherical; 0.24 to 0.34 mm. high, 0.3 mm. wide; wall thin, finely perforated; surface of test covered by numerous thornlike tubes, irregularly arranged, open, with aperture mostly surrounded by "collar" of heterogeneous shape; aperture of very different size and shape. [Perhaps related to rhizopods.] *L.Tert.(up.Eoc.),* Eu.(SE.Pol.).——Fig. 99,7. *S. multituba; 7a,b,* apertural view, ×80, ×225 (Szczechura, 1969).

Stylodiniopsis EISENACK, 1954, p. 75 [*S. maculatum;* OD]. Small oval or pear-shaped integument, thin-walled, 80 to 100 microns in size; with thin pedicle; covered with numerous, uniformly distributed, circular or oval spots; somewhat similar to *Palaeoperidinium spinosissimum* DEFLANDRE and *Stylodinium* KLEBS (dinoflagellate). [Systematic position uncertain.] *L.Tert.(low.Eoc.),* Eu.(N.Ger., Isle of Fehmarn); *L.Tert.(low.Oligo.),* Eu.(Samland, formerly NE. Ger., now USSR).——Fig. 99,3. *S. maculatum,* Samland; *ca.* ×250 (Eisenack, 1954).

Tetrabrachiophora EISENACK, 1954, p. 76 [*T. natans;* M]. Approximately globular integument (chitinous?), very thin, diameter about 100 microns; with 4 thin cylindrical branches of un-

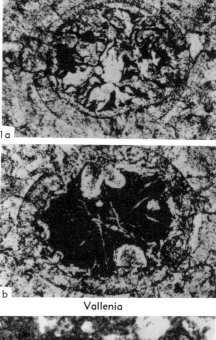

Vallenia

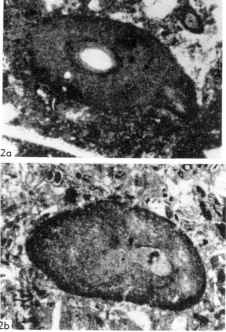

Tubiphytes

FIG. 100. Microproblematica (p. *W167*).

known length, diameter 5 microns. [Very probably integument of planktonic organism.] *L.Tert.(low.Oligo.)*, Eu.(Samland, formerly NE.Ger., now USSR).——Fig. 99,2. **T. natans*; ×290 (Eisenack, 1954).

Tholella Flower, 1961, p. 105 [**T. idiotica*; OD]. Thin-walled, high-arched test consisting of numerous plates, 1 mm. wide, 1.2 mm. high; main cavity supplemented by an accessory one; attached to corals; observed only in thin sections. [Systematic position doubtful.] *Ord.*, USA(N. Mex.-Texas-Utah).

Triangulina Quilty, 1970, p. 180 [**T. aequilateralis*; OD]. Test triangular or subtriangular, at least one angle quite sharp, diameter 0.5 to 0.7 mm., surface smooth; subdivided into 4 to 6 chambers arranged bilaterally; external wall perforate, 2 to 5 times as thick as intercameral walls, which are commonly bilamellar; walls usually composed of fibrous calcite; "aperture" a narrow conical opening. [Foraminiferal relationship is excluded because of wall structure; some similarities to Crisiidae (Bryozoa) exist, but affinities improbable.] *Tert.(up.Eoc.)*, SW.Australia; *Tert.(up.Oligo., Janjukian-low.Mio., Longford.)*, Australia(N.Tasmania-Victoria).——Fig. 97,3. **T. aequilateralis*, low.Mio., Tasmania(Cape Grim); *3a,b*, ×70 (Quilty, 1970).

Tubiphytes Maslov, 1956, p. 82 [**T. obscurus*; M] [*=Nigriporella* Rigby, 1958, p. 584 (type, *N. magna*); for discussion see Konishi, 1959]. Pear- or quince-shaped symmetrical ovoid bodies, about 2 mm. long (max. 6 mm.); composed of fibrous calcareous material; outer margins sharply delimited; usually with 1 or 2 small circular inclusions suggesting an initial tube or vacuole; rock-forming organism in Permian of USSR. [Systematic position questionable; described by Maslov as alga; interpreted by Rigby as hydrozoan.] *Carb.*, USA;-*Perm.*, Eu.(Aus.)-USA (Texas-N.Mexico-Wash.)-Asia (SE.Arabia-N.Iraq-USSR-Afghan.-Burma-S.China-Japan)-C.Am.(Guatem.)-N.Am.(Mexico).——Fig. 100,2. *T.* sp. (*=Nigriporella*), L.Perm.(Zinnar F.), Iraq(Mosul Liwa); *2a,b*, thin section, ×30, ×21 (Elliott, 1962).

Umbella Maslov in Bykova & Polenova, 1955, p. 40 [**U. bella*; OD] [*=Umbellina* Loeblich & Tappan, 1961, p. 284; obj.]. Globular to ovoid, hollow calcite bodies, long diameter as much as 0.45 mm., short diameters 0.40 mm.; shell composed of radially oriented calcite fibers, ranging from 10 to 150 microns in thickness. [Originally regarded as foraminifer; by others believed to be related to Charophyta or of uncertain affinities; name possibly applied to two different kinds of organisms (see *Treatise*, p. *C*322; Teichert, 1965, p. 103; Poyarkov, 1966; Veevers, 1970, p. 180; Peck, 1974)]. *Dev.*, N.Am.(USA), Eu.(Belg.), USSR), Australia. [Description prepared by Curt Teichert.]

Urnulella Szczechura, 1969, p. 90 [**U. costata*; OD]. Scoop-shaped calcareous test, fibrous, imperforate; about 0.4 mm. high, 0.3 mm. wide; surface ornamented by narrow ribs; distinct lateral ridge along longer axis; aperture approximately triangular; surrounded by a "collar"; shape and ornamentation of test rather varied. [Perhaps allied to rhizopods.] *L.Tert.(up.Eoc.)*, Eu.(SE. Pol.).——Fig. 97,2. **U. costata; 2a,b*, side view, apertural view, ×43 (Szczechura, 1969).

Vallenia Raunsgaard Pedersen, in Bondesen, Raunsgaard Pedersen, & Jørgensen, 1967, p. 20 [**V. erlingi*; OD]. Regular globular to flat elliptical structures of complex and uniform nature, diameter 0.25 to 1.5 mm., consisting of outer and inner spherical layer, both 3 to 5 microns thick, and 30 to 120 microns apart, occasionally connected by small indistinct radial "pillars"; in inner part of globules inside inner layers lies a dark to opaque carbonaceous core with irregular outer limitation; in a few specimens a special cellular structure occurs in outer part of core. [Abundant in dark grey dolomite; photosynthetic ?planktonic organisms, of ?vegetal origin, systematic position and phylogenetic affinity uncertain.] *L.Proteroz.(Ketilid., Vallen Gr.)*, SW.Greenl.—— Fig. 100,1. **V. erlingi*, Grænsesø F.; *1a,b*, thin sec., ×37.5 (Bondesen, Pedersen, & Jørgensen, 1967).

Voorthuyseniella Szczechura, 1969, p. 82 [**V. lageniformae*; OD] [*=Lagena-x* Voorthuysen, 1949, p. 31; for discussion see Szczechura, 1969, p. 83)]. Test calcareous, thin, hyaline, imperforate surface smooth, glossy; test consisting of globular or compressed main part and elongated horizontal gutterlike part at its base; in globular part a round or slitlike aperture, in basal part 2 lateral openings and one in the middle of bottom, latter sometimes prolonged into short internal tube; 0.25 to 0.3 mm. high, 0.27 to 0.37 mm. wide. [Similar to *Bicornifera* Lindenberg, 1965; systematic position unknown, assignment to Foraminifera excluded by Szczechura, 1969.] *Tert. (low.Eoc., Ypres.-up.Plio.)*, Eu.(France-Belg.-Neth.-Ger.-Port.-Italy)-S.USA (Ala.)-E.Asia(Taiwan), *Rec.*(S.China Sea-Gulf Mexico).——Fig. 99,4. **V. lageniformae*, up.Eoc., Pol.; *4a,b*, viewed from different sides, ×225 (Szczechura, 1969).

Warthinites Flower, 1961, p. 117 [**W. adhaerens*; OD]. Low-spired, widely umbilicate shells; spire slightly convex; 1 mm. in diameter, 0.6 mm. high; whorl gently rounded; attached by surface of spire on *Catenipora*; observed only in thin sections. [Systematic position unknown.] *Ord.*, USA(N.Mex.).

Xenokalymma Eisenack, 1968, p. 306 [**X. trematophora*; OD]. Very small lid-shaped cases, flat-arched, bean-shaped in outline, with circular frontal opening; consisting of black organic (chitinous?) material; about 1.1 mm. long, 0.5

mm. wide, diameter of the opening *ca.* 0.1 mm. [Affinities to *Xenotheka* Eisenack, 1937, likely.] *U.Ord.(Caradoc., Keila beds, D₂),* USSR(Est.). ——Fig. 97,6. **X. trematophora;* ×30 (Eisenack, 1968).

Xenotheka Eisenack, 1937, p. 239 [**X. klinostoma;* M]. Small, approximately loaf-shaped integument, consisting of dark chitinous material, 0.5 mm. long, 0.2 mm. high; flat-bottomed, with annulated tube tending obliquely upward, tube

with circular opening; somewhat similar to *Xenokalymma* Eisenack, 1968. [Foraminifer, according to Loeblich & Tappan (1964, p. *C*183); Eisenack (1970) suggested relationship to group Graptoblasti Kozłowski, 1949, and doubted (Eisenack, pers. commun., 1971) foraminifer relationship.] *Ord.*(Pleist. drift), Eu.(S.Finl.-Samland, formerly NE.Ger., now USSR).——Fig. 97,8. **X. klinostoma,* S.Finl.; *8a,b,* ×47 (Eisenack, 1970).

PSEUDOFOSSILS

This chapter deals with structures that in one way or another are suggestive of being "fossils," but are certainly or most probably of inorganic origin.

Concretions, clay galls, various trail-like markings, and even mud cracks and structures of diagenetic origin have been described and named as plant or animal fossils. Errors of this type occurred frequently when paleontology was a new field, but more recent examples may also be found (e.g., markings and structures of tectonic or diagenetic origin described by Fucini (1936, 1938) from the Verrucano of Florence and published in two voluminous books accompanied by many plates). Such structures are best grouped under the name "pseudofossils" (Hofmann, 1971).

["Tonrollen" or clay rolls are cylindrical or cornet-shaped bodies of varying sizes that are formed when thin layers of clay break up and the fragments curl up in rolls during desiccation. Such bodies have frequently been mistaken for fossils, or at least as possible fossils. This subject has been discussed by Voigt (1972a) who felt that "Tonrollen" could easily be mistaken for segments of an arthropod carapace. Examples would be "Protadelaidea howchini" Tillyard (in David & Tillyard, 1936, p. 64) and the peculiar bedding plane structures from the Jotnian Sandstone in Sweden reported by Lannerbro (1954). Recently, Elston and Clark discovered "fossil-like objects" in the Precambrian of Arizona (Fig. 101). Their picture of one of the Arizona specimens on the cover of *Geotimes* (December, 1972) launched a

lively discussion by Glaessner (1973a), Chowns (1973), Teichert (1973), and Lindström (1973) who seriously doubted its organic origin, but Elston (written communication, July, 1973) suggested that a unique preservational situation occurred whereby the clay layers might have been covered by thin algal mats or films.—Curt Teichert & W. G. Hakes.]

Obviously, the names listed below are invalid. They are included here at the request of the Editor for their historical interest and for the sake of completeness. Naming of "type species" is, of course, unnecessary. Nevertheless, in those cases in which "type species" have been formally designated or in which the "genera" are "monospecific," they have been cited.

Recently Hofmann (1971) has discussed, in detail, the Precambrian "macro-pseudofossil," Eozoon canadense. This name has no taxonomic status in paleontology. It is proposed that the existing name of such inorganic shapes be retained for discussion and summarization. Pseudofossils should continue to be identified but their names should not be printed in italics. Such names could be used as lithological terms such as oolite, styolite, and similar structures. Thus, Martinsson (1965) called a certain type of ripple mark "Kinneyan ripple marks," with reference to supposed algae described by Walcott (1914), under the "generic" name Kinneyia. In contrast, Osgood (1970, p. 388) rejected "quasilegal names for inorganic forms." They appear in his work, it is true, in italic print, but also in quotation marks.

Fig. 101. Casts of Precambrian pseudofossils in hyporelief from arkose member of Troy Quartzite, central Arizona, formed by curling and desiccation of a thin film of sediment, ×0.15 (from Geotimes, Dec. 1972; photo courtesy U. S. Geological Survey).

[It should be noted that, according to Art. 2(b) of the *International Code of Zoological Nomenclature,* formal names given to objects here regarded as pseudofossils,

if originally described as animal remains, "continue to compete in homonymy with names in the animal kingdom," that is, they cannot validly be used again for animal taxa.—Ed.]

Aenigmichnus HITCHCOCK, 1865, p. 20 [*A. multiformis; M]. Parallel lines, commonly changing to rows of dots or to moniliform lines, covering wide spaces; highly variable. [Surely inorganic (markings of drifting or rolling bodies).] *Trias.,* USA(Mass.).

Aequorfossa NEVIANI, 1925, p. 148 [*A. farnesinae; M]. Pseudofossil, interpreted by NEVIANI (1925) as medusa; see KIESLINGER (1939) and HARRINGTON & MOORE, 1956e, p. F159. *U.Tert. (up.Plio.),* Eu.(Italy).

Ammosphaeroides CUSHMAN, 1910, p. 51. See LOEBLICH & TAPPAN (1964), p. C786.

Anellotubulata WETZEL, 1967, p. 343 ("group"): see Mikrocalyx WETZEL, 1967, p. W176.

Antholithina CHOUBERT, TERMIER, & TERMIER, 1951, p. 28 [*A. rosacea; M]. Almost circular cross sections with radially disposed structures ("septa"), observed in thin sections. [Regarded by authors as calcareous algae. According to SCHINDEWOLF (1956, p. 468), grains with external covering of iron-oxyhydrate which in part has penetrated radially into the interior.] *Precam.,* Afr.(Morocco).

Archaeophyton BRITTON, 1888, p. 123 [*A. newberryanum; M]. Thin films of graphite lying parallel to bedding planes of limestones; at first regarded as "the most ancient plant yet discovered." *Precam.,* USA(N.J.).

Archaeospherina DAWSON, 1875, p. 139 [proposed without species name]. Small globular grains of serpentine distributed homogeneously throughout ophicalcite in highly metamorphic rocks, associated with Eozoon canadense. [Regarded by DAWSON as chamber fillings or germs or buds of Eozoon or as distinct organisms (Foraminifera?); for discussion see HOFMANN (1971, p. 12).] *Precam.,* N.Am.(Can.).

Aristophycus MILLER & DYER, 1878, p. 3 [*A. ramosum; M]. Branching structures; main "stem" dividing into secondary, tertiary and quaternary branches, forming a regular anastomosing raised pattern; main branches 2 to 6 mm. in diameter; secondary branches bifurcating consistently from main branch from below at angle of 30 to 60° off horizontal; preserved as convex epireliefs on the upper surface of the beds. [Described by MILLER & DYER (1878b) as a "fucoid"; interpreted by NATHORST (1881a), JAMES (1884, 1885), DAWSON (1888), and MILLER (1889) as inorganic ("mud washing," rill marks); by SEILACHER (1955, pers. commun.) as *"figures de viscosité";* by OSGOOD (1970) as *incertae sedis* but probably inorganic (?"some form of diagenetic flow pattern"); a seemingly plausible explanation as rill marks is not possible owing to mode of preservation of pattern as convex epire-

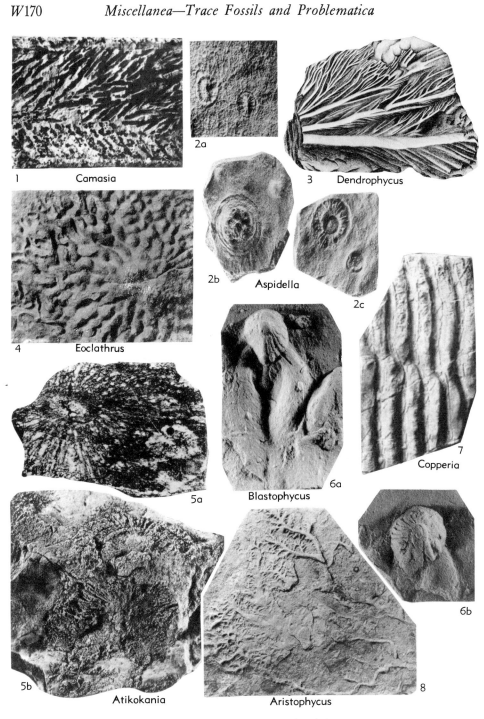

FIG. 102. Pseudofossils (p. *W*169-171, 173).

lief.]   *U.Ord.(Cincinnat.)*,   USA(Ohio).——FIG.   convex epirelief, holotype, ×0.4 (Osgood, 1970).
102,*8*. *A. ramosum, Maysville beds, Cincinnati;   **Aspidella** BILLINGS, 1872, p. 478 [*A. terranovica;

M]. Ovate structures, up to 3 by 4 cm. in size; rooflike ridge in central area of ellipse, with fine radial ridges and grooves extending to periphery; narrow ringlike border; mostly on bedding planes all oriented in one direction; having general aspect of small *Patella* flattened by pressure. [BILLINGS (1872) regarded Aspidella as fossil; MATTHEW (in PACKARD, 1898) interpreted it as slickensided mud concretions striated by pressure; WALCOTT (1899) and VAN HISE & LEITH (1909) were doubtful whether organic or inorganic; regarded by SCHINDEWOLF (1956) as inorganic and identical with Guilielmites GEINITZ; according to GOLDRING (1969), partly attributable to water- or gas-escape structures and interpreted by CLOUD (1968) as compaction and spall marks; according to HOFMANN (1971), inorganic, focused surfaces of rupture; for detailed discussion, complete summary of references, and various interpretations, see HOFMANN (1971, p. 16).] *Precam.*, N.Am.(Can., Newf.).——FIG. 102,2. *A. terranovica, St. John's F., Newf. (near St. John's); *2a-c*, ×1.3 (Hofmann, 1971).

Astrorhiza cretacea FRANKE, 1928, p. 7. Very small tubes, hollow, about 3 mm. long, 0.25 mm. in diameter, consisting of sandy particles bound together by calcareous cement. [Erroneously ascribed to agglutinated Cretaceous foraminifers from North Germany; same valid for "?Astrorhiza laguncula" (BORNEMANN, 1854) in FRANKE (1936, p. 11), from the "L.Jur." of North Germany; according to HILTERMANN (1952, p. 424), representing calcareous integuments around small roots of plants (German, *"Wurzel-Röhrchen"*).]

Atikokania WALCOTT, 1912, p. 17 [*A. lawsoni; OD] [=Attikokania METZGER, 1927, p. 6 (*nom. null.*)]. Pearshaped or cylindrical bodies, silicified in limestones, 3 to 35 cm. in diameter; with 1 or 2 "central cavities" and with radially arranged canals of irregular cross section and a concentric pattern of quartz in limestone. [At the time of its discovery Atikokania was regarded as the most ancient fossil, with varied interpretations: WALCOTT (1912a) compared it with a genus of Archaeocyatha and considered it related to sponges; ROTHPLETZ (1916, p. 73) regarded it as a lithistid sponge similar to *Aulocopium*; inorganic origin proposed by WALCOTT (1914), ABBOTT & ABBOTT (1914), SEWARD (1931), SCHINDEWOLF (1956), GLAESSNER (1962), CLOUD (1968) and others; for a complete list of references and summary of various interpretations see HOFMANN (1971, p. 26); according to HOFMANN (1971, p. 26), chemical, radial crystal growth, diffusion and replacement are involved; see also OKULITCH (1955, p. *E*20) and DE LAUBENFELS (1955, *E*33, *E*103).] *Precam.*, N.Am.(Can.).——FIG. 102,5. *A. lawsoni, Steeprock Gr., former Steep Rock Lake, Ont.; *5a,b*, ×1.7, ×1.3 (Hofmann, 1971).

Batrachoides HITCHCOCK, 1858, p. 121 [jr. hom.; *non* LACEPÈDE, 1800] [*B. nidificans; OD]

[=Batrachioides WEIGELT, 1927; Batracoides ILIE, 1937, *nom. null.*]. Shallow contiguous pits on bedding planes; about 2.5 cm. wide, depth about 1 cm.; compared with similar Recent excavations made by small fishes and tadpoles (SILLIMAN, 1851; HITCHCOCK, 1858). [Reasonably explained by KINDLE (1914) as interference ripples; see *Benjaminichnus* BOEKSCHOTEN, 1964, p. *W*189.] *Sil.*, USA(N.Y.); *Trias.*, USA(Mass.).

Blastophycus MILLER & DYER, 1878, p. 24 [*B. diadematus; M]. Bilobate structure with a budlike attachment at larger end covering junction of branches. [Originally described as "fucoid"; regarded by NATHORST (1881a, p. 97) as probably inorganic in origin; interpreted as cast of an enrolled trilobite associated with scour markings ("current crescent casts" POTTER & PETTIJOHN, 1964) by OSGOOD (1970, p. 390), whose explanation was based on laboratory experiments with *Flexicalymene* specimens; this "fossil," thus, is part body fossil, part inorganic.] *U.Ord.(Cincinnat.)*, USA(Ohio).——FIG. 102,6. *B. diadematus, Eden Gr., Cincinnati; *6a,b*, ×1, ×1.2 (Osgood, 1970).

Camasia WALCOTT, 1914, p. 115 [*C. spongiosa; OD]. Cross sections of compact layerlike bodies of spongioid appearance, numerous irregular tubelike openings. [Regarded as algae by FENTON & FENTON (1936); comparable structures from the Permian of England convincingly proved by HOLTEDAHL (1921) to be inorganic in origin; according to SCHINDEWOLF (1956), most probably diagenetic structures.] *Precam.(Belt Ser.)*, N.Am.(USA, Mont.), *Perm.*, Eu.(Eng.).——FIG. 102,1. *C. spongiosa, Belt Ser. (Newland Ls.), USA(Mont.); vert. sec., ×0.4 (Walcott, 1914).

Cayeuxina GALLOWAY, 1933, p. 156. See LOEBLICH & TAPPAN (1964), p. *C*786.

Chloephycus MILLER & DYER, 1878, p. 3 [*C. plumosum (=Buthotrephis filciformis U. P. JAMES, 1878, p. 9); M] [=Cloephycus DAWSON, 1888, p. 33; *nom. null.*]. Featherlike pattern; "stem" (0.5-5 mm. wide) with fine "filaments" issuing from it at angle of 20 to 30°. [Originally described as "fucoid," but doubtlessly inorganic as recognized by J. F. JAMES (1884), who described the form as "nothing more than a mark or series of marks . . . produced by the running of water down a sloping bank," and by NATHORST (1881a) and DAWSON (1888) (rill marks); according to OSGOOD (1970), modified groove casts.] *U.Ord.(Cincinnat.)*, USA(Ohio).——FIG. 103,5. C. plumosum, Eden Gr., Cincinnati; ×0.94 (Osgood, 1970).

Chondrus(?) binneyi KING, 1850, p. 2. Circular structures, irregularly scattered, each about 2 mm. in diameter, in form of raised ring with central depression. [Originally interpreted by KING as of plant origin; inorganic, according to STONELEY (1958, p. 332), comparable to pit and mound structures.] *U.Perm.*, Eu.(Eng.).

FIG. 103. Pseudofossils (p. *W*171, 173, 175-176).

Collinsia BAIN, 1927, p. 282 [*C. mississagiense; M]. Structure composed of quartz and sericite containing series of ellipsoids consisting of sericitized clay cemented by silica; irregularly grouped around a layered core of oval cross section; walls reputedly showing "cellular structure" [not observed by HOFMANN, 1971, p. 29]; found in massive quartzite; similar structures described as *Vallenia* PEDERSEN, 1966. [Interpreted as colonies of algal cells; according to HOFMANN (1971, p. 29) inorganic, "chemical."] *Precam.,* N.Am. (Can.).

Copperia WALCOTT, 1914, p. 109 [*C. tubiformis; OD] [=Cooperia CHOUBERT, TERMIER & TERMIER, 1951 *(nom. null.)*]. Differs from Greysonia WALCOTT, 1914, in greater irregularity of "growth" and more nearly cylindrical nature of tubes. [According to FENTON & FENTON (1936), identical with Greysonia and both "genera" of inorganic origin; C. ?minima CHOUBERT, TERMIER, & TERMIER, 1951, from the Precambrian of Morocco described as calcareous alga; according to SCHINDEWOLF (1956), type "species" and African "species" originated by diagenetic and tectonic processes.] *Precam.,* USA(Mont.)-?N.Afr. (Morocco).——FIG. 102,7. *C. tubiformis, Belt Ser.(Newland Ls.), Mont.; surface of group of tubes formed in horiz. position, ×0.7 (Walcott, 1914).

Ctenichnites MATTHEW in SELWYN, 1890, p. 147 [no species named]. Straight and parallel striae in sets interfering with each other; very similar to glacial striae. [Inorganic; for discussion see HOFMANN (1971, p. 20); interpreted by him as perhaps a combination of tool and flute marks.] *Precam.-Cam.,* N.Am.(Can.).

Cupulicyclus QUENSTEDT, 1879, p. 577 [no species designated]. Pressure cone, recognized as inorganic by QUENSTEDT himself. *M.Trias.(Muschelkalk)-L.Jur., Tert.,* Eu.(Ger.).

Cyathospongia(?) eozoica MATTHEW, 1890, p. 42. Needles interpreted as sponge spicules [RAUFF (1893) expressed strong doubt about affinity with sponges; according to CLOUD (1968), "probably crystals"; for discussion see HOFMANN (1971, p. 21).] *Precam.,* N.Am.(Can., N.B.).

Dendrophycus LESQUEREUX, 1884, p. 699 [*D. desorii; M]. Characteristic featherlike patterns of long straight, branching ridges, closely spaced, dichotomous, anastomosis lacking, organic material absent. [Originally described as sea weeds (LESQUEREUX, 1844; NEWBERRY, 1888, 1890), interpreted as rill marks (DAWSON, 1888, 1890; JAMES, 1889, SEWARD, 1898, FUCHS, 1894a, and LULL, 1915); more recent interpretation as "dendritic surge marks" by HIGH & PICARD (1968).] *Carb.-Trias.,* N.Am.(USA-Can.).——FIG. 102,3. D. triassicus NEWBERRY, U.Trias., Conn.; ?ca. ×0.2 (Newberry, 1888).

Dexiospira EHRENBERG, 1858, p. 309 *(non Dexiospira* CAULLERY & MESNIL, 1897). [Two species,

no type species designated]. "Fossil" preserved as ?glauconite grains. [According to LOEBLICH & TAPPAN (1964, p. C786), inorganic (small concretionary bodies).] *?L.Sil.,* USSR.

Dinocochlea WOODWARD, 1922, p. 246 [*D. ingens; M]. Very large horizontal bodies, spirally twisted to right or left. [Erroneously described as gastropod steinkerns (WOODWARD, 1922); interpreted by THOMAS (1935) as spiral concretion.] *L.Cret.,* G.Brit.(Eng.).

Dystactophycus MILLER & DYER, 1878, p. 2 [*D. mamillanum; M]. Resembling a small truncated cone, composed of flattened rings, larger ones overlapping smaller ones. [Originally described as alga; interpreted by NATHORST (1881a) as inorganic in origin; according to JAMES (1884), impression of coral base that left its mark in concentric rings; explained by OSGOOD (1970) as casts of markings made by sweeping crinoid stems.] *U.Ord.,* USA(Ohio).——FIG. 104,2. *D. mamillanum, Richmond beds, loc. unknown; ×0.5 (Osgood, 1970).

Eoclathrus SQUINABOL, 1887, p. 552 [*E. fenestratus; M]. Irregular, elongate, ridgelike structures nearly parallel to each other. [Originally described as alga (e.g., E. insignis FUCINI, 1936); doubtlessly of inorganic origin (markings on bedding planes).] *L.Dev., ?L.Perm., Tert.,* Eu. (Italy)-N.Afr.——FIG. 102,4. E. balboi DESIO, L.Dev., N.Afr.; ×0.3 (Desio, 1940).

Eophyton TORELL, 1868, p. 36 [*E. Linnaeanum; M] [=Rabdichnites DAWSON, 1873 *(partim)*; Rhabdichnites DAWSON, 1888 *(nam. van.)*; Eoichnites MATTHEW, 1891, p. 148 *(nom. van.)*; Aspidiaria silurica VLČEK, 1902]. Straight, parallel or curved drag markings on bedding planes, produced by organisms or inorganic objects. [Originally interpreted as plant origin (monocotyledons); eponymous for the Lower Cambrian Eophyton Ss. of Sweden; for short description of various interpretations, see KIESLINGER, 1939.] *Precam.-Rec.,* cosmop.——FIG. 103,1. E. sp., L. Cam.(Mickwitzia Ss.), Swed.; ×0.3 (Regnéll, 1962).

Eopteris morierei DE SAPORTA, ?1878. "Fossil" similar to *Cardiopteris* SCHIMPER; according to GOTHAN (1909) a ferric sulphide dendritic marking; description of "genus" and "species" not found. *Ord.,* Eu.(France).

Eospicula DE LAUBENFELS, 1955, p. E33 [*E. cayeuxi; M]. Needles resembling spicules of calcisponges; lumpy and crooked. [Believed by CAYEUX (1895) to be sponge; certainly inorganic in origin as shown by RAUFF (1896) and SCHINDEWOLF (1956).] *Precam.,* Eu.(France).

Eozoon DAWSON, 1865, p. 54 [*E. canadense; M] [=Eophyllum HAHN, 1880, p. 71 *(nom. van.)*]. Banded structures of coarsely crystalline calcite and serpentine. [Originally interpreted as gigantic Foraminifera; doubtlessly inorganic; for detailed discussion of 5 various types differing by texture

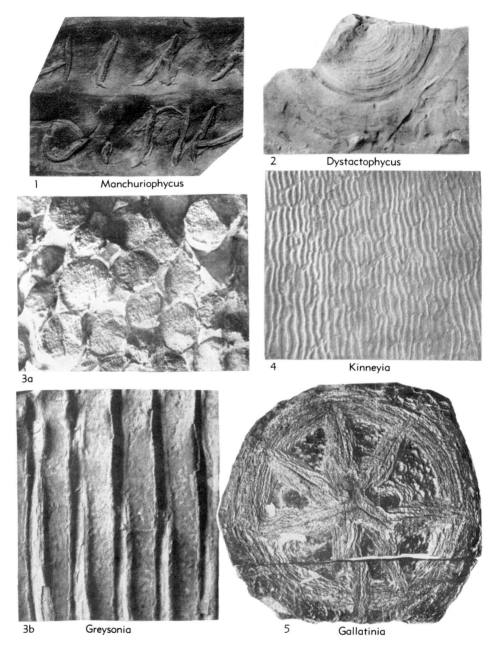

FIG. 104. Pseudofossils (p. *W*173, 175-176).

and mineral assemblage due to different metamorphic facies see HOFMANN (1971, p. 6); for a thorough historical summary of Eozoon papers see HOFMANN (1971, p. 6, fig. 4, pl. 1,2).] *Precam.,* N.Am.(Can.).

Flabellaria johnstrupi HEER, 1883, p. 70. Ripple marks, according to SCHENK (1890), not palm leaves as believed by HEER. *Tert.,* Greenl.

Forchhammera GÖPPERT, 1860, p. 438 [*F. silurica;* M]. Originally interpreted as alga; according to C. POULSEN and A. ROSENKRANTZ (pers. commun. to HÄNTZSCHEL, 1956), inorganic;

now interpreted as probably dendritic markings. *L.Ord.,* Eu.(Denm., Bornholm).

Gallatinia WALCOTT, 1914, p. 116 [*G. pertexa; OD]. Discoid, flattened, circular "fossil" (20 cm. in diameter), with several raylike "arms" more or less irregularly arranged. [Originally described as of plant origin; according to RAYMOND (1935) and SCHINDEWOLF (1956), inorganic, to be interpreted as separation concretion.] *Precam.,* USA (Mont.).——FIG. 104,5. *G. pertexa, Belt Ser. (Newland Ls.); upper surface, ×0.3 (Walcott, 1914).

Gloeocapsomorpha tazenakhtensis CHOUBERT, TERMIER, & TERMIER, 1951, p. 30. "Organisms" observed in thin sections of limestones; interpreted as calcareous algae. [According to SCHINDEWOLF (1956), certainly inorganic structures produced by combination of tectonic movements and metamorphic recrystallization.] *Precam.,* Afr.(Morocco).

Gothaniella FUCINI, 1936, p. 69 [*G. sphenophylloides; M]. Small rosettes, occurring together with bigger and more pronounced ones called Sewardiella FUCINI, 1936. [Interpreted by FUCINI (1936) as algae, by SACCO (1940) as *?Sphenophyllum;* according to PIA (1937) and GOTHAN (1942), doubtlessly inorganic.] *L.Cret.("Verrucano"),* Eu. (Italy).——FIG. 103,3. *G. sphenophylloides, Verrucano, Italy; ×2 (Fucini, 1936).

Grammichnus HITCHCOCK, 1865, p. 19 [*G. alpha; M]. Series of elongate impressions, repeated serially. [According to HITCHCOCK (1865), origin doubtful; interpreted by BROWN (1912) and LULL (1915) as probably roll or drag markings.] *Trias.,* USA(Mass.).

Greysonia WALCOTT, 1914, p. 108 [*G. basaltica; OD]. Large "tubes," irregularly rhomboidal or pentagonal in cross section; ends of group of "tubes" similar to group of very small basaltic columns. [Originally described as alga; very similar forms from Permian of England discussed by HOLTEDAHL (1921) and considered inorganic in origin; according to RAYMOND (1935), shrinkage cracks; interpreted by FENTON & FENTON (1936) as results of segregation of CaCO₃ and dolomite by percolating water; according to SCHINDEWOLF (1956), partly resembling ripples transformed by tectonic and diagenetic processes.] *Precam.,* USA(Mont.); *Perm.,* Eu.(Eng.).——FIG. 104,3. *G. basaltica, Precam.(BeltSer., Newland Ls.), USA(Mont.); 3a, view of end of tubes, ×0.7; 3b, sec. of mass of basalt-like columns, ×0.7 (Walcott, 1914).

Guilielmites GEINITZ, 1858, p. 19 [no type species designated] [=Calvasia sp. VON STERNBERG, 1820; Carpolites umbonatus VON STERNBERG, 1825; Cardiocarpum umbonatum BRONN, 1837; Carpolites clipeiformis GEINITZ, 1856; Gulielmites QUENSTEDT, 1867, *nom. null.*; Gulielmites DAWSON, 1873 *(nom. null.)*; ?Gaussia CHACHLOF, 1934 *(partim)*; ?Gaussia NEUBURG, 1934; Guilel-

mites FUCINI, 1936 *(nom. null.)*; Verrucania FUCINI, 1936]. Discoidal or ellipsoidal bodies, 1 to 5 cm. in diameter, with central depression or raised middle part; surface shining, weakly radially striated; occurring only in very fine-grained sediments such as shales and similar rocks. [Many different interpretations (fruits or seed of plants; especially palms; cones of conifers or *Araucaria*; concretions; diagenetic structures; burrows of pelecypods or soft-bodied animals) have been completely discussed in detail by ALTEVOIGT (1968b), who explained Guilielmites as the result of a special kind of slipping in the rock. He described it as a round or oval plastic clay body originated by rolling movement around a nuclear body of plant or animal origin which was compressed during diagenesis to the final discoidal or ellipsoidal shape, interpreting the weak striation on surface as slickensides. Disclike objects with fine radiating ridges have been described by WEBBY (1970c) who rejected a slickenside origin for them.] *Carb.-Tert.,* Eu.-Am.-Asia-Australia.——FIG. 103,2. *G. umbonatus (VON STERNBERG), L.Perm., Ger.; 2a,b, ×1 (Geinitz, 1856).

Halichondrites graphitiferus MATTHEW, 1890, p. 43. Long, thin "spicules" in graphitic shales and graphite lenses. [Interpreted by MATTHEW as sponge spicules; regarded by RAUFF (1893) as inorganic (systems of striae on graphite flakes?); according to CLOUD (1968), probably crystals; for other interpretations as crystal striations on cleavage planes or scratch markings or striations made by mineral impurities, see HOFMANN (1971, p. 22).] *Precam.,* Can.(N.B.).

Halleia FUCINI, 1936, p. 81 [*H. penicillata; M]. Not of plant origin; certainly inorganic; probably very slender flow markings. *L.Cret.("Verrucano"),* Eu.(Italy).

Hirmeria FUCINI, 1936, p. 103 [*H. notabilis; M]. Small parallel wrinkles, somewhat resembling Eoclathrus SQUINABOL, 1887; doubtlessly inorganic. *L.Cret.("Verrucano"),* Eu.(Italy).

Hurdia ?davidi CHAPMAN, 1926, p. 79. Reexamination of this supposed phyllocarid by BANKS (1962) proved it to be marking of tectonic origin. *Cam.,* Australia(Tasmania).

Interconulites DESIO, 1941, p. 83. Suggested as international name for cone-in-cone structures.

Kempia BAIN, 1927, p. 281 [*K. huronense; M]. Structures composed of rhythmic, curved, and regularly branching laminae of silica and less resistant weathering material; between resistant laminae to a pattern of "cellular" structure (fine tubuli or platelets, 0.2 mm. wide); observed in massive quartzite or argillite; somewhat similar to Newlandia WALCOTT. [Structures originally regarded as walls of colonial organisms, partly resembling stromatoporoids; according to HOFMANN (1971, p. 29), physiochemical phenomenon with diffusion banding resulting from rhythmic precipitation.] *Precam.,* N.Am.(Can.).

Kinneyia WALCOTT, 1914, p. 107 [*K. simulans; OD]. Reliefs reminiscent of very small ripple marks; 1 to 3 mm. wide, approximately parallel; similar to *Furchensteine* (furrow-stones) or corroded limestone flags. [Originally described as algae; regarded by RAYMOND (1935) and FENTON & FENTON (1936) as inorganic in origin; according to SCHINDEWOLF, probably ripplemarks, perhaps somewhat deformed by diagenetic or tectonic processes; Kinneyia dubia and K. labyrintica DESIO (1940) from the Lower Devonian of North Africa certainly of inorganic origin; generic name has been used adjectively by MARTINSSON (1965) for characterizing minute ripplelike structures observed in the Cambrian of Sweden ("kinneyian ripples").] *Precam.,* USA(Mont.); *Cam.,* Eu. (Swed.); *?Sil.,* N.Afr.——FIG. 104,4. *K. simulans, Precam.(Belt Ser., Newland Ls.), USA (Mont.); upper surface, ×0.7 (Walcott, 1914).

Kraeuselia FUCINI, 1936, p. 82 [*K. verrucana; M]. Narrow, long, tapering swellings, apparently screwshaped, twisted. [Inorganic.] *L.Cret.("Verrucano"),* Italy.

Lithodictuon CONRAD, 1837, p. 167 [*L. beckii; M] [=Dictuolites CONRAD, 1838 *(nom. van.);* Dictyolites DAWSON, 1888 *(nom. null.)]*. Mud cracks, at first interpreted as plants. [Plant origin first questioned by HALL (1852).] *Sil.,* USA(N.Y.).

Manchuriophycus ENDO, 1933, p. 47 [*M. yamamotoi; OD]. Shrinkage cracks, in part (e.g., M. yamamotoi, M. inexpectans) in normal form of polygons. [Erroneously interpreted by ENDO (1933) as fillings of soft cylindrical stems of algae; explained by LEE (1939) as worm burrows; M. sawadai YABE (1939) and M. sibiricus MASLOV (1947) are flexuous or even curved spindleshaped sand-bodies with tapering ends, occurring mostly in troughs of simple or interference ripples; M. sawadai regarded by YABE (1939) as cylindrical organism without hard external crust *(inc. sed.),* and M. sibiricus interpreted by MASLOV (1947, 1956) with some doubt as of plant origin. All curved forms according to HÄNTZSCHEL (1949) are sinusoidal contraction cracks; the same or similar structures have repeatedly been described (not all being named Manchuriophycus) from sediments of various ages; more recent papers (discussion in SCHINDEWOLF, 1956) are mostly in agreement with HÄNTZSCHEL (1949) as inorganic in origin, partly as organic (even metazoan) in origin; for later descriptions and discussions of such structures, see FRAREY & MCLAREN (1963), BARNES & SMITH (1964), YOUNG (1967), HOFMANN (1967), DONALDSON (1968), LAUERMA & PIISPANEN (1967), CLOUD (1968), GLAESSNER (1969).] *Precam.-Trias.,* Eu.-Asia-N.Am.-Greenl.——FIG. 104,1.

M. sawadai YABE, Precam., Asia; ×0.4 (Yabe, 1939).

Matthewina GALLOWAY, 1933, p. 157. See LOEBLICH & TAPPAN (1964), p. C786.

Medusichnites MATTHEW, 1891, p. 143 [No species named] [=Taonichnites MATTHEW in SELWYN, 1890, p. 146]. Group of striae, more or less parallel; converging from furrowed margin. [Interpreted by MATTHEW as drag markings made by numerous tentacles of medusoid; doubtlessly inorganic; regarded by HOFMANN (1971, p. 19), particularly "Medusichnites Form γ" MATTHEW, 1891, as sole markings; similar structures reproduced in the laboratory; for discussion see HOFMANN, 1971, p. 20.] *Precam.,* N.Am.(Can.).

Membranites FUCINI, 1938, p. 216 [Three "species," no "type species" designated]. Probably inorganic. *L.Cret.("Verrucano"),* Eu.(Italy).

Mikrocalyx WETZEL, 1967, p. 344 [*M. pullulans forma syringata; OD]. Very small cylindrical tubes of chitin-like or calcareous material, finely annulated, about 250 microns long, 50 microns thick; "anterior" end funnel-shaped, expanding, "posterior" end open and bent like a hook; found only in fragments; "varieties" very heterogeneous; many "formae" distinguished. [Originally interpreted to possibly have been parts of some larger organisms settling in colonies; "genus" representative of WETZEL's "group" Annellotubulata (p. W169); similar forms have been described from the Ecca Series by MCLACHLAN (1973) who considered that they formed inorganically (MCLACHLAN, written commun. to W. G. HAKES, 1973). Recently PICKETT & SCHEIBNEROVA (1974, p. 100) described "anellotubulates" resulting from reaction of hydrogen peroxide with certain iron minerals. An inorganic origin has also been described by RICHARDSON et al. (1973).—W. G. HAKES.] *Perm.,* S.Afr.; *L.Jur.(up.Lias.),* Eu.(N. Ger., from boreholes in Holstein).

Neantia LEBESCONTE, 1887, p. 786 [Four "species," no "type species" designated]. Wrinklelike structures. [Interpreted by LEBESCONTE as sponge; organic origin first doubted by DEWALQUE (1887); according to SEILACHER (1956), rill marks or ripplemarks; regarded by CLOUD (1968) as ripplemarks.] *Precam.,* Eu.(France).——FIG. 105,2. N. rhedonensis LEBESCONTE; ×0.8 (Lebesconte, 1887).

Newlandia WALCOTT, 1914, p. 104 [*N. frondosa; OD]. Irregular hemispherical or bowl-shaped bodies; diameter up to 80 cm.; consisting of concentric, subparallel, subequidistant layers; similar to *Collenia* or *Cryptozoon*. [Interpreted as algae by WALCOTT (1914), FENTON & FENTON (1936), and EDGELL (1964); according to PIA (1936) (describing and discussing similar Triassic specimens from Spain called "Newlandien"), inorganic in origin (rhythmical precipitates); other similar forms found by HOLTEDAHL (1921) in Permian of England and explained as inorganic structures;

regarded by SCHINDEWOLF (1956) and GLAESSNER (1962, p. 471) as formed by diagenetic processes.] *Precam.,* USA(Mont.)-W.Australia; *Perm.,* G.Brit. (Eng.); *Trias.,* Eu.(Spain).——FIG. 103,4. **N. frondosa,* Precam. (Belt Ser., Newland Ls.), USA (Mont.); upper surface, large frond, ✕0.5 (Walcott, 1914).

Nipterella paradoxica HINDE, 1889, p. 144 [=Calathium paradoxicum BILLINGS, 1865, p. 358]. Regarded by BILLINGS (1865) and several later authors as sponge; on reexamination of holotype, NITECKI (1968) recognized this form as a cherty concretion riddled with dendrites of pyrolusite; see also NITECKI's list of synonyms. *L.Ord.,* N. Am.(E.Can.).

Palaeotrochis EMMONS, 1856 [Two "species," no type species designated]. Double cone, with grooved surface; cones juxtaposed base to base. [Regarded by EMMONS (1856) as a coral ("the oldest organic body yet discovered"); according to HALL (1857), "nothing but concretions"; compared by MARSH (1868) with cone-in-cone-structures; interpreted by DILLER (in WALCOTT, 1899) as biconical spherulites in an acid volcanic rock.] *Precam.,* USA(N.Car.).——FIG. 105,*1a,b,e.* P. minor; mag. unknown (Emmons, 1856).——FIG. 105,*1c,d,f.* P. major; mag. unknown (Emmons, 1856).

Palmacites martii HEER, 1855, p. 97 [=Palmanthium martii SCHIMPER, 1870, p. 506]. "Fossil," interpreted by HEER (1855) as ?fruit or flower of a palm; according to SCHIMPER & SCHENK (1885), most probably inorganic. [Found in molasse deposits.] *U.Tert.,* Eu.(Switz.).

Panescorsea DE SAPORTA, 1882, p. 25 [*P. glomerata; M] [=Panescorsaea FUCHS, 1895; Panescorea ANDREWS, 1955 *(nom. null.)*]. Long parallel ridges on bedding planes. [Erroneously explained by DE SAPORTA as seaweed; interpreted as of inorganic origin by NEWBERRY (1885), NATHORST (1886) and FUCHS (1895) (ripple marks or a special kind of current marks or flute casts).] *Cret.-Tert.,* Eu.(France-Italy).

Phyllitites FUCINI, 1936, p. 78 [*P. rugosus; M]. Markings on bedding planes, certainly inorganic. [Erroneously explained as of plant origin.] *L. Cret.("Verrucano"),* Eu.(Italy).

Phytocalyx BORNEMANN, 1886, p. 13 [*P. antiquus; M]. Structureless conical or hemispherical bodies. [Originally regarded as algae; according to HINDE (1887), inorganic concretions or fillings of burrows.] *Cam.,* Eu.(Italy, Sardinia).

Piaella FUCINI, 1936, p. 95 [*P. biformis; M]. Doubtlessly inorganic, not of plant origin as suggested by FUCINI, 1936. *L.Cret.("Verrucano"),* Eu.(Italy).

Polygonolites DESIO, 1941, p. 81. Suggested as international designation for mud cracks.

Protadelaidea TILLYARD in DAVID & TILLYARD, 1936, p. 64 [*P. howchini; OD] [=Protoadelaidea SEILACHER, 1956 *(nom. null.)*]. Fragments in

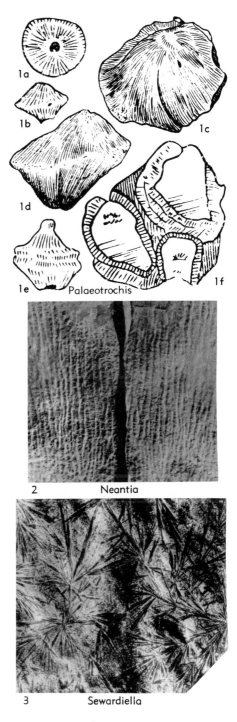

FIG. 105. Pseudofossils (p. *W*176-177, 179).

Protadelaidea

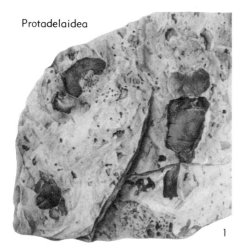

Fig. 105A. Pseudofossils (p. *W*177-178).

form of ochreous to black crusts in quartzites, with vaguely regular outlines. [Erroneously believed to represent body segments of giant arthropods; doubtlessly inorganic in origin as interpreted by Teichert in Hupé (1952), especially for very similar forms from the Precambrian of Morocco (mud flakes or flattened clay pellets); same opinion held by Schindewolf (1956), Seilacher (1956), and Cloud (1968); according to Glaessner (1959b), possibly formed also by pyritized soft plant tissue.] *Precam.(Adelaide System),* S. Australia.——Fig. 105A,*1.* *P. howchini; ✕0.45 (David, 1950).

Pseudopolyporus Hollick, 1910 [*P. carbonicus; M]. "Fossil" closely resembling fungus (especially *Polyporus*) and originally described as such; according to Pia (1927), probably inorganic (concretion). *Penn.,* N.Am.(USA,W.Va.).

Reynella David, 1928, p. 200 [*R. howchini; M]. Small fragments of exceedingly irregular shape. [Erroneously explained by David (1922, 1928) as belonging to problematical crustaceans; according to Glaessner (1959b, p. 525) not recognizable as animal remains, certainly inorganic; interpreted by Cloud (1968) as mud flakes.] *Precam.(Brighton Ls.),* S.Australia.

Rhysonetron Hofmann, 1967, p. 504 [*R. lahtii; OD]. Long curved cylindrical rods or spindles, vermiform, rarely with branchings; occasionally with faint, distinctly oblique crescentic corrugations along sides; distinct median longitudinal markings; end of spindles tapering; largest specimen 14 cm. long, 7 mm. wide; constant in morphology; preserved as sand casts in troughs of ripple marks; cleanly separated from matrix. [Originally regarded as Metazoa of unknown systematic position, compared with tubes of modern annelids; interpreted by Donaldson

(1967) as originating through deformation of an algal mat; recently explained by Hofmann (1971, p. 39) as diagenetic structure resulting from shrinkage crack filling modified by compaction and injecting processes followed by impression into substrate and superstrate, passing through a Manchuriophycus stage to the final Rhysonetron stage; similar structures observed in Precambrian quartzites in Finland (Lauerma & Piispanen,

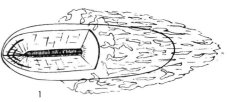

1

Telemarkites

2
Rhysonetron

Rivularites

3

Fig. 106. Pseudofossils (p. *W*178-179).

1967).] *Precam.*, N.Am.(Can.).——Fɪɢ. 106,2. *R. lahtii, Huron.(Bar River F.), Ont.(Flack Lake); ×0.3 (Hofmann, 1971). [Courtesy Geol. Survey Canada, Photo 200446-B.]

Rivularites Fʟɪᴄʜᴇ, 1906, p. 46 [*R. repertus; M]. Type species (U.Trias., Lorraine) rather unrecognizable, pustulated surfaces of bedding planes?; neither holotype nor other specimens of type could be located. [Erroneously explained by Fʟɪᴄʜᴇ (1906) as algal in origin; American "species" R. permiensis Wʜɪᴛᴇ (1929) (Hermit Shale, Ariz.) interpreted as alga, but bedding plane features doubtlessly inorganic in origin, very similar to mud flow markings; compared by C. L. Fᴇɴᴛᴏɴ (1946) with small symmetrical ripple marks; very similar pitted surfaces of bedding planes of dolomite (L. Trias., W.Pakistan) compared by Kᴜᴍᴍᴇʟ & Tᴇɪᴄʜᴇʀᴛ (1970) with Rivularites and interpreted as systems of capped interference ripples or as wrinkle marks (German, *Runzelmarken*) by Tᴇɪᴄʜᴇʀᴛ (1970).] *Perm.*, USA (Ariz); *U.Trias.(Keuper)*, Eu.(France).——Fɪɢ. 106,3. R. permiensis Wʜɪᴛᴇ, Perm.(Hermit Sh.), Ariz.; ×0.34 (White, 1929).

Rutgersella Jᴏʜɴsᴏɴ & Fox, 1968, p. 119 [*R. truexi; OD]. Bell shaped, elliptical to nearly round, strongly bilaterally symmetric structure (5 to 10 cm.) divided by a prominent "dorsal" ridge from which radiate numerous (40 to 52) convex segments; "lappets" occur around margin. [Originally considered a dipleurozoan and assigned to the family Dickinsoniidae by the authors; undoubtedly inorganic, considered by Cʟᴏᴜᴅ (1973, p. 125) as the imprints of spokelike, radiate growths of pyrite which grew under pressure and prior to the lithification of the sediment.] *L.Sil.*, N.Am.(Pa.). [Description supplied by W. G. Hᴀᴋᴇs.]

Schafferia Fᴜᴄɪɴɪ, 1938, p. 133 [*S. verrucana; M]. Apparently markings on bedding planes [Originally interpreted as of plant origin.] *L. Cret.("Verrucano")*, Eu.(Italy).

Sewardiella Fᴜᴄɪɴɪ, 1936, p. 47 [*S. verrucana; M] [=Baieropsis (?)verrucana Fᴜᴄɪɴɪ, 1928, p. CVII (*non Baieropsis* Fᴏɴᴛᴀɪɴᴇ, 1889, p. 205, a plant)]. Sharply stamped impressions of rosettes on bedding planes resembling *Annularia* or tiny palm branches or fans. [Originally interpreted by Fᴜᴄɪɴɪ (1928, 1936) as algae and by Sᴀᴄᴄᴏ (1940) as belonging to Sphenophyllales, doubtlessly molds of radiate crystal aggregates (?gypsum, ?ice) as recognized by Gᴏᴛʜᴀɴ (1933, 1942), Hᴀ̈ɴᴛᴢsᴄʜᴇʟ (1935b), Rᴇᴅɪɴɪ (1938), ᴠᴏɴ Hᴜᴇɴᴇ (1941).] *L.Cret.("Verrucano")*, Eu. (Italy).——Fɪɢ. 105,3. *S. verrucana; ca. ×0.8 (Fucini, 1936).

Sickleria Mᴜ̈ʟʟᴇʀ, 1846, p. 83 [*S. labyrinthiformis; M]. Originally regarded by Rᴜ̈ᴘᴘᴇʟʟ (1845) and Mᴜ̈ʟʟᴇʀ (1846) as plants, but immediately after Mᴜ̈ʟʟᴇʀ's publication recognized by Sᴄʜɪᴍᴘᴇʀ (1846) as shrinkage cracks in sandstone. *L.Trias.(Buntsandstein)*, Eu.(Ger., Thuringia).

Sidneyia groenlandica Cʟᴇᴀᴠᴇs in Cʟᴇᴀᴠᴇs & Fox, 1935, p. 485. Somewhat distorted and poorly preserved "fossil" originally interpreted as abdominal region of an arachnid of Middle Cambrian age. [According to Eʜᴀ (1953, p. 15-16), of Precambrian age and probably a group of damaged ripplemarks partly removed by erosion; regarded by Sᴄʜɪɴᴅᴇᴡᴏʟғ (1956) as "at least very uncertain fossil."] *Precam.*, E.Greenl.

Spirocerium Eʜʀᴇɴʙᴇʀɢ, 1858, p. 310 [*S. priscum; M]. "Microfossil," according to Lᴏᴇʙʟɪᴄʜ & Tᴀᴘᴘᴀɴ (1964, p. C786), inorganic, globular mass of ?glauconite. *?L.Sil.*, USSR.

Stylolithes Kʟᴏ̈ᴅᴇɴ, 1828, p. 58 [*S. sulcatus; M]. Regarded by Kʟᴏ̈ᴅᴇɴ as problematical fossil; actually stylolites. ("Genus" described from *M. Trias.(Muschelkalk)*, Eu.(Ger.).

Tazenakhtia Cʜᴏᴜʙᴇʀᴛ, Tᴇʀᴍɪᴇʀ & Tᴇʀᴍɪᴇʀ, 1951, p. 31 [*T. aenigmatica; M]. "Organisms" observed in thin sections of limestones. [Interpreted as questionable foraminifers, but also compared with calcareous algae (*Nubecularites* Mᴀsʟᴏᴠ); according to Sᴄʜɪɴᴅᴇᴡᴏʟғ (1956), inorganic structures due to combination of tectonic movements and metamorphic recrystallization.] *Precam.*, Afr.(Morocco).

Telemarkites Dᴏɴs, 1959, p. 262 [*T. enigmaticus; M]. Ellipsoidal nodules with inner structure composed of concentric and radial elements and long axis parallel to bedding planes; central tube lying parallel to long axis; 2 to 4 cm. long, 1 to 2 cm. across; composed mainly of fine-grained quartz and feldspars (mostly albite), together with muscovite and calcite; many globular algae, three-dimensionally preserved, in cores of nodules, appearing to be arranged in colonies; probably silicified by gelatinous silica during lifetime; size and shape of nodules very similar to *Botswanella* Pғʟᴜɢ & Sᴛʀᴜ̈ʙᴇʟ, 1969, but different in nature and origin. [According to Dᴏɴs (1963), sponges or of organic-controlled inorganic origin (concretions formed by intervention of algae); regarded by Cʟᴏᴜᴅ (1968, p. 54) as doubtful concretions; according to Pғʟᴜɢ & Sᴛʀᴜ̈ʙᴇʟ (1969), concretions of algae-controlled synsedimentary origin.] *U.Precam.(Telemark Suite, Bandak Gr.)*, Eu.(S. Nor.).——Fɪɢ. 106,1. *T. enigmaticus; simplified reconstr. showing internal structures; ×1.3 (Dons, 1959).

Tubiphyton Cʜᴏᴜʙᴇʀᴛ, Tᴇʀᴍɪᴇʀ & Tᴇʀᴍɪᴇʀ, 1951, p. 29 [*T. taghdoutensis; M]. Supposed "organisms" observed in thin sections of limestones. [Interpreted as calcareous algae; according to Sᴄʜɪɴᴅᴇᴡᴏʟғ (1956), inorganic structures due to combination of tectonic movement and metamorphic recrystallization.] *Precam.*, Afr. (Morocco).

UNRECOGNIZED AND UNRECOGNIZABLE "GENERA"

Numerous "genera," mostly based on badly preserved fossils, are included in this group, because descriptions are insufficient and illustrations inadequate. The majority of them are so nondescript that they do not deserve to be named. Many of these fossils will remain unrecognizable for a long time. In only a few cases are investigations of new and better material likely to clarify their systematic position.

Some of the unrecognizable "genera" mentioned in the first edition of the *Treatise,* Part W (HÄNTZSCHEL, 1962), have since been reinterpreted as trace fossils or have been found to be of inorganic origin. Such names are here transferred to the appropriate sections of this contribution. On the other hand, additional names, mostly of monospecific "genera," have been based on insufficiently described lebensspuren. These useless names should under no circumstances be revived. After this listing, many of these genera should never be mentioned or discussed again in the literature.

Amanlisia LEBESCONTE, 1891, p. 4 [*A. simplex*; M]. Interpreted as alga resembling *Palaeophycus simplex* HALL; according to SEILACHER (1956a, p. 167), uncharacteristic trail. *Precam.,* Eu.(France).

Amansites BRONGNIART, 1849, p. 58 ("Genus" introduced for the "group" of *Fucoides dentatus* BRONGNIART, 1828, p. 70). [Interpreted as plant in origin; according to SCHIMPER (1869, p. 214), graptolites.] *"Calcaire de transition,"* N.Am. (Can.).

Amaralia KEGEL, 1967, p. 5, 7 [*A. paulistana*; M]. Rather poorly figured trail composed of 2 different elements but probably belonging to each other: 1) narrow or wide network consisting of small round ribs, 1 to 2 mm. in breadth and height, with median furrow and occasionally with fine transverse annulation, and 2) elliptical or circular trails with or without connection with the networks, up to 15 mm. long and 7 mm. wide, somewhat comparable to resting trails like the "coffee-beans" *Isopodichnus,* but differing by commonly lacking median furrow; both components regarded as belonging to Bilobites by KEGEL. *Perm.(IratíF.),* S.Am.(Brazil, São Paulo).

Ampelichnus HITCHCOCK, 1865, p. 19 [*Grammepus uniordinatus* HITCHCOCK, 1858; M] [=*Ampelichnus sulcatus* HITCHCOCK, 1865, p. 19]. According to HITCHCOCK (1865, p. 19), of doubtful origin, track or plant. *Trias.,* USA (Mass.).

Archaeorrhiza TORELL, 1870, p. 7 [*A. tuberosa*; M]. "Plant, *radicibus similis*"; never figured. *L.Cam.,* Eu.(S.Swed.).

Archaeoscolex MATTHEW, 1889, p. 59 [*A. corneus*; M]. Dubious fossil, interpreted as insect larva; according to HANDLIRSCH (1906-08, p. 338-339), perhaps a myriapod; no specimens located in Canadian collections *U.Carb.* (age stated by MATTHEW: *Dev.*), Can.(N.B.).

Armelia LEBESCONTE, 1891, p. 5 [*A. barrandei*; M]. Interpreted as perhaps belonging to cystoids; according to SEILACHER (pers. commun., 1956), problematic body fossil. *Precam.,* Eu.(France).

Asabellarifex KLÄHN, 1932, p. 14. Poorly founded, rather superfluous "genus" proposed for vertical burrows resembling *Sabellarifex* RICHTER, but believed to be burrowed in downward direction, not built upward as tubes like *Sabellarifex* (HÄNTZSCHEL, 1965). *L.Cam.*(Pleist. drift), Eu.(Ger.-Swed.).

Astropolithon DAWSON, 1878, p. 83 [*A. Hindii*; M]. Oval or circular ridge, raised and arched or (depending on preservation) more or less compressed; diameter 3 to 7 cm.; articulated by numerous (about 30) "rays"; ridge surrounding central area, apparently smooth depression (with central ?axis). (Description based on Spanish specimens; no type or other specimens from Canada located; "genus" only once figured for about a century.) [Originally explained as of plant origin but later (by DAWSON, 1890, p. 605-606) as possible mouths of large burrows with radiating trails or as organisms; also compared by DAWSON (1878) with *Astylospongia radiata* LINNARSSON; Spanish specimens regarded as Scyphomedusae by VAN DER MEER MOHR & OKULITCH (1967) and VAN DEN BOSCH (1969); interpretation as trace fossils seems next to impossible, for some smaller Spanish specimens are preserved as sharply delimited ellipsoidal bodies, lying crowded and in part obliquely to one another.] *Cam.,* Eu.(Spain)-N.Am.(Can., N.Scotia).——FIG. 107,3. *A. hindii,* L.Cam.; ×?0.7 (Dawson, 1890).

Atlantaia BORRELLO, 1966, p. 26 [*A. argentina*; OD]. ?Inorganic. *Ord.(F.La Tinta),* S.Am. (Arg.).

Balanulina RZEHAK, 1888, p. 265 [*B. ḳittlii*; M]. Microfossil, interpreted as foraminifer; according to LOEBLICH & TAPPAN (1964, p. C786), unrecognizable. *L.Tert.(up.Oligo.),* Eu.(Aus.).

Beaumontia DAVID, 1928, p. 203 [*non* MILNE-EDWARDS & HAIME, 1851; *nec* EUDES-DESLONGCHAMPS, 1856] [*B. eckersleyi*; M] [=*Beaumontella* DAVID, 1928, p. 208 *(nom. null.)*]. Nodular bodies. [Interpreted by DAVID (1928) as various parts of eurypterids; according to GLAESSNER (1959b, p. 525), not recognizable as

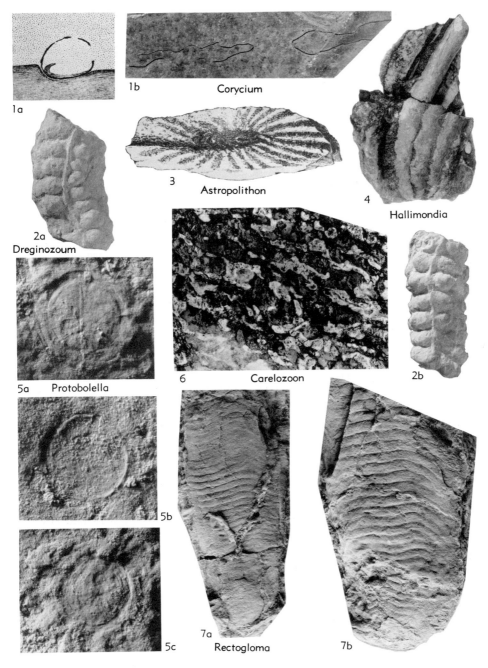

1a

1b   Corycium

3   Astropolithon

4

Hallimondia

2a
Dreginozoum

2b

5a   Protobolella

6   Carelozoon

5b

7a   Rectogloma

7b

5c

Fig. 107. Unrecognizable "genera" (p. *W*180, 182, 184-185, 187-188).

animal remains; regarded by Cloud (1968) as mud flakes.] *Precam.(Beaumont Dol., Adelaide System)*, S.Australia.

Beltina Walcott, 1899, p. 238 [**B. danai*; M]. Angular fragments of thin chitinous or carbonaceous films, from approximately 1 to several cm. in size, commonly much distorted and compressed; without distinctive ornamentation. [Regarded by WALCOTT (1899) as fragmentary remains of Merostomata and compared with *Pterygotus* or *Eurypterus* fragments; considered by WHITE (1929) and FENTON & FENTON (1937a) as probably noncalcareous algae, if partly not inorganic ("segregated carbon"); according to CLOUD (1968), inorganic or algal in origin; for discussion of various interpretations, see HOFMANN (1971, p. 23).] *Precam.,* USA-Can.——FIG. 108,2. **B. danai,* Belt Ser.(Greyson Sh.), USA (Mont.); *2a,* body segment, ×?; *2b,* appendage with 2 large basal? joints and 2 smaller terminal joints, ×2; *2c,* unidentified fragment with terminal curved spine, ×4; *2d,* portion of jointed appendage, ×3 (Walcott, 1899).

Bipezia MATTHEW, 1910, p. 121 [**B. bilobata;* M] [=*Bipesia* MATTHEW, 1910, p. 125 *(nom. null.)*]. Spindle-shaped "footprints," pointed at both ends, in pairs opposite each other, coalescing laterally; 10 mm. long, 3 mm. wide. [Interpretation doubtful, certainly not of vertebrate origin, as MATTHEW believed; according to GLAESSNER (1957), possibly synonymous with *Isopodichnus* BORNEMANN, 1889.] *U.Carb.*(MATTHEW reported *Dev.*), Can.(N.B.).

Bisulcus HITCHCOCK, 1865, p. 18 [**B. undulatus;* M]. Continuous paired grooves separated by single ridge; poorly figured. [Doubtful whether trail or of inorganic origin; interpreted by HITCHCOCK (1865) as annelid trail, by LULL (1915) as ?mollusk trail; according to BROWN (1912), probably drag marks.] *Trias.,* USA(Mass.).

Bitubulites BLUMENBACH, 1803, p. 23 [**B. problematicus;* M]. "Genus" (especially the "species" *B. irregularis* VON SCHLOTHEIM, 1820, p. 376) possibly synonymous with *Rhizocorallium* ZENKER, 1836; name apparently not used again during the last century. *M.Trias.,* Eu.(Ger.).

Boliviana SALTER, 1861, p. 71 [Three species, no type species designated]. *?Sil.,* S.Am.(Bol.).

Bonariensia BORRELLO, 1966, p. 27 [**B. nuda;* OD]. ?Inorganic. *Ord.(F.La Tinta),* S.Am. (Arg.).

Bucinella FUCINI, 1936, p. 82 [**B. verrucana;* M]. ?Uncharacteristic trail. *L.Cret.("Verrucano"),* Eu. (Italy).

Calcinema BORNEMANN, 1886, p. 290 [**C. triasinum;* M]. Thin-walled tube, straight or gently curved, cross section nearly circular, diameter 0.15 to 0.20 mm. [Imperfectly described, never again investigated; systematic position uncertain; interpreted by BORNEMANN as alga, this explanation questioned by FRANTZEN (1888) and PIA (1927).] *Low.M.Trias.(Muschelkalk, Schaumkalk),* Eu.(Ger., Thuringia).

Camptocladus FENTON & FENTON, 1937, p. 1081 [**C. intertextus;* OD]. Superfluous name for branched, flexuous, intertwined burrows; "genus" proposed on assumption that burrows are of crustacean origin. *Penn.,* USA(Texas).

Carelozoon METZGER, 1924, p. 50 [**C. jatulicum;* M]. Irregularly ramifying annd branching, irregularly shaped structures about 0.5 mm. in diameter; circular in cross section, forming network in rock; with crustal layer and possible tabulae; reminiscent of stromatoporoids. [Affinities unknown; ?coelenterate, ?calcareous alga; according to SEILACHER (1956a) and CLOUD (1968), concretionary in origin.] *Precam.,* Eu.(Finl.)-?USSR. ——FIG. 107,6. **C. jatulicum,* Finl.; cross sec., ×1.1 (Häntzschel, 1962, photo courtesy Geol. Survey Finland).

Caridolites NICHOLSON, 1873, p. 289 [**C. wilsoni;* M]. Tracks, not described in detail; thought to be made by *Ceratiocaris.* [Name apparently not used since 1873.] *L.Paleoz.,* G.Brit.(Eng.).

Ceraospongites MAYER, 1964, p. 108 [**C. lotzae;* M]. Gently curved, cylindrical bodies, some forming T-shaped structures, internally contain randomly oriented cylindrical cavities. [Considered a sponge by MAYER (1964) and established only on fragmentary material.] *M.Trias.(up.Muschelkalk),* Eu.(S.Ger.).

Chapadamlidium BORRELLO, 1966, p. 28[**C. robustum;* OD]. ?Inorganic. *Ord.(F.La Tinta),* S.Am.(Arg.).

Charruia RUSCONI, 1955. See KNIGHT *et al.,* 1960, p. 1324, and FISHER, 1962, p. W140.

Chauviniopsis DE SAPORTA, 1872, p. 119 [**C. pellati;* M]. Interpreted as algae. *U.Jur.(low. Portland.),* Eu.(France).

Chordophyllites cicatricosus TATE, 1876, p. 474. Cylindrical "stem" of rather great length on bedding planes; interpreted as plant in origin ("fucoid"); ?burrows. *L.Jur.,* G.Brit.(Eng.).

Clematischnia WILSON, 1948, p. 10 [**Buthotrephis succulens* HALL, 1847, p. 62; OD]. Irregularly bifurcating burrows, about 5 mm. in diameter; surface ringed by undulating ridges at 2 to 5 mm. intervals. [Interpreted as alga, but certainly trace fossil; "genus introduced with some hesitation"; chondritid-like burrows, according to OSGOOD (1970, p. 333), who proposed to include *Clematischnia* provisionally within *Chondrites s.l.*] *M.Ord.,* Can.

Climacodichnus HITCHCOCK, 1865, p. 20 [**C. corrugatus;* M]. Small, ladderlike rows of impressions, trackway 15 mm. wide, resembling steps of *Acanthichnus. Trias.,* USA(Mass.).

Codites VON STERNBERG, 1833, p. 20 [**C. serpentinus;* M]. Originally interpreted as plant, later regarded as ?sponge. *U.Jur.,* Eu.(France-Ger.).

Conchyophycus DE SAPORTA, 1872, p. 150 [**C. marcygnianus;* M]. Interpreted as alga with reservation; very doubtful. *U.Trias.,* Eu.(France).

Confervites BRONGNIART, 1828, p. 86 [No type species designated] [=*Confervides* SCHIMPER, 1869 *(nom. null.)*]. Most forms placed here, especially those from Tertiary beds, are remains of threadlike algae (PIA, 1927), or tissue residues of higher plants. [According to NATHORST

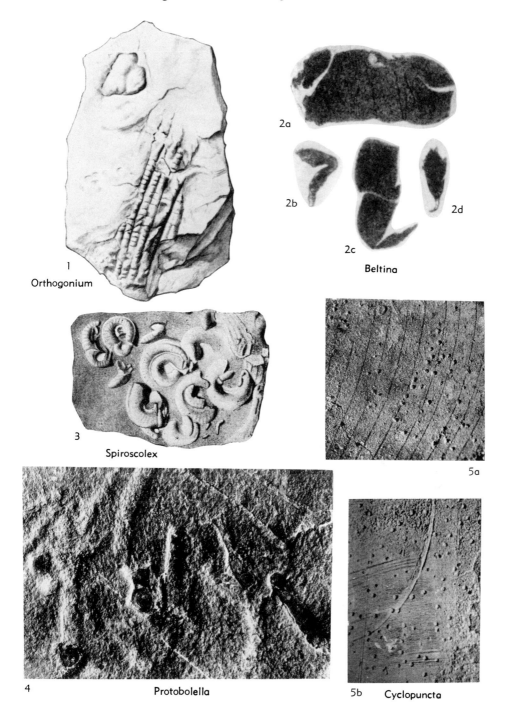

1
Orthogonium

2a

2b

2c

2d

Beltina

3
Spiroscolex

5a

4
Protobolella

5b Cyclopuncta

FIG. 108. Unrecognizable "genera" (p. W181-182, 184, 186-188).

(1881a), some "species," such as *C. padellae* HEER (1877, p. 103) from Jurassic of Switzerland, are probably trace fossils resemblng *Chondrites; Confervites* has been regarded by DE LAUBENFELS (1955, p. E104) as representing unrecognizable supposed sponges.] *Jur.-Tert.,* Eu.

Cophinus KOENIG in MURCHISON, 1839, p. 697 [*C. dubius*; M]. Problematical structure resembling inverted 4-sided pyramid with columnlike rounding at each corner; always found in vertical position. [Tentatively explained by SOWERBY and SALTER (see MURCHISON, 1859, p. 147) as impressions of rooted crinoid stems which produced observed pattern by wavy and somewhat rotatory motion; possibly inorganic.] *U.Sil.(Ludlov.),* G. Brit.(Eng.).

Corycium SEDERHOLM, 1911, p. 28 [*C. enigmaticum*; M] [=*Corycinium* C. L. FENTON, 1946, p. 259 *(nom. null.)*]. Saclike structures with carbonaceous walls occurring in sandy beds; filling mass commonly shows concentric internal structure. [The "fossil" or at least its carbon regarded as of organic origin by SEDERHOLM (1911), METZGER (1927), SEWARD (1931), RANKAMA (1948, 1950) and MATISTO (1963); compared by OHLSON (1961) with Recent lake balls; considered as inorganic in origin by KREJCI (1924, 1925), VAN STRAATEN (1949), SCHINDEWOLF (1956), and CLOUD (1968). *Precam.,* Eu.(Finl.). ——FIG. 107,1. *C. enigmaticum; 1a,* vert. sec., ×0.7 (Sederholm, 1911); *1b,* ×0.2 (Häntzschel, 1962, courtesy Geol. Survey Finland).

Crenobaculus FRITSCH, 1908, p. 7 [*C. Draboviensis*; M]. Rod-shaped ?structure with series of small nodes on external surface; circular in cross section; up to 17 cm. long. [?Body fossil; original illustration nondescript.] *M.Ord.,* Eu.(Czech.). [Description supplied by W. G. HAKES.]

Cunicularis HITCHCOCK, 1858, p. 163 [*C. retrahens*; M]. ?Ramified trails. *Trias.,* USA (Conn.-Mass.).

Cyclopuncta ELIAS, 1958, p. 50 [*C. girtyi*; M]. Shallow subhemispherical holes; diameter 0.1 to 0.3 mm.; generally irregularly scattered on cephalopod shells, in some specimens tending to follow growth lines. [Such structures explained by GIRTY (1909) as perforations in shells probably made by small gregarious animals (e.g., the lorica-secreting infusorian *Folliculina*), scar being produced by prolonged passive attachment.] *Miss.,* USA(Okla.).——FIG. 108,5. *C. girtyi; 5a,* on *Cravenoceras* sp., ×8 (Elias, 1958); *5b,* on *Bactrites? smithianus,* ×4.6 (Girty, 1909).

Dasycladites FUCINI, 1936, p. 74 [*D. subclavaeformis*; M]. Nondescript form with trifid, pointed "branches," similar in outline to a dasycladacean alga. According to PIA (1937), so nondescript that it should never have been named. *L.Cret.* (*"Verrucano"),* Eu.(Italy).

Dazeodesma BORRELLO, 1966, p. 28 [*D. symmetrica*; OD]. ?Inorganic. *Ord.(F.La Tinta),* S.Am.(Arg.).

Digitolithus FRITSCH, 1908, p. 23 [*D. rugatus*; M]. Main structure "as large as a finger," covered with tubercles; flat branches originate from it. [Origin uncertain, description based on a single discovery that was associated with a fucoid.] *Ord.,* Eu.(Czech.). [Description applied by W. G. HAKES.]

Discoidina TERQUEM & BERTHELIN, 1875, p. 15. See LOEBLICH & TAPPAN (1964, p. C786).

Discophycus WALCOTT, 1883, p. 19 [*D. typicalis*; M]. Discoid, slightly convex bodies; diameter 4 to 12 cm.; outline varying from circular to orbicular, substance ?coriaceous. [Interpreted by JAMES (1884) as inorganic (produced by air bubbles in mud); by RUEDEMANN (1925) as "actual remains of organisms" (seaweeds, sponges, and fragments of eurypterids).] *U.Ord.,* USA(N.Y.).

Dreginozoum VON DER MARCK, 1894, p. 6 [*D. nereitiforme*; M]. Narrow curving median ridge, about 1 mm. wide, with small oval disclike round-edged appendages on both sides, closely spaced like roll of coins; whole fossil up to several decimeters long, 10 to 15 mm. wide. [Somewhat obscure; apparently not a trail; variously regarded as resembling *Nereites*, algae, *Serpula*, or mollusks; similar fossils are *Oncophorus* GLOCKER, 1850 (?U.Cret., Czech.), *Platyrhynchus* GLOCKER, 1850 (*non* LEUCKART, 1816) (?U.Cret., Czech.) and particularly *Gyrochorte bisulcata* GEINITZ, 1883 (Oligo., Ger.); tentatively compared by HÄNTZSCHEL (1964a, p. 298) with egg capsules of some marine prosobranchs attached to a string as observed on the Recent genus *Busycon* (east coast USA).] *U.Cret.,* Eu.(Ger.).——FIG. 107,2. *D. nereitiforme,* U.Campan., Ger.(Beckum); *2a,* ×0.9; *2b,* ×1.1 (Häntzschel, 1964a).

Dryalus BARRANDE, 1872, p. 585 [*D. obscurus*; M]. ?Fragment of body fossil (of fish or crustacean, according to BARRANDE); interpreted by FRITSCH (1908) as belonging to a genus similar to *Acanthodes*. *Sil.,* Eu.(Czech.).

Duovestigia BUTTS, 1891, p. 19 [*D. scala*; M]. Described as amphibian footprint, but apparently of invertebrate origin; according to O. KUHN (pers. commun., 1960), probably limuloid. *U. Carb.,* USA(Mo.).

Durvillides SQUINABOL, 1887, p. 560 [*D. eocenicus*; M]. ?Meandering trail. *Eoc.,* Eu.(Italy).

Eocladophora FUCINI, 1936, p. 79 [Several "species," no type species designated]. Long, narrow, threadlike pads or ridges; probably inorganic. *L.Cret.("Verrucano"),* Eu.(Italy).

Eurypterella MATTHEW, 1889, p. 60 [*E. ornata*; M]. Dubious fossil interpreted as peculiar small crustacean; no specimens could be located in Canadian collections. *U.Carb.*(age stated by MATTHEW: *Dev.*), Can.(N.B.).

Fengtienia ENDO & RESSER, 1937, p. 326 [*F. peculiaris*; M]. Unrecognizable genus of "trilobite" founded on impression of 2 individuals lying side by side, perhaps in copulation. [According to ÖPIK (1959), probably only a *"Rusophycus"*;

see also HARRINGTON *et al.*, 1959, p. O102 and O525.] *M.Cam.*, China (Manchuria).

Flabellichnus KARASZEWSKI, 1971, p. 105 [**F. lewinskii*; M]. "Inflorescence"-shaped imprints, consisting of several "petals" rapidly narrowing and tapering; particular "petals" spindle-shaped. [Photos and description not sufficient for an interpretation.] *L.Jur.(low.Lias., Hettang.),* Eu. (Pol.).

Fruticristatum WEBSTER, 1920, p. 288 [Three "species," no type species designated]. Originally described as an alga, never figured; apparently fillings of uncharacteristic burrows. *M.Dev.,* USA (Iowa).

Furca BARRANDE in FRITSCH, 1908, p. 8 [**F. bohemica*; M]. Structure with a straight-line anterior truncation and 4 gently recumbent posterior lobes, 2 "lateral" and 2 "medial," about 25 mm. long; external surface composed of numerous subrectangular plates. [?Pluteus larva of crinoid.] *Ord.,* Eu.(Czech.). [Description supplied by W. G. HAKES.]

Gleichenophycus MASSALONGO, 1884 [**G. granulosus* MASSALONGO in CAPELLINI, 1884, p. 541; SD ANDREWS, 1955, p. 162]. MASSALONGO's first description of the "genus" not found; according to FUCHS (1895, p. 406), *G. italicus* MASSALONGO identical with *Caulerpa lehmanni* HEER, 1877; ?screwlike burrow. [Found in flysch deposits.] *U.Cret.,* Eu.(Switz.-Italy).

Gracilerectus WEBSTER, 1920, p. 288 [**G. hackberryensis*; M]. Straight or curved, cylindrical "stems," similar to *Fruticristatum* WEBSTER. [Originally regarded as algae ("fucoids"), but most probably uncharacteristic burrows.] *M.Dev.,* USA(Iowa).

Grammepus HITCHCOCK, 1858, p. 155 [**G. erismatus*; SD LULL, 1953, p. 48]. Doubtful (?arthropod) trail. *Trias.,* USA(Mass.).

Granifer FRITSCH, 1908, p. 7 [**G. stolatus*; M]. Nodules with various morphologies; may be round or rod-shaped, several centimeters in dimension; covered with tiny nodes about 1 mm. in diameter. [?Concretion.] *Ord.,* Eu.(Czech.). [Description supplied by W. G. HAKES.]

Guttolithus FRITSCH, 1908, p. 7 [**G. Strasseri*; M]. Subconical structure with blunt apical end; 14 cm. long, diameter 2.2 to 4 cm.; external surface weakly nodose. [Origin uncertain; insignificant and unnecessary name based on single discovery.] *Ord.,* Eu. (Czech.). [Description supplied by W. G. HAKES.]

Hallimondia CASEY, 1961, p. 600 [**H. fasciculata*; M]. Straight, parallel, trough- or tubelike structures, unbranched, about 40 cm. long, 1 cm. in diameter, cross section approximately semicircular; tube walls built up of concentric wavy layers; "tubes" apparently commencing at a common point, grouped together at first, then gradually diverging, not interlaced; in limestones forming nuclei of phosphate nodules. [?Organic in origin.] *L.Cret.(up.Apt.),* G.Brit.(Eng.).——FIG.

107,4. **H. fasciculata*, Sandgate Beds, Eng.; ×0.7 (Casey, 1961).

Halysichnus HITCHCOCK, 1858, p. 162 [**H. laqueatus*; SD LULL, 1953, p. 51]. Repeatedly looped, chainlike trail with ridges on each side. *Trias.,* USA(Mass.).

Harpagopus HITCHCOCK, 1848, p. 247 [**H. dubius*; SD LULL, 1953, p. 54]. Rather obscure tracks; obliquely placed elliptical impressions. *?M. Dev.,* USA(N.Y.); *Trias.,* USA(Mass.); *Jur.,* USA (N.J.).

Hauthaleia BORRELLO, 1966, p. 29 [**H. concava*; OD]. ?Inorganic. *Ord.(F.La Tinta),* S.Am. (Arg.).

Helviensia DE LIMA, 1895, p. 94 [**H. delgadoi*; M]. Originally interpreted as alga; ?uncharacteristic burrows. *?U.Cam.,* Eu.(Port.).

Hippodophycus HALL & WHITFIELD, 1872, p. 203 [**H. cowlesi*; M]. Described as marine plant with swelling root. [According to M. GOLDRING (pers. commun., 1953), holotype (only described specimen) probably lost, perhaps inorganic]. *U. Dev.,* USA(N.Y.).

Hoplichnus HITCHCOCK, 1848, p. 230 [**H. quadrupedans*; M] [=*H. poledrus* HITCHCOCK, 1858, p. 136, *nom. van.*] [=*Chelichnus gigas* JARDINE, 1850, p. 208]. Hoofshaped, semioval reliefs resembling impressions of horseshoes; diameter about 5 cm. [Perhaps markings or (particularly the "species" *H. equus* HITCHCOCK, 1858) lebensspuren; for interpretation of similar hoofshaped structures from the German Buntsandstein (Thuringia) as U-shaped dwelling tubes with spreite similar to *Rhizocorallium*, see W. QUENSTEDT (1932a, p. 93).] *Penn.-Trias.,* USA; *?Penn.-Trias.,* Eu.(Eng.-Ger.).

Hylopus(?) variabilis MATTHEW, 1910, p. 120. Very doubtful "footprints," referred to vertebrates. [According to ABEL (1935, p. 78), not a vertebrate track, but an unrecognizable form.] *Dev.,* Can.(N.B.).

Ichnophycus HALL, 1852, p. 26 [**I. tridactylus*; M]. Doubtful tridactyl impressions, similar to foot of bird in outline. [According to OSGOOD (1970, p. 345), "probably a portion of a burrow," but not comparable with *Dactylophycus* MILLER & DYER as pointed out by MILLER (1889).] *Sil.,* USA(N.Y.).

Isnardia BORRELLO, 1966, p. 30 [**I. aenigmatica*; OD]. ?Inorganic. *Ord.(F.La Tinta),* S.Am. (Arg.).

Keidelasma BORRELLO, 1966, p. 32 [**K. bonariensis*; OD]. ?Inorganic. *Ord.(F.La Tinta),* S.Am. (Arg.).

Krishnania SAHNI & SHRIVASTAVA, 1954, p. 40 [**K. acuminata*; M]. Somewhat similar to *Fermoria* CHAPMAN but differing from it by its acuminately ovate shape, in outline resembling *Lingula*; longest axis 7.5 mm., maximum width 4 mm., narrowing abruptly at one end, rounded at other; deep marginal furrow with fine median rib dividing it. [Rather poorly figured, "need restudy"

(CLOUD, 1968).] *U.Precam.(Vindhyan)*, C.India (Neemuch distr.).

Laminarites BRONGNIART, 1828, p. 54 [no type species designated]. "Genus" comprising very heterogenous "species"; similar to *Laminaria*; straight and parallel structures on bedding planes. [Seemingly in part of plant origin (e.g., *L. antiquissimus* EICHWALD, 1856), in part probably inorganic in origin (e.g., *L. lagrangei* DE SAPORTA & MARION, 1883), and partly, according to MESCHINELLI & SQUINABOL, 1892, also trails.]

Laminopsis FUCINI, 1938, p. 204 [*L. insignis*; M]. Probably inorganic. *?M.Trias.*, Eu.(Italy).

Lepidotruncus FRITSCH, 1908, p. 23 [*L. fortis*; M] [=*Lipidotruncus* FRITSCH, 1908, p. 28, *nom. null.*]. Large subcylindrical structure, length and width 17 cm.; may branch; surface covered with transverse, irregular striations. [Listed among "Problematica botanica" by author.] *Ord.*, Eu. (Czech.). [Description supplied by W. G. HAKES.]

Leptophycus FRITSCH, 1908, p. 21 [*Fucoides papyraceus* BARRANDE in FRITSCH, 1908, p. 21); M] [*non Leptophycus* JOHNSON, 1940 (stromatolite from the Pennsylvanian of Colorado, USA)]. Leaf-shaped structure rolled in the form of a cornet about 18 cm. long. [Erroneously regarded by FRITSCH (1908, p. 20) as belonging to *Alectorurus* SCHIMPER, 1869.] *Ord.*, Eu.(Czech.). [Description of structure supplied by W. G. HAKES.]

Lingulella montana FENTON & FENTON, 1936, p. 616. Small linguloid-shaped structures with concentric wrinkles paralleling anterior margin. [Considered to be brachiopod by FENTON & FENTON (1936); thought to resemble stromatolites by GLAESSNER (1962); and interpreted as inorganic by ROWELL (1971), who believed wrinkles formed as result of deformation by slippage.] *Precam.*, USA(Mont.). [Description supplied by CURT TEICHERT and W. G. HAKES.]

Lithodictyon TORELL, 1870, p. 7 [*L. fistulosum*; M]. Not figured; ?inorganic. *L.Cam.*, Eu. (Swed.).

Lithostachys FISCHER-OOSTER, 1858, p. 59 [*L. alpina*; M]. ?Plant. *?L.Cret.*, Eu.(Switz.).

Macrocystites FUCINI, 1936, p. 75 [*M. similis*; M]. Trail or inorganic (see PIA, 1937, p. 1098). *?M.Trias.*, Eu.(Italy).

Mastocarpites TREVISAN in DE ZIGNO, 1856, p. 22 [*non* TREVISAN, 1849 *(nom. nud.)*] [*Algacites erucaeformis* VON STERNBERG, 1833, p. 36; OD]. ?Coprolite (Andrews, 1955). *U.Jur.*, Eu.(Ger.).

Micrapium TORELL, 1870, p. 11 [*M. erectum*; M]. Never figured; according to NATHORST, 1881a, p. 50, burrows or of inorganic origin; see also WESTERGÅRD, 1931, p. 12. *L.Cam.*, Eu. (Swed.).

Myriodocites MARCOU (before 1880). *Fide* ZITTEL, 1880, p. 568, resembling *Nereites*; MARCOU's description not seen.

Naites GEINITZ, 1867, p. 8 [*N. priscus*; M].

Rather valueless name for a trail somewhat resembling that made by the Recent genus *Nais* MÜLLER, 1771. [Interpreted by GEINITZ as a bodily preserved annelid; according to PFEIFFER (1968, p. 693), an uncharacteristic burrow.] *L. Carb.*, Eu.(Ger.).

Nanopus? **vetustus** MATTHEW, 1910, p. 121. Doubtful "footprints," referred to vertebrates; according to ABEL (1935, p. 78), not a vertebrate track, but unrecognizable. *Dev.*, Can.(N.B.).

Nematolites KEEPING, 1882, p. 489 [No type species designated]. Poorly preserved, "curious irregular branching structures." *Sil.*, G.Brit.(Eng.).

Octoia BORRELLO, 1967, p. 4 [*B. subandina*; OD]. From description and figures doubtful whether or not of inorganic origin. *Dev.*, S.Am. (S.Bol.).

Oncophorus GLOCKER, 1850, p. 937 [*non* RUDOW, 1870; *nec* EPPELSHEIM, 1885] [*O. beskidensis*; M]. Sinuous trail. [Originally interpreted as body fossil; placed provisionally by GEINITZ (1852, p. 28) in his "genus" *Nereograpsus;* recognized by NATHORST (1881a, p. 85) as trace fossil; related to or identical with *Dreginozoum*, resembling *Gyrochorte bisculcata* E. GEINITZ; name apparently not used again for nearly a century.] *?U.Cret.*, Eu.(Czech.).

Orthocaris FRITSCH, 1908, p. 12 [*O. splendens*; M]. Rod-shaped structure with oblique plications; 18 mm. long, 8 mm. wide; external surface has a high luster. [Insignificant ?body fossil described from a fragment.] *Ord.*, Eu.(Czech.). [Description supplied by W. G. HAKES.]

Orthogonium GÜRICH, 1933, p. 146 [*O. parallelum*; M]. Consisting of several articulated rows suggestive of crinoid arms; row 3 or 4 mm. wide, about 6 cm. long; spongelike body similar to dictyospongiids (HÄNTZSCHEL, 1965). [Type of *Orthogonium* has been lost and the form not seen since (GLAESSNER written commun. to TEICHERT, 1972). Questionable fossil.] *Precam.(Nama Syst., Kuibis Quartzite)*, SW.Afr.——FIG. 108,*1.* *O. parallelum*; ×0.6 (Gürich, 1933).

Ostrakichnites PACKARD, 1900, p. 66. Name proposed for trails insufficiently described by DAWSON (1873, p. 55) as *Protichnites carbonarius;* according to PACKARD, not belonging to *Protichnites*. [Name apparently not used since 1900.] *Carb.*, Can.(N.S.).

Palaeonereis EICHWALD, 1856, p. 409 [*non* HUNDT, 1940, p. 214]. ?Trails. [Notation "*Palaeonereis* m." [mihi] appears in EICHWALD, 1860, p. 680, with *P. prisca*, a species already mentioned by EICHWALD, 1856, p. 409.] *Ord.*, Eu.(Est.).

Palaeonereis HUNDT, 1940, p. 214 [*non* EICHWALD, 1856]. Small crawling trail or burrow. *L.Dev.*, Eu.(Ger., Thuringia).

Petromonile CASEY, 1961, p. 600 [*Siphonia* (or *Spongites*) *benstedii* BENSTED, 1862, p. 335; OD]. "Stems," about 10 mm. in diameter; irregularly branching, periodically lobed. [Originally described as "sponges"; according to CASEY, probably infilled feeding burrows (written commun., 1973).] *L.Cret.(up.Apt.),* G.Brit.(Eng.).——Fig. 109,*1.* **P. benstedii* (Bensted), low. Greensand, Kent (Maidstone, Iguanodon Quarry); *1a,b,* syntypes, ×0.7 (Casey, n; Inst. Geol. Sci. London Repos. no. GSM 114818,114819).

Phycoidella Matthew, 1890, p. 144 [**P. stichidifera*; M]. Strap-shaped "fronds" showing irregular rows of dark spots or granules transversely arranged on "stem." [According to Matthew, related to *Fucoides circinnatus* Brongniart and regarded as alga; perhaps trace fossil.] *Cam.,* N.Am.(Can.).

Potiria Borrello, 1966, p. 33 [**P. trunciformis;* OD.] ?Inorganic. *Ord.(F.La Tinta),* S.Am. (Arg.).

Protobolella Chapman, 1935, p. 117 [*Fermoria minima Chapman, 1935, p. 114; SM Sahni, 1936] [=*Fermoria* Sahni, 1936, p. 465; obj.]. Small disc-shaped carbonaceous structures, 2 to 4 mm. in diameter, concentrically wrinkled. [Interpreted as ?atremate brachiopod (Chapman, 1935); compared with algae by Sahni & Shrivastava (1954); Howell (1956); and others; according to Misra & Dube (1952) probably inorganic; regarded by Cloud (1968) as "possibly algal, but need restudy"; for account of the nomenclature of genus and type species see Rowell (1971); considered synonymous with *Chuaria circularis* Walcott, 1899.] *Precam. (VindhyanF.),* India.——Fig. 108,*4. P.* sp.; attached to filament-like bodies; ×2 (Sahni, in Häntzschel, 1962).——Fig. 107,*5.* **P. minima; 5a,* traces of concentric ornament on marginal brim; *5b,* severely exfol. disc, with marginal brim; *5c,* concentric traces probably due to crushing, ×10 (Rowell, 1971). [Considerably modified and edited version of text originally prepared by author.—Ed.]

Protostigma Lesquereux, 1878, p. 169 [**P. sigillarioides*; M]. Originally regarded as the oldest lycopod; according to Osgood (1970, p. 395), possibly cast of a burrow or internal mold of a nautiloid or a stromatoporoid; "must remain *Inc. sed.*" *U.Ord.,* USA(Ohio).

Pseudotaeniopteris Sze, 1951, p. 81 [**P. piscatorius*; M]. Oval impressions with thick median vein; similar to *Taeniopteris* Brongniart, but not of plant origin. *L.Cret.(Wealden),* China(Manchuria).

Ptilichnus Hitchcock, 1858, p. 144 [4 "species," no type species designated]. Finlike impressions, arranged in rows; others consisting of parallel slightly curved grooves. [Regarded by Hitchcock as swimming trails of fishes; according to Brown (1912, p. 544-546), more likely markings of rolling or dragging objects.] *Trias.,* USA(Mass.).

Ptychoplasma Fenton & Fenton, 1937, p. 1080 [**P. excelsum*; OD]. Poorly figured trails. [Considered by Fenton & Fenton as having been made by bivalves.] *Penn.,* USA(Texas).

Pucksia Sollas, 1895, p. 302 [**P. machenryi*; M].

FIG. 109. Unrecognizable "genera" (p. W186-187).

Long, narrow threadlike markings. [According to Sollas (1900, p. 278), indubitably organic in origin but of unknown systematic position.] *Cam.,* Eu.(Ire.).

Punctatumvestigium Butts, 1891, p. 44 [**P. circuliformis*; M]. Described as amphibian footprint, but obviously of invertebrate origin. *U.Carb.,* USA(Mo.).

Pyrophyllites. Star-shaped ?trace fossil; cited by Dawson, 1890 (p. 604), together with *Scolithus* and *Asterophycus*; author unknown; no species described. *Sil., Can.*(Ont.).

Quallites Fritsch, 1908, p. 10 [**Q. graptolitarum* (=*Q. problematicus* Fritsch, 1908, explan. pl. 9, fig. 6); M]. Disc-shaped structure (about 3 cm. in diameter) possessing numerous arms with crenulate edges (about 0.5 cm. wide and several centimeters long). *Sil., Eu.*(Czech.). [Description supplied by W. G. Hakes.]

Radicites Fritsch, 1908, explan. pl. 6, fig. 8 [**R. rugosus*; M] [=*Radix* Fritsch, 1908, p. 8 (*non* de Montfort, 1810) (type, *R. corrugatus*)]. Ramifying, branching structure with numerous transverse striations. *Ord., Eu.* (Czech.). [Description supplied by Curt Teichert and W. G. Hakes.]

Radicopsis Fucini, 1938, p. 179 [Many "species," no type species designated]. Probably inorganic; name perhaps not meant as genus. ?*M.Trias., Eu.*(Italy).

Radiophyton Meunier, 1887, p. 59 [**R. sixii*; M]. Tetraradiate, probably accidental structure. *U.Jur., Eu.*(France).

Rectogloma van Tuyl & Berckhemer, 1914, p. 275 [**R. problematica*; M]. Body shaped like an orthoconic cephalopod shell; elliptical in transverse section; apex terminating in spiral coil; closely placed sinuous sutures on surface which disappear completely on apical coil. [Coprolite, according to Knight et al. (1960, p. 1324); this is a doubtful interpretation.] *U.Dev., USA*(Pa.).
——Fig. 107,7. **R. problematica; 7a,b,* ×1.2 (Häntzschel, 1962, courtesy Am. Museum Nat. History).

Rhizomorpha Hernandez-Pacheco, 1908, p. 86 [2 species, no type species designated]. Superfluous name for bulging structures on bedding planes; 3 to 12 mm. in diameter; irregularly branched. *L.Sil., Eu.*(Spain).

Saccophycus U. P. James, 1879, p. 17 [**S. inortus*; M]. Possibly burrows, smooth or striated longitudinally (see J. F. James, 1885, p. 157); never figured. According to Osgood (1970, p. 299), the single specimen not located. *U.Ord., USA* (Ohio).

Schilleria Borrello, 1966, p. 34 [*non* Dahl, 1907; *nec* Girault, 1932] [**S. acuta*; OD]. ?Inorganic. *Ord.(F.La Tinta), S.Am.*(Arg.).

Shikamaia Ozaki, 1968, p. 28 [**S. akasakaensis*; OD]. Flat, dislike "fossil" with undulated "dorsal" and flat "ventral" side, both traversed by median longitudinal "canal"; shell walls composed of calcite, varying in thickness from 0.5 to 2 cm.; transverse section of middle part of "fossil" showing elongate rhomboidal outlines with large inner cavity; systematic position problematic, all phyla of invertebrates, vertebrates and plants are excluded from consideration [!]. [Undoubtedly of inorganic origin.] *L.Perm.*, Japan.

Solicyclus Quenstedt, 1879, p. 578 [Published without species]. Elliptical reliefs, smooth internally; marginal seam divided by numerous radial rays. *L.Jur., Eu.*(Ger.).

Sphaerapus Hitchcock, 1858, p. 164 [**S. larvalis*; SD Lull, 1953, p. 47] [=*Sphaeropus* Lull, 1953, p. 47; *nom. null.*]. ?Trackway consisting of 2 rows of small (diam. 3-5 mm.) hemispherical impressions. *Trias., USA*(Mass.).

Spirochorda Schimper in Schimper & Schenk, 1879, p. 51 [**Dictyota spiralis* Ludwig, 1869, p. 114; OD]. Possibly braided trail. *U.Dev., Eu.* (Ger.).

Spiroscolex Torell, 1870, p. 12 [**Arenicolites spiralis* Torell, 1868, p. LXII; OD]. Transversely ribbed, strongly curved, spiral structures 2 cm. in diameter; transverse ribs slightly elevated. [Originally interpreted as worms; regarded by Nathorst (1881a, p. 28) as impressions of tentacles of medusae; for discussion see also Hofmann (1971, p. 18).] ?*Precam.*, Can.; *Cam.*, Eu.(Swed.-Est.).
——Fig. 108,3. **S. spiralis* (Torell), Cam., Swed.; ×0.5 (Walcott, 1890).

Spongolithus Fritsch, 1908, p. 14 [12 species, no type species designated]. ?Cylindrical structure 14 mm. thick with smaller numerous branches like the "leaves of a willow." Very heterogeneous group of ridgelike and tracklike structures. *Ord., Eu.*(Czech.). [Description supplied by W. G. Hakes.]

Squamopsis Fucini, 1938, p. 182 [Two "species," no type species designated]. Probably inorganic. *L.Cret.("Verrucano")*, Eu.(Italy).

Squamularia Rothpletz, 1896, p. 892 [*non* Gemmellaro, 1899] [**Caulerpa cicatricosa* Heer, 1877, p. 153; SD Häntzschel, 1962, p. W242]. Possibly small fucoids. *Tert.*, Eu.

Striocyclus Quenstedt, 1879, p. 577 [Without species names]. Reliefs on bedding planes with radial, wormlike ornament and central hollow. *L.Jur., Eu.*(Ger.).

Tandilinia Borrello, 1966, p. 34 [**T. mesoconica*; OD]. ?Inorganic. *Ord.(F. La Tinta), S.Am.*(Arg.).

Thinopus antiquus Marsh, 1896, p. 374. Single "footprint" with 3 "toe-impressions," described as earliest record of a terrestrial vertebrate; according to Abel (1935, p. 77), and others, not a vertebrate footprint; in Abel's opinion, a "fossil" that can be interpreted in various ways; possibly fish coprolite. *U.Dev., USA*(Pa.).

Triadonereites Mayer, 1954, p. 227 [**T. mesotriadica*; M]. General name for burrows of varying shape, annulated in part; believed to be made by *Triadonereis* Mayer, 1954. [No clear diagnosis.] *M.Trias.(Muschelkalk)*, Eu. (S.Ger.).

Trianisites Rafinesque, 1821, p. 286 [**T. cliffordi*; M]. See Harrington & Moore (1956e, p. F159).

Trichoides HARKNESS, 1855, p. 474 [*T. ambiguus; M*]. Hairlike bodies, generally straight, some slightly curved; length irregular, about 1 inch; never figured. [Perhaps inorganic.] *Ord.*, Eu. (Scot.).

Tropidaulus FENTON & FENTON, 1937, p. 1080 [*T. magnus*; OD]. Burrows, undersurface with transverse wrinkles and median keel or ridge, 1.5 to 2 cm. wide; rather poorly figured. [According to the authors, made by arthropods or large annelids.] *Penn.*, USA(Texas).

Truncus FRITSCH, 1908, p. 8 [*T. ramifer*; M]. Subcylindrical structure about 10 cm. long, exterior covered with dimples; bent branch extending to "left" side of structure. [?Inorganic.] *Ord.*, Eu.(Czech.). [Description supplied by W. G. HAKES.]

Valonites SORDELLI, 1873, p. 367 [*V. utriculosus*; M]. Small hemispherical, poorly figured forms, belonging to ?algae. *Tert.(Plio.)*, Eu.(Italy).

Vesicolithus FRITSCH, 1908, only figure: pl. 3, fig. 5 [*V. guttalis*; M]. Hemispheres of different size occurring in quartzite; up to 1 cm. in diameter. [?Inorganic.] *Ord.*, Eu.(Czech.). [Description supplied by W. G. HAKES.]

Vucetichia BORRELLO, 1966, p. 35 [*V. psamma*; OD]. ?Inorganic. *Ord.(F. La Tinta)*, S.Am. (Arg.).

Walcottia MILLER & DYER, 1878, p. 39 [*W. rugosa*; M]. "Genus" including 3 different "species" of long, tapering, rugose, flexuous impressions of wormlike shape. [*W. sulcata* U. P. JAMES, 1881, p. 44, was never figured, no type specimens or other material located; according to OSGOOD (1970, p. 398), perhaps small trails similar to *Cruziana*; ?*W. cookana* MILLER & DYER, 1878, p. 11, according to OSGOOD (1970, p. 380), "impossible to make any interpretation of the species."] *U.Ord.*, USA(Ohio).

Yaravidium BORRELLO, 1966, p. 36 [*Y. coniformis*; OD]. ?Inorganic. *Ord.(F. La Tinta)*, S.Am.(Arg.).

Zearamosus WEBSTER, 1920, p. 286 [*Z. elleria*; M]. Originally described as of plant origin; ?burrows similar to *Gracilerectus* WEBSTER, 1920, and *Fruticristatum* WEBSTER, 1920. *Dev.*, USA (Iowa).

GENERA OF RECENT LEBENSSPUREN

"Generic" names have been proposed for three types of Recent lebensspuren. In a fourth case, an important Paleozoic trace fossil (*Nemapodia* EMMONS) has been proven to be present in the Recent environment. Names for Recent lebensspuren are neither necessary nor justifiable. Often their producer can be determined, in which case description of the traces are sufficient. Examples are "star-shaped feeding traces of *Corophium*" or a "*Paraonis* meander." When the producer is unknown, it is sufficient to give a morphological-ecological description such as "branching grazing trace of supposedly polychaete origin." Also, it is sufficient to indicate similarity in shape of the Recent form with that of a known fossil ichnogenus.

1

FIG. 110. Recent lebensspuren (p. *W*189).

Benjaminichnus BOEKSCHOTEN, 1964, p. 423 [*nom. subst. pro* Batrachoides HITCHCOCK, 1858, p. 121 (*non* LACEPÈDE, 1800)]. Proposed for possible fossil and Recent tadpole nests or similar traces. [Batrachoides antiquior and B. nidificans HITCHCOCK, 1858, p. 122 (Sil. and Trias., USA) interpreted by SHEPARD (1867) and KINDLE (1914, p. 160) as interference ripple marks. Recent tadpole nests should not be named; all supposed fossil tadpole nests are of inorganic origin as proved by CAMERON & ESTES (1971), thus Batrachoides HITCHCOCK and *Benjaminichnus* BOEKSCHOTEN are to be abandoned; for discussion and nearly complete references of papers dealing with tadpole nests see CAMERON & ESTES (1971).]
——FIG. 110,1. Tadpole structures from Tenn., in dry condition, ×0.12 (Maher, 1962).

Corophites ABEL, 1935, p. 463 [*nom. nud.*]. Suggested as name for burrows made by Recent amphipod *Corophium,* especially for (rare) simple shafts with sidewise branchings. *Rec.*

Ephemerites ABEL, 1935 [*non* GEINITZ, 1865]. Horizontal U-shaped burrows produced by larvae of ephemerids; occurring in fresh-water deposits. [Shown by SEILACHER (1951), to be spreiten burrows.] *Rec.*

Mystichnis VYALOV & ZENKEVICH, 1961, p. 58

[**M. pacificus*; M]. Crawling trails on bottom of Pacific at depth of 3,000 m.; 10 cm. wide; producer unknown. *Rec.*

Nemapodia EMMONS, 1844, expl. pl. 2, fig. 1. [**N. tenuissima*; M]. Described as fossil trails, but according to RICHTER (1924), trail of Recent gastropod feeding in meanders on surface of slabs of Paleozoic slates (as shown by RICHTER for *N. tenuissima,* described by GEINITZ (1852) from bedding planes of L.Carb. slates of Saxony).

INVALID NAMES

The following group contains a list of names considered to be invalid. These names are either available or unavailable.

Acanthus GROSSHEIM, 1946, p. 116 [*non* BLOCH, 1795; *nec* DUMONT, 1816; *nec* GISTL, 1834; *nec* LOCKINGTON, 1876)] [**A. dodecimanus*; M] (HÄNTZSCHEL, 1965). *L.Tert.,* USSR.

Agarites DE SAPORTA, 1890, p. 313 [*non* AGASSIZ, 1841] [**A. fenestratus*; M] (HÄNTZSCHEL, 1965). *U.Jur.,* Eu.(France).

Aglaopheniolites. According to SEILACHER (pers. commun., 1956), name used in Italian paleontological collections for trace fossil from Italian flysch. [Very probably manuscript name.]

Anapaleodictyon TANAKA, 1970 [MS. name, *nom. null.* for *Protopaleodictyon?*].

Arabesca VYALOV, 1972, p. 76. Two species described, *A. caucasica* and *A. daghestanica,* but no type designated, hence name not available. *Tert. (low. Paleoc.),* USSR (N. Daghestan). [Description supplied by CURT TEICHERT.]

Boteillites. According to SEILACHER (pers. commun., 1956), name used in Italian paleontological collections for trace fossils from the Italian flysch. [Probably manuscript name.]

Cochlea HITCHCOCK, 1858, p. 162 [jr. hom.; *non* MARTYN, 1874; *nec* GRAY, 1847] [**C. archimedea*; M]. Trackway resembling an archimedean screw (HITCHCOCK, 1858). *Trias.,* USA (Mass.).

Corticites FUCINI, 1938, p. 170 [*nom. nud.*] [jr. hom.; *non* ROSSMAESSLER, 1840]. ?Inorganic (HÄNTZSCHEL, 1965). *L.Cret.("Verrucano"),* Eu. (Italy).

Cyclophycus ULRICH, 1880, *nom. nud.* (*fide* OSGOOD, 1970, p. 399).

Cylindrites GÖPPERT, 1842 (?1841), p. 115 [*non* GESNER, 1758; *nec* GMELIN, 1793; *nec* SOWERBY, 1824] [=*Spongites* GEINITZ, 1842 (*partim*) (*non* OKEN, 1815); ?*Astrocladia furcata* GERSTER, 1881; *Goniophycus* DE SAPORTA, 1884]. Like *Palaeophycus,* used as general term for cylindrical and

not vertical fillings of burrows (HÄNTZSCHEL, 1965). [Jr. hom.] *Mesoz.,* Eu.

Gordioides FRITSCH, 1908, pl. 11 [*nom. nud.*] [**G. spiralis*; M]. *Sil.,* Eu.(Czech.).

Itieria DE SAPORTA, 1872, p. 122 [jr. hom.; *non* MATHERON, 1842] (ANDREWS, 1955). *U.Jur.,* Eu. (France).

Leuconoe BOGACHEV, 1930, p. 73 [jr. hom.; *non* BOIE, 1830] [**L. paradoxa*; M]. Larva of arthropod of unknown systematic position (HÄNTZSCHEL, 1965). [Found in flysch deposits.] *L.Tert. (low. Eoc.),* USSR.

Lunula HITCHCOCK, 1865, p. 17 [*non* KOENIG, 1825, *nec* LAMARCK, 1812] [**L. obscura*; M]. Trail consisting of narrow axis, with laterally extended lunate impressions on both sides (HITCHCOCK, 1865). [Possibly made by phyllopod or myriapod; jr. hom.] *Trias.,* USA(Mass.).

Montfortia LEBESCONTE, 1887, p. 782 [jr. hom.; *non* RECLUZ, 1843]. Small horizontal, oblique, or perpendicular burrows, 1 to 2 mm. wide, occasionally showing annulation; very similar to *Planolites* (SEILACHER, 1956a). [Probably worm trails; not a sponge as interpreted by LEBESCONTE.] *Precam.,* Eu.(France).

Nisea DE SERRES, 1840, p. 13 [*non* RAFINESQUE, 1815 (*nom. nud.*)] [=*Nemausina* DUMAS, 1876]. Irregularly shaped globular or ellipsoidal bodies which give off 2 or more long, transversely striped or slightly segmented tubes (HÄNTZSCHEL, 1965). [Interpreted as annelids, mollusks, or coelenterates.] *L.Cret.,* Eu.(S.France).

Palaeocrista HUNDT, 1941, p. 70 [*nom nud.*; diagnosis and designation of type species missing] (HÄNTZSCHEL, 1965). *L.Ord.,* Eu.(Ger.).

"?Palaeoscolex ratcliffei" ROBISON, 1969, p. 1172. Burrows, 7 to 10 mm. wide, mostly flattened; filled with ovoid pellets, 1.5 mm. long, 0.5 mm. wide; burrows sometimes containing bundles of pellets and thus similar to pelletal clusters known as *Tomaculum* GROOM. [In one of these burrows a body remain of the annelid *Palaeoscolex ratcliffei* ROBISON (1969, p. 1171) has been found; ROBISON believed these burrows were made by that annelid but he wrongly used the name for the burrow of its supposed producer.] *M.Cam.(Spence Sh.)*, N. Am.(USA, N.Utah).

Palaeotenia guilleri Crié, 1883, p. 49. Name proposed by Crié for *Fraena goldfussi* Rouault but obviously not used since 1885. *Ord.*, Eu.(France).

Parinassa Hundt, 1941, p. 124 [*P. pennaeformis*; M] [*nom. nud.*; no diagnosis]. (Häntzschel, 1965). *L.Ord.*, Eu.(Ger.).

Phyllonia Hundt, 1941, p. 53 [*nom. nud.*; diagnosis and designation of type species missing] (Häntzschel, 1965). *L.Ord.*, Eu.(Ger.).

Platyrhynchus Glocker, 1850, p. 940 [jr. hom.; *non* Leuckart, 1816, *nec* Swainson, 1820; *nec* Cuvier, 1826; *nec* Wagler, 1830; *nec* Agassiz, 1846; *nec* van Beneden, 1876; *nec* Chevrolat, 1882] [*P. problematicus*; M] (Häntzschel, 1965). Probably a track; similar to *Dreginozoum*. *?U.Cret.*, Eu.(Ger.).

Portelia Boursault, 1889, p. 728 [jr. hom.; *non* de Quatrefages, 1850] [*P. meunieri*; M]. Nondescript, branched cylindrical fillings of tunnels; very poorly figured (Andrews, 1955). *U.Jur.*, Eu.(France).

Sagittarius Hitchcock, 1865, p. 16 [jr. hom.; *non* Vosmaer, 1767; *nec* Hermann, 1783] [*S. alternans*; M]. Two parallel rows of delicately curved tracks, with concave sides toward each other, resembling many small bows alternating with one another (Hitchcock, 1865). [Insect trail.] *Trias.*, USA(Mass.) (See Häntzschel, 1962, fig. 129,3).

Saltator Hitchcock, 1858, p. 137 [jr. hom.; *non* Vieillot, 1816]. Inorganic markings or tracks made by animals moving by leaps; 2 "species" having little in common (Hitchcock, 1858). *Trias.*, USA(Mass.).

Schaderthalis Hundt, 1931, p. 51, 56 [*nom. nud.*, no description nor diagnosis, 3 poor figures only] [*S. bruhmii*; M] [=*Schaderthalia* Hundt, 1931, p. 67 *(nom. null.)*]. Very numerous tiny furrows, arranged parallel and closely adjacent, smooth and sharply incised; similar to finger impressions. ["*Schaderthalia*" regarded by Seilacher (1960,

p. 49) as identical to *Lophoctenium globulare* Gümbel, 1879, p. 469 (*nom. nud.*; no description nor diagnosis, figure only); Pfeiffer (1968, p. 672) ascribed "*Schaderthalis*" to his ichnogenus *Agrichnium* as *A. bruhmi* (Pfeiffer, 1968) though being much smaller than the type species *A. fimbriatum* (Ludwig) and differing from it in much more regular arrangement and parallelism of the furrows.] *Low. (?) M. Dev. (Nereiten-Quarzit)*, Eu.(Ger., Thuringia).

Sphenopus Fritsch, 1908, p. 11, 12 [jr. hom.; *non* Steenstrup, 1856] [*S. pectinatus*; M] (Fritsch, 1908). *Ord.*, Eu.(Czech.).

Tubotomaculum Richter in Gómez de Llarena, 1949, p. 117, 127 [=*nom. nud.*, used in title of announced but never published paper] (see under *Tomaculum* Groom, 1902, p. W143)].

Tubulites H. D. Rogers, 1838 [*nom. nud.*, provided for *Skolithos* Haldeman, not published; preoccupied by *Tubulites* Gesner, 1758].

Vermiculites Rouault, 1850, p. 744 [jr. hom.; *non* Bronn, 1848] [*V. panderi*; M]. Poorly described and never figured (Andrews, 1955). *Ord.*, Eu.(France).

Wellerites Flower, 1961, p. 115 [*non* Plummer & Scott, 1937] [*W. gracilis*; OD]. Long, slender, calcareous tubes, somewhat widening distally, 1 mm, long, 0.3 mm. wide; at bases forming small colonies attached to *Catenipora*; known only from single thin section. [Systematic position unknown.] *Ord.*, USA(N.Mex.).

Zonarites von Sternberg, 1833, p. 34 [jr. hom.; *non* *Zonarites* Rafinesque, 1831] [*Fucoides flabellaris* Brongniart, 1823, p. 311; SD Andrews, 1955, p. 262] [Probably = *Zonarides striatus* Squinabol, 1887 (*Saportia* Squinabol, 1891), as well as plants (e.g., *Z. digitatus* von Sternberg, 1833, =*Zonarides* Schimper, 1869)]. "Genus" comprising starlike trace fossils (e.g., *Z. alcicornis* Fischer-Ooster, 1858) (Andrews, 1955). [According to Seilacher (1955), branched feeding burrows with fecal pellets stuffed transversely into them. Fuchs (1895, p. 408) considered *Zonarites alcicornis* to belong to *Phymatoderma* Brongniart, 1849.] *?Perm., Tert.*, Eu.

REFERENCES

Abbott, G., & Abbott, C. P.

1914, *Is Atikokania a concretion?*: Nature, v. 94, p. 477-478.

Abel, Othenio

1912, *Grundzüge der Palaeobiologie der Wirbeltiere*: 708 p., 470 text-fig., E. Schweizerbart (Stuttgart).

1926a. *Amerikafahrt: Eindrücke, Beobachtungen und Studien eines Naturforschers auf einer Reise nach Nordamerika und Westindien*: 462 p., 273 text-fig., G. Fischer (Jena).

1926b, *Die Lebensspuren in der oberen Trias des Connecticut-Tales in Connecticut und Massachusetts*: Zool.-botan. Gesell. Wien, Verhandl., v. 74/75, p. 145-150.

1927, *Lebensbilder aus der Tierwelt der Vorzeit*: 2nd. edit., 714 p., 557 text-fig., Gustav Fischer (Jena).

1929, *Aufklärung der Kriechspuren im Greifen-*

steiner Sandstein bei Kierling im Wiener-wald: Akad. Wiss. Wien, math.-nat. Kl., Anz., v. 66, p. 240-242.

1935, *Vorzeitliche Lebensspuren:* 644 p., 530 text-fig., Gustav Fischer (Jena).

Adegoke, O. S.

1966, *Silicified sand-pipes belonging to Chaceia (?) (Pholadidae: Martesiinae) from the late Miocene of California:* Veliger, v. 9, p. 233-235, pl. 21.

Adkins, W. S.

1928, *Handbook of Texas Cretaceous fossils:* Univ. Texas, Bull., no. 2838, 385 p., 37 pl.

Agassiz, Louis

1833, *Briefliche Mitteilung an Prof. Bronn vom. 8.11.1833:* Neues Jahrb. Mineralogie, Geognosie, Geologie u. Petrefaktenkd., 1833, p. 675-677.

Ager, D. V.

1963, *Principles of paleoecology:* 371 p., McGraw-Hill (New York).

———, & Wallace, Peigi

1970, *The distribution and significance of trace fossils in the uppermost Jurassic rocks of the Boulonnais, northern France:* in Trace fossils, T. P. Crimes & J. C. Harper (eds.), Geol. Jour., spec. issue no. 3, p. 1-18, text-fig. 1-7, pl. 1, Seel House Press (Liverpool).

Åhman, E., & Martinsson, Anders

1965, *Fossiliferous Lower Cambrian at Äspelund on the Skäggenäs Peninsula:* Geol. Fören. i Stockholm, Förhandl., v. 87, pt. 1, p. 139-151, illus.

Ahr, W. M., & Stanton, R. J., Jr.

1973, *The sedimentologic and paleoecologic significance of Lithotrya, a rock-boring barnacle:* Jour. Sed. Petrology, v. 43, p. 20-23, text-fig. 1-5.

Alberti, Helmut

1968, *Trilobiten (Proetidae, Otarionidae, Phacopidae) aus dem Devon des Harzes und des Rheinischen Schiefergebirges:* Geol. Jahrb., Beihefte, v. 73, p. 1-147, text-fig. 1-31, table 1-8, pl. 1-21.

Alessandri, Giulio de

1900, *Appunti di geologia e di paleontologia sui dintorni di Acqui:* Soc. Italiana Sci. Nat. e Museo Civile Storia Nat., Atti, v. 39, p. 173-348, 1 pl.

Allasinaz, Andrea

1968, *Revisione ed interpretazione del genere Bactryllium Heer:* Riv. Ital. Paleontologia e Stratigrafia, v. 74, no. 4, p. 1065-1146, text-fig. 1-20, pl. 68-83, 2 tables.

Allen, A. T., & Lester, J. G.

1953, *Animal tracks in an Ordovician rock of*

northwest Georgia: Georgia Geol. Survey, Bull. 60, p. 205-214, text-fig. 1-5.

Alloiteau, James

1952, *Sous-classe des Alcyonaria Milne-Edwards 1857:* in Traité de Paléontologie, Jean Piveteau (ed.), v. 1, p. 408-417, Masson et Cie (Paris).

Alpert, Stephen

1973, *Bergaueria Prantl (Cambrian and Ordovician), a probable actinian trace fossil:* Jour. Paleontology, v. 47, p. 919-924, text-fig. 1-3, pl. 1.

Altevogt, Gustav

1968a, *Erste Asterosoma-Funde (Problem.) aus der oberen Kreide Westfalens:* Neues Jahrb. Geologie, Paläontologie, Abhandl., v. 132, p. 1-8, text-fig. 1,2, 2 pl.

1968b, *Das Problematikum Guilielmites Geinitz, 1858. Ein Deutungsversuch:* Same, Abhandl., v. 132, p. 9-21, 2 pl.

Ameron, H. W. J. van

1966, *Phagophytichnus ekowskii nov. ichnogen. & nov. ichnosp., eine Missbildung infolge von Insektenfrass, aus dem spanischen Stephanien (Provinz Leon):* Leidse Geol. Meded., v. 38, p. 181-184, text-fig. 1-3, pl. 1-3.

———, & Boersma, M.

1971, *A new find of the ichnofossil Phagophytichnus ekowskii van Ameron:* Geologie en Mijnbouw, v. 50, p. 667-670, text-fig. 1.

Ami, H. M.

1903, *Description of tracks from the fine-grained siliceous mudstones of the Knoydart formation (Eo-Devonian) of Antigonish County, Nova Scotia:* Nova Scotia Inst. Sci., Proc. Trans., v. 10 (1898-1902), p. 330-332, pl. 2.

1905, *Preliminary list of the fossils collected by Professor L. W. Bailley from various localities in the Province of New Brunswick during 1904:* Canada Geol. Survey, Dept. Summ. Rept. for 1904, p. 289-292.

Ammon, Ludwig von

1900, *Über das Vorkommen von "Steinschrauben" (Daemonhelix) in der oligocänen Molasse Oberbayerns:* Geognost. Jahresh., v. 13, p. 55-69, text-fig. 1-5, 1 pl.

Amstutz, G. C.

1958, *Coprolites: A review of the literature and a study of specimens from southern Washington:* Jour. Sed. Petrology, v. 28, p. 498-508, text-fig. 1-3.

Anderheggen, F.

1927, *Actualités illustrés:* Nature, v. 55, pt. 2, p. 240, text-fig. 1, 2 (Paris).

Anderson, Ann

1972, *An analysis of supposed fish trails from interglacial sediments in the Dwyka Series, near Vryheid, Natal:* Internatl. Union Geol. Sci. Comm. Stratigraphy, Second Gondwana Symposium (July-Aug., 1970), Proc. & Papers, p. 637-647, text-fig. 1-5, 3 pl. Council Sci. Industr. Res. (Pretoria).

Andrée, Karl

1920, *Über einige fossile Problematika. I. Ein Problematikum aus dem Paläozoikum von Battenberg an der Eder und das dasselbe beherbergende Gestein:* Neues Jahrb. Mineralogie, Geologie, Paläontologie, 1920, pt. I, p. 55-88, 1 pl.

1927, *Bedeutung und zeitliche Verbreitung von Arenicoloides Blanckenhorn und verwandten Formen:* Paläont. Zeitschr., v. 8, p. 120-128.

Andrews, H. N., Jr.

1955, *Index of generic names of fossil plants, 1820-1950:* U. S. Geol. Survey, Bull. 1013, 262 p.

1970, *Index of generic names of fossil plants, 1820-1965:* Same, Bull. no. 1300, 354 p.

Anelli, M.

1935, *Appunti paleontologi a proposito delle cosidette "argille scagliose":* Riv. Ital. Paleontologia, v. 41, p. 33-44, pl. 1.

Antoniazzi, Alberto

1966, *Alcuni Palaeodictyon rinvenuti nei terreni Miocenici dell'Appennino Forlivese:* Museo Civico Storia Naturale Verona, Mem., v. 14, p. 455-463, pl. 1.

Antun, P.

1950, *Sur les Spirophyton de l'Emsien de l'Oesling (Grand-Duché de Luxembourg):* Soc. Géol. Belgique, Ann., v. 73, Bull., ser. B, p. 241-262, text-fig. 1-3, pl. 1.

Arai, M. N., & McGugan, Alan

1968, *A problematical coelenterate (?) from the Lower Cambrian, near Moraine Lake, Banff Area, Alberta:* Jour. Paleontology, v. 42, p. 205-209, text-fig. 1, 2, pl. 36.

Avnimelech, Moshè

1955, *Occurrence of fossil Phoronidea-like tubes in several geological formations in Israel:* Res. Council Israel, ser. B, v. 5, no. 2, p. 174-177.

Azpeitia Moros, Florentino

1933, *Datos para el estudio paleontólogico del Flysch de la Costa Cantábrica y de algunos otros puntos de España:* Inst. Geol. Min. España, Bol., v. 53, p. 1-65, pl. 1-19.

Babin, Claude, Glemarec, Michel, Termier, Henri, & Termier, Geneviève

1971, *Rôle des Maldanes (Annélides Polychètes)* *dans certains types de bioturbation:* Soc. Géol. Nord, Ann., v. 91, no. 3, p. 203-206, pl. 33-35.

Baily, W. H.

1865, *The Cambrian rocks of the British Islands, with especial reference to the occurrence of this formation and its fossils in Ireland:* Geol. Mag., v. 2, p. 385-400, text-fig. 1-6.

Bain, G. W.

1927, *Huronian stromatoporid-like masses:* Pan-American Geologist, v. 47, p. 281-284.

Ballance, P. F.

1964, *The sedimentology of the Waitemata Group in the Takapuna section, Auckland:* New Zealand Jour. Geology & Geophysics, v. 7, p. 466-499, text-fig. 1-26.

Bambach, R. K.

1971, *Adaptations in Grammysia obliqua:* Lethaia, v. 4, p. 169-183, text-fig. 1-11.

Bandel, Klaus

1967a, *Trace fossils from two Upper Pennsylvanian sandstones in Kansas:* Univ. Kansas Paleont. Contrib., Paper 18, p. 1-13, text-fig. 1-3, pl. 1-5.

1967b, *Isopod and limulid marks and trails in Tonganoxie Sandstone (Upper Pennsylvanian) of Kansas:* Same, Paper 19, p. 1-10, pl. 1-3.

1973, *A new name for the ichnogenus Cylindrichnus Bandel, 1967:* Jour. Paleontology, v. 47, no. 5, p. 1002.

Banks, M. R.

1962, *On Hurdia? davidi Chapman from the Cambrian of Tasmania:* Austral. Jour. Sci., v. 25, no. 5, p. 222-223.

Banks, N. L.

1970, *Trace fossils from the Late Precambrian and Lower Cambrian of Finnmark, Norway:* in Trace fossils, T. P. Crimes & J. C. Harper (eds.), Geol. Jour., spec. issue no. 3, p. 19-34, text-fig. 1-2b, pl. 1-3, Seel House Press (Liverpool).

1973, *Trace fossils in the Halkkavarre section of the Dividal Group (?Late Precambrian—Lower Cambrian), Finnmark:* Norges Geol. Unders. 288, p. 1-6, text-fig. 1-4.

Barbour, I. H.

1892a, *Notice of new gigantic fossils:* Science, v. 19, p. 99-100, text-fig. 1-3.

1892b, *On a new order of gigantic fossils:* Univ. Nebraska, Univ. Studies, v. 1, no. 4, p. 301-335, text-fig. 1-18, 6 pl. (July).

1895, *Is Daemonelix a burrow?:* Am. Naturalist, v. 29, p. 517-527, text-fig. 1-5.

Barnes, W. C., & Smith, A. G.

1964, *Some markings associated with ripple-marks*

from the Proterozoic of North America: Nature, v. 201, no. 4923, p. 1018-1019.

Barrande, Joachim

1872, *Système Silurien de la centre de la Bohême. Suppl. vol. 1, Trilobites, Crustacés divers et poissons:* 647 p., 35 pl., publ. by author (Prague, Paris).

Barrois, Ch.

1882, *Recherches sur les terrains anciens des Asturies et de la Galice:* Soc. Géol. Nord, Mém., v. 2, pt. 1, 630 p., 20 pl.

1888, *Note sur l'existence du genre Oldhamia dans les Pyrénées:* Same, v. 15 (1887-88), p. 154-157, pl. 3.

Barsanti, Leopoldo

1902, *Considerazioni sopra il genere Zoophycos:* Atti Soc. Toscana Sci. Nat., Mem., v. 18, p. 68-95, pl. 3.

Barthel, K. W.

1969, *Die obertithonische, regressive Flachwasser-Phase der Neuburger Folge in Bayern:* Bayer. Akad. Wiss., math-naturw. Kl., Abhandl., no. 142, p. 1-174, text-fig. 1-39, 14 pl.

———, **& Barth, Walter**

1972, *Palecologic specimens from the Devonian of Bolivia:* Neues. Jahrb. Geologie, Paläontologie, Monatsh. 1972, p. 573-581, text-fig. 1-5.

Bartrum, J. A.

1948, *Two undetermined New Zealand Tertiary fossils:* Jour. Paleontology, v. 22, p. 488-489, pl. 76.

Bassler, R. S.

1915, *Bibliographic index of American Ordovician and Silurian fossils:* U. S. Natl. Museum, Bull. 92, v. 1, 718 p.

1941, *A supposed jellyfish from the pre-Cambrian of the Grand Canyon:* Same, Proc., v. 89, no. 3104, p. 519-522, illus.

1952, *Taxonomic notes on genera of fossil and Recent Bryozoa:* Washington Acad. Sci., Jour., v. 42, p. 381-385.

1953, *Bryozoa:* in Treatise on invertebrate paleontology, R. C. Moore (ed), Part G, 253 p., 175 text-fig., Geol. Soc. America & Univ. Kansas Press (New York; Lawrence, Kans.).

Bather, F. A.

1909, *Fossil representation of the lithodomous worm Polydora:* Geol. Mag., ser. 5, v. 6, p. 108-110.

1910, *Some fossil annelid burrows:* Same, ser. 5, v. 7, p. 114-116.

1911, *Upper Cretaceous terebelloids from England:* Same, ser. 5, v. 8, p. 481-487, 549-556, pl. 24.

1924, *Tiosa siphonalis Marcel de Seres, a supposed Liassic annelid:* The Naturalist, 1924, p. 7-10, text-fig. 1-3.

Bayer, F. M.

1955, *Remarkably preserved fossil seapens and their recent counterparts:* Washington Acad. Sci., Jour., v. 45, p. 294-300, text-fig. 1, 2.

Beaudoin, B., & Gigot, P.

1971, *Figures de courant et traces de pattes d'oiseaux associées dans la molasse Miocène de Digne, Basses Alpes (France):* Sedimentology, v. 17, p. 241-256, text-fig. 1-3, 5 pl.

Becker, H. F., & Donn, William

1952, *A new occurrence and description of the fossil Arthrophycus:* Science, v. 115, no. 2982, p. 214-215, text-fig. 1, 2.

Bein, Georg

1932, *Die Stellung des Richelsdorfer Gebirges zum Thüringer Walde und Rheinischen Schiefergebirge:* Deutsch. Geol. Gesell., Zeitschr., v. 84, p. 786-829, text-fig. 1-8, pl. 25-27.

Belford, D. J.

1967, *Occurrence of the genus Draffania Cummings in Western Australia:* Australia Bur. Min. Res., Geology & Geophysics, Bull. 92, Palaeont. Papers 1966, p. 49-54, text-fig. 1, 2, pl. 6.

Bell, B. M., & Frey, R. W.

1969, *Observations on ecology and the feeding and burrowing mechanisms of Mellita quinquiesperforata (Leske):* Jour. Paleontology, v. 43, p. 553-560.

Bender, Friedrich

1963, *Stratigraphie der "Nubischen Sandsteine" in Süd-Jordanien. (Auswertung der paläozoischen Fossilien von Reinhold Huckriede):* Geol. Jahrb., v. 81, p. 237-276, text-fig. 1-11, 6 pl., 1 table.

Bengtson, Stefan

1968, *The problematic genus Mobergella from the Lower Cambrian of the Baltic area:* Lethaia, v. 1, p. 325-351, text-fig. 1-18.

Bentz, A.

1929, *Fossile Röhrenbauten im Unterneokom des Isterberges bei Bentheim:* Preuss. Geol. Landesanst., Jahrb., v. 49, pt. 2 (1928), p. 1173-1183, pl. 1.

Berger, Walter

1957, *Eine spiralförmige Lebensspur aus dem Rupel der bayrischen Beckenmolasse:* Neues Jahrb., Geologie, Paläontologie, Monatsh., 1957, p. 538-540, text-fig. 1.

Bergström, Jan

1969, *Remarks on the appendages of trilobites:* Lethaia, v. 2, p. 395-414.

1970, *Rusophycus as an indication of early Cambrian age:* in Trace fossils, T. P. Crimes & J. C. Harper (eds.), Geol. Jour., spec. issue no. 3, p. 35-42, text-fig. 1-3, pl. 1, Seel House Press (Liverpool).

1972, *Appendage morphology of the trilobite Cryptolithus and its implications:* Lethaia, v. 5, p. 85-94, text-fig. 1-3.

1973, *Organization, life, and systematics of trilobites:* Fossils and Strata, no. 2, p. 1-69, text-fig. 1-16, pl. 1-5.

Bernhauser, A.

1953, *Über Mycelites ossifragus Roux. Auftreten und Formen im Tertiär des Wiener Beckens:* Österr. Akad. Wiss., Sitzungsber., math.-nat. Kl., pt. 1, v. 162, p. 119-127, text-fig. 1-6.

1962, *Über Mycelites ossifragus Roux und Palaeomycelites lacustris Bystrow:* Mikroskopie, v. 17, p. 187-193, text-fig. 1-5.

Beyer, K.

1943, *Neue Fundpunkte von Tomaculum problematicum Groom im Ordovicium des Sauerlandes:* Reichsstelle Bodenf., Jahrb., v. 63 (1942), p. 124-133, pl. 3.

Biernat, Gertruda

1961, *Diorygma atrypophilia n. gen. n. sp.—a parasitic organism of Atrypa zonata Schnur:* Acta Palaeont. Polonica, v. 6, p. 17-28, 4 pl.

Bigot, A. P. D.

1886, *Quelques mots sur les Tigillites:* Soc. Linnéenne Normandie, Bull., sér. 3, v. 10, p. 161-165.

Billings, Elkanah

1861-65, *Palaeozoic fossils, Vol. 1: containing descriptions and figures of new or little known species of organic remains from the Silurian rocks:* Canada, Geol. Survey, 426 p., 401 text-fig.

1866, *Catalogue of the Silurian fossils of the Island of Anticosti, with descriptions of some new genera and species:* Same, Publ. 427, 93 p., 28 text-fig.

1872, *On some fossils from the primordial rocks of Newfoundland:* Canad. Naturalist, n. ser., v. 6, p. 465-479, 14 text-fig.

1874, *Palaeozoic fossils. Vol. 2:* Canada, Geol. Survey, 144 p., 85 text-fig., 10 pl.

Binney, E. W.

1852, *On some trails and holes found in rocks of the Carboniferous strata, with remarks on the Microconchus carbonarius:* Manchester Lit. Philos. Soc., Mem. & Proc., ser. 2, v. 10, p. 181-201, pl. 1, 2.

Birkenmajer, Krzysztof

1959, *Fucusopsis angulatus Palibin (Problematica) z warstw pstrych (Dan-Paleocen) Osłony Pienińskiego pasa skałkowego:* Polsk. Towarzyst. Geol., v. 29 (1959), no. 2, p. 227-232, text-fig. 1, pl. 22. [*Fucusopsis angulatus Palibin (Problematica) from the variegated beds (Danian—Paleocene) of the Pieniny Klippenbelt mantle (Central Carpathians).*]

——, & Bruton, D. L.

1971, *Some trilobite resting and crawling traces:* Lethaia, v. 4, p. 303-319, 14 text-fig.

Bischoff, Bernhard

1968, *Zoophycos, a polychaete annelid, Eocene of Greece:* Jour. Paleontology, v. 42, p. 1439-1443, text-fig. 1, pl. 179, 180.

Blake, J. A., & Evans, J. W.

1973, *Polydora and related genera as borers in mollusk shells and other calcareous substrates:* Veliger, v. 15, p. 235-249, text-fig. 1-4.

Blanckenhorn, M.

1902, *Über drei interessante geologische Erscheinungen in der Gegend von Mellrichstadt und Ostheim vor der Rhön.:* Deutsch. Geol. Gesell., Zeitschr., v. 54, Monatsber., p. 102-106, text-fig. 1.

1916, *Organische Reste im mittleren Buntsandstein Hessens:* Gesell. z. Beförderg. gesamten Naturwiss. Marburg, Sitzungsber., 1916, p. 21-43.

1936, *Natürliche Erklärung der "Runensteine" von Willinghausen:* Hessenland, v. 47, p. 5-10, text-fig. 1-4.

Blumenbach, J. F.

1803, *Specimen Archaeologicae Telluris Terrarumque imprimis Hannoverarum:* 28 p., 3 pl., H. Dieterich (Göttingen).

Boekschoten, G. J.

1964, *Tadpole structures again:* Jour. Sed. Petrology, v. 34, p. 422-423.

1966, *Shell borings of sessile epibiontic organisms as palaeoecological guides (with examples from the Dutch coasts):* Palaeogeography, Palaeoclimatology, Palaeoecology, v. 2, p. 333-379, text-fig. 1-16.

1967, *Palaeoecology of some mollusca from the Tielrode Sands (Pliocene, Belgium):* Same, v. 3, p. 311-362, text-fig. 1-40.

1970, *On bryozoan borings from the Danian at Fakse, Denmark:* in Trace fossils, T. P. Crimes & J. C. Harper (eds.), Geol. Jour., spec. issue no. 3, p. 43-48, text-fig. 1, Seel House Press (Liverpool).

Bogachev [Bogatschew], V. V.

1908, *Problematicheskaya vodorosly Taonurus v Russkom Paleogen:* Ezhegodnik po Geologii i Mineralogii, v. 10, no. 7-8, p. 221-223, text-fig. 1-3 (Ger. text p. 224-226). [*The problematic alga Taonurus from the Russian Paleogene.*]

1930, *Fukoidy i ieroglify kavkazskogo flisha:* Azerbaydzhan. Neft. Khosiaist. (=Aserbeidschanische Erdölwirtschaft), 1930, no. 7-8 (103-104), p. 71-73, 1 pl. [*Fucoids and hieroglyphs of the Caucasian flysch.*]

Bondesen, E., Raunsgaard Pedersen, K., & Jørgensen, O.

1967, *Precambrian organisms and the isotopic composition of organic remains in the Ketilidian of south-west Greenland:* Meddel. Grønland, v. 164, no. 4, p. 1-41, text-fig. 1-8, pl. 1-13, tables 1-5.

Bonet, Federico

1956, *Zonificación microfaunística de las calizas Cretácicas del este de México:* Asoc. Mexicana Geólogos Petroleros, Bol., v. 8, no. 7-8, p. 389-488, 31 pl.

Bornemann, J. G.

1886-91, *Die Versteinerungen des cambrischen Schichtensystems der Insel Sardinien. Teil 1:* Nova Acta Kaiserl. Leopold.-Carol. Akad. Deutsch. Naturf., Part 1 (1886), v. 51, p. 1-147, 33 pl.; Part 2 (1891), v. 56, p. 425-510, pl. 19-28.

1889, *Über den Buntsandstein in Deutschland und seine Bedeutung für die Trias:* Beiträge Geologie, Paläontologie, v. 1, 61 p., 3 pl.

Borrello, A. V.

1966, *Paleontografia Bonaerense. Fasc. 5. Trazas, restos tubiformes y cuerpos problematicos de la Formacion La Tinta, Sierras Septentrionales-Provincia de Buenos Aires:* Prov. Buenos Aires, Gobernac., Comis. Investig. Cient., 42 p., 46 pl., text-fig. 1-3, pl. 1.

1967, *Octoia subandina gen. et sp. nov. del Devonico del Sur de Bolivia:* Same, Notas 4, no. 3, p. 1-8, pl. 1, 2.

Bosc, L. A. G.

1802, *Histoire naturelle des vers, contenant leur description et leur moeurs:* 3 vol., illus., Roret (Paris). (2nd edit., 1830).

Bosch, W. J. van den

1969, *Geology of the Luna-Sil region, Cantabrian Mountains (NW Spain):* Leidse Geol. Meded., v. 44, p. 137-225, 116 text-fig. (Span. resumé).

Bouček, Bedřich

1938, *Über "Skolithen"-artige Grabspuren aus den Drabover Quartziten des böhmischen Ordoviziums:* Paläont. Zeitschr., v. 19, p. 244-253, text-fig. 1, pl. 17-19.

———, & Eliáš, Mojmír

1962, *O novém zajímavém bioglyfu z Paleogénu Československých flyšových Karpat:* Geol. Práce Zprávy, v. 25-26, p. 145-151, pl. 7, 8 (Ger. resumé). [*On an interesting trace fossil from the Paleogene of the Czechoslovakian Carpathian flysch.*]

Bouma, A. H.

1972, *Distribution of sediments and sedimentary structures in the Gulf of Mexico:* in Contributions on the geological and geophysical oceanography of the Gulf of Mexico, Richard Rezak & V. J. Henry (eds.), Texas A & M Univ. Oceanographic Studies, v. 3, p. 35-65, text-fig. 1-21.

Bourne, D. W., & Heezen, B. C.

1965, *A wandering enteropneust from the abyssal Pacific, and the distribution of spiral tracks on the sea floor:* Science, v. 150, p. 60-63, 5 text-fig.

Boursault, M. H.

1889, *Nouvelles empreintes problématiques des couches boloniennes du Portel (Pas-de-Calais):* Soc. Géol. France, Bull., sér. 3, v. 17, p. 725-728, 2 pl.

Boyd, D. W.

1966, *Lamination deformed by burrowers in Flathead Sandstone (Middle Cambrian) of central Wyoming:* Wyoming Univ., Contrib. Geology, v. 5, no. 1, p. 45-53, text-fig. 1-7.

———, & Ore, H. T.

1963, *A Wyoming specimen of Dendrophycus:* Wyoming Univ., Contrib. Geology, v. 2, p. 63-86.

Bradley, John

1973, *Zoophycos and Umbellula (Pennatulacea): their synthesis and identity:* Palaeography, Palaeoclimatology, Palaeoecology, v. 13, p. 103-128, text-fig. 1-11.

Brady, L. F.

1939, *Tracks in the Coconino sandstone compared with those of small living arthropods:* Plateau, v. 12, p. 32-34, text-fig. 1-4.

1947, *Invertebrate tracks from the Coconino sandstone of Northern Arizona:* Jour. Paleontology, v. 21, p. 466-472, text-fig. 1, 2, pl. 66-69.

1949, *Oniscoidichnus, new name for Isopodichnus Brady 1947 not Bornemann 1889:* Same, v. 23, p. 573.

1961, *A new species of Palaeohelcura Gilmore from the Permian of northern Arizona:* Same, v. 35, p. 201-202, pl. 1.

Braithwaite, C. J. R., & Carter, D. J.

1972, *Crustacean burrows in the Seychelles, Indian Ocean:* Palaeogeography, Palaeoclimatology, Palaeoecology, v. 11, p. 265-285, pl. 1-3.

Branson, C. C.

1959, *Some problematical fossils:* Oklahoma Geology Notes, v. 19, no. 4, p. 82-87, text-fig. 1-6.

1960, *Conostichus:* Same, v. 20, no. 8, p. 195-207, 4 pl.

1961, *New records on the scyphomedusan Cono-*

stichus: Same, v. 21, p. 130-138, text-fig. 1, 3 pl.

1962, *Conostichus, a scyphomedusan index fossil:* Same, v. 22, p. 251-253, pl. 1.

1964, *Sole trails in an Atoka siltstone:* Same, v. 24, p. 180-184.

Brattström, Hans

1936, *Ulophysema öresundense, n. gen. et sp., eine neue Art der Ordnung Cirripedia Ascothoracica:* Arkiv Zoologi, v. 28A, no. 23, p. 1-10, text-fig. 1-4.

Braun, F.

1847, *Die fossilen Gewächse aus den Gränzschichten zwischen dem Lias und Keuper des neu aufgefundenen Pflanzenlagers in dem Steinbruche von Veitlahm bei Culmbach:* Flora oder allg. botan. Zeitung, n.r. 5, v. 1 (Jahrg. 30, Bd. 1), p. 81-87.

Britton, N. L.

1888, *On an Archean plant from the white crystalline limestone of Sussex County, N. J.:* New York Acad. Sci., Ann., v. 4, p. 123-124, pl. 7.

Brönnimann, Paul

1955, *Microfossils incertae sedis from the Upper Jurassic and Lower Cretaceous of Cuba:* Micropaleontology, v. 1, p. 28-51, text-fig. 1-10.

1972, *Remarks on the classification of fossil anomuran coprolites:* Paläont. Zeitschr., v. 46, p. 99-103, text-fig. 1.

———, Caron, J. -P., & Zaninetti, L.

1972a, *New galatheid anomuran (Crustacea, Decapoda) coprolites from the Rhetian of Provence, southern France:* Gesell. Geologie & Bergbaustud., Mitteil., v. 21, p. 905-920, text-fig. 1-5, pl. 1, 2.

1972b, *Parafavreina, n. gen., a new thalassinid anomuran (Crustacea, Decapoda) coprolite form-genus from the Triassic and Liassic of Europa and North Africa:* Same, Mitteil., v. 21, p. 941-956, text-fig. 1-4, pl. 1, 2.

———, Cros, P., & Zaninetti, L.

1972, *New thalassinid anomuran (Crustacea, Decapoda) coprolites from infraliasic limestones of the Dolomites, Italy:* Gesell. Geologie & Bergbaustud., Mitteil., v. 21, p. 921-928, text-fig. 1-5, 1 pl.

———, & Masse, J. P.

1968, *Thalassinid (Anomura) coprolites from Barremian Aptian passage beds, Basse Provence, France:* Rev. Micropaléontologie, v. 11, no. 3, p. 153-160, text-fig. 1-3, pl. 1, 2.

———, & Norton, Peter

1960, *On the classification of fossil faecal pellets and description of new forms from Cuba,* Guatemala and Libya: Eclogae Geol. Helvetiae, v. 53, no. 2, p. 832-842, text-fig. 1-5.

———, & Zaninetti, L.

1972a, *New names for favreine and parafavreine thalassinid anomuran (Crustacea, Decapoda) coprolites from the Jurassic of Greece and Algeria:* Paläont. Zeitschr., v. 46, p. 221-224, text-fig. 1, 2.

1972b, *Revision of the micro-coprolite Palaxius ? triasicus (Elliot), 1962, and description of a new Triassic thalassinid anomuran (Crustacea, Decapoda) coprolite from France, Austria and Libya:* Gesell. Geologie & Bergbaustud., Mitteil., v. 21, p. 929-940, text-fig. 1-3, pl. 1, 2.

———, ———, & Baud, A.

1972, *New thalassinid (Crustacea, Decapoda) coprolites from the Préalpes médianes rigides of Switzerland and France (Chablais):* Gesell. Geologie & Bergbaustud., Mitteil., v. 21, p. 885-904, text-fig. 1-8, pl. 1.

Broili, Ferdinand

1924, See von Zittel.

Bromley, R. G.

1967, *Some observations on burrows of thalassinidean Crustacea in chalk hardgrounds:* Geol. Soc. London, Quart. Jour., v. 123, p. 157-182.

1970, *Borings as trace fossils and Entobia cretacea Portlock, as an example:* in Trace fossils, T. P. Crimes & J. C. Harper (eds.), Geol. Jour., spec. issue no 3, p. 49-90, text-fig. 1-4, pl. 1-5, Seel House Press (Liverpool).

1972, *On some ichnotaxa in hard substrates, with a redefinition of Trypanites Mägdefrau:* Paläont. Zeitschr., v. 46, p. 93-98, text-fig. 1, pl. 18.

———, & Asgaard, Ulla

1972, *Notes on Greenland trace fossils: 1. Freshwater Cruziana from the Upper Triassic of Jameson Land, East Greenland (p. 7-13, text-fig. 1-4); II. The burrows and microcoprolites of Glyphea rosenkrantzi, a Lower Jurassic palinuran crustacean from Jameson Land, East Greenland (p. 15-21, text-fig. 5-9); III. A large radiating burrow-system in Jurassic micaceous sandstones of Jameson Land, East Greenland (p. 23-30, text-fig. 12-14);* Grønlands Geol. Undersøgelse, Rapport no. 49, p. 1-30, text-fig. 1-14.

———, & Surlyk, F.

1973, *Borings produced by brachiopod pedicles, fossil and Recent:* Lethaia, v. 6, p. 349-365, text-fig. 1-14.

Brongniart, A. T.

1823, *Observations sur les Fucoides:* Soc. Histoire

Nat. Paris, Mém., v. 1, p. 301-320, pl. 19-21.

1828-38, *Histoire des végétaux fossiles ou recherches botaniques et géologiques sur les végétaux renfermés dans les diverses couches du globe:* v. 1, p. 1-136 (1828a); p. 137-208 (1829); p. 209-248 (1830); p. 249-264 (1831?); p. 265-288 (1832?); p. 289-336 (1834); p. 337-368 (1835?); p. 369-488 (1836); v. 2, p. 1-24 (1837); p. 25-72 (1838); plates appeared irregularly: v. 1, pl. 1-166; v. 2, pl. 1-29; G. Dufour & E. d'Ocagne (Paris).

1849, *Tableau des genres de végétaux fossiles considérés sous le point de vue de leur classification botanique et de leur distribution géologique:* Dictionnaire Univ. Histoire Nat., v. 13, p. 1-127 (52-176).

Bronn, H. G.

1837-38, *Lethaea Geognostica oder Abbildungen und Beschreibungen der für die Gebirgsformationen bezeichnendsten Versteinerungen:* v. 1, p. 1-672, atlas, pl. 1-47 (1837); v. 2, p. 673-1350 (1838), E. Schweizerbart (Stuttgart).

1848-49, *Index palaeontologicus oder Übersicht der bis jetzt bekannten fossilen Organismen 1. Abt., 1.: Hälfte:* 1384 p., E. Schweizerbart (Stuttgart).

1848-49, *Handbuch einer Geschichte der Natur. 3 Bd. Organisches Leben (Fortsetzung und Schluss). Index palaeontologicus oder Uebersicht der bis jetzt bekannten fossilen Organismen, bearb. unter mitwirkung der Herren Prof. H. R. Göppert und H. v. Meyer:* 2 vol. in 3; v. 1, 1381 p.; v. 2, 980 p., E. Schweizerbart (Stuttgart).

1853-56, *Lethaea Geognostica;* 3rd edit. by H. G. Bronn & C. F. Roemer (eds.), *Bd. 3: Caeno-Lethaea. VI. Theil: Molasse-Periode,* by H. G. Bronn, 124 pl., E. Schweizerbart (Stuttgart). [Not seen by the editors.]

Brotzen, F.

1941, *Några bidrag till Visingsöformationens stratigrafi och tektonik:* Geol. Fören. Stockholm, Förhandl., v. 63, p. 245-261, text-fig. 1-5.

Brown, A. P.

1912, *The formation of ripple marks, tracks and trails:* Acad. Nat. Sci. Philadelphia, Proc., v. 63, p. 536-547.

Brown, Barnum, & Vokes, H. E.

1944, *Fossil imprints of unknown origin. II. Further information and a possible explanation:* Am. Jour. Sci., v. 242, p. 656-672, text-fig. 1, 4 pl.

Brunton, Howard

1966, *Predation and shell damage in a Viséan*

brachiopod fauna: Palaeontology, v. 9, p. 355-359, pl. 60.

Bryson, A.

1865, *Surface-markings on sandstone:* Geol. Mag., v. 2, p. 189-191, text-fig. 1.

Buckland, William

1837, *Geology and mineralogy considered with reference to natural theology:* 2 vol., 2nd. edit., 87 pl.; v. 1, 599 p., v. 2, 128 p., 69 pl., Wm. Pickering (London).

Buller, A. T., & McManus, John

1972, *Corophium burrows as environmental indicators of Quaternary estuarine sediments of Tayside:* Scot. Jour. Geology, v. 8, p. 145-150, text-fig. 1, 2, 1 pl.

Buoi, Luigi De

1932, *Si di alcune impronte fossili problematiche:* Soc. Naturalisti e Matem. Modena, Atti, v. 63, p. 145-149, pl. 4.

Burling, L. D.

1917, *Protichnites and Climatichnites: a critical study of some Cambrian trails:* Am. Jour. Sci., ser. 4, v. 44, p. 387-398, text-fig. 1-5.

Butts, Edward

1891a, *Recently discovered foot-prints of the amphibian age, in the Upper Coal Measure Group of Kansas City, Missouri:* Kansas City Scientist, v. 5, p. 17-19, text-fig. 1, 2.

1891b, *Foot-prints of new species of amphibians in the Upper Coal Measure Group of Kansas City, Missouri:* Same, v. 5, p. 44, text-fig. 1, 2.

Bykova, E. V., & Polenova, E. N.

1955, *Foraminiferi, radiolarii i ostrakody devona Volgo-Uralskoi oblast:* Vses. Nefty. Nauchno-issledovatel. Geol.Razved. Inst. (VNIGRI), Trudy, n. ser., v. 87, 141 p., 24 pl. [*Foraminifers, radiolarians, and ostracodes from Devonian of the Volga-Urals region.*]

Byrne, Frank, & Branson, Jack

1941, *Permian organic burrows:* Kansas Acad. Sci., Trans., v. 44, p. 257-261, text-fig. 1-6.

Bystrov [Bystrow], A. P.

1956, *O razrushenii skeltnykh elementov iskopaemykh zhivotnykh gribami:* Leningrad Univ. Vestnik, v. 6, (ser. Geol. i Geogr. vyp. 1), 30-46. [*On destruction of skeletal elements of fossil animals by fungi.*]

Callison, George

1970, *Trace fossils of trilobites from the Deadwood Formation (Upper Cambrian) of western South Dakota:* Southern California Acad. Sci., Bull., v. 69, p. 20-26, text-fig. 1-3.

Cameron, Barry

1967, *Fossilization of an ancient (Devonian) soft-bodied worm:* Science, v. 155, p. 1246-1248.

1968, *Commensalism of new serpulid worm from the Hamilton Group (Middle Devonian) of New York:* Jour. Paleontology, v. 42, p. 850-852.

1969a, *New name for Palaeosabella prisca (Mc-Coy), a Devonian worm-boring, and its preserved probable borer:* Same, v. 43, p. 189-192, text-fig. 1, 2.

1969b, *Paleozoic shell-boring annelids and their trace fossils:* Am. Zoologist, v. 9, p. 689-703, text-fig. 1-8.

———, & **Estes, Richard**

1971, *Fossil and Recent "tadpole nests": a discussion:* Jour. Sed. Petrology, v. 41, p. 171-178, text-fig. 1.

Campbell, A. S.

1954, *Tintinnina:* in Treatise on invertebrate paleontology, R. C. Moore (ed.), p. D166-D180, text-fig. 88-92, Geol. Soc. America & Univ. Kansas Press (New York; Lawrence, Kans.).

Capellini, Giovanni

1884, *Il Cretaceo superiore e il gruppo di Priabona nell'Appennino settentrionale e in particolare nel Bolognese e loro rapporti col Grès de Celles in parte en con gli strati a Clavulina Szaboli:* Accad. Sci. Ist. Bologna, Mem., ser. 4, v. 5, p. 535-550.

Carriker, M. R., & Smith, E. H.

1969, *Comparative calcibiocavitology: Summary and conclusions:* in Penetration of calcium carbonate substrates by lower plants and invertebrates, M. R. Carriker, E. H. Smith, & R. T. Wilce (eds.),: Am. Zoologist, v. 9, p. 1011-1020.

———, ———, & **Wilce, R. T. (eds.)**

1969, *Penetration of calcium carbonate substrates by lower plants and invertebrates (a symposium):* Am. Zoologist, v. 9, no. 3, edit. 2, 1020 p.

———, & **Yochelson, E. L.**

1968, *Recent gastropod boreholes and Ordovician cylindrical borings:* U. S. Geol. Survey, Prof. Paper 593-B, 26 p., text-fig. 1, 2, 5 pl.

Carruthers, William

1871, *On some supposed vegetable fossils:* Geol. Soc. London, Quart. Jour., v. 27, p. 443-449, pl. 19.

Carter, H. J.

1887, *Note on "Tubulations Sableuses" of the Etage Bruxellien in the environs of Brussels:* Ann. Mag. Nat. History, ser. 4, v. 19, p. 382-393, pl. 18.

Casey, Raymond

1961, *The stratigraphical palaeontology of the Lower Greensand:* Palaeontology, v. 3, p. 487-621, text-fig. 1-14, pl. 77-84.

Caster, K. E.

1938, *A restudy of the tracks of Paramphibius:* Jour. Paleontology, v. 12, p. 3-60, text-fig. 1-9, 13 pl.

1939, *Were Micrichnus scotti Abel and Artiodactylus sinclairi Abel of the Newark Series (Triassic) made by vertebrates or limuloids?:* Am. Jour. Sci., v. 237, p. 786-797.

1940, *Die sogenannten "Wirbeltierspuren" und die Limulus-Fährten der Solnhofener Plattenkalke:* Paläont. Zeitschr., v. 22, p. 12-29.

1942a, *A laotirid from the Upper Cambrian of Wyoming:* Am. Jour. Sci., v. 240, p. 104-112, 1 pl.

1942b, *Two siphonophores from the Paleozoic:* Palaeont. Americana, v. 3, no. 14, p. 1-30, text-fig. 1-5, pl. 1, 2.

1944, *Limuloid trails from the Upper Triassic (Chinle) of the Petrified Forest National Monument, Arizona:* Am. Jour. Sci., v. 242, p. 74-84, text-fig. 1, 2, pl. 1.

1945, *A new jellyfish (Kirklandia texana Caster) from the Lower Cretaceous of Texas:* Palaeont. Americana, v. 3 (no. 18), p. 173-220, text-fig. 1-8, pl. 16-20 (Separate: p. 1-52, pl. 1-5).

1957, *Problematica:* in Treatise on marine ecology and paleoecology, v. 2, H. S. Ladd (ed.), Geol. Soc. America, Mem. 67, p. 1025-1032.

———, & **Brooks, H. K.**

1956, *New fossils from the Canadian-Chazyan (Ordovician) hiatus in Tennessee:* Bull. Am. Paleontology, v. 36, no. 157, p. 157-199, pl. 12-23.

———, & **Macke, W. B.**

1952, *An aglaspid merostome from the Upper Ordovician of Ohio:* Jour. Paleontology, v. 26, p. 753-757, pl. 109.

Caterini, F.

1925, *Che cosa sono i Nemertiliti?:* Atti Soc. Toscana Sci. Nat., Mem., v. 36, p. 309-321, pl. 10.

Cayeux, Lucien

1895, *De l'existence de nombreux débris de spongiaires dans le Précambrien de Bretagne:* Soc. Géol. Nord, Ann., v. 23, p. 52-65, pl. 1, 2.

Chachlof, W. A.

1934, *Eine neue Gattung Gaussia n. gen. aus dem Oberkarbon von Sibirien:* Centralbl. Mineralogie, Geologie, Paläontologie, 1934, ser. B, p. 346-351, text-fig. 1-4.

Chakrabarti, A.

1972, *Beach structures produced by crab pellets:* Sedimentology, v. 18, p. 129-134, text-fig. 1-7.

Chamberlain, C. K.

1971a, *Morphology and ethology of trace fossils from the Ouachita Mountains, southeastern Oklahoma:* Jour. Paleontology, v. 45, p. 212-246, text-fig. 1-8, pl. 29-32.

1971b, *Biogenic mounds in the Dakota Sandstone of northwestern New Mexico:* Same, v. 45, p. 641-644, pl. 74.

1971c, *Bathymetry and paleoecology of Ouachita geosyncline of southeastern Oklahoma as determined by trace fossils:* Am. Assoc. Petroleum Geologists, Bull., v. 55, p. 34-50.

————, & Baer, James

1973, *Ophiomorpha and a new thallassinid burrow from the Permian of Utah:* Brigham Young Univ., Geology Studies, v. 20, p. 79-94, text-fig. 1-5, pl. 1-3.

————, & Clark, D. L.

1973, *Trace fossils and conodonts as evidence for deep-water deposits in the Oquirrh Basin of Central Utah:* Jour. Paleontology, v. 47, no. 4, p. 663-682, text-fig. 1-7, pl. 1-3.

Chapman, E. J.

1878, *On the probable nature of the supposed fossil tracks known as Protichnites and Climactichnites:* Canad. Jour. Sci. Lit. and History, n.ser., v. 15, p. 486-490.

Chapman, Frederick

1913, *Note on tracks made by a common gastropod, Bittium cerithium Qu. and G., sp.:* Victorian Naturalist, v. 29, p. 139-140.

1926, *On a supposed phyllocarid from the older Paleozoic of Tasmania:* Royal Soc. Tasmania, Papers & Proc. for 1925, p. 79-80, pl. 10.

1929, *On some remarkable annelid remains from Arthur River, N.W. Tasmania:* Same, Papers & Proc. for 1928, p. 1-5, pl. 1.

1935, *Primitive fossils, possibly atremateous and neotremateous Brachiopoda, from the Vindhyans of India:* Geol. Survey India, Records, v. 69, p. 109-120, pl. 1, 2.

Cheng, Ying-Min

1972, *On some lebensspuren from Taiwan:* Acta Geol. Taiwanica, no. 15, p. 13-22, text-fig. 1-5, 4 pl.

Chenoweth, P. A.

1960, *Starfish impressions from the Hilltop Shale:* Oklahoma Geology Notes, v. 20, p. 35-36, text-fig. 1, 2.

Chiplonkar, G. W., & Badwe [Badve], R. M.

1970, *Trace fossils from the Bagh Beds:* Palaeont.

Soc. India, Jour., v. 14 [1969], p. 1-10, pl. 1-3.

1972, *Trace fossils from the Bagh Beds—part II:* Same, v. 15 [1970], p. 1-5, pl. 1.

Chisholm, J. I.

1968, *Trace-fossils from the Geological Survey boreholes in East Fife, 1963-4:* Great Britain Geol. Survey, Bull. no. 28, p. 103-119, 1 text-fig., pl. 5-7.

1970a, *Lower Carboniferous trace-fossils from the Geological Survey boreholes in West Fife (1965-6):* Same, Bull. no. 31, p. 19-35, text-fig. 1, pl. 1-4.

1970b, *Teichichnus and related trace-fossils in the Lower Carboniferous of St. Monace, Scotland:* Same, Bull. no. 32, p. 21-51, text-fig. 1-3, pl. 6-8.

Choubert, G., Termier, Henri, & Termier, Geneviève

1951, *Les calcaires précambriens de Tahgdout et leurs organismes problématiques:* Serv. Géol. Maroc, Notes et Mém., v. 85 (Notes du Serv. Géol., 5), p. 9-34, pl. 1-5.

Chowns, T. M.

1973, *Fossils vs mud rolls:* Geotimes, v. 18, no. 3, p. 11 (letter to editor).

Churkin, Michael, Jr., & Brabb, E. E.

1965, *Occurrence and stratigraphic significance of Oldhamia, a Cambrian trace fossil, in east-central Alaska:* U. S. Geol. Survey, Prof. Paper 525-D, p. D120-D124, text-fig. 1-4.

Clark, T. H.

1923, *New fossils from the vicinity of Boston. Aspidella-like markings from the Cambridge slate:* Boston Soc. Nat. History, Proc., v. 36, p. 482-485, text-fig. 1.

————, & Usher, J. L.

1948, *The sense of Climactichnites:* Am. Jour. Sci., v. 246, p. 251-253, text-fig. 1, 2.

Clarke, J. M.

1908, *The beginnings of dependent life:* New York State Museum, Bull., v. 121, p. 146-169, pl. 1-13.

1921, *Organic dependence and disease: their origin and significance:* Same, v. 221-222 (1919), p. 1-113, text-fig. 1-105.

1924, *Rosetted trails of the Palaeozoic:* Same, v. 251, p. 128-130, 1 pl.

————, & Swartz, C. K.

1913, *Systematic paleontology of the Upper Devonian deposits of Maryland:* in Maryland Geol. Survey, Middle and Upper Devonian vol., text, p. 535-701.

Clarke, R. H.

1968, *Burrow frequency in abyssal sediments:* Deep-Sea Research, v. 15, p. 397-400.

Claypole, E. W.

1895, *Daemonelix or what?:* Am. Geologist, v. 16, p. 113.

Cleaves, A. B., & Fox, E. F.

1935, *Geology of the west end of Ymer Island, East Greenland:* Geol. Soc. America, Bull., v. 46, p. 463-488, text-fig. 1, 4 pl.; discussion by Fox, reply by Cleaves, p. 2018-2021.

Cloud, P. E., Jr.

1960, *Gas as a sedimentary and diagenetic agent:* Am. Jour. Sci., v. 258-A (Bradley Vol.), p. 35-45.

1968, *Pre-metazoan evolution and the origins of the Metazoa:* in Evolution and environment, E. T. Drake (ed.), p. 1-72, text-fig. 1-9, Yale Univ. Press (New Haven).

1973, *Pseudofossils—a plea for caution:* Geology, v. 1, no. 3, p. 123-127, text-fig. 1-7.

———, **& Bever, J. E.**

1973, *Trace fossils from the Flathead Sandstone, Fremont County, Wyoming, compared with Early Cambrian forms from California and Australia:* Jour. Paleontology, v. 47, p. 883-885, 1 pl.

———, **& Nelson, C. C.**

1966, *Phanerozoic—Cryptozoic and related transitions: new evidence:* Science, v. 154, p. 766-770, text-fig. 1-3.

———, **& Semikhatov, M. A.**

1969, *Proterozoic stromatolite zonation:* Am. Jour. Sci., v. 267, p. 1017-1061, text-fig 1-15, 7 pl.

Codez, J., & Saint-Seine, Roseline

1958, *Révision des Cirripèdes Acrothoraciques fossiles:* Soc. Géol. France, Bull., sér. 6, v. 7, p. 699-719, text-fig. 1-4, 3 pl., tables 1-3.

Cole, G. A. J.

1901, *Recent observations on Oldhamia and Histioderma:* Irish Naturalist, v. 10, p. 81-86.

Colom, G.

1945, *Nannoconus steinmanni Kamptner y Lagena colomi Lapparent:* Inst. Geol. Barcelona, Publ., v. 7 ("Miscelanea Almera"), pt. 1, p. 123-132.

Colton, G. W.

1967, *Late Devonian current directions in western New York with special reference to Fucoides graphica:* Jour. Geology, v. 75, p. 11-22, text-fig. 1-6, 2 pl.

Condra, G. E., & Elias, M. K.

1944, *Carboniferous and Permian ctenostomatous Bryozoa:* Geol. Soc. America, Bull. 55, no. 5, p. 517-568, pl. 1-13.

Conkin, J. E., & Conkin, B. M.

1968, *Scalarituba missouriensis and its strati-*

graphic distribution: Univ. Kansas Paleont. Contrib., Paper 31, p. 1-7, pl. 1-4.

Conrad, T. A.

1837, *First annual report on the geological survey of the third district of the State of New York. Organic remains of the Red Sandstone:* New York Geol. Survey, Ann. Rept. 1, 1837, p. 155-186.

1838. *Report on the paleontological department of the survey:* Same, Ann. Rept. 2, 1838, p. 107-119.

Conybeere, C. E. B., & Crook, K. A. W.

1968, *Manual of sedimentary structures:* Australia Bur. Min. Resources, Geology, Geophysics, Bull. 102, 322 p., 5 text-fig., 108 pl., 12 tables.

Cook, D. O.

1971, *Depressions in shallow marine sediment made by benthic fish:* Jour. Sed. Petrology, v. 41, p. 577-578.

Cotta, Bernhard

1839, *Notiz über Thierfährten im Bunten Sandstein bei Pölzig zwischen Ronneburg und Weissenfels:* Neues Jahrb. Mineralogie, Geologie, Paläontologie, 1839, p. 10-15, pl. 1.

Cotter, Edward

1973, *Large Rosselia in the Upper Cretaceous Ferron Sandstone, Utah:* Jour. Paleontology, v. 47, p. 975-978, text-fig. 1, pl. 1, 2.

Coulter, H. W.

1955, *Fucoidal markings in the Swan Peak Formation, southeastern Idaho:* Jour. Sed. Petrology, v. 25, p. 282-284.

Cowie, J. W., & Spencer, A. M.

1970, *Trace fossils from the Late Precambrian/ Lower Cambrian of East Greenland:* in Trace fossils, T. P. Crimes & J. C. Harper (eds.), Geol. Journal, spec. issue no. 3, p. 91-100, text-fig. 1, 2, pl. 1, 2, Seel House Press (Liverpool).

Cox, L. R.

1929, *A spiral puzzle:* Nat. History Mag. (Brit. Museum), v. 2, p. 16-26, text-fig. 1-8.

Cragin, F. W.

1893, *A contribution to the invertebrate paleontology of the Texas Cretaceous:* Texas Geol. Survey, 4th Ann. Rept., pt. 2, p. 139-246.

1894, *New and little known Invertebrata from the Neocomian of Kansas:* Am. Geologist, v. 14, p. 1-12, pl. 1.

Cramer, H. R.

1958, *Additions to the Hamilton biota at Rockville, Dauphin County, Pennsylvania:* Pennsylvania Acad. Sci., Proc., v. 32, p. 184-187.

1961, *Suppression of the name Tasmanites rockvillensis:* Jour. Paleontology, v. 35, p. 1087.

Crié, L. A.

1878, *Les Tigillites siluriennes:* Acad. Sci. [Paris], Comptes Rendus, v. 86, p. 687-689.

1883, *Les origines de la vie. Essai sur la flore primordiale: organisation, développement, affinités, distribution géologique et géographique:* 75 p., 20 text-fig., O. Doin (Paris).

Crimes, T. P.

1968, *Cruziana: a stratigraphically useful trace fossil:* Geol. Mag., v. 105, p. 360-364, text-fig. 1, 2, pl. 9-11.

1969, *Trace fossils from the Cambro-Ordovician rocks of North Wales and their stratigraphic significance:* Geol. Jour., v. 6, p. 333-338, text-fig. 1-3.

1970a, *A facies analysis of the Arenig of Western Lleyn, North Wales:* Geol. Assoc., Proc., v. 81, pt. 2, p. 221-239, text-fig. 1-5.

1970b, *Trilobite tracks and other trace fossils from the Upper Cambrian of North Wales:* Geol. Jour., v. 7, p. 47-68, text-fig. 1-7, pl. 5-13.

1970c, *The significance of trace fossils in sedimentology, stratigraphy, and palaeoecology with examples from lower Paleozoic strata:* in Trace fossils, T. P. Crimes & J. C. Harper (eds.), Geol. Jour., spec. issue no. 3, p. 101-126, text-fig. 1-9, pl. 1-5, Seel House Press (Liverpool).

1973, *From limestones to distal turbidites: A facies and trace fossil analysis in the Zumaya flysch (Paleocene—Eocene), North Spain:* Sedimentology, v. 20, p. 105-131, text-fig. 1-15.

————, & Harper, J. C. (eds.)

1970, *Trace fossils:* Geol. Jour., spec. issue no. 3, 547 p., illus., Seel House Press (Liverpool).

Crookall, R.

1931, *On some curious fossils from the Downtonian and Lower Old Red Sandstone of Scotland:* Royal Soc. Edinburgh, Proc., v. 50 (1930), pt. II, p. 175-178, 1 pl.

Cummings, R. H.

1957, *A problematic new microfossil from the Scottish Lower Carboniferous:* Micropaleontology, v. 3, no. 4, p. 407-409, text-fig. 1-9.

Cushman, J. A.

1910, *A monograph of the Foraminifera of the North Pacific Ocean. Pt. I. Astrorhizidae and Lituolidae:* U. S. Natl. Museum, Bull. 71, 134 p., 203 text-fig.

Cuvillier, Jean

1954, *Niveaux à coprolithes de Crustacés:* Soc. Géol France, Bull., sér. 6, v. 4, p. 51-53, pl. 3.

————, & Sacal, V.

1951, *Corrélations stratigraphiques par microfaciès en Aquitaine:* 23 p., 90 pl., Brill (Leiden).

Dahmer, Georg

1937, *Lebensspuren aus dem Taunusquarzit und den Siegener Schichten (Unterdevon):* Preuss. Geol. Landesanst., Jahrb., 1936, v. 57, p. 523-539, text-fig. 1, 2, pl. 31-35.

Daley, Brian

1968, *Sedimentary structures from a non-marine horizon in the Bembridge Marls (Oligocene) of the Isle of Wight, Hampshire, England:* Jour. Sed. Petrology, v. 38, p. 114-127, text-fig. 1-13.

Dames, W. B.

1881, *Vorlegung eines Exemplares von Aspidorhynchus acutirostris Ag. aus den lithographischen Schiefern von Solnhofen:* Gesell. Naturforsch. Freunde Berlin, Sitzungsber. 1881, p. 48-49.

Darder Pericás, Bartolomé

1945, *Estudio geologico del sur de la provincia de Valencia y norte de la Alicante:* Geol. Min. España, Bol., v. 57, p. 59-362, text-fig. 1-86, pl. 1-11.

David, T. W. E.

1922, *Occurrence of remains of small Crustacea in the Proterozoic (?) or Lower Cambrian (?) rocks of Reynella, near Adelaide:* Royal Soc. South Australia, Trans. Proc., v. 46, p. 6-8, pl. 2.

1928, *Notes on newly-discovered fossils in the Adelaide Series (Lipalian ?), South Australia:* Same, Trans. & Proc., v. 52, p. 191-209, pl. 13-18a.

1950, *The geology of the Commonwealth of Australia:* v. 1, 747 p., 208 text-fig., 58 pl., Edward Arnold & Co. (London).

————, & Tillyard, R. J.

1936, *Memoir on fossils of the Late Pre-Cambrian (newer Proterozoic) from the Adelaide Series, South Australia:* 122 p., 13 pl., Angus & Robertson (Sydney).

Davitashvili, L. Sh.

1945, *Tsenozy zhivykh organizmov i organicheskikh ostatkov:* Akad. Nauk Gruzin. SSR, Soobshch., v. 6, no. 7, p. 527-534. [*Biocoenoses of living organisms and of organic remains.*] (Traduction no. 2077, Bureau de Recherches géologiques, géophysique et minières, Paris).

Dawson, J. W.

1862, *Notice on the discovery of additional remains of land animals in the coal-measures of the South-Joggins, Nova Scotia:* Geol. Soc. London, Quart. Jour., v. 18, p. 5-7.

1864, *On the fossils of the genus Rusophycus:* Canadian Naturalist and Geologist, n. ser., v. 1, p. 363-367, 458, text-fig. 1-4.

1865, *On the structure of certain organic remains*

in the Laurentian limestones of Canada:
Geol. Soc. London, Quart. Jour., v. 21, p.
51-59, pl. 6, 7.

1868, *Acadian geology:* 2nd edit., 694 p., Macmillan & Co. (London).

1873, *Impressions and footprints of aquatic animals and imitative markings on Carboniferous rocks:* Am. Jour. Sci., ser. 3, v. 5, p. 16-24, text-fig. 1-5.

1875, *The dawn of life:* 239 p., Dawson Bros. (Montreal).

1878, *Supplement to the 2. edit. of Acadian Geology:* in Acadian geology, 3rd. edit., 102 p., Macmillan & Co. (London).

1888, *The geological history of plants:* The International Scientific Series, v. 61, 290 p., 79 text-fig., D. Appleton & Co. (New York).

1890, *On burrows and tracks of invertebrate animals in Palaeozoic rocks, and other markings:* Geol. Soc. London, Quart. Jour., v. 46, p. 595-617, text-fig. 1-19.

Debey, M. H.

1849, *Entwurf zu einer geognostisch-geogenetischen Darstellung der Gegend von Aachen:* Gesell. Deutscher Naturf. Aerzte, Amtliche Ber., 25 Vers., p. 269-328.

———, & Ettinghausen, C. R. von

1859, *Die urweltlichen Thallophyten des Kreidegebirges von Aachen und Maastricht:* Akad. Wiss. Wien, Denkschr., math. -nat. Kl., v. 16, p. 131-214, pl. 1-3.

Deecke, Wilhelm

1895, *Eocäne Kieselschwämme als Diluvialgeschiebe in Vorpommern und Mecklenburg:* Naturwiss. Ver. Neu-Vorpommern u. Rügen, Mitteil., v. 26 (1894), p. 166-170, pl. 1.

Deflandre, Georges

1955, *Palaeocryptidium n. g. cayeuxi n. sp., microorganismes incertae sedis des phtanites briovériens bretons:* Soc. Géol. France, Compte Rendu somm. des séanc., no. 9, p. 182-185.

1957, *Remarques sur deux genres de Protistes du Précambrien (Arnoldia Hovasse 1956, Cayeuxipora Graindor 1957):* Acad. Sci. [Paris], Comptes Rendus, séanc. hebd., v. 244, pt. 2, p. 2640-2641.

1959, *Sur les nannofossils calcaires et leur systématique:* Rev. Micropaléontologie, v. 2, p. 121-152, 4 pl.

1961, *Étude micropaléontologique:* in Edgar Aubert de la Rûe & Georges Deflandre, Sur un calcaire à microorganismes enclavé dans un basalte du Val Studer, Archipel de Kerguelen: Muséum Natl. Histoire Nat. Paris, Bull., sér. 2, v. 33, p. 123-127, 3 pl.

1963, *Sur les microrhabdulidés, famille nouvelle de nannofossiles calcaires:* Acad. Sci. [Paris],

Comptes Rendus, v. 256, no. 16, p. 3484-3486.

———, & Dangeard, L.

1938, *Schizosphaerella, un nouveau microfossil méconnu du Jurassique moyen et supérieur:* Acad. Sci. [Paris], Comptes Rendus, v. 207, p. 1115-1117, text-fig. 1-6.

———, & Ters, Mireille

1966, *Sur la présence d' acritarches ordoviciens dans les schistes subardoidiers de la région de la Mothe-Achard (Vendée). Extension du silurien (grès armoricain et schistes d' Angers) en Vendée littorale:* Acad. Sci. [Paris], Comptes Rendus, ser. D, v. 262, no. 2, p. 237-240.

Dekay, J. E.

1824, *Note on the organic remains, termed Bilobites, from the Catskill Mountains:* Lyceum Nat. History New York, Ann., v. 1, p. 45-49, pl. 5.

Delgado, J. F. N.

1884, *Note sur les échantillons de Bilobites envoyés à l'exposition géographique de Toulouse:* Soc. Histoire Nat. Toulouse, Bull., v. 18, p. 126-131, 2 pl.

1885, *Estudo sobre os bilobites e outros fosseis das quartzites da base do systema silurico de Portugal:* 113 p., 42 pl., Academia real das sciencias (Lisboa) (incl. Fr. transl.).

1910, *Terrains paléozoiques du Portugal. Étude sur les fossiles des schistes à néréites de San Domingos et des schistes à néréite et à graptolites de Barrancos:* Commis. Serv. Geol. Portugal, v. 56, 68 p., 51 pl.

Demanet, Felix, & Van Straelen, V.

1938, *Faune Houillère de la Belgique:* Edit. Musée Royal Histoire Nat. Belgique, 317 p., 144 pl. (Bruxelles).

Derichs, Franz

1928, *Über Flysch-Chondriten:* Senckenbergiana, v. 9, p. 214-219, text-fig. 1-3.

Derville, Henry

1931, *Les marbres du Calcaire carbonifère en Bas-Boulonnais:* 322 p., 24 pl., O. Boehm (Strasbourg).

1950, *Contribution à l'étude des calcisphères du calcaire de Bachant:* Soc. Géol. Nord, Ann., v. 70, p. 273-285.

Desio, Ardito

1923, *Sopra una Lorenzinia del Flysch dei dintorni di Firenze:* Riv. Ital. Paleontologia, v. 29, p. 7-10, 1 pl.

1940, *Vestigia problematiche paleozoiche della Libia:* Ann. Museo Libico Storia Naturale, v. 2, p. 47-92, 13 pl. (=Publ. Ist. Geol. Pal. Geogr. fis.r. Univ. Milano, ser. pal., v. 20, Milano).

1941, *Un nuovo reperto di Lorenzinia carpatica (Zuber) nel Flysch dell Albania settentrionale:* Riv. Ital. Paleontologia, v. 47, p. 7-8, 1 text-fig.

Deslongchamps, Eugène

1856, *Notice sur des empreintes ou traces d'animaux existant à la surface d'une roche des grès, au lieu dit les Vaux-d'Aubin, près Argentan, département de l'Orne, et connus dans le pays sous le nom de pas de boeufs:* Soc. Linnéenne Normandie, Mém., v. 10 (1854-55), p. 19-44, pl. 3.

Destombes, Jacques

1964, *A propos de Leckwyckia aenigmatica H. et G. Termier de l'Ordovicien moyen du Maroc central:* Serv. Géol. Maroc, Notes et Mém., v. 172 (1963), p. 67.

Dettmer, Friedrich

1915, *Neues zum Fucoidenproblem:* Centralbl. Mineralogie, Geologie, Paläontologie, 1915, p. 285-287, 1 text-fig.

Deunff, Jean

1957, *Ampelitocystis, genre nouveau de microorganisme; chitinoïde du gothlandien armoricain:* Soc. Géol. et Minér. Bretagne, Bull., n. sér., v. 2, p. 1-3.

Dewalque, G. J. G.

1882, *Fragments paléontologiques:* Soc. Géol. Belgique, Ann., v. 8 (1880-81), Mém., p. 43-54, pl. 1-3.

1887 [not seen by the editors].

DeWindt, J. T.

1973, *Occurrence of Rusophycus in the Poxono Island Formation (Upper Silurian) of eastern Pennsylvania:* Jour. Paleontology, v. 47, p. 999-1000, 1 text-fig.

Dike, E. F.

1972, *Ophiomorpha nodosa Lundgren: Environmental implications in the Lower Greensand of the Isle of Wight:* Geologists' Assoc., Proc., v. 83, p. 165-178, pl. 13-14.

Dillon, W. P.

1964, *Flotation technique for separating fecal pellets and small marine organisms from sand:* Limnology & Oceanography, v. 9, p. 601-602.

———, & Zimmerman, H. B.

1970, *Erosion by biological activity in two New England submarine canyons:* Jour. Sed. Petrology, v. 40, p. 542-547, text-fig. 1-9.

Dimian, Mihai, & Dimian, Elena

1964, *Recherches sédimentologiques sur la zone du Flysch Crétacé supérieur-Paléogène et de la Molasse miocène entre Valea Zăbalei et Valea Buzănluè:* Comit. Geol. Inst. Geol. Dări de Seamă ale Séd'ntelov, v. 49 (1961-

62), pt. 1-a; p. 361-382, 12 pl. (Ruman. text.; Russ., French resumés).

Dineley, D. L.

1960, *The Old Red Sandstone of Eastern Ekmanfjorden, Vestspitsbergen:* Geol. Mag., v. 97, p. 18-32, text-fig. 1-5.

Dionne, Jean-Claude

1969, *Tadpole holes: A true biogenic sedimentary structure:* Jour. Sed. Petrology, v. 39, p. 358-360.

1969, *Une structure sédimentaire alvéolaire d'origine biologique:* Rev. Géographie Montréal, v. 23, p. 197-199.

1972, *Ribbed grooves and tracks in mud tidal flats of cold regions:* Jour. Sed. Petrology, v. 42, p. 848-851, text-fig. 1-7.

Dörjes, Jürgen

1972, *Georgia coastal region, Sapelo Island, U. S. A.: sedimentology and biology. VII. Distribution and zonation of macrobenthic animals:* Senckenbergiana Maritima, v. 4, p. 183-216, text-fig. 1-5, 2 pl.

Donahue, Jack

1971, *Burrow morphologies in north-central Pacific sediments:* Marine Geology, v. 11, p. M1-M7, text-fig. 1-4.

Donaldson, Douglas, & Simpson, Scott

1962, *Chomatichnus, a new ichnogenus and other trace-fossils of Wegber Quarry:* Liverpool & Manchester Geol. Jour., v. 3, pt. 1, p. 73-81, text-fig. 1-3, pl. 3, 4.

Donaldson, J. A.

1967, *Precambrian vermiform structures: A new interpretation:* Canad. Jour. Earth Sci., v. 4, p. 1273-1276, 1 pl.

Dons, Johannes

1959, *Fossils (?) of Precambrian age from Telemark, Southern Norway:* Norsk Geol. Tidsskrift, v. 39, no. 2/3, p. 249-262, text-fig. 1-8.

1963, *The Precambrian Telemark area in south central Norway:* Geol. Rundschau, v. 52 (1962), no. 1, p. 261-268, text-fig. 1-4.

Doughty, P. J.

1965, *Trace fossils of the Liassic rocks of north west Lincolnshire:* The Mercian Geologist, v. 1, p. 143-152, pl. 6, 7.

Douvillé, Henri

1908, *Perforations d Annélides:* Soc. Géol. France, Bull., v. 7, p. 361-370, pl. 12.

Driscoll, E. G.

1969, *Animal-sediment relationships of the Coldwater and Marshal formations of Michigan:* in Stratigraphy and paleontology: Essays in honour of Dorothy Hill, K. S. W. Campbell (ed.), p. 337-352, A. N. U. Press (Canberra).

Dubois, Paul, & Lessertisseur, Jacques
1965, *Note sur Bifungites, trace problematique du Devonien du Sahara:* Soc. Géol. France, Bull., sér. 7, v. 61 (1964), p. 626-634, text-fig. 1, 7 pl.

Dudich, Endre
1962, *Ein neues Anneliden-Wohnrohr aus dem helvetischen Schotterkomplex in der Nähe von Budapest:* Földtani Közlöny, v. 92, p. 107-109, text-fig. 1-3.

Duff, Joseph
1865, *Carboniferous sandstone with surface marks:* Geol. Mag., v. 2, p. 136-137, pl. 4.

Dumas, Emilien
1876, *Statistique géologique, minéralogique, métallurgique et paléontologique de Département du Gard:* pt. 2, 735 p., 9 pl., A. Bertrand (Paris et Nimes).

Dumortier, Eugène
1861, *Note sur le calcaire à Fucoides, base de l'Oolithe inférieure dans le bassin du Rhône:* Soc. Géol. France, Bull., sér. 2, v. 18, p. 579-587, pl. 12.

Duncan, P. M.
1876, *On some unicellular algae parasitic within Silurian and Tertiary corals, with a notice of their presence in Calceola sandalina and other fossils:* Geol. Soc. London, Quart. Jour., v. 32, p. 205-216, pl. 16.

———, & Jenkins, H. M.
1870, *On Palaeocoryne, a genus of tubularine Hydrozoa from the Carboniferous Formation:* Royal Soc. London, Philos. Trans., v. 159 (for 1869), p. 693-699, pl. 66.

Duns, J.
1877, *On an unnamed Palaeozoic annelid:* Royal Soc. Edinburgh, Proc., v. 9, p. 352-359, pl. 4.

Durkin, M. K.
1968, *Notes on the trace fossils at Bean, Kent:* Geologists' Assoc., Proc., v. 79, p. 215-218, text-fig. 1, 2 pl.

Edgell, H. S.
1964, *Precambrian fossils from the Hamersley Range, Western Australia, and their use in stratigraphic correlation:* Geol. Soc. Australia, Jour., v. 11, p. 235-259.

Eha, Silvio
1953, *The pre-Devonian sediments on Ymers Ø, Suess Land, and Ella Ø (East Greenland) and their tectonics:* Meddel. Grønland, v. 111, p. 1-105, illus.

Ehlers, Ernst
1868, *Über eine fossile Eunicee aus Solenhofen, Eunicites avitus, nebst Bemerkungen über fossile Würmer überhaupt:* Zeitschr. Wiss. Zoologie, v. 18, p. 421-443, pl. 29.

Ehrenberg, C. G.
1858a, *Über fortschreitende Erkenntnis massenhafter mikroskopischer Lebensformen in den untersten silurischen Thonschichten bei Petersburg:* Akad. Wiss. Berlin, Monatsber., 1858, p. 295-311.
1858b, *Über massenhafte mikroskopische Lebensformen der ältesten silurischen Grauwacken-Thone bei Petersburg:* Same, Monatsber., 1858, p. 324-337.

Ehrenberg, Kurt
1938, *Bauten von Decapoden (Callianassa sp.) aus dem Miozän (Burdigal) von Burgschleinitz bei Eggenburg im Gau Nieder-Donau (Niederösterreich):* Paläont. Zeitschr., v. 20, p. 263-284, pl. 27-29.
1941, *Über einige Lebensspuren aus dem Oberkreideflysch von Wien und Umgebung:* Palaeobiologica, v. 7, p. 282-313, text-fig. 1-10.
1944, *Ergänzende Bemerkungen zu den seinerzeit aus dem Miozän von Burgschleinitz beschriebenen Gangkernen und Bauten dekapoder Krebse:* Paläont. Zeitschr., v. 23, p. 354-359.

Eichwald, Eduard
1846, *Geognoziya Preimushchestvenio v "otnoshenii k" Rossii:* 572 p., publ. by author (St. Petersburg). [*Geology, particularly in relation to Russia.*]
1856, *Beitrag zur geographischen Verbreitung der fossilen Thiere Russlands. Alte Periode:* Soc. Impér. Naturalistes Moscou, Bull., v. 29, p. 406-453.
1860-68, *Lethaea Rossica ou paléontologie de la Russie:* v. 1, 1657 p. (1860); v. 2, 1304 p. (1865-68), E. Schweizerbart (Stuttgart).

Eisenack, Alfred
1934, *Über Bohrlöcher in Geröllen baltischer Obersilurgeschiebe:* Zeitschr. Geschiebeforsch., v. 10, p. 89-94, text-fig. 1, 2.
1954, *Mikrofossilien aus Phosphoriten des samländischen Unteroligozäns und über die Einheitlichkeit der Hystrichosphaerideen:* Palaeontographica, v. 105A, p. 49-95, text-fig. 1-8, pl. 7-12.
1955, *Chitinozoen, Hystrichosphären und andere Mikrofossilien aus dem Beyrichia-Kalk:* Senckenbergiana Lethaea, v. 36, p. 157-188, text-fig. 1-13.
1962, *Neue problematische Mikrofossilien:* Neues Jahrb. Geologie, Paläontologie, Abhandl., v. 114, p. 135-141, text-fig. 1, pl. 5.

1966, *Über Chuaria wimani Brotzen:* Same, Monatsh., v. 1, p. 52-56, text-fig. 1, 2.

1968, *Problematika aus baltischem Ordovizium und Silur:* Same, Abhandl., v. 131, p. 305-309, pl. 22, fig. 1, 2.

1970, *Xenotheka klinostoma und ihre systematische Stellung:* Same, Monatsh., 1970, p. 449-451, text-fig. 1, 2.

1971, *Weitere Mikrofossilien aus dem Beyrichien-kalk (Silur):* Same, Monatsh., 1971, p. 449-460, text-fig. 1-34.

Eldredge, Niles

1970, *Observations on burrowing behavior in Limulus polyphemus (Chelicerata, Merostomata), with implications on the functional anatomy of trilobites:* Am. Museum Novitates, no. 2436, p. 1-17, text-fig. 1-4.

Elias, M. K.

1957, *Late Mississippian fauna from the Redoak Hollow Formation of southern Oklahoma:* Jour. Paleontology, v. 31, p. 370-427, pl. 39-50.

1958, *Late Mississippian fauna from the Redoak Hollow Formation of Southern Oklahoma, Pt. 4. Gastropoda, Scaphopoda, Cephalopoda, Ostracoda, Thoracica, and Problematica:* Same, v. 32, p. 1-57, text-fig. 1-45, pl. 1-4.

Ellenberger, François

1948, *Le probléme lithologique de la craie durcie de Meudon. Bancs-limites et "contacts par racines": lacune sous-marine ou émersion?:* Soc. Géol. France, Bull., sér. 5, v. 17 (1947), p. 255-274, illus.

Ellenor, D. W.

1970, *The occurrence of the trace fossil Zoophycos in the Middle Devonian of northeastern New South Wales, Australia:* Palaeogeography, Palaeoclimatology, Palaeoecology, v. 7, p. 69-78.

Elliott, G. F.

1958, *Fossil microproblematica from the Middle East:* Micropaleontology, v. 4, p. 419-428, pl. 1-3.

1960, *Fossil calcareous algal floras of the Middle East with a note on a Cretaceous problematicum, Hensonella cylindrica gen. et sp. nov.:* Geol. Soc. London, Quart. Jour., v. 115, pt. 3, p. 217-232, pl. 8.

1962, *More microproblematica from the Middle East:* Micropaleontology, v. 8, p. 29-44, 6 pl.

1963, *Problematical microfossils from the Cretaceous and Palaeocene of the Middle East:* Palaeontology, v. 6, p. 293-300, pl. 46-48.

Emery, K. O.

1953, *Some surface features of marine sediments*
made by animals: Jour. Sed. Petrology, v. 23, p. 202-204.

Emmons, Ebenezer

1844, *The Taconic System; based on observations in New York, Massachusetts, Maine, Vermont, and Rhode Island:* 68 p., 6 pl., Caroll & Cook, printers (Albany).

1856, *On new fossil corals from North Carolina:* Am. Jour. Sci., ser. 2, v. 22, p. 389-390, text-fig. 1, 2.

Endo, Ruiji

1933, *Manchuriophycus, n.g., from a Sinian formation of South Manchuria:* Japan. Jour. Geology, Geography, v. 11, p. 43-48, pl. 6, 7.

1951a, *Stratigraphical and paleontological studies of the later Paleozoic calcareous algae in Japan, 1. Several new species from the Sakamoto-zawa section in the Kitakami Mountainous Land:* Palaeont. Soc. Japan, Trans. Proc., n. ser., no. 4, p. 121-129, pl. 10, 11.

1951b, *Stratigraphical and paleontological studies of the later Palaeozoic calcareous algae in Japan, V. Several species from the Iwaizaki limestone, Motoyoshi-gun, in the Kitakami Mountainous Land:* Japan. Jour. Geology, Geography, v. 23, p. 120-126, pl. 11, 12.

——, & Resser, C. E.

1937, *The Sinian and Cambrian formations and fossils of southern Manchukuo:* Manchurian Sci. Museum, Bull., v. 1, 474 p., 73 pl.

Etheridge, Robert

1876, *Appendix A. Description of new fossils occurring in the Arenig or Skiddaw slates, in J. Ward, The geology of the northern part of the English Lake District:* Geol. Survey, England and Wales, Mem., 1876, p. 108-112.

Etheridge, Robert Jr.

1891, *On the occurrence of microscopic fungi, allied to the genus Palaeoachlya, Duncan, in the Permo-Carboniferous rocks of N.S. Wales and Queensland:* Geol. Survey, New South Wales, Records, v. 2, pt. 3, p. 95-99, pl. 7.

1899, *On two additional perforating bodies, believed to be thallophytic cryptogams, from the lower Palaeozoic rocks of N. S. Wales:* Austral. Museum, Records, v. 3, no. 5, p. 121-127, 1 pl.

1904, *An endophyte (Stichus mermisoides) occurring in the test of a Cretaceous bivalve:* Same, Records, v. 5, no. 4, p. 255-257, pl. 30, 31.

Ettinghausen, C. R. von

1863, *Die fossilen Algen des Wiener und des Karpathen-Sandsteines:* K. Akad. Wiss.

Wien, math.-nat. Kl., Sitzungsber., Abt. 1, v. 48, p. 444-467, pl. 1, 2.

Evans, J. W.

1970, *Palaeontological implications of a biological study of rock-boring clams (Family Pholadidae):* in Trace fossils, T. P. Crimes & J. C. Harper (eds.), Geol. Jour., spec. issue no. 3, p. 127-140, pl. 1-7, table 1, Seel House Press (Liverpool).

Ewing, Maurice, & Davis, R. A.

1967, *Lebensspuren photographed on the ocean floor:* in Deep-sea photography, J. B. Hersey (ed.), Johns Hopkins Univ., Oceanographic Studies no. 3, p. 259-294, 104 text-fig.

Eyerman, John

1890, *Notes on geology and mineralogy. 1. Fossil footprints from the Jura (?)—Trias of New Jersey:* Acad. Nat. Sci. Philadelphia, Proc. 1889, p. 32-35.

Farrés [Farrés Malian], Francisco

1963, *Observaciones paleoicnologicas y estratigraficas en el Flysch Maestrichtiense de la Pobla de Segur (Prov. de Lérida):* Inst. Geol. Min. España, Notas & Comun., v. 71, p. 95-136, text-fig. 1-7, pl. 1-9.

1967, *Los "Dendrotichnium" de España:* Same, Notas & Comun., v. 94, p. 29-36, 3 pl.

Farrow, G. E.

1966, *Bathymetric zonation of Jurassic trace fossils from the coast of Yorkshire, England:* Palaeogeography, Palaeoclimatology, Palaeoecology, v. 2, p. 103-151, 11 text-fig., pl. 1-7.

1971, *Back-reef and lagoonal environments of Aldabra Atoll distinguished by their crustacean burrows:* Zool. Soc. London, Symposium, no. 28, p. 455-500.

Faul, Henry

1950, *Fossil burrows from the Precambrian Ajubik Quartzite of Michigan:* Jour. Paleontology, v. 24, p. 102-106.

1951, *The naming of fossil footprint "species":* Same, v. 25, p. 409.

———, & Roberts, W. A.

1951, *New fossil footprints from the Navajo (?) Sandstone of Colorado:* Jour. Paleontology, v. 25, p. 266-274, text-fig. 1-5, pl. 40-43.

Fauvel, A.

1868, *Compte-rendu de l'excursion linnéenne à Bagnoles-de-l'Orne:* Soc. Linnéenne Normandie, Bull., sér. 2, v. année 1867, p. 523-534, pl. 5.

Felix, J.

1913, *Über ein cretaceisches Geschiebe mit Rhizocorallium Gläseli n. sp. aus dem Diluvium bei Leipzig:* Naturforsch. Gesell. Leipzig,

Sitzungsber., v. 39 (1912), p. 19-25, 37, pl. 1.

Fenton, C. L.

1946, *Algae of the Pre-Cambrian and early Paleozoic:* Am. Midland Naturalist, v. 36, p. 259-263.

———, & Fenton, M. A.

1924, *The stratigraphy and fauna of the Hackberry stage of the Upper Devonian:* Univ. Michigan Museum Geology, Contrib., v. 1, 260 p., text-fig. 1-7, pl. 45.

1931a, *Apparent gastropod trails in the Lower Cambrian:* Am. Midland Naturalist, v. 12, p. 401-405.

1931b, *Algae and algal beds in the Belt Series of Glacier National Park:* Jour. Geology, v. 39, p. 670-686, text-fig. 1, 10 pl.

1932, *Boring sponges in the Devonian of Iowa:* Am. Midland Naturalist, v. 13, p. 42-54, pl. 6-9.

1933, *Oboloid brachiopods in the Belt Series of Montana* (abstr.): Geol. Soc. America, Bull., v. 44, p. 190.

1934a, *Traces of invertebrates and plants:* Same, Proc. for 1933, p. 369.

1934b, *Arthraria-like markings made by annelids and snails:* Pan-Am. Geologist, v. 61, p. 264-266, text-fig. 1.

1934c, *Lumbricaria, a holothuroid casting?:* Same, v. 61, p. 291-292, text-fig. 1, pl. 28.

1934d, *Scolithus as a fossil phoronid:* Same, v. 61, p. 341-348, text-fig. 1, pl. 1.

1936, *Walcott's "Pre-Cambrian Algonkian algal flora" and associated animals:* Geol. Soc. America, Bull., v. 47, p. 609-620, text-fig. 1, pl. 1-3.

1937a, *Belt Series of the North: Stratigraphy, sedimentation, paleontology:* Same, v. 48, p. 1873-1970, text-fig. 1-20, 19 pl.

1937b, *Olivellites, a Pennsylvanian snail burrow:* Am. Midland Naturalist, v. 18, p. 452-453, text-fig. 1.

1937c, *Archaeonassa, Cambrian snail trails and burrows:* Same, v. 18, p. 454-456, text-fig. 1, 1 pl.

1937d, *Burrows and trails from Pennsylvanian rocks of Texas:* Same, v. 18, p. 1079-1084, 3 pl.

1937e, *Trilobite "nests" and feeding burrows:* Same, v. 18, p. 446-451, text-fig. 1-6.

Ferguson, John

1961, *Claviradix, a new genus of the family Palaeocorynidae from the Carboniferous rocks of County Durham:* Yorkshire Geol. Soc., Proc., v. 33, no. 2, p. 135-148, text-fig. 1-5, 2 pl.

Fiege, Kurt

1944, *Lebensspuren aus dem Muschelkalk Nordwestdeutschlands:* Neues Jahrb. Mineralogie,

Geologie, Paläontologie, Abhandl. B, v. 88, p. 401-426, text-fig. 1-6.

1951, *Eine Fisch-Schwimmspur aus dem Culm bei Waldeck, mit Bemerkungen über die Lebensräume und die geographische Verbreitung der karbonischen Fische Nordwest-Europas:* Same, Monatsh., Jahrg. 1951, v. 1, p. 9-31, text-fig. 1-9.

Firtion, Fridolin

1958, *Sur la présence d'ichnites dans le Portlandien de l'Ile d'Oléron (Charente maritime):* Ann. Univ. Saraviens. (Naturw.), v. 7, p. 107-112, text-fig. 1-3, 2 pl.

————, **Schömer, R., Schröder, H., & Schröder, K.**

1959, *Guilielmiten im Westfal C des Saarlandes:* Ann. Univ. Saraviens. (Naturw.), v. 8, p. 233-236, text-fig. 1, 1 pl.

Fischer, A. G.

1962, *Fossilien aus Riffkomplexen der alpinen Trias: Cheliosporites Wähner, eine Foraminifere?:* Paläont. Zeitschr., v. 36, p. 118-124.

Fischer, P. H.

1866, *Étude sur les Bryozoaires perforants de la famille des Térébriporides:* Muséum Histoire Nat., Nouv. Arch., v. 2, p. 293-313, pl. 11.

Fischer, Peter, & Paulus, Bruno

1969, *Spurenfossilien aus den oberen Nohn-Schichten der Blanckenheimer Mulde (Eifelium, Eifel):* Senkenbergiana Lethaea, v. 50, p. 81-101, text-fig. 1, pl. 1-3.

Fischer-Ooster, Carl von

1858, *Die fossilen Fucoiden der Schweizer Alpen, nebst Erörterungen über deren geologisches Alter:* 72 p., 18 pl., Huber (Bern).

Fischer de Waldheim, G. F.

1811, *Notice des fossiles du Gouvernement de Moscou. III. Recherches sur les encrinites, les polycères et les ombellulaires:* 32 p., 2 pl. (Moscou).

1837, *Oryctographie du Gouvernement de Moscou:* 202 p., 51 pl., A. Semen (Moscou).

Fisher, D. W.

1962, *Small conoidal shells of uncertain affinities:* in Treatise on invertebrate paleontology, R. C. Moore (ed.), Part W, p. W98-W143, text-fig. 50-84, Geol. Soc. America & Univ. Kansas Press (New York & Lawrence, Kans.).

Fitch, Asa

1850, *A historical, topographical and agricultural survey of the County of Washington. Pt. 2-5:* New York Agric. Soc., Trans., v. 9 (1849), p. 753-944.

Fliche, Paul

1906, *Flore fossile du Trias en Lorraine et en Franche-Comté:* Soc. Sci. Nancy, Bull. Séanc., sér. 3, v. 6 (1905), p. 1-66, pl. 1-5.

Flores, R. M.

1972, *Delta front-delta plain facies of the Pennsylvania Haymond Formation, northeastern Marathon Basin, Texas:* Geol. Soc. America, Bull., v. 83, p. 3415-3424, text-fig. 1-6.

Flower, R. H.

1955, *Trails and tentacular impressions of orthoconic cephalopods:* Jour. Paleontology, v. 29, p. 857-867, text-fig. 1-4.

1961, *Part I. Montoya and related colonial corals. Part II. Organisms attached to Montoya corals:* New Mexico State Bur. Mines & Min. Res., Mem. 7, 229 p., 10 text-fig., 52 pl.

Flügel, Erik

1959, *Fossile Hydrozoen, eine wenig bekannte Gruppe riffbildender Meerestiere:* Universum, Jahrg. 14, no. 1, p. 19-24, illus.

1964, *Mikroproblematika aus den rhätischen Riffkalken der Nordalpen:* Paläont. Zeitschr., v. 38, p. 74-87, text-fig. 1, pl. 8, 9.

1972, *Mikroproblematika in Dünnenschliffen von Trias-Kalken:* Gesell. Geologie & Bergbaustud., Mitteil., v. 21, p. 957-988, text-fig. 1, 2, pl. 1-5.

————, **& Hötzl, Heinz**

1970, *Foraminiferen, Calcisphaeren und Kalkalgen aus dem Schwelmer Kalk (Givet) von Letmathe im Sauerland:* Neues Jahrb. Geologie, Paläontologie, Abhandl., v. 137, p. 358-395, text-fig. 1-5, tables 1-16.

Flügel, Helmut

1963, *Algen und Problematica aus dem Perm Süd-Anatoliens und Irans:* Österr. Akad. Wiss., Sitzungsber., math.-naturw. Kl., pt. 1, v. 172, p. 85-95, 2 pl.

Folk, R. L.

1965, *On the earliest recognition of coprolites:* Jour. Sed. Petrology, v. 35, p. 272-273.

Forbes, Edward

1849?, *On Oldhamia, a new genus of Silurian fossils:* Geol. Soc. Dublin, Jour., v. 4 (1848-50), p. 20 [1851]. (Exact date of pt. 1, vol. 4, not determined; probably Feb., 1849.)

Forchhammer, J. G.

1845, *On the influence of fucoidal plants upon the formations of the earth, on metamorphism in general, and particularly the metamorphosis of the Scandinavian alum slate:* Brit. Assoc. Advanc. Sci., Rept. 14th mtg., York (1844), p. 155-169.

Ford, T. D.

1958, *Pre-Cambrian fossils from Charnwood Forest:* Yorkshire Geol. Soc., Proc., v. 31, p. 211-217, pl. 13.

————, **& Breed, W. J.**

1970, *Tadpole holes formed during desiccation of overbank pools:* Jour. Sed. Petrology, v. 40, p. 1044-1045.

1972, *The problematic Precambrian fossil Chuaria:* 24th Internatl. Geol. Congress, Montreal, sec. 1, p. 11-18, 2 pl.

1973a, *Late Precambrian Chuar Group, Grand Canyon, Arizona:* Geol. Soc. America, Bull., v. 84, p. 1243-1260, 12 text-fig.

1973b, *The problematical Precambrian fossil Chuaria:* Paleontology, v. 16, p. 535-550, pl. 61-63.

————, ————, **& Downie, Charles**

1969, *Preliminary geologic report of the Chuar Group, Grand Canyon, Arizona:* in Geology and natural history of the Grand Canyon region, Fifth Field Conf., Powell Centennial River Exped., Four Corners Geol. Soc. Grand Canyon Guidebook, p. 114-122 (Durango, Colo.).

Forti, Achille

1926, *Alghe del Paleogene di Bolca (Verona) et loro affinità con tipi oceanici viventi:* 19 p., 5 pl., Soc. Cooperativa tip. (Padova).

Fraipont, Ch.

1912, *Empreinte néreitiforme du marbre noir de Denée:* Soc. Géol. Belgique, Ann., v. 38 (1910-1911), p. M31-M36, pl. 3.

1915, *Essais de paléontologie expérimentale:* Geol. Fören. Stockholm, Förhandl., v. 37, p. 435-451.

Franke, Adolf

1928, *Die Foraminiferen der oberen Kreide Nord- und Mitteldeutschlands:* Preuss. Geol. Landesanst., Abhandl., n. ser., v. 111, 207 p., 18 pl.

1936, *Die Foraminiferen des deutschen Lias:* Same, Abhandl., n. ser., no. 169, 138 p., 2 text-fig., 12 pl.

Frantzen, W.

1888, *Untersuchungen über die Gliederung des unteren Muschelkalks in einem Theile von Thüringen und Hessen und über die Natur der Oolithkörner in diesen Gebirgsschichten:* Preuss. Geol. Landesanst., Jahrb. 1887, p. 1-93, 3 pl.

Frarey, M. J., & McLaren, D. J.

1963, *Possible metazoans from the early Proterozoic of the Canadian Shield:* Nature, v. 200, no. 4905, p. 467-462, text-fig. 1.

Frey, R. W.

1968, *The lebensspuren of some common marine invertebrates near Beaufort, North Carolina. 1. Pelecypod burrows:* Jour. Paleontology, v. 42, p. 570-574, text-fig. 1-4.

1970a, *Trace fossils of Fort Hays Limestone Member, Niobrara Chalk (Upper Cretaceous), west-central Kansas:* Univ. Kansas Paleont. Contrib., Art. 53 (Cretaceous 2), 41 p., 5 text-fig., 10 pl.

1970b, *Environmental significance of Recent marine lebensspuren near Beaufort, North Carolina:* Jour. Paleontology, v. 44, p. 507-519, pl. 89-90.

1971, *Ichnology—the study of fossil and recent lebensspuren:* in Trace fossils, B. F. Perkins (ed.), Louisiana State Univ., Misc. Publ. 71-1, p. 91-125, text-fig. 1-21.

1972, *[Discussion] in Walter Häntzschel & O. Kraus, Names based on trace fossils (ichnotaxa): request for a recommendation.* Z. N. (S.) 1973, Bull. Zool. Nomenclature, v. 29, p. 141.

1973, *Concepts in the study of biogenic sedimentary structures:* Jour. Sed. Petrology, v. 43, p. 6-19, text-fig. 1-6.

1974, *The study of trace fossils:* Springer-Verlag New York, Inc. (New York) (in press).

————, **& Chowns, T. M.**

1972, *Trace fossils from the Ringgold road cut (Ordovician and Silurian), Georgia:* Georgia Geol. Survey, Guidebook 11, p. 25-55, text-fig. 1-4, pl. 1-5 (Athens).

————, **& Cowles, J. G.**

1969, *New observations on Tisoa, a trace fossil from the Lincoln Creek Formation (mid-Tertiary) of Washington:* The Compass, v. 47, p. 10-22, text-fig. 1, 4 pl.

1972, *The trace fossil Tisoa in Washington and Oregon:* Ore Bin, v. 34, no. 7, p. 113-119, text-fig. 1-4.

————, **& Howard, J. D.**

1969, *Profile of biogenic sedimentary structures in a Holocene barrier island-salt marsh complex, Georgia:* Gulf Coast Assoc. Geol. Soc., Trans., v. 19, p. 427-444.

1970, *Comparison of Upper Cretaceous ichnofaunas from siliceous sandstones and chalk, Western Interior Region, U. S. A.:* in Trace fossils, T. P. Crimes & J. C. Harper (eds.), Geol. Jour., spec. issue no. 3, p. 141-166, text-fig. 1-8, table 1-3, Seel House Press (Liverpool).

1972, *Georgia coastal region, Sapelo Island, U. S. A.: sedimentology and biology. VI. Radiographic study of sedimentary structures made by beach and offshore animals in aquaria:* Senckenbergiana Maritima, v. 4, p. 169-182, text-fig. 1-8.

————, **& Mayou, T. V.**

1971, *Decapod burrows in Holocene barrier island beaches and washover fans, Georgia:*

Senckenbergiana Maritima, v. 3, p. 53-77, pl. 1-4.

Frischmann, Ludwig

1853, *Versuch einer Zusammenstellung der bis jetzt bekannten fossilen Thier- und Pflanzen-Überreste des lithographischen Kalkschiefers in Bayern. Ein Programm:* 46 p. (Eichstätt).

Fritel, P. H.

1925, *Végétaux paléozoiques et organismes problématiques de l'Ouedai:* Soc. Géol. France, Bull., sér. 4, v. 25, p. 33-48, text-fig. 1-6, pl. 2, 3.

Fritsch, Anton

1908, *Problematica Silurica:* in Joachim Barrande, Système Silurien du centre de la Bohême, 28 p., text-fig. 1-7, 12 pl., publ. from Barrande Fund, by author and editor (Prague).

Fritz, M. A.

1925. *The stratigraphy and paleontology of Toronto and vicinity. IV. Hydrozoa, Echinodermata, Trilobita, markings:* Ontario Dept. Mines, Rept., v. 32, pt. 7, p. 1-46, text-fig. 1-5, 4 pl.

1965, *Bryozoan fauna from the Middle Ordovician of Mendoza, Argentina:* Jour. Paleontology, v. 39, p. 141-142, pl. 19.

Fuchs, Theodor

1894a, *Über pflanzenähnliche "Fossilien," durch rinnendes Wasser hervorgebracht:* Naturw. Wochenschr., v. 9, p. 229-231.

1894b, *Über einige von der österreichischen Tiefsee-Expedition S. M. Schiffes "Pola" in bedeutenden Tiefen gedredschten Cylindrites-ähnlichen Körper und deren Verwandtschaft mit Gyrolithes:* Akad. Wiss. Wien. math-nat. Kl., Denkschr., v. 61, p. 11-21, 3 pl.

1894c, *Über eine fossile Halimeda aus dem eocänen Sandstein von Greifenstein:* Same, Sitzungsber., v. 103, pt. 1, p. 200-204, 1 pl.

1895, *Studien über Fucoiden und Hieroglyphen:* Same, Denkschr., v. 62, p. 369-448, 9 pl.

1901, *Über Medusina geryonoides von Huene:* Centralbl. Mineralogie, Geologie, Paläontologie, 1901, p. 166-167.

1905, *Kritische Besprechung einiger im Verlaufe der letzten Jahre erschienenen Arbeiten über Fucoiden:* K. K. Geol. Reichsanst. Wein, Jahrb., v. 54 (1904), p. 359-388, pl. 10.

1909, *Über einige neuere Arbeiten zur Aufklärung der Natur der Alectoruriden:* Geol. Gesell. Wien, Mitteil., v. 2, p. 335-350, text-fig. 1-12.

Fucini, Alberto

1928, *Sulla scoperta di una flora Wealdiana nel Mt. Pisano:* Accad. Gioenia Sci. Nat. Catania, Boll., v. 58, 4 p., 1 pl.

1936. *Problematica verrucana, Parte I:* Palaeont. Italica, Append. 1, p. 1-126, text-fig. 1-26, pl. 1-76.

1938, *Problematica verrucana. Parte II:* Same, Append. 2, p. 127-258, pl. 77-148D.

Fürsich, F. T.

1973, *A revision of the fossils Spongeliomorpha, Ophiomorpha and Thalassinoides:* Neues Jahrb. Geologie, Paläontologie, Monatsh., Jahrg., 1973, Heft 12, p. 719-735, text-fig. 1-6.

Gabelli, L. da

1900, *Sopra un' interessante impronte medusoide:* Il Pensiero Aristotelico nella Scienza moderna, v. 1, no. 2, p. 74-78, 1 pl. (Bologna).

1927, *Noticia sobre el hallazgo de la Lorenzinia apenninica Da Gabelli en el eoceno de Guipuzcoa:* Soc. Española Historia Nat., Bol., v. 27, p. 46-56.

Gaillard, Christian

1972, *Paratisoa contorta n. gen., n. sp. trace fossile nouvelle de l'Oxfordian du Jura:* Archiv. Sci., v. 25, fasc. 1, p. 149-160.

Galloway, J. J.

1922, *Nature of Taonurus and its use in estimating geologic time:* Geol. Soc. America, Bull., v. 33, p. 199.

1933, *A manual of Foraminifera:* James Furman Kemp memorial series, Pub. 1, 483 p., 42 pl., Principia Press (Bloomington, Ind.).

Gardet, Gustav, Laugier, R., & Lessertisseur, Jacques

1957, *Sur un Problématicum du Lias inférieur de la Haute-Marne: Siphonites heberti de Sap.:* Soc. Géol. France, Bull., sér. 6, v. 6, p. 997-1000, text-fig. 1, 2.

Garrett, Peter

1970, *Phanerozoic stromatolites: noncompetitive ecologic restriction of grazing and burrowing animals:* Science, v. 169, p. 171-173.

Gatrall, Michael, & Golubic, Stjepko

1970, *Comparative study of some Jurassic and Recent endolithic fungi using scanning electron microscope:* in Trace fossils, T. P. Crimes & J. C. Harper (eds.), Geol. Jour., spec. issue no. 3, p. 167-178, pl. 1-3, Seel House Press (Liverpool).

Geinitz, Eugen

1883, *Die Flötzformationen Mecklenburgs:* Verein Freunde Naturgesch. Mecklenburg, Archiv, v. 37, p. 1-151, 246-250, pl. 1-6.

1888, *IX. Beitrag zur Geologie Mecklenburgs:* Same, Archiv, v. 41 (1887), p. 143-216, pl. 4-6.

1895, *Über einige räthselhafte Fossilien:* Naturwiss.

Wochenschr., v. 10, p. 213-216, text-fig. 1, 2.

Geinitz, H. B.

1839-42, *Charakteristik der Schichten und Petrefacten des sächsisch-böhmischen Kreidegebirges:* 116 p., 24 pl., Arnold (Dresden & Leipzig).

1846, *Grundriss der Versteinerungskunde:* viii + 815 p., plates, Arnold (Dresden & Leipzig).

1849-50, *Das Quadersandsteingebirge oder Kreidegebirge in Deutschland:* 292 p., 12 pl., Craz & Gerlach (Freiberg).

1852-53, *Die Versteinerungen der Grauwackenformation in Sachsen und den angrenzenden Länder-Abteilungen: Heft I. Die Graptolithen . . . sowie der silurischen Formation überhaupt:* 58 p., 6 pl. (1852); *Heft II. Geologische Verhältnisse der Grauwackenformation in Sachsen. . . :* 95 p., 20 pl. (1853), W. Engelmann (Leipzig).

1858, *Die Leitpflanzen des Rothliegenden und des Zechsteingebirges oder der permischen Formation in Sachsen:* K. Polytech. Schule u. kgl. Baugewerkschule, p. 1-27, pl. 1, 2 (Dresden).

1862, *Dyas oder die Zechsteinformation und das Rothliegende. II. Die Pflanzen der Dyas und Geologisches:* p. 131-342, pl. 24-42, Wilhelm Engelmann (Leipzig).

1863, *Über zwei neue dyadische Pflanzen:* Neues Jahrb. Mineralogie, Geologie, Paläontologie, 1863, p. 525-530, pl. 6, 7.

1864a, *Über organische Überreste in dem Dachschiefer von Wurzbach bei Lobenstein:* Same, 1864, p. 1-9, pl. 1, 2.

1864b, *Zwei Arten von Spongillopsis Geinitz:* Same, 1864, p. 517-519.

1867a, *Die organischen Überreste im Dachschiefer von Wurzbach bei Lobenstein:* in H. B. Geinitz & K. Th. Liebe, Über ein Äquivalent der takonischen Schiefer Nordamerikas in Deutschland und dessen geologische Stellung, Nova Acta Acad. Caes. Leopold.-Carol. German. Natur. Curios., 33, pt. 3, p. 1-24, pl. 1-8.

1867b, *Über Dictyophyton ? Liebeanum Gein. aus dem Culmschiefer vom Heersberge zwischen Gera und Weyda:* Neues Jahrb. Mineralogie, Geologie, Paläontologie, 1867, p. 286-288, text-fig. 3, pl. 3.

1879, *Zur Nereitenfrage:* Deutsch. Geol. Gesell., Zeitschr., v. 31, p. 621-623.

Gekker [Hecker], R. F.

1957, *Vvedenie v paleoekologiyu:* 126 p., 20 pl., Gosudar. Nauch.-Tekh. Izd. Lit. Geol. i Okhrane Nedr (Moskva). [*Introduction to paleoecology.*]

1960, *Bases de la Paléoécologie:* Bur. Rech. Géol. Min., Ann. serv. inform. géol., Paris, v. 44,

98 p., text-fig. 1-30, 19 pl. (transl. by J. Roger of 1957 publ.).

1965, *Introduction to paleoecology:* transl. and edited by M. K. Elias and R. C. Moore, 166 p., 31 text-fig., 17 pl., American Elsevier Publ. Co. (New York).

1970, *Palaeoichnological research in the palaeontological institute of the Academy of Sciences of the USSR:* in Trace fossils, T. P. Crimes & J. C. Harper (eds.), Geol. Jour., spec. issue no. 3, p. 215-226, text-fig. 1-5, Seel House Press (Liverpool).

———, Osipova. A. I., & Belskaya, T. N.

1962, *Ferganskiy zaliv paleogenovogo morya Srednei Azii; ego istoriya, osadki, fauna, flora, usloviya ikh obitaniya i razvitie:* kniga 1, 335 p., kniga 2, 332 p., illus., Akad. Nauk SSSR, Paleont. Inst. (Moskva). [*Ferghanian gulf of the Paleogene sea in Central Asia; its history, sedimentation, fauna, flora, conditions of their environment and evolution.*]

———, & Ushakov, B. V.

1962, *Vermes. Chervi. Tip Plathelminthes. Ploskie chervi. Tip Nemathelminthes. Kruglye chervi. Tip Nemertini. Hemertiny. Tip Annelida Kol'chatye chervi:* in Osnovy Paleontologii, Yu. A. Orlov (ed.), Gubki, Arkheotsiaty, Kishechnopolostnye, Chervi, p. 433-464, text-fig. 1-46, pl. 1-5, Izdatelstvo Akad. Nauk SSSR (Moskva).

Germar, E. F.

1827, *Über die Versteinerungen von Solenhofen:* in Ch. Keferstein, Teutschland, geognostisch-geologisch dargestellt, v. 4, no. 2, p. 89-110, pl. 1a (Weimar).

Germs, G. J. B.

1968, *Discovery of a new fossil in the Nama System, South West Africa:* Nature, v. 219, p. 53-54 (London).

1972, *Trace fossils from the Nama Group, Southwest Africa:* Jour. Paleontology, v. 46, p. 864-870, text-fig. 1, 2, pl. 1, 2.

1973, *Possible spригginid worm and a new trace fossil from the Nama Group, South West Africa:* Geology, v. 1, no. 2, p. 69-70, text-fig. 1-5.

Gernant, R. E.

1972, *The paleoenvironmental significance of Gyrolithes (lebensspur):* Jour. Paleontology, v. 46, p. 735-741, text-fig. 1, 2, 1 pl.

Gerster, Carl

1881, *Die Plänerbildungen um Ortenburg bei Passau:* Nova Acta Acad. Caes. Leop.-Carol. German. Natur. Curios., v. 42, p. 1-60, 1 pl.

Gevers, T. W.

1973, *A new name for the ichnogenus Arthro-*

podichnus Gevers, 1971: Jour. Paleontology, v. 47, no. 5, p. 1002.

——, Frakes, L. A., Edwards, L. N., & Marzolf, J. E.

1971, *Trace fossils in the Lower Beacon sediments (Devonian), Darwin Mountains, South Victoria Land, Antarctica:* Jour. Paleontology, v. 45, p. 81-94, pl. 18-20.

——, & Twomey, A.

1974, *Trace fossils in lower Beacon sediments (Devonian), Upper Wright Valley, Victoria Land, Antarctica:* (in prep.).

Ghent, E. D., & Henderson, R. A.

1966, *Petrology, sedimentation, and paleontology of Middle Miocene graded sandstones and mudstone, Kaiti Beach, Gisborne:* Royal Soc. New Zealand, Trans., Geol., v. 4, p. 147-169, text-fig. 1-4, 2 pl.

Giebel, C. G.

1853, *Beitrag zur Paläontologie des Texanischen Kreidegebirges:* Naturw. Ver. Sachsen u. Thüringen, Jahresber., v. 5 (1852), p. 358-375, pl. 7.

1857, *Zur Fauna des lithographischen Schiefers von Solnhofen. 4. Holothurienreste im Lithographischen Schiefer:* Zeitschr. f.d. Gesamt. Naturwiss, v. 9, p. 385-388, pl. 5.

Gilchrist, J. D. F.

1908, *New forms of the Hemichordata from South Africa:* S. Afr. Philos. Soc. (Royal Soc. South Afr.), Trans., v. 17, p. 151-176, pl. 16-17.

Gilmore, C. W.

1926, *Fossil footprints from the Grand Canyon:* Smithson. Misc. Coll., v. 77, no. 9, 41 p., text-fig. 1-23, 12 pl.

1927, *Fossil footprints from the Grand Canyon, 2d Contribution:* Same, v. 80, no. 3, 78 p., text-fig. 1-37, pl. 1-21.

Girotti, Odoardo

1970, *Echinospira pauciradiata g. n., sp. n., ichnofossil from the Serravallian-Tortonian of Ascoli Piceno (central Italy):* Geologia Romana, v. 9, p. 59-62, text-fig. 1-3 (Italian resumé).

Glaessner, M. F.

1947, *Decapod Crustacea (Callianassidae) from the Eocene of Victoria:* Royal Soc. Victoria, Proc., v. 59, p. 1-7, pl. 1, 2.

1957, *Palaeozoic arthropod trails from Australia:* Paläont. Zeitschr., v. 31, p. 103-109, pl. 10, 11.

1958, *New fossils from the base of the Cambrian in South Australia:* Royal Soc. South Australia, Trans., v. 81, p. 185-188, 1 pl.

1959a, *Fauna:* in M. F. Glaessner & B. Daily, The geology and Late Precambrian fauna

of the Ediacara Reserve, S. Austral. Museum, Records, v. 13, p. 377-398, text-fig. 1, 2, pl. 42-47.

1959b, *The oldest fossil faunas of South Australia:* Geol. Rundschau, v. 47 (1958), no. 2, p. 522-531, text-fig. 1-5.

1959c, *Precambrian Coelenterata from Australia, Africa and England:* Nature, v. 183, no. 4673, p. 1472-1473.

1962, *Pre-Cambrian fossils:* Biol. Reviews, v. 37, p. 467-494, 1 pl.

1963, *Zur Kenntnis der Nama-Fossilien Südwest-Afrikas:* Naturhist. Mus. Hofmuseums, Wien, Ann., v. 66, p. 113-120, 3 pl.

1969, *Trace fossils from the Precambrian and basal Cambrian:* Lethaia, v. 2, p. 369-393.

1971, *Die Entwicklung des Lebens im Praekambrium und seine geologische Bedeutung:* Geol. Rundschau, v. 60, no. 4, p. 1323-1339, text-fig. 1-8.

1972, *Precambrian fossils—a progress report:* 23rd Internatl. Congress, Internatl. Paleont. Union, Proc. 1968, p. 277-384 (Warszawa).

1973a, *Pseudo(?) fossils:* Geotimes, v. 18, no. 3, p. 11 (letter to editor).

1973b, *Trace fossils and the base of the Cambrian:* Ichnology Newsletter, Winter 1972-73, no. 6, p. 7-8.

——, & Wade, Mary

1966, *The Late Precambrian fossils from Ediacara, South Australia:* Palaeontology, v. 9, p. 599-628, text-fig. 1-3, pl. 97-103.

1971, *Praecambridium—a primitive arthropod:* Lethaia, v. 4, p. 71-77.

Glazek, Jerzy, Marcinowski, Ryszard, & Wierzbowski, Andrzej

1971, *Lower Cenomanian trace fossils and transgressive deposits in the Crakow upland:* Acta Geol. Polonica, v. 21, no. 3, p. 433-448, text-fig. 1-3, 2 pl.

Glocker, F. E.

1841, *Über die kalkführende Sandsteinformation auf beiden Seiten der mittleren March, in der Gegend zwischen Kwassitz und Kremsier:* Nova Acta Acad. Caes. Leop.-Carol. German. Natur. Curios., v. 19, suppl. 2, p. 309-334, pl. 4.

1850, *Über einige neue fossile Thierformen aus dem Gebiete des Karpathensandsteins:* Same, v. 22, pt. 2, p. 935-946, pl. 73.

Göppert, H. R.

1842 [1841 ?], *Über die fossile Flora der Quadersandsteinformation in Schlesien, als erster Beitrag zur Flora der Tertiärgebilde:* Nova Acta Caes. Leop.-Carol. German. Natur. Curios., v. 19, pt. 2, p. 97-134, pl. 46-53.

1848, *Zur Flora des Quader-Sandsteins in Schlesien:* Neues Jahrb. Mineralogie, Geognosie, Geologie, Petrefaktenkd., 1848, p. 269-278.

1851, *Über die Flora des Übergangsgebirges:* Deutsch. Geol. Gesell., Zeitschr., v. 3, p. 185-207.

1852, *Fossile Flora des Übergangsgebirges:* Nova Acta Caes. Leop.-Carol. German. Natur. Curios., v. 22, suppl., 299 p., 44 pl.

1860, *Über die fossile Flora der silurischen, der devonischen und unteren Kohlenformation oder des sogenannten Übergangsgebirges:* Same, v. 27, p. 425-606, pl. 34-45.

1854-65, *Die fossile Flora der permischen Formation:* Palaeontographica, v. 12, 316 p., 64 pl.

Götzinger, Gustav, & Becker, Helmut

1932, *Zur geologischen Gliederung des Wienerwaldflysches (Neue Fossilfunde):* Geol. Bundesanst. Wien, Jahrb., v. 82, p. 343-396, text-fig. 1-5, pl. 7-11.

1934, *Neue Fährtenstudien im ostalpinen Flysch:* Senckenbergiana, v. 16, p. 77-94, text-fig. 1-13.

Goldfuss, G. A.

1831, *Petrefacta Germaniae:* 1. Theil, 252 p., 71 pl., Arnz & Co. (Düsseldorf).

1862, *Petrefacta Germaniae:* 2nd. edit., pt. 1, 234 p.; pt. 2, 298 p.; pt. 3, 120 p., 201 pl., List & Francke (Leipzig). [Not seen by the editors.]

Goldring, Roland

1962, *The trace fossils of the Baggy Beds (Upper Devonian) of North Devon, England:* Paläont. Zeitschr., v. 36, p. 232-251, text-fig. 1-5, pl. 22, 23.

1964, *Trace-fossils and the sedimentary surface in shallow-water marine sediments:* in Deltaic and shallow marine deposits, L. M. J. U. van Straaten (ed.), Developments in sedimentology, v. 1, p. 136-143, Elsevier (Amsterdam).

1969, *Criteria for recognizing Precambrian fossils:* Nature, v. 223, p. 1076.

———, & Seilacher, Adolf

1971, *Limulid undertracks and their sedimentological implications:* Neues Jahrb. Geologie, Paläontologie, Abhandl., v. 137, p. 422-442, text-fig. 1-9.

———, & Stephenson, D. G.

1970, *Did Micraster burrow?:* in Trace fossils, T. P. Crimes, and J. C. Harper (eds.), Geol. Jour., spec. issue no. 3, p. 179-184, text-fig. 1, Seel House Press (Liverpool).

Gómez de Llarena, Joaquin

1946, *Revision de algunos datos paleontologicos del Flysch Cretaceo y Numulitico de Guipuzcoa:* Inst. Geol. Min. España, Notas y Comun., v. 15, p. 113-165, text-fig. 1-5, 8 pl.

1949, *Datos paleoicnologicos:* Same, v. 19, p. 115-127, text-fig. 1-8.

Gortani, M.

1920, *Osservazioni sulle impronte medusoidi del Flysch (Lorenzinia e Atollites):* Riv. Ital. Paleontologia, v. 26, p. 56-72, pl. 2, 3.

Gothan, Walther

1909, *Vermeintliche und zweifelhafte Versteinerungen:* Himmel u. Erde, v. 21, p. 472-486.

1933, *Über die fossilen Problematika der Monti Pisani bei Pisa:* Gesell. Naturforsch. Freunde Berlin, Sitzungsber. 1933, p. 250-256, text-fig. 1-4.

1942, *Pflanzen und Pseudofossilien:* Deutsch. Botan. Gesell., Ber., v. 60, p. 93-97, text-fig. 1, 2.

———, & Weyland, Hermann

1954, *Lehrbuch der Paläobotanik:* 535 p., 1 pl., Akad.-Verlag (Berlin).

Gottis, Charles

1954, *Sur un Tisoa très abondant dans le Numidien de Tunisie:* Soc. Sci. Nat. Tunisie, Bull., v. 7, p. 183-192, pl. 25-27.

Grabau, A. W.

1913, *Early Paleozoic delta deposits of North America:* Geol. Soc. America, Bull., v. 24, p. 399-528, pl. 12.

Graindor, M. J.

1957, *Cayeuxidae nov. fam., organismes à squelette du Briovérien:* Acad. Sci. [Paris], Comptes Rendus, v. 244, no. 15, p. 2075-2077.

Grant, R. E.

1826, *Notice of a new zoophyte (Cliona celata, Gr.) from the Firth of Forth . . . :* Edinburgh New Philos. Jour., v. 15, p. 78-81.

Green, Upfield

1899, *On some new and peculiar fossils from the Lower Devonian of the South Coast of Cornwall:* Royal Geol. Soc. Cornwall, Trans., v. 12, p. 227-228, pl. F.

Greensmith, J. T.

1956, *Sedimentary structures in the Upper Carboniferous of north and central Derbyshire, England:* Jour. Sed. Petrology, v. 26, p. 343-355, text-fig. 1-3, pl. 1-3.

Gregory, M. R.

1969, *Trace fossils from the turbidite facies of the Waitemata Group, Whangaparoa Peninsula, Auckland:* Royal Soc. New Zealand, Trans., Earth Sci., v. 7, p. 1-20, text-fig. 1-6, 8 pl.

Greiner, Hugo

1972, *Arthropod trace fossils in the Lower Devonian Jacquet River Formation of New Brunswick:* Canad. Jour. Earth Sci., v. 9, p. 1772-1777, fig. 1-10.

Grier, N. M. (ed.)

1927, *The Hitchcock lecture upon ichnology, and the Dartmouth college ichnological collection:* Am. Midland Naturalist, v. 10, p. 161-197.

Gripp, Karl

1927, *Über einen "geführte Mäander" erzeugenden Bewohner des Ostsee-Litorals:* Senckenbergiana, v. 9, p. 93-99, text-fig. 1-6.

1967, *Polydora biforans n. sp., ein in Belemniten-Rostren bohrender Wurm der Kreide-Zeit:* Meyniana, v. 17, p. 8-10, text-fig. 1-3, pl.

Groom, Theodore

1902, *The sequence of the Cambrian and associated beds of the Malvern Hills:* Geol. Soc. London, Quart. Jour., v. 58, p. 89-149, text-fig. 1-35.

Grossgeim [Grossheim], V. A.

1946, *O znachenii i metodike izucheniya ieroglifov (na materiale Kavkazskogo flisha):* Akad. Nauk SSSR, Izvestiya, ser. geol., no. 2, p. 111-120, text-fig. 1-7. [*On the significance and methods of study of hieroglyphs on material of the Caucasian flysch.*]

Grubić, Aleksander

1961, *Lorencinije iz eocenskoy fliša Crne Gore:* Sedimentologija, v. 1, p. 51-58. [*Lorenziniae from the Eocene flysch of Montenegro.*]

1970, *Rosetted trace fossils: a short review:* in Trace fossils, T. P. Crimes & J. C. Harper (eds.), Geol. Jour., spec. issue no. 3, p. 185-188, text-fig. 1, Seel House Press (Liverpool).

Grunau, H. R.

1959, *Mikrofazies und Schichtung ausgewählter, jungmesozoischer Radiolaritführender Sedimentserien der Zentral-Alpen:* Internatl. Sed. Petrogr. Ser., v. 4, 179 p., text-fig. 1-90, 11 pl.

Gümbel, C. W.

1861, *Geognostische Beschreibung des bayrischen Alpengebirges und seines Vorlandes. (=Geognostische Beschreibung des Königreiches Bayern, Abt. 1):* 950 p., 36 pl., J. Perthes (Gotha).

1863, *Über Clymenien in den Übergangsgebilden des Fichtelgebirges:* Palaeontographica, v. 11, p. 85-165, pl. 15-21.

1879, *Geognostische Beschreibung des Fichtelgebirges mit dem Frankenwalde und dem westlichen Vorlande:* v. 3, 698 p., Perthes (Gotha).

Gürich, Georg

1930a, *Über den Kuibis-Quarzit in Südwestafrika:* Deutsch Geol. Gesell., Zeitschr., v. 82, p. 637.

1930b, *Die bislang ältesten Spuren von Organismen*

in Südafrika: 15th Internatl. Geol. Congr. South Africa 1929, Comptes Rendus, p. 670-680, text-fig. 1-5.

1933, *Die Kuibis-Fossilien der Nama-Formation von SW-Afrika. Nachträge und Zusätze:* Paläont. Zeitschr., v. 15, p. 137-154, text-fig. 1-6.

Gulline, A. B.

1967, *The first proved Carboniferous deposits in Tasmania:* Australian Jour. Sci., v. 29, p. 369-393.

Gussow, W. C.

1973, *Chuaria circularis Walcott from the Precambrian Hector Formation, Banff National Park, Alberta, Canada:* Jour. Paleontology, v. 47, p. 1108-1112, text-fig. 1, 2.

Guthörl, Paul

1934, *Die Arthropoden aus dem Carbon und Perm des Saar-Nahe-Pfalz-Gebietes:* Preuss. Geol. Landesanst., Abhandl., new ser., v. 164, 219 p., text-fig. 1-116, 30 pl.

Haas, Otto

1954, *Zur Definition des Begriffs "Lebensspuren":* Neues Jahrb. Geologie, Paläontologie, Monatsh., v. 8, p. 379.

Hadding, Assar

1929, *The pre-Quaternary sedimentary rocks of Sweden. III. The Paleozoic and Mesozoic sandstones of Sweden:* Lunds Univ. Årsskr., n. ser., pt. 2, v. 25, no. 3, 287 p., text-fig. 1-138.

Häntzschel, Walter

1930, *Spongia ottoi Geinitz, ein sternförmiges Problematikum aus dem sächsischen Cenoman:* Senckenbergiana, v. 12, p. 261-274, text-fig. 1-3.

1934, *Schraubenförmige und spiralige Grabgänge in turonen Sandsteinen des Zittauer Gebirges:* Same, v. 16, p. 313-324, text-fig. 1-4.

1935a, *Erhaltungsfähige Schleifspuren von Gischt am Nordseestrand:* Natur u. Volk, v. 65, p. 461-465, text-fig. 1-4.

1935b, *Rezente Eiskristalle in meerischen Sedimenten und fossile Eiskristall-Spuren:* Senckenbergiana, v. 17, p. 151-167, text-fig. 1-12.

1938, *Quer-Gliederung bei Littorina-Fährten, ein Beitrag zur Deutung von Keckia annulata Glocker:* Same, v. 20, p. 297-304, text-fig. 1-6.

1939, *Die Lebensspuren von Corophium volutator (Pallas) und ihre paläontologische Bedeutung:* Same, v. 21, p. 215-227, text-fig. 1-7.

1949, *Zur Deutung von Manchuriophycus Endo und ähnlichen Problematika:* Geol. Staatsinst. Hamburg, Mitteil., v. 19, p. 77-84, text-fig. 1-5.

1952, *Die Lebensspur Ophiomorpha Lundgren im Miozän bei Hamburg, ihre weltweite Verbreitung und Synonymie:* Same, v. 21, p. 142-153, pl. 13, 14.

1955, *Lebensspuren als Kennzeichen des Sedimentationsraumes:* Geol. Rundschau, v. 43, p. 551-562, text-fig. 1, 2.

1958, *Oktokoralle oder Lebensspur?:* Geol. Staatsinst. Hamburg, Mitteil., v. 27, p. 77-87, text-fig. 1-7.

1962, *Trace fossils and Problematica:* in Treatise on invertebrate paleontology, R. C. Moore (ed.), Part W, p. W177-W245, text-fig. 109-149, Geol. Soc. America & Univ. Kansas Press (New York & Lawrence, Kans.).

1964a, *Spurenfossilien und Problematika im Campan von Beckum (Westf.):* Fortschr. Geol. Rheinld. Westfal., v. 7, p. 295-308, pl. 1-4.

1964b, *Die Spuren-Fauna, bioturbate Texturen und Marken in unterkambrischen Sandstein-Geschieben Norddeutschlands und Schwedens:* Der Aufschluss, Sonderheft, v. 14, p. 88-102, text-fig. 1-9.

1965, *Vestigia invertebratorum et Problematica:* Fossilium Catalogus. 1: Animalia, Pars 108, 142 p., W. Junk (s'Gravenhage).

1966, *Recent contributions to knowledge of trace fossils and Problematica:* Univ. Kansas Paleont. Contrib., Paper 9, p. 10-17, text-fig. 1-19.

1970, *Star-like trace fossils:* in Trace fossils, T. P. Crimes & J. C. Harper (eds.), Geol. Jour., spec. issue no. 3, p. 201-214, pl. 1, 2, Seel House Press (Liverpool).

1972, *Lebensspuren in den Kulm-Tonschiefern von Neustadt a. d. Weinstrasse:* Oberrhein. Geol. Abh., v. 21, p. 107-115, text-fig. 1-4.

————, El-Baz, Farouk, & Amstutz, G. C.

1968, *Coprolites: An annotated bibliography:* Geol. Soc. America, Mem., v. 108, 132 p., text-fig. 1-6, 11 pl., 3 tables.

————, & Kraus, O.

1972, *Names based on trace fossils (ichnotaxa): request for a recommendation.* Z. N. (S.) 1973: Bull. Zool. Nomenclature, v. 29, p. 137-141.

————, & Reineck, H. -E.

1968, *Fazies-Untersuchungen im Hettangium von Helmstedt (Niedersachsen):* Geol. Staatsinst. Hamburg, Mitteil., v. 37, p. 5-39, text-fig. 1-3, 6 pl.

Häusel, Wilhelm

1965, *Hinterlassenschaften einstiger "wurmförmiger" Organismen auf unterdevonischen Fossilien:* Natur u. Museum, v. 95, p. 388-398, text-fig. 1-6.

Hagenow, K. F. von

1840, *Monographie der Rügenschen Kreideversteinerungen II. Abth. Radiarien u. Annulaten:* Neues Jahrb. Mineralogie, Geognosie, Geologie, Petrefaktenkd., 1840, p. 631-672, 1 pl.

Hakes, W. G.

1974, *Trace fossil analysis of two Pennsylvanian shales in Kansas—cyclic sedimentation or continual mud flat deposition?:* Am. Assoc. Petroleum Geologists & Soc. Econ. Paleontologists & Mineralogists, Ann. Mtgs. Abstracts, San Antonio, Texas, v. 1, p. 41-42.

Haldeman, S. S.

1840, *Supplement to number one of "A monograph of the Limniades, and other freshwater univalve shells of North America," containing descriptions of apparently new animals in different classes, and the names and characters of the subgenera in Paludina and Anculosa:* 3 p. (Philadelphia).

Hall, James

1847-52, *Palaeontology of New York:* v. 1, 338 p., 87 pl. (1847); v. 2, 362 p. (1852); State of New York (Albany, N. Y.).

1850, *On the trails and tracks in the sandstones of the Clinton group of New York; their probable origin etc.: and a comparison of some of them with Nereites and Myrianites:* Am. Assoc. Advanc. Sci., Proc., v. 2, p. 256-260.

1857, *Palaeotrochis of Emmons:* Am. Jour. Sci., pt. 2, v. 23, p. 278.

1863, *Observations upon some spiralgrowing fucoidal remains of the Paleozoic rocks of New York:* New York State Cabinet, 16th Ann. Rept., p. 76-83, text-fig. 1-4, pl.

1865, *Figures and descriptions of Canadian organic remains; Dec. II, Graptolites of the Quebec Group:* Geol. Survey Canada, 151 p., 21 pl.

1886, *Note on some obscure organisms in the roofing slate of Washington County, New York:* Trustees New York State Museum Nat. History, 39th Ann. Rept., v. 160, pl. 11.

————, & Whitfield, R. P.

1872, *Remarks on some peculiar impressions in sandstone of the Chemung group, New York:* New York State Museum, Ann. Rept., v. 24, p. 201-204, text-fig. 1.

Hallam, Anthony

1960, *Kulindrichnus langi, a new trace-fossil from the Lias:* Palaeontology, v. 3, p. 64-68, pl. 15.

1970, *Gyrochorte and other trace fossils in the Forest Marble (Bathonian) of Dorset, England:* in Trace fossils, T. P. Crimes & J. C. Harper (eds.), Geol. Jour., spec. issue no. 3, p. 189-200, text-fig. 1, 2, pl. 1, 2, Seel House Press (Liverpool).

————, & Swett, K.

1966, *Trace fossils from the Lower Cambrian pipe rocks of the north-west Highlands:* Scot. Jour. Geology, v. 2, p. 101-106, 1 pl.

Hamblin, W. K.

1962, *X-ray radiography in the study of structures in homogeneous sediments:* Jour. Sed. Petrology, v. 32, p. 201-210, text-fig. 1-6.

1965, *Internal structures of "homogeneous" sandstones:* Kansas State Geol. Survey, Bull. 175, pt. 1, 37 p.

Hamm, Fritz

1929, *Über Rhizocoralliden im Kreidesandstein der Umgegend von Bentheim:* Provinzialstelle Naturdenkmalpflege Hannover, Mitteil., v. 2, p. 101-107, text-fig. 1-6.

Handlirsch, Anton

1906-08, *Die fossilen Insekten und die Phylogenie der rezenten Formen:* 1430 p., text-fig. 1-14, 51 pl., W. Engelmann (Leipzig).

Hanley, J. H., Steidtmann, J. R., & Toots, Heinrich

1971, *Trace fossils from the Casper Sandstone (Permian) South Laramie Basin, Wyoming and Colorado:* Jour. Sed. Petrology, v. 41, p. 1065-1068, text-fig. 1-5.

Hanor, J. S., & Marshall, N. F.

1971, *Mixing of sediment by organisms:* in Trace fossils, B. F. Perkins (ed.), Louisiana State Univ., Misc. Publ. 71-1, p. 127-135.

Hardy, C. T.

1956, *Fucoidal markings in the Swan Peak Formation, southwestern Idaho:* Jour. Sed. Petrology, v. 26, p. 369.

Hardy, P. G.

1970, *New xiphosurid trails from the Upper Carboniferous of Northern England:* Palaeontology, v. 13, p. 188-190, pl. 40.

Harkness, Robert

1855a, *On annelid tracks in the equivalent of the Millstone grits in the South-West of the County of Clare:* Edinburgh New Philos. Jour., n. ser., v. 1, p. 278-284, pl. 5.

1855b, *Notes on the fossil fucoids, zoophytes, and annelids of the flags and sandstones at Barlae:* Geol. Soc. London, Quart. Jour., v. 11, p. 473-476.

Harrington, H. J., & Moore, R. C.

1955, *Kansas Pennsylvanian and other jellyfishes:* Kansas Geol. Survey, Bull. 114, pt. 5, p. 153-162, 2 pl.

1956a, *Protomedusae:* in Treatise on invertebrate paleontology, R. C. Moore (ed.), Part F, p. F21-F23, text-fig. 11, 12, Geol. Soc. America & Univ. Kansas Press (New York; Lawrence, Kans.).

1956b, *Scyphomedusae:* in Treatise on invertebrate paleontology, R. C. Moore (ed.), Part F, p. F38-F53, text-fig. 29-41, Geol. Soc. America & Univ. Kansas Press (New York; Lawrence, Kans.).

1956c, *Trachylinida:* in Treatise on invertebrate paleontology, R. C. Moore (ed.), Part F, p. F68-F76, text-fig. 53-61, Geol. Soc. America & Univ. Kansas Press (New York; Lawrence, Kans.).

1956d, *Siphonophorida:* in Treatise on invertebrate paleontology, R. C. Moore (ed.), Part F, p. F145-F152, text-fig. 115-121, Geol. Soc. America & Univ. Kansas Press (New York; Lawrence, Kans.).

1956e, *Medusae incertae sedis and unrecognizable forms:* in Treatise on invertebrate paleontology, R. C. Moore (ed.), Part F, p. F153-F161, text-fig. 122-131, Geol. Soc. America & Univ. Kansas Press (New York; Lawrence, Kans.).

————, et al.

1959, *Arthropoda 1:* in Treatise on invertebrate paleontology, R. C. Moore (ed.), Part O, 560 p., 416 text-fig., Geol. Soc. America & Univ. Kansas Press (New York; Lawrence, Kans.).

Hartman, W. D.

1957, *Ecological niche differentiation in the boring sponges (Clionidae):* Evolution, v. 11, p. 294-297.

Hary, Armand

1969, *Recherches biostratigraphiques et pétrographiques dans les couches à entroques au "Heselberg" près de Moersdorf (Basse Sûre):* Soc. Naturalistes Luxembourgeois, Bull., v. 70 (1965), n. sér., v. 59, p. 109-138, text-fig. 1-17.

Hatai, Kotora

1968, *A pipy structure from the Lower Cretaceous Miyako Group, Iwate Prefecture, northeast Honshu, Japan:* Japan. Jour. Geology, Geography, v. 39 (no. 2-4), p. 125-137, text-fig. 1, 2.

————, Kotaka, Tamio, & Noda, Hiroshi

1970, *Supplementary note on the faecal pellets from the early Miyagian Kogota Formation, Kogota-Machi, Miyagi Prefecture, Northeast Honshu, Japan:* Saito Ho-on Kai Museum, Res. Bull., no. 39, p. 7-11, text-fig. 1.

————, & Murata, Masafumi

1971, *Two trace fossils from the southern part of the Kitakimi Massif, northeastern Honshu, Japan:* Saito Ho-on Kai Museum, Res. Bull., no. 40, p. 9-12, pl. 2.

————, & Noda, Hiroshi

1971a, *Peculiar markings on a sandstone layer of*

the Hagino Formation, Nagano Prefecture: Palaeont. Soc. Japan, Trans. Proc., no. 83, p. 162-165.

1971b, *A plantlike fossil from the Maekawa Formation Isawa-Gun, Iwate Prefecture, Japan:* Saito Ho-on Kai Museum, Res. Bull., no. 40, p. 1-6, pl. 1.

1972, *A problematica from the Mizuho-To of Nigata Prefecture:* Palaeont. Soc. Japan, Trans. Proc., n. ser., no. 86, p. 319-324, pl. 39.

Hattin, D. E.

1971, *Widespread synchronously deposited, burrow-mottled limestone beds in Greenhorn Limestone (Upper Cretaceous) of Kansas and southeastern Colorado:* Am. Assoc. Petrol. Geologists, Bull., v. 55, p. 412-432.

————, **& Frey, R. W.**

1969, *Facies relations of Crossopodia sp., a trace fossil from the Upper Cretaceous of Kansas, Iowa, and Oklahoma:* Jour. Paleontology, v. 43, p. 1435-1440.

Hauff, Bernhard

1921, *Untersuchung der Fossilfundstätten von Holzmaden im Posidonienschiefer des oberen Lias Württembergs:* Palaeontographica, v. 64, p. 1-42, pl. 1-21.

Haug, Émile

1907-11, *Traité de Géologie. II. Les périodes géologiques:* no. 1, p. 539-2024, 290 text-fig., 64 pl., A. Colin (Paris).

Haughton, S. H.

1956, *The Naukluft Mountains (S. W. Africa):* in S. H. Haughton & H. Martin, *The Nama System in south-west Africa:* 20th Internatl. Geol. Congress Mexico, El Sistema Cambrico, Sympos., p. 323-339, text-fig. 1-4 (Mexico City).

1960, *An archaeocyathid from the Nama System:* Royal Soc. South Africa, Trans., v. 36, p. 57-59, pl. 3-5.

1963-64, *Two problematic fossils from the Transvaal system:* Geol. Survey South Africa, Ann., v. 1 (1962), p. 257-260, 2 pl.

Hauptfleisch, Paul

1897, *Die als fossile Algen (und Bacterien) beschriebenen Pflanzenreste oder Abdrücke:* in A. Engler & K. Prantl, *Die natürlichen Pflanzenfamilien:* pt. 1, no. 2, p. 545-569, Stahel (Leipzig).

Hayasaka, Ichiro

1935, *The burrowing activities of certain crabs and their geologic significance:* Am. Midland Naturalist, v. 16, p. 99-103.

Hecht, Günter

1960, *Über Kalkalgen aus dem Zechstein Thü-* *ringens:* Freiberger Forschungshefte, v. 89, p. 125-176, text-fig. 1-58.

Hedström, Hermann

1923, *On "Discinella Holsti MBG." and Scapha antiquissima (Markl.) of the division Patellacea:* Sver. Geol. Undersök., ser. C, Avh. Upps., no. 313 (=Årsbok 16 [1922], no. 3), 13 p., 1 pl.

Heer, Oswald

1853, *Beschreibung der angeführten Pflanzen und Insekten:* in A. Escher v.d. Linth, Geologische Bemerkungen über das nördliche Vorarlberg und einige angrenzenden Gegenden, Allg. Schweiz. Gesell. f.d. gesamten Naturwiss., Neue Denkschr., v. 13, p. 115-135, pl. 6-8.

1855, *Flora Tetiaria Helvetiae. 1:* 117 p. 50 pl., J. Würster & Co. (Winterthur).

1864-65, *Die Urwelt der Schweiz:* 622 p., 368 text-fig., 11 pl., F. Schulthess (Zürich).

1876-77, *Flora Fossilis Helvetiae. Die vorweltliche Flora der Schweiz:* 182 p., 70 pl., J. Würster & Co. (Zürich).

1883, *Die fossile Flora der Polarländer:* in Flora fossilis arctica, v. 7, 275 p., pl. 48-110, J. Würster & Co. (Zürich).

Heezen, B. C.

1970, *Modern abyssal ichnology* (abstr.): Geol. Soc. America, Abstracts with Programs, ann. mtg., p. 574, Milwaukee, Wis.

————, **& Hollister, C. D.**

1971, *The face of the deep:* 659 p., illus., Oxford Univ. Press (New York).

————, **Tharp, Marie, & Bentley, C. R.**

1972, *Morphology of the earth in the Antarctic and Subantarctic:* in V. C. Bushnell (ed.), Antarctic map folio series, Am. Geogr. Soc., folio 16, 16 p., text-fig. 1-9, 8 pl.

Heim, Albert

1921, *Geologie der Schweiz. 2. Die Schweizer Alpen, 1. Hälfte:* 476 p., 160 text-fig., 27 pl., Chr. Herm. Tauchnitz (Leipzig).

Heinberg, Claus

1970, *Some Jurassic trace fossils from Jameson Land (East Greenland):* in Trace fossils, T. P. Crimes & J. C. Harper (eds.), Geol. Jour., spec. issue no. 3, p. 227-234, text-fig. 1-4, Seel House Press (Liverpool).

1973, *The internal structure of the trace fossils Gyrochorte and Curvolithus:* Lethaia, v. 6, p. 227-238, text-fig. 1-12.

Heinzelin, Jean de

1965, *Pogonophores fossiles?:* Soc. Belge Géologie, Paléontologie, Hydrologie, Bull., v. 73, no. 3 (1964), p. 501-510, text-fig. 1, 2, 2 pl.

Heller, Florian

1929, *Geologische Untersuchungen im Bereiche des fränkischen Grundgipses:* Naturhist. Gesell. Nürnberg, Abhandl., v. 23, p. 49-114, pl. 1-6.

Helwig, James

1972, *Stratigraphy, sedimentation, paleogeography, and paleoclimates of Carboniferous ("Gondwana") and Permian of Bolivia:* Am. Assoc. Petroleum Geologists, Bull., v. 56, p. 1008-1033, text-fig. 1-17.

Henbest, L. G.

1960, *Fossil spoor and their environmental significance in Morrow and Atoka Series, Pennsylvanian, Washington County, Arkansas:* U. S. Geol. Survey, Prof. Paper, 400-B, p. 383-385, 1 pl.

Hernandez-Pacheco, Eduardo

1908, *Consideraciones respecto à la organizacion género de vida y manera de fossilizarse algunos organismos dudosos de la época silurica y estudio de las especies de algas y huellas de gusa nos arenicolas des silúrico inferior de Alcuéscar (Cáceres):* [R.] Soc. Española Historia Nat., Bol., v. 8, p. 75-91, 4 pl.

Hersey, J. B. (ed.)

1967, *Deep-sea photography:* 310 p., Johns Hopkins Press (Baltimore).

Hertweck, Günther

1970, *The animal community of a muddy environment and the development of biofacies as effected by the life cycle of the characteristic species:* in Trace fossils, T. P. Crimes & J. C. Harper (eds.), Geol. Jour., spec. issue no. 3, p. 235-242, text-fig. 1, pl. 1, Seel House Press (Liverpool).

1972, *Georgia coastal region, Sapelo Island, U. S. A.: sedimentology and biology. V. Distribution and environmental significance of lebensspuren and in-situ skeletal remains:* Senckenbergiana Maritima, v. 4, p. 125-167, text-fig. 1-14, 4 pl.

———, & Reineck, H. -E.

1966, *Untersuchungsmethoden von Gangbauten und anderen Wühlgefügen mariner Bodentiere:* Natur u. Museum, v. 96, no. 11, p. 429-438.

Hester, N. C.

1970, *A detailed study of lithified specimens of Ophiomorpha* (abstr.): Geol. Soc. America, Abstracts with Programs, annual mtg., p. 576 (Milwaukee, Wis.).

———, & Pryor, W. A.

1972, *Blade-shaped crustacean burrows of Eocene age: a composite form of Ophiomorpha:* Geol. Soc. America, Bull., v. 83, p. 677-688.

Heymons, R.

1928, *Über Morphologie und verwandtschaftliche Beziehungen des Xenusion auerswaldae Pomp. aus dem Algonkium:* Zeitschr. Morphologie, Ökologie der Tiere, v. 10, p. 307-329, text-fig. 1-7.

High, L. R., & Picard, D. M.

1968, *Dendritic surge marks ("Dendrophycus") along modern stream banks:* Univ. Wyoming, Contrib. Geology, v. 7, no. 1, p. 1-6, text-fig. 1-6.

Hildebrand, Erich

1924, *Geologie und Morphologie der Umgebung von Wertheim a.M.:* 79 p., text-fig., 8 pl., F. Wagner (Freiburg i. Br.).

Hill, Dorothy, & Wells, J. W.

1956, *Hydroida and Spongiomorphida:* in Treatise on invertebrate paleontology, R. C. Moore (ed.), Part F, p. F81-F89, text-fig. 65-74, Geol. Soc. America & Univ. Kansas Press (New York; Lawrence, Kans.).

Hill, G. W., & Hunter, R. E.

1973, *Burrows of the ghost crab Ocypode quadrata (Fabricius) on the barrier islands, south-central Texas coast:* Jour. Sed. Petrology, v. 43, p. 24-30, text-fig. 1-6.

Hill, R. T.

1890, *Occurrence of Goniolina in the Comanche series of the Texas Cretaceous:* Am. Jour. Sci., ser. 3, v. 40, p. 64-65.

1893, *Paleontology of the Cretaceous formations of Texas: the invertebrate paleontology of the Trinity division:* Biol. Soc. Washington, Proc., v. 8, p. 9-40, pl. 1.

Hillmer, Gero, & Schulz, M. -G.

1973, *Ableitung der Biologie und Ökologie eines Polychaeten der Oberkreide durch Analyse des Bohrganges Ramosulcichnus biforans (Gripp) nov. ichnogen.:* Geol.-Paläont. Inst. Univ. Hamburg, Mitteil., v. 42, p. 5-24, text-fig. 1-9, pl. 1-3.

Hiltermann, Heinrich

1952, *Astrorhiza cretacea Franke 1928 als Scheinfossil und ähnliche Wurzelröhrchen (Rhizosolenien):* Geol. Jahrb., v. 66, p. 421-424, text-fig. 1-21.

———, & Schmitz, H. -H.

1968, *Problematische Apatit-Körperchen in limnischem Jungtertiär auf der Schwäbischen Alb.:* Geol. Jahrb., v. 85, p. 299-314, text-fig. 1-4, pl. 32-34, 2 tables.

Hinde, G. J.

1887, *Review of J. G. Bornemann: Die Versteinerungen des cambrischen Schichtensystems der Insel Sardiniens . . . (1886):* Geol. Mag., ser. 3, v. 4, p. 226-229.

1889, *On Archaeocyathus Billings and on other genera allied to or associated with it from the Cambrian strata of North America, Spain, Sardinia, and Scotland:* Geol. Soc. London, Quart. Jour., v. 45, p. 125-148, illus.

Hise, Ch. R. van, & Leith, Ch. K.
1909, *Precambrian geology of North America:* U. S. Geol. Survey, Bull., v. 360, 939 p.

Hisinger, Wilhelm
1837, *Lethaea svecica seu petrificata Sveciae, iconibus et characteribus illustrata:* 124 p., 36 pl., Norstedt & Söner (Holmiae [Stockholm]).

Hitchcock, Ch. H.
1898, *Recent progress in ichnology:* U. S. Geol. Survey, Mon. 29, p. 400-406.

Hitchcock, Edward
1837, *Fossil footsteps in sandstone and graywacke:* Am. Jour. Sci., v. 32, p. 174-176.
1841, *Final report on the geology of Massachusetts:* v. 2, p. 301-831, 55 pl., J. H. Butler (Northampton).
1844, *Report on ichnolithology or fossil footmarks:* Am. Jour. Sci., v. 47, p. 292-322, 2 pl.
1848, *An attempt to discriminate and describe the animals that made the fossil footmarks of the United States, and especially of New England:* Am. Acad. Arts Sci., Mem., n. ser., v. 3, p. 129-256, 24 pl.
1858, *Ichnology of New England. A report on the sandstone of the Connecticut Valley, especially its footprints:* 220 p., 60 pl., W. White (Boston).
1865, *Supplement of the ichnology of New England:* 96 p., 20 pl., Wright & Porter (Boston).

Högbom, A. A.
1915a, *Zur Deutung der Scolithus-Sandsteine und "Pipe Rocks":* Geol. Inst. Uppsala, Bull., v. 13, p. 45-60, text-fig. 1-5.
1915b, *Om djurspår in den uppländska ishafsleran:* Geol. Fören. Stockholm, Förhandl., v. 37, p. 33-44, pl. 1.
1926, *Om problematiska fossil från Närkes underkambrium:* Same, Förhandl., v. 48, p. 135-142, text-fig. 1.

Hölder, Helmut
1972, *Endo- und Epizoen von Belemniten-Rostren (Megateuthis) im nordwestdeutschen Bajocium (Mittlerer Jura):* Paläont. Zeitschr., v. 46, p. 199-220, 16 text-fig., pl. 28.

Hoernes, Rudolf
1904, *Über Koprolithen und Enterolithen:* Biolog. Zentralbl., v. 24, p. 566-576.

———, & Hollmann, R.
1969, *Bohrgänge mariner Organismen in jurassi-schen Hart- und Felsböden:* Neues Jahrb. Geologie, Paläontologie, Abhandl., v. 133, p. 79-88, text-fig. 1-4, 1 pl.

Hofmann, H. J.
1967, *Precambrian fossils (?) near Elliot Lake, Ontario:* Science, v. 156, p. 500-504, text-fig. 1-8.
1971, *Precambrian fossils, pseudofossils and problematica in Canada:* Geol. Survey Canada, Bull., v. 189, 146 p., text-fig. 1-10, 25 pl.
1972a, *Precambrian remains in Canada: fossils, dubiofossils, and pseudofossils:* 24th Internatl. Geol. Congress, sec. 1, p. 20-30, 5 text-fig. (Montreal).
1972b, *Systematically branching burrows from the Lower Ordovician (Quebec Group) near Quebec, Canada:* Paläont. Zeitschr., v. 46, p. 186-198, text-fig. 1-7, pl. 27.

Hollick, C. A.
1910, *A new fossil polypore:* Mycologia, v. 2, p. 93-94, text-fig.

Holtedahl, Olaf
1921, *On the occurrence of structures like Walcott's Algonkian algae in the Permian of England:* Am. Jour. Sci., ser. 5, v. 1, p. 195-206, text-fig. 1-8.

Horne, R. R., & Gardiner, P. R. R.
1973, *A new trace fossil from non-marine upper Paleozoic red beds in County Wexford and County Kerry, Ireland:* Geologie en Mijnbouw, v. 52, no. 3, p. 125-131, text-fig. 1, 2, pl. 1-5.

Hosius, August
1893, *Über marine Schichten im Wälderthon von Gronau (Westfalen) und die mit denselben vorkommenden Bildungen (Rhizocorallium Hohendali, sog. Dreibeine):* Deutsch. Geol. Gesell., Zeitschr., v. 45, p. 34-53, pl. 2, 3.

Hovasse, Raymond
1956, *Arnoldia antiqua, gen. nov., sp. nov., foraminifère probable du précambrien de la Côte-d'Ivoire:* Acad. Sci. [Paris], Comptes Rendus, v. 242, no. 21, p. 2582-2584.

———, & Couture, R.
1961, *Nouvelle découverte dans l'antecambrien de las Côte-d'Ivoire, de Birrimarnoldia antiqua (gen. nov.) =Arnoldia antiqua Hovasse 1956:* Acad. Sci. [Paris], Comptes Rendus, v. 252, no. 7, p. 1054-1056.

Howard, J. D.
1966, *Characteristic trace fossils in Upper Cretaceous sandstones of the Book Cliffs and Wasatch Plateau:* Utah Geol. Mineral. Survey, Central Utah, Coal Bull., v. 80, p. 35-53, text-fig. 1-19.
1968, *X-ray radiography for examination of burrowing in sediments by nearshore inverte-*

brate organisms: Sedimentology, v. 11, p. 249-258.

1969, *Radiographic examination of variations in barrier island facies, Sapelo Island, Georgia:* Gulf Coast Assoc. Geol. Soc., Trans., v. 19, p. 217-232.

————, & **Dörjes, Jurgen**

1972, *Animal-sediment relationships in two beach-related tidal flats; Sapelo Island, Georgia:* Jour. Sed. Petrology, v. 42, text-fig. 1-14.

————, & **Elders, C. A.**

1970, *Burrowing patterns of haustorid amphipods from Sapelo Island, Georgia:* in Trace fossils, T. P. Crimes & J. C. Harper (eds.), Geol. Jour., spec. issue no. 3, p. 243-262, text-fig. 1-3, pl. 1-9, table 1, Seel House Press (Liverpool).

————, **Frey, R. W., & Kingery, F. A.**

1973, *Physical and biogenic characteristics of sediments from outer Georgia continental shelf* (abstr.): Am. Assoc. Petroleum Geologists, Bull., v. 57, p. 784.

————, ————, & **Reineck, H. -E.**

1972, *Georgia coastal region, Sapelo Island, U. S. A.: sedimentology and biology. 1. Introduction:* Senckenbergiana Maritima, v. 4, p. 3-14, text-fig. 1, 2.

————, & **Reineck, H. -E.**

1972a, *Georgia coastal region, Sapelo Island, U. S. A.: sedimentology and biology. IV. Physical and biogenic sedimentary structures of the nearshore shelf:* Senckenbergiana Maritima, v. 4, p. 81-123, text-fig. 1-12, pl. 1-5.

1972b, *Georgia coastal region, Sapelo Island, U. S. A.: sedimentology and biology. VIII. Conclusions:* Same, v. 4, p. 217-222, text-fig. 1.

Howell, B. F.

1934, *Bovicornellum vermontense, a peculiar new Cambrian fossil from Vermont:* Wagner Free Inst. Sci. Philadelphia, Bull., v. 9, p. 112-113, 1 pl.

1943, *Burrows of Skolithos and Planolites in the Cambrian Hardyston sandstone at Reading, Pennsylvania:* Same, Publ., v. 3, p. 3-33, pl. 1-8.

1945, *Skolithos, Diplocraterion, and Sabellidites in the Cambrian Antietam sandstone of Maryland:* Same, Bull., v. 20, p. 33-39, pl. 1, 2.

1946, *Silurian Monocraterion clintonense burrows showing the aperture:* Same, Bull., v. 21, p. 29-37, pl. 1-3.

1956, *Evidence from fossils of the age of the Vindhyan system:* Palaeont. Soc. India, Jour., v. 1, p. 108-112.

1957a, *New Cretaceous scoleciform annelid from Colorado:* Same, v. 2, p. 149-152, pl. 16.

1957b, *Stipsellus annulatus, a Skolithos-like Cambrian fossil from Arizona:* Wagner Free

Inst. Sci., Bull., v. 32, no. 2, p. 17-19, text-fig. 1, 2.

1958, *Skolithos Woodi Whitfield in the Upper Cambrian of Minnesota and Wisconsin:* Same, Bull., v. 33, p. 17-24, pl. 1, 2.

1962, *Worms:* in Treatise on invertebrate paleontology, R. C. Moore (ed.), Part W, p. W144-W177, text-fig. 85-108, Geol. Soc. America, Univ. Kansas Press (New York; Lawrence, Kans.).

Huckriede, Reinhold

1952, *Eine spiralförmige Lebensspur aus dem Kulmkieselschiefer von Biedenkopf an der Lahn (Spirodesmos archimedeus n. sp.):* Paläont. Zeitschr., v. 26, p. 175-180, text-fig. 1-3.

Hülsemann, Jobst

1966, *Spiralfährten und "geführte Mäander" auf dem Meeresboden:* Natur u. Museum, v. 96, p. 449-455, text-fig. 1-4.

Huene, Friedrich von

1901a, *Kleine paläontologische Mittheilungen. 1. Medusina geryonides:* Neues Jahrb. Mineralogie, Geologie, Paläontologie 1901, v. 1, p. 1-12, text-fig. 1, 2.

1901b, *Nochmals Medusina geryonides v. Huene:* Centralbl. Mineralogie, Geologie, Paläontologie, 1901, p. 167.

1941, *Die Tetrapoden-Fährten im toskanischen Verrucano und ihre Bedeutung:* Neues Jahrb. Mineralogie, Geologie, Paläontologie, Beil. Bd. 86 B, p. 1-34, text-fig. 1-8, pl. 1-8.

Hundt, Rudolf

1931, *Eine Monographie der Lebensspuren des unteren Mitteldevons Thüringens:* 68 p., 128 text-fig., Weg (Leipzig).

1939, *Das mitteldeutsche Graptolithenmeer:* 395 p., 565 text-fig., Martin Boerner Verlag (Halle).

1940, *Neue Lebensspuren aus dem Ostthüringer Paläozoikum:* Zentralbl. Mineralogie, Geologie, Paläontologie, 1940, B, p. 210-216.

1941a, *Das mitteldeutsche Phycodesmeer:* 136 p., 124 text-fig., Fischer (Jena).

1941b, *Beiträge zur Kenntnis der Phycodesschichten Ostthüringens:* Geol. Thüringen, Beiträge, v. 6, p. 124-131.

Hunger, Richard

1947, *Isopodichnus tritylotos nov. spec. aus dem unteren Muschelkalk von Köllme bei Halle (Saale):* Biolog. Zentralbl., v. 66, p. 416-420, text-fig. 1.

Hupé, Pierre

1952, *Sur des Problématica du Précambrien III:* Serv. Géol. Maroc, Divis. Mines et Geologie, Notes et Mém., v. 103, p. 297-333, text-fig. 72-99, pl. 12-24.

Hyde, J. E.

1953, *The Mississippian formations of central and southern Ohio:* Ohio Geol. Survey, Bull., v. 51, 355 p.

Ilie, M. D.

1937, *Note sur l'origine du genre Palaeodictyon (Batracioides nidificans):* Inst. Géol. Roumanie, Comptes Rendus, v. 21, p. 62-64.

Issel, A.

1890, *Impressions radiculaires et figures de viscosité ayant l'apparence de fossiles:* Soc. Belge de Géologie, Bull., v. 3, Mém., p. 450-455, pl. 14.

Jablonský, Eduard

1973, *Mikroproblematika aus der Trias der West-karpaten:* Geol. Carpathica, Geol. Zborník, v. 14, p. 415-423, text-fig. 1, pl. 1-3.

Jacob, K.

1938, *Fossil algae from Waziristan:* Indian Botan. Soc., Jour., v. 17, p. 173-176, 1 pl.

Jacobsen, V. H.

1970, *A simple tool for collecting burrowing animals in submerged areas:* Limnology & Oceanography, v. 15, p. 646-648, 1 text-fig.

Jaekel, Otto

1929, *Die Spur eines neuen Urvogels (Protornis bavarica) und deren Bedeutung für die Urgeschichte der Vögel:* Paläont. Zeitschr., v. 11, p. 201-238, text-fig. 1-20, pl. 7.

James, J. F.

1884-85, *The fucoids of the Cincinnatian Group:* Cincinnati Soc. Nat. History, Jour., v. 7, p. 124-132, pl. 5, 6 (pt. 1, 1884); p. 151-166, pl. 8, 9 (pt. 2, 1885).

1886, *Remarks on some markings on the rocks of the Cincinnatian group, described under the names of Ormathichnus and Walcottia:* Same, Jour., v. 8, p. 160-163.

1889, [not seen by the editors].

1890, *Fucoids and other problematic organisms:* Am. Naturalist, v. 24, pt. 2, p. 1222.

1892a, *Manual of the palaeontology of the Cincinnati group. Part II:* Cincinnati Soc. Nat. History, Jour., v. 14, p. 149-163.

1892b, *The preservation of plants as fossils:* Same, Jour., v. 15, p. 75-78.

1892c, *Studies in problematic organisms. The genus Scolithus:* Geol. Soc. America, Bull., v. 3, p. 32-44, text-fig. 1-15.

1893, *Remarks on the genus Arthrophycus Hall:* Cincinnati Soc. Nat. History, Jour., v. 16, p. 82-86.

1894, *Studies in problematic organisms. Nr. 2. The genus Fucoides:* Same, Jour., v. 16, p. 62-81.

James, U. P.

1879, *Description of new species of fossils and remarks on some others, from the Lower and Upper Silurian rocks of Ohio:* The Paleontologist, no. 3, p. 17-24.

1881, *Contributions to paleontology: fossils of the Lower Silurian Formation: Ohio, Indiana and Kentucky:* Same, no. 5, p. 33-44.

1883, *Descriptions of fossils from the Cincinnati Group:* Cincinnati Soc. Nat. History, Jour., p. 235-236, 1 pl.

Janicke, Volkmar

1967, *Fossil-Sediment-Structuren in untertithonischen Plattenkalken der südlichen Frankenalb:* Diss. Univ. München, 116 p., 24 text-fig., 15 pl.

1970, *Lumbricaria—ein Cephalopoden-Koprolith:* Neues Jahrb. Geologie, Paläontologie, Monatsh. 1970, p. 50-60, text-fig. 1-7.

Jansa, Lubomir

1972, *Depositional history of the coal-bearing Upper Jurassic-Lower Cretaceous Kootenay Formation, southern Rocky Mountains, Canada:* Geol. Soc. America, Bull., v. 83, p. 3199-3222, text-fig. 1-16.

Jardine, William

1850, *Note to Mr. Harkness's paper on "The position of the Impressions of Footsteps in the Bunter Sandstone of Dumfrieshire":* Ann. Mag. Nat. History, ser. 2, v. 6, p. 208-209.

1853, *The ichnology of Annandale, or illustrations of footmarks impressed on the New Red sandstone of Cornockle Muir:* 17 p., text-fig. 1, 13 pl., W. H. Lizars (Edinburgh).

1858, [not seen by the editors].

Jarvis, M. M.

1905, *On the fossil genus *Porocystis Cragin:* Biol. Bull. (Marine Biol. Lab. Woods Hole), v. 9, p. 388-390, text-fig. 1-6.

Jessen, Werner

1950a, *"Augenschiefer" -Grabgänge, ein Merkmal für Faunenschiefer-Nähe im westfälischen Oberkarbon:* Deutsch. Geol. Gesell., Zeitschr. 1949, v. 101, p. 23-43, text-fig. 1-6.

1950b, *Die Augenschiefer, ihre Bedeutung für die Auffindung mariner Horizonte und ihre Stellung im oberkarbonischen Sedimentationsrhythmus des Ruhrgebietes:* Glückauf, v. 86, p. 731-733.

Johnson, Helgi, & Fox, S. K., Jr.

1968, *Dipleurozoa from Lower Silurian of North America:* Science, v. 162, p. 119-120, text-fig. 1-3.

Johnson, R. G.

1971, *Animal-sediment relations in shallow water benthic communities:* Marine Geology, v. 11, p. 93-104, text-fig. 1-6.

————, & Richardson, E. S., Jr.

1969, *The morphology and affinities of Tulli-monstrum:* Fieldiana, Geol., v. 12, no. 8, p. 119-149, text-fig.

1970a, *Fauna of the Francis Creek shale in the Wilmington Area:* in Depositional environments in parts of the Carbondale Formation, Western and northern Illinois, Illinois Geol. Survey, Guidebook Series, no. 8, p. 53-60, illus.

1970b, *Pennsylvanian invertebrates of the Mazon Creek area, Illinois; the morphology and affinities of Tullimonstrum:* Fieldiana, Geol., v. 12, no. 8, p. 119-149, illus.

Jordan, Reiner

1969, *Deutung der Astrorhizen der Stromatoporoiden (?Hydrozoa) als Bohrspuren:* Neues Jahrb. Geologie, Paläontologie, Monatsh. 1969, p. 705-711, text-fig. 1-5.

Joukowsky, Étienne, & Favre, Jules

1913, *Monographie géologique et paléontologique du Salève:* Soc. Phys. Histoire Nat. Genève, Mém., v. 37, no. 4, p. 295-523, text-fig. 1-56, 29 pl.

Joysey, K. A.

1959, *Probable cirripede, phoronid, and echiuroid burrows within a Cretaceous echinoid test:* Palaeontology, v. 1, p. 397-400, pl. 70.

Jux, Ulrich

1964, *Kommensalen oberdevonischer Atrypen aus Bergisch Gladbach (Rheinisches Schiefergebirge):* Neues Jahrb. Geologie, Paläontologie, Monatsh., 1964, p. 675-687, text-fig. 1-7.

Kamptner, Erwin

1931, *Nannoconus steinmanni nov. gen., nov. spec., ein merkwürdiges gesteinsbildendes Mikrofossil aus dem jüngeren Mesozoikum der Alpen:* Paläont. Zeitschr., v. 13, p. 288-298, text-fig. 1-3.

Karakasch, N. J.

1910, *Les restes problématiques du Cephalites maximus Eichw.:* Soc. Imp. Nat. St. Pétersbourg, Travaux, v. 35, no. 5 (Sec. Géol. Minéral.), p. 154-155, pl. 8.

Karaszewski, Władysław

1967, *Konkrecje związane z kanalikami U-kształtnyme robaków w w spągowych warstwach aalenu Świętokrzyskiego:* Kwartal. Geol., v. 11, no. 3, p. 632-636, 6 pl. [*Concretions connected with U-shaped worm burrows in the Aalenian deposits of the Świętokrzyskie Mts.*]

1971a, *Some fossil traces from the lower Liassic of the Holy Cross Mts, Central Poland:* Acad. Polonaise Sci., Bull., sér. Sci. Terre, v. 19, no. 2, p. 101-105, 1 text-fig., pl. 1-7.

1971b, *Ślady nieznanego organizmu zwierzęcego z serii gielniowskiej (dolnego pliensbachu) liasu Świętokrzyskiego:* Kwartal. Geol., v. 15, no. 4, p. 885-889, text-fig. 1, 5 pl. [*Traces of an unknown animal in the Gielniów Series (lower Pliensbachian) of the Świętokrzyskie Mountains Lias.*]

1973a, *A star-like trace fossil in the Jurassic of the Holy Cross Mts:* Acad. Polon. Sci., Bull., sér. Sci. Terre, v. 21, p. 157-160, text-fig. 1, photo 1-3 (Russ. summ.).

1973b, *Rhizocorallium, Gyrochorte and other problematics from the Middle Jurassic of the Inowłódz region:* Same, Bull., sér. Sci. Terre, v. 21. [Not seen by the editors.]

1973c, *O skamieniałościach śladowych w jurze Świętokrzyskiej:* Przegląd Geologiczny, nr. 11(247), p. 598-599, text-fig. 1-4. [*Trace fossls in the Jurassic of the Swiętokrzyskie Mts.*]

Katto, Jiro

1960, *Some Problematica from the so-called unknown Mesozoic strata of the southern part of Shikoku, Japan:* Tohoku Univ., Sci. Rept., ser. 2 (geol.), spec. vol. 4, p. 323-334, 2 pl.

Katzer, Friedrich

1896, *Beiträge zur Paläontologie des älteren Palaeozoicums in Mittelböhmen:* Böhm. Gesell. Wiss., Sitzungsber., math.-nat. kl., 1895, no. 14, 17 p., 2 pl.

Kauffman, E. G.

1969, *Form, function, and evolution:* in Treatise on invertebrate paleontology, R. C. Moore (ed.), Part N, p. N129-N205, text-fig. 87-99, Geol. Soc. America & Univ. Kansas (Boulder, Colo.; Lawrence, Kans.).

Kayser, F. H. E.

1872, *Neue Fossilien aus dem rheinischen Devon:* Deutsch. Geol. Gesell., Zeitschr., v. 24, p. 691-700, pl. 27, 28.

Kazmierczak, Jósef, & Pszczółkowski, Andrzej

1969, *Burrows of Enteropneusta in Muschelkalk (Middle Triassic) of the Holy Cross Mountains, Poland:* Acta Palaeont. Polonica, v. 14, p. 299-318, text-fig. 1-9, pl. 1-5. (Pol. & Russ. summ.).

Keen, A. M.

1969a, *Veneracea:* in Treatise on invertebrate paleontology, R. C. Moore (ed.), Part N, p. N670-N690, text-fig. E142-E152, Geol. Soc. America & Univ. Kansas (Boulder, Colo.; Lawrence, Kans.).

1969b, *Gastrochaenacea:* in Treatise on invertebrate paleontology, R. C. Moore (ed.), Part N, p. N699-N700, text-fig. E160, Geol. Soc.

America & Univ. Kansas (Boulder, Colo.; Lawrence, Kans.).

Keeping, Walter

1882, *On some remains of plants, Foraminifera and Annelida, in the Silurian rocks of central Wales:* Geol. Mag., ser. 2, v. 9, p. 485-491, pl. 11.

Kegel, Wilhelm

1966, *Rastos do Devoniano da Bacia do Parnaiba:* Divis. Geol. Mineral., Depart. Nac. Prod. Mineral., Bol. 233, 32 p., text-fig. 1-4, 11 fig.

1967, *Rastos do grupo dos Bilobites da formação Irati, São Paulo:* Divis. Geologia, Mineralogia Brasil, Notas prelim. e estudos, v. 136, 9 p., text-fig. 1, 2.

Keij, A. J.

1965, *Miocene trace fossils from Borneo:* Paläont. Zeitschr., v. 39, p. 220-228, text-fig. 1-3, 2 pl.

1969a, *Bicornifera lindenbergi n. sp. from the upper Oligocene of Escornebéon, S. W. France:* Neues Jahrb. Geologie, Paläontologie, Monatsh. 1969, p. 241-246, text-fig. 1-7.

1969b, *Dicasignetella, a bryozoan-like problematicum from the Bartonian of Belgium:* Rev. Micropaléontologie, v. 12, no. 1, p. 21-24, text-fig. 1, 2, 1 pl.

1970, *Taxonomy and stratigraphic distribution of Voorthuyseniella (Problematica). I:* K. Nederl. Akad. Wetenschappen, Verhandel., ser. B, v. 73, p. 479-499, text-fig. 1-7, 8 pl.

Kemper, Edwin

1968, *Einige Bemerkungen über die Sedimentationsverhältnisse und die fossilen Lebensspuren des Bentheimer Sandsteins (Valanginium):* Geol. Jahrb., v. 86, p. 49-106, text-fig. 1-13, pl. 2-9.

Kennedy, W. J.

1967, *Burrows and surface traces from the Lower Chalk of southern England:* Brit. Museum (Nat. History), Bull., Geol., v. 15, p. 125-167, text-fig. 1-7, 9 pl.

1970, *Trace fossils in the Chalk environment:* in Trace fossils, T. P. Crimes & J. C. Harper (eds.), Geol. Jour., spec. issue no. 3, p. 263-282, pl. 1-5, Seel House Press (Liverpool).

———, Jakobson, M. E., & Johnson, R. T.

1969, *A Favreina-Thalassinoides association from the Great Oolite of Oxfordshire:* Palaeontology, v. 12, p. 549-554, text-fig. 1, 2, 1 pl.

———, & MacDougall, J. D. S.

1969, *Crustacean burrows in the Weald Clay (Lower Cretaceous) of south-eastern England and their environmental significance:* Palaeontology, v. 12, p. 459-471, 1 text-fig., 1 pl.

———, & Sellwood, B. W.

1970, *Ophiomorpha nodosa Lundgren, a marine indicator from the Sparnacian of southeast England:* Geol. Assoc., Proc., v. 81, p. 99-110, text-fig. 1-3, 2 pl.

Kiaer, Johan

1924, *The Downtonian fauna of Norway:* Vidensk. Selsk. Kristiania, Skrifter, v. 1, 1924, no. 6, 139 p.

Kieslinger, Alois

1924, *Medusae fossiles:* Fossilium Catalalogus, I: Animalia, Pars 26, 20 p., W. Junk (Berlin).

1939, *Scyphozoa:* in Handbuch der Paläozoologie, O. H. Schindewolf (ed.), pt. 5, v. 2A, p. A69-A109, 42 text-fig., Gebrüder Borntraeger (Berlin).

Kilian, C.

1931, *Sur l'âge des grès à Harlania et sur l'extension du Silurien dans le Sahara oriental:* Acad. Sci. [Paris], Comptes Rendus Séances, v. 192, p. 1742-1743.

Kilpper, Karl

1962, *Xenohelix Mansfield 1927 aus der miozänen niederrheinischen Braunkohlenformation:* Paläont. Zeitschr., v. 36, p. 55-58, pl. 7.

Kinahan, J. R.

1858, *On the organic relations of the Cambrian rocks of Bray (County of Wicklow) and Howth (County of Dublin); with notices of the most remarkable fossils:* Geol. Soc. Dublin, Jour., v. 8 (1857-60), p. 68-72, pl. 6, 7.

1859 (?), *On Haughtonia (Kinahan), a new genus of Cambrian fossil from Bray Head, County of Wicklow:* Same, Jour., v. 8 (1857-1860), p. 116-120, text-fig. 1, 2.

1887, *Oldhamia:* Royal Geol. Soc. Ireland, Jour., v. 17 (n. ser. 7, pt. 2) (1885-87), p. 166-170.

Kindelan, V.

1919, *Nota sobre el Cretaceo y el Eoceno de Guipuzcoa:* Inst. Geol. Minero España, Bol., ser. 2, v. 20, p. 163-198, 25 pl.

Kindle, E. M.

1914, *An inquiry into the origin of "Batrachioides the Antiquor" of the Lockport Dolomite of New York:* Geol. Mag., ser. 6, v. 1, p. 158-161, pl. 8, 9.

King, A. F.

1965, *Xiphosurid trails from the Upper Carboniferous of Bude, north Cornwall:* Geol. Soc. London, Proc., no. 4626, p. 162-165, 1 text-fig.

King, R. H.

1955, See under Harrington & Moore, 1955.

King, Wm.

1850, *A monograph of the Permian fossils of England:* Palaeontograph. Soc., Mon., v. 38, 258 p., 28 pl.

Kirtley, D. W., & Tanner, W. F.

1968, *Sabellariid worms: Builders of a major reef type:* Jour. Sed. Petrology, v. 38, p. 73-78.

Klähn, Hans

1932, *Erhaltungsfähige senkrechte Gänge im Dünensand und die Scolithus-Frage:* Zeitschr. Geschiebeforschg., v. 8, 1-18.

Klöden, K. F.

1828, *Beiträge zur mineralogischen und geognostischen Kenntniss der Mark Brandenburg. 1. Stück:* Progr. z. Prüfg. d. Zöglinge d. Gewerbeschule Ostern 1828, p. 1-82, W. Dieterici (Berlin).

1834, *Die Versteinerungen der Mark Brandenburg, insonderheit diejenigen, welche sich in den Rollsteinen und Blöcken der südbaltischen Ebene finden:* 378 p., 10 pl., Lüderitz (Berlin).

Knight, J. B., Batten, R. L., Yochelson, E. L., & Cox, L. R.

1960, *Supplement. Paleozoic and some Mesozoic Caenogastropoda and Opisthobranchia:* in Treatise on invertebrate paleontology, R. C. Moore (ed.), Part I, p. I310-I331, text-fig. 206-216, Geol. Soc. America and Univ. Kansas Press (New York; Lawrence, Kans.).

Knox, R. W. O'B.

1973, *Ichnogenus Corophiodes:* Lethaia, v. 6, p. 133-146, text-fig. 1-7.

Kobayashi, Teiichi

1945, *Notakulites toyomensis, a new trail found in the Upper Permian Toyoma Series in Nippon:* Japan. Jour. Geology, Geography, v. 20, p. 13-18, pl. 2.

Kochansky, V., & Herak, Mílan

1960, *On the Carboniferous and Permian Dasycladaceae of Yugoslavia:* Geol. Vjesnik (Zagreb), v. 13, p. 65-94, pl. 1-9. (English; summary in Serbian.)

Kochansky-Devidé, Vanda

1958, *Die Neoschwagerinenfauna der südlichen Crna Gora (Jugoslavien):* Geol. Vjesnik (Zagreb), v. 11, p. 45-76, pl. 1-6.

———, **& Ramovš, A.**

1955, *Neoschwagerinski skladi in njih fuzulinidna favna pri Bohinjski Beli v Bledu—Die Neoschwagerinenschichten und ihre Fusulinidenfauna bei Bohinjska Bela und Bled (Julische Alpen, Slowenien, NW-Jugoslawien):* Slovensk. Akad. Znan. in Umet.

(Ljubljana), Razprave, ser. 4, v. 3, p. 361-424, pl. 1-7.

Kolbe, H. J.

1888, *Zur Kenntnis von Insektenbohrgängen in fossilen Hölzern:* Deutsch. Geol. Gesell., Zeitschr., v. 40, p. 131-137, pl. 11.

Kolesch, Karl

1921, *Beitrag zur Stratigraphie des mittleren Buntsandsteins im Gebiete des Blattes Kahla (S.-A.):* Preuss. Geol. Landesanst., Jahrb., v. 40, pt. II (1919), p. 307-382, text-fig. 1-15 [in no. 2 of pt. II: appeared 1921].

Konishi, Kenji

1958, *Devonian calcareous algae from Alberta, Canada, pt. 2 of Studies of Devonian algae:* Colorado School Mines, Quart., v. 53, no. 2, p. 85-109, 4 pl.

1959, *Identity of algal Tubiphytes Maslov, 1956, and hydrozoan genus Nigriporella Rigby, 1958:* Palaeont. Soc. Japan, Trans. & Proc., n. ser., no. 35, p. 142.

Korde, K. B.

1959, *Problematicheskie ostatki iz Kembriyskikh otlozheniy Yugo-vostoka Sibirskoy Platformy:* Akad. Nauk SSSR, Doklady, v. 125, no. 3, p. 625-627, pl. 1. [*Problematic fossils from the Cambrian deposits of the southeast of the Siberian Platform.*]

Korn, Hermann

1929, *Fossile Gasblasenbahnen aus dem Thüringer Palaeozoikum. Eine neue Deutung von Dictyodora:* Zeitschr. Naturwiss., v. 89, p. 25-46, text-fig. 1-3.

Kozłowski, Roman

1959, *Un microfossile énigmatique:* Acta Palaeont. Polonica, v. 4, p. 273-277, text-fig. 1, 2.

1965, *Oeufs fossiles des céphalopodes?:* Same, v. 10, p. 3-9, text-fig. 1, 2.

Kräusel, Richard, & Weyland, Hermann

1932, *Pflanzenreste aus dem Devon. II:* Senckenbergiana, v. 14, p. 185-190.

1934, *Lennea schmidti, eine pflanzenähnliche Tierspur aus dem Devon:* Paläont. Zeitschr., v. 16, p. 95-102, text-fig. 1-3, pl. 7-11.

Kraus, E.

1930, *Über rhizocorallide Bauten im ostbaltischen Devon:* Naturforscher-Vereins zu Riga, Korrespondenzblatt, v. 60, p. 171-185, 1 text-fig.

Kraus, Olaf, & Ott, Ernst

1968, *Eine ladinische Riff-Fauna im Dobratsch-Gipfelkalk (Kärnten, Österreich) und Bemerkungen zum Faziesvergleich von Nordalpen und Drauzug:* Bayer. Staatssamml. Paläontologie, Hist. Geologie, Mitteil., p. 263-290, text-fig. 1-3, pl. 17-20.

Krejci, Karl

1924, *Über Corycium-ähnliche Bildungen im rumänischen Salzgebirge:* Centralbl. Mineralogie, Geologie, Paläontologie, 1924, p. 59-60, text-fig. 1.

1925, *Über Corycium und tektonisch entstandene ähnliche Gebilde:* Same, 1925, B, p. 315-320, 2 text-fig.

Krejci-Graf, Karl

1932, *Definition der Begriffe Marken, Spuren, Fährten, Bauten, Hieroglyphen und Fucoiden:* Senckenbergiana, v. 14, p. 19-39.

1937, *Über Fährten und Bauten tropischer Krabben:* Geologie der Meere und Binnengewässer, v. 1, p. 177-182, text-fig. 1-5.

Krestew, Krestoe

1928, *Über das Carbon des Iskur-Défilés in Bulgarien und seine Altersstellung:* Preuss. Geol. Landesanst., Jahrb., v. 49, pt. 1 (1928), p. 551-579, text-fig. 1-7, pl. 37-39.

Książkiewicz, Marian

1954, *Uwarstwienie frakcjonalne i laminowane we fliszu karpackim:* Polsk. Towarzst. Geol., Rocznik (Soc. Géol. Pologne, Ann.), v. 22 (1952), p. 399-449. [*Graded and laminated bedding in the Carpathian flysch.*]

1958, *Stratigrafia serii magurskiej w Beskidzie Średnim:* Pánst. Instyt. Geol., Biulet., v. 135, p. 43-96, text-fig. 1-7, 5 pl. [*Stratigraphy of the Magura series in the Średni Beskid (Carpathians).*]

1960, *O niektórych problematykach z fliszu Karpat Polskich. Część I:* Kwartal. Geol., v. 4, no. 3, p. 735-747, text-fig. 1, pl. 1-4. [*On some problematic organic traces from the flysch of the Polish Carpathians. Part 1.*]

1961, *O niektórych problematykach z fliszu Karpat Polskich. Część II:* Same, v. 5, p. 882-890, pl. 1, 2. [*On some problematic organic traces from the flysch of the Polish Carpathians. Part 2.*]

1968, *O niektórych problematykach z fliszu Karpat Polskich (Część III):* Polsk. Towarzyst. Geol. (Ann. Soc. Géol. Pologne), Rocznik, v. 38, no. 1, p. 3-17, text-fig. 1-6, 6 pl. [*On some problematic organic traces from the flysch of the Polish Carpathians (Part III).*]

1970, *Observations on the ichnofauna of the Polish Carpathians:* in Trace fossils, T. P. Crimes & J. C. Harper (eds.), Geol. Jour., spec. issue no. 3, p. 283-322, text-fig. 1-8, pl. 1-4, table 1, Seel House Press (Liverpool).

Kuenen, Ph. H.

1957, *Sole markings of graded graywacke beds:* Jour. Geology, v. 65, p. 231-258.

Kuhn, Oskar

1937, *Neue Lebensspuren von Würmern aus der deutschen Obertrias:* Gesell. Naturforsch. Freunde Berlin, Sitzungsber., 1937, p. 363-373, text-fig. 1-5.

1952, *Eine neue Perlkettenfährte aus dem Lias Oberfrankens:* Neues Jahrb. Geologie, Paläontologie, Monatsh., 1952, p. 224-229, text-fig. 1, 2.

1958, *Die Fährten der vorzeitlichen Amphibien und Reptilien:* 64 p., 13 pl., Meisenbach (Bamberg).

1966, *Die Tierwelt des Solnhofener Schiefers:* 40 p., 144 text-fig., Z. Zeimsen (Wittenberg Lutherstadt).

Kummel, Bernhard, & Teichert, Curt

1970, *Stratigraphy and paleontology of the Permian-Triassic boundary beds, Salt Range and Trans-Indus Ranges, West Pakistan:* in Stratigraphic boundary problems, Permian and Triassic of West Pakistan, Bernhard Kummel & Curt Teichert (eds.), Geology Dept. Univ. Kansas, Spec. Publ. 4, 110 p., 38 text-fig.

Kurr, J. G.

1845, *Beiträge zur fossilen Flora der Juraformation Württembergs:* 21 p., 3 pl., Guttenbergsche Buchdr. (Stuttgart).

Lamouroux, J. V. F.

1816, *Histoire des polypiers coralligènes flexibles, vulgairement nommés zoophytes:* 560 p., F. Poisson (Caen).

Lange, F. W.

1942, *Restos vermiformes do "Arenito das Furnas":* Arquivos Museu Paranaense, v. 2, p. 3-8, 1 pl.

Lange, Werner

1932, *Über spirale Wohngänge, Lapispira bispiralis n.g. et n. sp., ein Leitfossil aus der Schlotheimien-Stufe des Lias Norddeutschlands:* Deutsch. Geol. Gesell., Zeitschr., v. 84, p. 537-543, 2 pl.

Lannerbro, Ragnar

1954, *Description of some structures, possibly fossils, in Jotnian sandstone from Mångsbodarna in Dalecarlia:* Geol. Fören. Stockholm, Förhandl., v. 76, no. 1, p. 46-50, text-fig. 1-7.

Laporte, L. F.

1969, *Paleoecology: fossils and their environments:* Jour. Geol. Education, v. 17, p. 75-80, text-fig. 1-5.

Lapparent, Jacques de

1924, *Les calcaires à globigérines du Crétacé supérieur et des couches de passage à l'Éocène dans les Pyrénées occidentales:* Soc. Géol. France, Bull., sér. 4, v. 24, p. 615-641, pl. 20.

1931, *Sur les prétendus "embryons de Lagena":*

Same, Comptes Rendus somm. seánc., sér. 5, v. 1, p. 222-223.

Laubenfels, M. W. de

1955, *Porifera:* in Treatise on invertebrate paleontology, R. C. Moore (ed.), Part E, p. E21-E112, text-fig. 14-89, Geol. Soc. America & Univ. Kans. Press (New York; Lawrence, Kans.).

Lauerma, Raimo, & Piispanen, Risto

1967, *Worm-shaped casts in Precambrian quartzite from Kunsamo, northeastern Finland:* Commiss. Géol. Finlande, Bull., v. 229 (C. R. Soc. Géol. Finlande, v. 39), p. 189-197, text-fig. 1-7.

Laughton, A. S.

1957, *A new deep-sea underwater camera:* Deepsea Research, v. 4 (1956-57), p. 120-125, text-fig. 1-14.

1959, *Die Photographie des Meeresbodens:* Endeavour (German edit.), v. 18, p. 178-185, text-fig. 1-17.

Lebesconte, Paul

1883a, *Oeuvres posthumes de Marie Rouault, . . . publiées par les soins de P. Lebesconte, suivies de: Les Cruziana et Rysophycus, connus sous le nom général Bilobites, sont-ils des végétaux ou des traces d'animaux?:* 73 p., 22 pl., Savy (Rennes-Paris).

1883b, *Présentation à la société des oeuvres posthumes de Marie Rouault par P. Lebesconte, suivies d'une note sur les Cruziana et Rysophycus:* Soc. Géol. France, Bull., sér. 3, v. 11, p. 466-472.

1887, *Constitution générale du Massif breton comparée à celle du Finisterre:* Same, sér. 3, v. 14(1886), p. 776-820, pl. 34-36.

1891, *Les Poudingues rouges de Montfort:* Revue Sci. Natur. de l'Ouest, 1891, no. 3, p. 1-8, text-fig. 1-9.

LeCalvez, Yolande

1959, *Étude de quelques Foraminifères nouveaux du Cuisien franco-belge:* Rev. Micropaléontologie, v. 2, no. 2, p. 88-94, pl. 1.

Lee, I. S.

1939, *The geology of China:* 528 p., 93 text-fig., Thos. Murby & Co. (London).

Lehner, Leonhard

1937, *Fauna und Flora der fränkischen albüberdeckenden Kreide. II. Fauna:* Palaeontographica, ser. A, v. 87, p. 158-230, 4 pl.

Lemche, Henning

1973, *Comments on the application concerning trace fossils. Z. N. (S.) 1973:* Bull. Zool. Nomenclature, v. 30, pt. 2, p. 70.

Lemoine, Marie

1960, *Comparaison de Distichoplax biserialis et*

des Rhabdopleura fossiles et actuels: Rev. Micropaléontologie, v. 3, no. 2, p. 95-102.

Leriche, Maurice

1931, *Les vestiges du Panisélien rejetés sur la côte flamande, le prolongement, sous la mer du Nord, les assises tertiaires de la Flandre:* Soc. Géol. Nord, Ann., v. 56, p. 254-262, text-fig. 1, 2.

Lesquereux, Leo

1869, *On Fucoides in the Coal Formations:* Am. Philos. Soc., Trans., n.ser., v. 13, p. 313-328, pl. 7.

1873, *Lignitic formation and fossil flora:* U. S. Geol. Survey Territories, Ann. Rept., v. 6, p. 317-427.

1876, *Species of fossil marine plants from the Carboniferous measures:* Indiana, Geol. Survey, Ann. Rept., v. 7 (1875), p. 134-145, 2 pl.

1878, *Land plants, recently discovered in the Silurian rocks of the United States:* Am. Philos. Soc., Proc., v. 17, p. 169-173, text-fig. 1-8, pl. 1.

1880-84, *Description of the coal flora of the Carboniferous Formation in Pennsylvania and throughout the United States:* Pennsylvania Geol. Survey, 2nd Rept. Progr., v. 1-3, 977 p., illus. (Atlas, 1879).

1883, *Principles of Paleozoic botany and the fauna of the Coal Measures:* Indiana Dept. Geol. Nat. History, 13th Ann. Rept., pt. 2, 188 p., pl. 1-39.

1887, *On the character and distribution of Paleozoic plants:* Pennsylvania Geol. Survey, Ann. Rept. for 1886, pt. 1, p. 452-522.

Lessertisseur, Jacques

1955, *Traces fossiles d'activité animale et leur significance paléobiologique:* Soc. Géol. France, Mém. n.sér., v. 74, p. 1-150, text-fig. 1-68, pl. 1-11.

Leuchs, Kurt

1928, *Beiträge zur Lithogenesis kalkalpiner Sedimente:* Neues Jahrb. Mineralogie, Geologie, Paläontologie, v. 59, p. 357-430, pl. 25-26.

Leutze, W. P.

1958, *Eurypterids from the Silurian Tymochtee Dolomite of Ohio:* Jour. Paleontology, v. 32, p. 937-942, pl. 122.

Lewarne, G. C.

1964, *Starfish traces from the Namurian of County Clare, Ireland:* Palaeontology, v. 7, pt. 3, p. 508-513, text-fig. 1, 2.

Lewis, D. W.

1970, *The New Zealand Zoophycos:* New Zealand Jour. Geology, Geophysics, v. 13, p. 295-315, text-fig. 1-10.

Leymerie, M. A.

1842, *Suite du mémoire sur le terrain Crétacé du Département de l'Aube:* Soc. Géol. France, Mém., v. 5, pt. 1, p. 1-34, pl. 1-13.

Lima, W. Del

1895, *Notice sur une alge paléozoique:* Direçao Trabalh. Geolog. Portugal, Commun., v. 3, p. 92-96, pl. 1-4.

Linck, Otto

1942, *Die Spur Isopodichnus:* Senckenbergiana, v. 25, p. 232-255, text-fig. 1-10.

1943, *Die Buntsandstein-Kleinfährten von Nagold. (Limuludichnulus nagoldensis n.g. n.sp., Merostomichnites triassicus n.sp.):* Neues Jahrb. Mineralogie, Geologie, Paläontologie, Monatsh., Abt. B, 1943, p. 9-27, text-fig. 1, 2 pl.

1949a, *Fossile Bohrgänge (Anobichnium simile n.g.n.sp.) an einem Keuperholz:* Same, 1949, ser. B, p. 180-185, text-fig. 1, 2.

1949b, *Lebens-Spuren aus dem Schilfsandstein (Mittl. Keuper km 2) NW-Württembergs und ihre Bedeutung für die Bildungsgeschichte der Stufe:* Verein Vaterl. Naturkd. Württemberg, Jahresh., v. 97-101, p. 1-100, text-fig. 1-5, 8 pl.

1954, *Schwänzel-Gruben von Kaulquappen:* Aus der Heimat, v. 62, pt. 1, p. 15-16, text-fig. 1, 2.

1956, *Drift-Marken von Schachtelhalm-Gewächsen aus dem Mittleren Keuper (Trias):* Senckenbergiana Lethaea, v. 37, p. 39-51, text-fig. 1-3, 2 pl.

1961, *Leben-Spuren niederer Tiere (Evertebraten) aus dem württembergischen Stubensandstein (Trias. Mittl. Keuper 4) verglichen mit anderen Ichnocoenosen des Keupers:* Stuttgarter Beitr. Naturkd., no. 66, 20 p., 5 pl.

Lindenberg, H. S.

1965, *Problematica aus dem inneralpinen Tertiär Pseudarcella Spandel, emend. und Bicornifera n.g.:* Neues Jahrb. Geologie, Paläontologie, Monatsh., 1965, p. 18-29, text-fig. 1-6.

Lindström, Maurits

1973, *Clay rolls as pseudofossils:* Geotimes, v. 18, no. 5, p. 11-12.

Linnarsson, J. G. O.

1869, *On some fossils found in the Eophyton sandstone at Lugnås in Sweden:* Geol. Mag., v. 6, pt. 1, p. 393-406, pl. 11-13. (Translated from: Öfvers. Kongl. Vetensk. Akad. Förhandl., 1869, p. 1-16, pl. 7-9).

1871, *Geognostiska och palaeontologiska iakttagelser öfver Eophytonsandstenen i Vestergötland:* K. Svenska Vetenskapsakad., Handl., v. 9, no. 7, 19 p., 5 pl.

Linné, Carl [Linnaeus, Carolus]

1766-68, *Systema naturae per regna tria naturae, secundum classes, ordines, genera, species, cum characteribus, differentiis synonymis, locis . . . :* edit. 12, v. 1, pt. 1(1766), 532 p.; pt. 2(1767), p. 533-1328; v. 2(1767), 736 p., 142 p. (list of genera and species); v. 3(1768), 235 p., 3 pl., Laurentii Salvii (Holmiae).

Lisson, C. I.

1904, *Los Tigillites del Salto del Fraile y algunes Sonneratia del Morro Solar:* Cuerpo Ingen. Minas del Perú, Bol., no. 17, 64 p., text-fig. 1-38.

Llueca, G. F.

1927, *Noticia sobre el hallazgo del la Lorenzinia apenninica da Gabelli en el Eoceno de Guipuzcoa:* Soc. Española Historia Nat., Bol., v. 27, p. 46-56.

Loeblich, A. R., Jr., & Tappan, Helen

1961, *Suprageneric classification of the Rhizopodea:* Jour. Paleontology, v. 35, p. 245-330.

1964, *Protista:* in Treatise on invertebrate paleontology, R. C. Moore (ed.), Part C, v. 1-2, 900 p., text-fig., Geol. Soc. America and Univ. Kansas Press (New York; Lawrence, Kans.).

1968, *Annotated index to genera, subgenera and suprageneric taxa of the ciliate order Tintinnida:* Jour. Protozoology, v. 15, p. 185-192.

Lörcher, Ernst

1931, *Eine neue fossile Qualle aus den Opalinusschichten und ihre paläogeographische Bedeutung:* Oberrhein. Geol. Verein, Jahresber. u. Mitteil., n. ser., v. 20, p. 44-46, text-fig. 1-3, pl. 1.

Logan, W. E.

1860, *On the track of an animal lately found in the Potsdam Formation:* Canad. Naturalist and Geologist, v. 5, no. 4, p. 279-285.

Lombard, Augustin

1938, *Microfossiles d'attribution incertaine du Jurassique supérieur alpin:* Eclogae Geol. Helvetiae, v. 30 (1937), p. 320-331.

1945, *Attribution de microfossiles du Jurassique supérieur alpin à des Chlorophycées. (Proto- et Pleurococcacées):* Same, v. 38, no. 1, p. 163-173.

Lomnicki, A. M.

1886, *Slodkowodny utwor trzeceorzedny na Podulu galicyjskiem:* Akad. Umiejet Krakow, Kom. Fizyogr., v. 20, no. 2, p. 48-119, pl. 1-3. [*The Tertiary freshwater formations in Galician Podolia.*]

Lorenz, Theodor

1903, *Geologische Studien im Grenzgebiete zwischen helvetischer und ostalpiner Fazies. II. Der südliche Rhaetikon:* Naturforsch. Gesell. Freiburg i. Br., Berichte, v. 12, p. 34-95, text-fig. 1-19, pl. 1-9.

Lorenz von Liburnau, J. R.

1897, *Eine fossile Halimeda aus dem Flysch von Muntigl (monticulus) bei Salzburg:* Akad. Wiss. Wien, math.-nat. Kl., Sitzungsber., v. 106, Part 1, p. 174-177, 2 pl.

1900, *Zur Deutung der fossilen Fucoiden-Gattungen Taenidium und Gyrophyllites:* Same, Denkschr., v. 70, p. 523-583, text-fig. 1-21, 4 pl.

1902, *Ergänzung zur Beschreibung der fossilen Halimeda fuggeri:* Same, Sitzungsber., v. 3, pt. 1, p. 685-712, pl. 1, 2.

Loring, A. D., & Wang, K. K.

1971, *Re-evaluation of some Devonian lebensspuren:* Geol. Soc. America, Bull., v. 82, p. 1103-1106.

Love, L. G.

1958, *Micro-organisms and the presence of syngenetic pyrite (with discussion):* Geol. Soc. London, Quart. Jour., v. 113, pt. 4, no. 452, p. 429-440, pl. 33.

Lucas, Gabriel, & Rech-Frollo, Marguerite

1965, *"Traces en rosette" du flysch éocène de Jaca (Aragon), Essai d'interprétation:* Soc. Géol. France, Bull., sér. 7, v. 6 (1964), p. 163-170, text-fig. 1, 1 pl.

Ludwig, Rudolf

1869, *Fossile Pflanzenreste aus den paläolithischen Formationen der Umgebung von Dillenburg, Biedenkopf und Friedberg und aus dem Saalfeldischen:* Palaeontographica, v. 17, p. 105-128, pl. 18-28.

Lugn, A. L.

1941, *The origin of Daemonelix:* Jour. Geology, v. 49, p. 673-696, text-fig. 1-6.

Lull, R. S.

1915, *Triassic life of the Connecticut valley:* State Connecticut Geol. Nat. History Survey, Bull., v. 24, 285 p., text-fig. 1-126, 12 pl.

1953, *Triassic life of the Connecticut valley:* Same, Bull., v. 81, 331 p., text-fig. 1-168, 12 pl.

Lundgren, S. A. B.

1891, *Studier öfver fossilförande lösa block:* Geol. Fören. Stockholm, Förhandl., v. 13, p. 111-121, text-fig. 1, 2.

Maas, Otto

1902, *Über Medusen aus dem Solenhofer Schiefer und der unteren Kreide der Karpathen:* Palaeontographica, v. 48, p. 297-322, pl. 22, 23.

Macarovici, Neculai

1969, *Observations sur la présence de certains lamellibranches lithophages fossiles du Miocène dans le Sud-Est de l'Europe et dans la Mer Noire:* Am. Zoologist, v. 9, p. 721-724, text-fig. 1.

McCoy, Frederick

1850, *On some genera and species of Silurian Radiata in the collection of the University of Cambridge:* Ann. Mag. Nat. History, ser. 2, v. 6, p. 270-290.

1851, *On some new Protozoic Annulata:* Same, ser. 2, v. 7, p. 394-396.

1851-55, *A systematic description of the British Palaeozoic fossils in the Geological Museum of the University of Cambridge:* in A. Sedgwick, A synopsis of the classification of the British Paleozoic rocks: 661 p., 25 pl., J. W. Parker (London, Cambridge). [p. 1-184 (1851); p. 185-406 (1852); p. 407-661 (1855)].

MacCulloch, John

1814, *Remarks on several parts of Scotland which exhibit quartz rocks, and on the nature and connexions of this rock in general:* Geol. Soc. London, Trans., v. 2, p. 450-487.

McGugan, Alan

1963, *Problematical "Zoophycos" from the Permian of Western Canada:* Ann, Mag. Nat. History, ser. 13, v. 6, p. 107-112.

McKee, E. D.

1947, *Experiments on the development of tracks in fine cross-bedded sand:* Jour. Sed. Petrology, v. 17, p. 23-28, pl. 1, 2.

MacKenzie, D. B.

1971, *Post-Lytle Dakota Group on west flank of Denver Basin, Colorado:* Mountain Geologist, v. 8, p. 91-131, text-fig. 1-10, 4 pl.

1972, *Tidal sand flat deposits in Lower Cretaceous Dakota Group near Denver, Colorado:* Same, v. 9, p. 269-277, text-fig. 1-8.

Mackinnon, D. J., & Biernat, Gertruda

1970, *The probable affinities of the trace fossil Diorygma atrypophilia:* Lethaia, v. 3, p. 163-172, text-fig. 1-7.

McLachlan, I. R.

1973, *Problematic microfossils from the Lower Karoo Beds in South Africa:* Palaeontologia Africana, v. 15, p. 1-21, text-fig. 1-3, plates (fig. 4-74).

MacLeay, W. S.

1839, *Note on the Annelida:* in R. I. Murchison, The Silurian System, pt. II, p. 699-701, J. Murray (London).

Macsotay, Oliver

1967, *Huellas problematicas y su valor paleo-*

ecologico en Venezuela: GEOS (Venezuela), v. 16, p. 7-79, pl. 1-18, 1 map.

Madsen, F. F., & Wolff, T.

1965, *Evidence of the occurrence of Ascothoracica (parasitic cirripeds) in Upper Cretaceous:* Dansk Geol. Foren., Medd., v. 15, p. 556-558, text-fig. 1, 1 pl.

Mägdefrau, Karl

1932, *Über einige Bohrgänge aus dem unteren Muschelkalk von Jena:* Paläont. Zeitschr., v. 14, p. 150-160, text-fig. 1-4, pl. 5.

1934, *Über Phycodes circinatum Reinh. Richter aus dem thüringischen Ordovicium:* Neues Jahrb. Mineralogie, Geologie, Paläontologie, Beil.-Bd. 72, B, p. 259-282, text-fig. 1-6, pl. 10, 11.

1937, *Lebensspuren fossiler "Bohr"-Organismen:* Beiträge Naturkdl. Forschg. Südwestdeutschl., v. 2, p. 54-67, 2 pl.

1941, *Review of R. Hundt: Das Mitteldeutsche Phycodes-Meer (Jena 1941):* Neues Jahrb. Mineralogie, Geologie, Paläontologie, Referate, 1941, v. 3, p. 525-527.

Maher, S. W.

1962, *Primary structures produced by tadpoles:* Jour. Sed. Petrology, v. 32, p. 138-139, text-fig. 1, 2.

Maillard, Gustave

1887, *Considérations sur les fossiles décrits comme Algues:* Soc. Paléont. Suisse, Mém., v. 14, 40 p., 5 pl.

Malaroda, Roberto

1947, *Segnalazione di nuove impronte nelle arenarie del flysch eocenico della conca di Trieste:* Museo Civico Storia Nat. Trieste, Atti, v. 16, no. 5, p. 57-64, 2 pl.

Malmgren, A. J.

1867, *Spetsbergens, Grönlands, Islands och den skandinaviska halföns hittils kända Annulata Polychaeta:* 127 p., 14 pl., J. C. Frenckell & son (Helsingfors).

Malz, Heinz

1964, *Kouphichnium walchi, die Geschichte einer Fährte und ihres Tieres:* Natur. u. Museum, v. 94, p. 81-97, text-fig. 1-15.

1968, *Climactichnites—die Kriechspur eines noch unbekannten kambrischen Tieres:* Same, v. 98, p. 369-373, text-fig. 1-5.

Mansfield, W. C.

1927, *Some peculiar fossils from Maryland:* U. S. Natl. Museum, Proc., v. 71, art. 16, p. 1-9, 5 pl.

1930, *Some peculiar fossil forms from California and Mexico:* Same, v. 77, art. 13, p. 1-3, 2 pl.

Marcinowski, Ryszard

1972, [not seen by editors].

Marck, W. van der

1863, *Fossile Fische, Krebse und Pflanzen aus dem Plattenkalk der jüngsten Kreide in Westphalen:* Palaeontographica, v. 11, p. 1-83, pl. 1-14.

1873, *Neue Beiträge zur Kenntnis der fossilen Fische und anderer Thierreste der jüngsten Kreide Westfalens:* Same, v. 22, p. 55-74, pl. 1, 2.

1894, *Dreginozoum nereitiforme, ein vergessenes Fossil der oberen Kreide Westfalens von Dolberg bei Hamm:* Naturh. Ver. Preuss. Rheinl. Westf., Verhandl., v. 51, p. 1-9, pl. 1.

Marple, M. F.

1956, *On the fossil Conostichus:* Ohio Jour. Sci., v. 56, p. 29-30.

Marsh, O. C.

1868, *On the Palaeotrochis of Emmons from North Carolina:* Am. Jour. Sci., ser. 2, v. 45, p. 217-219, text-fig. 1.

1896, *Amphibian footprints from the Devonian:* Same, ser. 4, v. 2, p. 374-375.

Martini, Erland, & Mentzel, Rolf

1971, *Lebensspuren und Nannoplankton aus dem Alzeyer Meeressand (Mittel-Oligozän):* Hess. Landesamt Bodenforsch., Notizbl., v. 99, p. 54-61, pl. 6, 7.

Martinsson, Anders

1965, *Aspects of a Middle Cambrian thanatotope on Öland:* Geol. Fören. Stockholm, Förhandl., v. 87, p. 181-230, text-fig. 1-35.

1970, *Toponomy of trace fossils:* in Trace Fossils, T. P. Crimes & J. C. Harper (eds.), Geol. Jour., spec. issue no 3, p. 323-330, text-fig. 1, 2, Seel House Press (Liverpool).

1972, [Discussion] in Walter Häntzschel & O. Kraus, Names based on trace fossils (ichnotaxa): request for a recommendation. Z. N. (S.) 1973, Bull. Zool. Nomenclature, v. 29, p. 140.

Maslov, V. P.

1947, *Geologiya verkhovev rek Leny i Kirengi:* Akad. Nauk SSSR, Trudy, Inst. Geol., no. 85, geol. ser. no. 24, p. 1-64, illus. [*The geology of the headwaters of the Lena and Kirenga river region.*]

1956, *Iskopaemye izvestkovie vodorosli SSSR:* Same, Inst. Geol., Trudy, v. 160, p. 3-301, text-fig. 1-136, pl. 1-86. [*Fossil calcareous algae from the U.S.S.R.*]

Massalongo, Abramo

1851, *Sopra le piante fossili dei terreni terziari del Vicentino:* 263 p., A. Bianchi (Padova).

1855a, *Monografia delle nereidi fossili del M. Bolca:* 35 p., 6 pl., G. Antonelli (Verona).

1855b, *Zoophycos, novum genus plantorum fossilium:* 52 p., 3 pl., Antonelli (Verona).

1856, *Studii Palaeontologici:* 53 p., 7 pl., Antonelli (Verona).

1859, *Syllabus plantarum fossilium hucusque in formationibus tertiariis agri veneti detectarum:* 179 p., A. Merlo (Veronae).

———, & Scarabelli, Guiseppe

1859, *Studii sulla flora fossile e geologia stratigraphica delle Senigalliese:* 504 p., 44 pl., Galeati e figlio (Imola).

Mathieu, Gilbert

1949, *Contribution à l'étude des Monts Troglodytes dans l'extrème Sud-Tunisien. Géologie régionale des environs de Matmata Medenine et Foum-Tatahouine:* Rég. Tunis, Dir. Trav. Publ., Ann. mines et géologie, v. 4, 82 p., text-fig. 1-11, 3 pl.

Matisto, Arvo

1963, *Über den Ursprung des Kohlenstoffs in Corycium:* Neues Jahrb. Geologie, Paläontologie, Monatsh., 1963, p. 433-441.

Matthew, G. F.

1888, *On Psammichnites and the early trilobites of the Cambrian rocks in eastern Canada:* Am. Geologist, v. 2, p. 1-9.

1889, *On some remarkable organisms of the Silurian and Devonian rocks of New Brunswick:* Royal Soc. Canada, Proc. Trans., v. 6 (for 1888), sec. 4, p. 49-62, pl. 4.

1890, *On Cambrian organisms in Acadia:* Same, Proc. Trans, v. 7 (1889), sec. 4, p. 135-162, pl. 5-9.

1891, *Illustrations of the fauna of the St. John Group, no. V:* Same, Trans., v. 8 (1890), sec. 4, p. 123-166, pl. 11-16.

1899, *Studies on Cambrian faunas, no. 4. Fragments of the Cambrian faunas of Newfoundland:* Royal Soc. Canada, Proc. & Trans., ser. 2, v. 5, sec. iv, Papers for 1899, p. 97-119, illus.

1901, *Monocraterion and Oldhamia:* The Irish Naturalist, v. 10, p. 135-136.

1903, *On Batrachian and other footprints from the coal measures of Joggings, Nova Scotia:* Nat. History Soc. New Brunswick, Bull., v. 5, no. 21, p. 103-108, 1 pl.

1910, *Remarkable forms of the Little River Group:* Royal Soc. Canada, Proc. & Trans., ser. 3, v. 3 (1909), sec. 4, p. 115-125, 4 pl.

Matyasovszky, Jakab-tól

1878, *Ein fossiler Spongit aus dem Karpathensandsteine von Kis-Lipnik im Sároser Comitate:* Termész. Fuzetek, v. 2, p. 262-266 (Hung.), p. 297-301 (Ger.), pl. 12.

Mayer, Gaston

1952, *Lebensspuren von Bohrorganismen aus dem*

unteren *Hauptmuschelkalk (Trochitenkalk) des Kraichgaues:* Neues Jahrb. Geologie, Paläontologie, Monatsh., 1952, p. 450-456.

1954, *Neue Beobachtungen an Lebensspuren aus dem unteren Hauptmuschelkalk (Trochitenkalk) von Wiesloch:* Same, Abhandl., v. 99, p. 223-229, pl. 14-18.

1955, *Kotpillen als Füllmasse in Hoernesien und weitere Kotpillenvorkommen im Kraichgauer Hauptmuschelkalk:* Same, Monatsh., 1955, p. 531-535, text-fig. 1-6.

1956, *Eine Schichtfläche mit Biocoenosen, Strömungsmarken und Lebensspuren aus dem mittleren Hauptmuschelkalk von Bruchsal:* Beiträge Naturkdl. Forschg. Südwestdeutschl., v. 15, p. 6-10, text-fig. 1, pl. 2, 3.

1958, *Rhizocorallien mit Wandkörperchen:* Der Aufschluss, 1958, p. 314-316, text-fig. 1, 2.

1964, *Noeh einmal: Spongeliomorphe Gebilde aus dem Muschelkalk:* Same, v. 15, p. 107-111.

Mayer, Karl

1878, *Zur Geologie des mittleren Ligurien:* Naturforsch. Gesell. Zürich, Vierteljahrsschr., v. 23, p. 74-94.

Mayr, F. X.

1966, *Sur Frage des "Auftriebes" und der Einbettung bei Fossilien der Solnhofener Schichten:* Geol. Blätter Nordost-Bayern, v. 16, no. 2-3, p. 102-107.

1967, *Paläobiologie und Stratinomie der Plattenkalke der Altmühlalb:* Erlanger Geol. Abh., no. 67, p. 1-40, text-fig. 1-7, pl. 1-16.

Meer Mohr, C. G. van der

1969, *The stratigraphy of the Cambrian Lancara Formation between the Luna River and Esla River in the Cantabrian Mts., Spain:* Leidse Geol. Mededel., v. 43, p. 233-316, text-fig. 1-61.

———, & Okulitch, V. J.

1967, *On the occurrence of Scyphomedusa in the Cambrian of the Cantabrian Mountains (NW Spain):* Geol. & Mijnbouw, v. 46, p. 361-362.

Meneghini, G. G. A.

See Savi, P. & Meneghini, G. G. A.

Mertin, Hans

1941, *Decapode Krebse aus dem subhercynen und Braunschweiger Emscher und Untersenon sowie Bemerkungen über einige verwandte Formen in der Oberkreide:* Nova Acta Leopoldina, v. 10, no. 68, 118 p., 8 pl.

Meschinelli, Luigi, & Squinabol, Senofonte

1892, *Flora tertiaria italica:* 575 p., Seminario (Padova).

Metzger, A. T. T.

1924, *Die jatulischen Bildungen von Suojärvi in*

Ostfinnland: Commiss. Géol. Finlande, Bull., v. 64, 86 p., text-fig. 1-39.

1927, *Zum Problem der präkambrischen Fossilien und Lebensspuren:* Nassau. Verein Naturkd. Wiesbaden, Jahrb., v. 79, p. 1-17.

Meunier, Stanislaus

1886, *Sur quelques empreintes problématiques des couches boloniennes du Pas-de-Calais:* Soc. Géol. France, Bull., sér. 3, v. 14, p. 564-568, pl. 29, 30.

1887, *Radiophyton Sixii:* Le Naturaliste, sér. 2, v. 9, p. 58-59, text-fig. 1.

1891, *Staurophyton bagnolensis Stan. Meunier. Nouveau fossile des grès armoricains de Bagnoles (Orne):* Same, sér. 2, v. 13, p. 134.

Michaelis, Hermann

1972, *Die Lebensspur einer Dipterenlarve im Dünensand:* Natur u. Museum, v. 102, p. 421-424, text-fig. 1-3.

Michelau, Paul

1956, *Belorhaphe kochi (Ludwig 1869), eine Wurmspur im europäischen Karbon:* Geol. Jahrb., v. 71 (1955), p. 299-330, text-fig. 1, 2, pl. 28-31.

Middlemiss, F. A.

1962, *Vermiform burrows and rate of sedimentation in the Lower Greensand:* Geol. Mag., v. 99, p. 33-40, text-fig. 1.

Miller, S. A.

1875, *Some new species of fossils from the Cincinnati group and remarks upon some described forms:* Cincinnati Quart. Jour. Sci., v. 2, no. 4, p. 349-355, text-fig. 1-3.

1877, *The American Paleozoic fossils, a catalogue of the genera and species:* 253 p., the author (Cincinnati, Ohio).

1880, *Silurian ichnolites, with definitions of new genera and species. Note on the habit of some fossil annelids:* Cincinnati Soc. Nat. History, Jour., v. 2, p. 217-229, 2 pl.

1889, *North American geology and palaeontology for the use of amateurs, students and scientists:* 664 p., Western Methodist Book Concern (Cincinnati, Ohio).

———, & Dyer, C. B.

1878a, *Contributions to paleontology, no. 1:* Cincinnati Soc. Nat. History, Jour., v. 1, p. 24-39.

1878b, *Contributions to paleontology, no. 2:* 11 p., pl. 3, 4, privately publ. (Cincinnati, Ohio).

Minato, Masao, & Suyama, Kunio

1949, *Kotfossilien von Arenicola-artigem Organismus aus Hokkaido, Japan:* Japan. Jour. Geology, Geography, v. 21, p. 277-279, pl. 11.

Miroshnikov, L. D.

1959, *Iskopaemye Stsifoidnye meduzy iz Kembriya Sibiri:* Priroda, Akad. Nauk SSSR, no. 11, p. 109-110, 1 text-fig. [*Fossil Scyphomedusae from the Siberian Cambrian.*]

———, & Kravtsov, A. G.

1965, *Pozdnekembriyskie stsifomeduzy sibirskoy platformy:* Vses. Paleont. Obshch., Ezhegodnik, v. 17, p. 46-66, text-fig. 1-20 [*Late Cambrian Scyphomedusae from the Siberian Platform.*]

Misra, R. C., & Dube, S. N.

1952, *A new collection and re-study of the organic remains from the Suket shales (Vindhyans) Rampura Madhya Bharat:* Science and Culture, v. 18, p. 46-48.

Mitzopoulos, Max

1939, *Ein Medusenvorkommen im Eozänflysch des Peleponnes:* Praktika Akad. Athen., v. 14, p. 258-259, 1 pl.

Moberg, J. C.

1892, *Om en nyupptäckt fauna i block af kambrisk sandsten, insamlade af dr N. O. Holst:* Sveriges Geol. Undersök., ser. C, Afhandl. och Upps., no. 125, 18 p., 1 pl.

Moodie, R. L.

1929, *Vertebrate footprints from the red beds of Texas:* Am. Jour. Sci., ser. 5, v. 17, p. 352-368.

Moore, D. G., & Scruton, P. C.

1957, *Minor internal structures of some Recent unconsolidated sediments:* Am. Assoc. Petroleum Geologists, Bull., v. 41, p. 2723-2751, text-fig. 1-16.

Morière, Jules

1879, *Sur les empreintes offertes par les grès siluriens dans le département de l'Orne et connu vulgairement sous le nom de "Pas de Boeuf":* Assoc. Franç. Advanc. Sci., Comptes Rendus, v. 7, Paris, 1878, p. 570-576.

Morningstar, Helen

1922, *Pottsville fauna of Ohio:* Geol. Survey Ohio, Bull., ser. 4, no. 25, 274 p., 16 pl.

Morris, John

1851, *Paleontological notes:* Ann. Mag. Nat. History, ser. 2, v. 8, p. 85-90, pl. 4.

Mortelmans, Georges

1957, *Traces fossiles de vie dans les argilites lukuguiennes de Vuele Nyoka, de Luena-Kisulu et de la Lovoy (Katanga):* Acad. Royal Sci. Colon., Bull. séanc., n. sér. 3, 1957-3, p. 607-627, text-fig. 1-5.

Müller, A. H.

1955a, *"Helminthoide" Lebensspuren aus der Trias*

von Thüringen: Geologie, v. 4, p. 407-415, 2 pl.

1955b, *Über die Lebensspur Isopodichnus aus dem oberen Buntsandstein (unt. Röt) von Göschwitz bei Jena und Abdrücke ihres mutmasslichen Erzeugers:* Same, v. 4, p. 481-489, text-fig. 1, pl. 1, 2.

1955c, *Das erste Benthos (Planolites ? vermiculare n. sp.) aus dem Stinkschiefer Mitteldeutschlands (Zechstein, Stassfurtserie):* Same, v. 4, p. 655-659, text-fig. 1, 2.

1956a, *Über problematische Lebensspuren aus dem Rotliegenden von Thüringen:* Geol. Gesell. DDR, Berichte, v. 1, p. 147-154, text-fig. 1, 2, pl. 5-8.

1956b, *Weitere Beiträge zur Ichnologie, Stratinomie und Ökologie der germanischen Trias:* Geologie, v. 5, p. 405-423, text-fig. 1-3, 5 pl.

1959, *Weitere Beiträge zur Ichnologie, Stratinomie und Ökologie der germanischen Trias. II:* Same, v. 8, p. 239-261, text-fig. 1-5, 7 pl.

1962, *Zur Ichnologie, Taxiologie und Ökologie fossiler Tiere:* Freiberger Forschungsh., C, v. 151, p. 5-49, text-fig. 1-21.

1963, *Lehrbuch der Paläozoologie. II. Invertebraten. Teil 3: Arthropoda 2-Stomachorda, Abschlusskapitel über die Ichnologie der Invertebraten:* xvii + 698 p., 854 text-fig., Verl. Fischer (Jena).

1966, *Neue Lebensspuren (Vestigia invertebratorum) aus dem Karbon und der Trias Mitteldeutschlands:* Geologie, v. 15, p. 712-725, text-fig. 1-5, pl. 1, 2.

1967, *Zur Ichnologie von Perm und Trias in Mitteldeutschland:* Same, v. 9, p. 1061-1071, text-fig. 1-6, 2 pl.

1969a, *Nautiliden-Kiefer (Cephalopoda) mit Resten des Cephalopodiums aus dem Muschelkalk des Germanischen Triasbeckens:* Deutsch. Akad. Wiss. Berlin, Monatsber., v. 11, no. 4, p. 307-316, text-fig. 1-3, 1 pl.

1969b, *Zum Lumbricaria-Problem (Miscellanea), mit einigen Bemerkungen über Saccocoma (Crinoidea, Echinodermata):* Same, Monatsber., v. 11, no. 10, p. 750-758.

1969c, *Über ein neues Ichnogenus (Tambia n. g.) und andere Problematica aus dem Rotliegenden (Unterperm) von Thüringen:* Same, Monatsber., v. 11, no. 11/12, p. 922-931, text-fig. 1-4, pl. 1, 2.

1969d, *Medusenartige Problematica (Miscellanea) und die Frage einer marinen Beeinflussung des tieferen Buntsandsteins:* Geologie, v. 18, no. 4, p. 441-445.

1969e, *Zur Kenntnis von Ophiomorpha (Miscellanea):* Same, v. 18, no. 9, p. 1102-1109, text-fig. 1, 2 pl.

1970a, *Neue Tetrapoden-Fährten aus dem terrestrischen Zechstein:* Deutsch. Akad. Wiss. Berlin, Monatsber., v. 12, no. 2-3, p. 197-207, text-fig. 1, 2, 2 pl.

1970b, *Über Ichnia vom Typ Ophiomorpha und Thalassinoides (Vestigia invertebratorum, Crustacea):* Same, Monatsber., v. 12, no. 10, p. 775-787, text-fig. 1-4, 2 tables.

1970c, *Aktuopaläontologische Beobachtungen an Quallen der Ostsee und des Schwarzen Meeres:* Natur u. Museum, v. 100, no. 7, p. 321-322, text-fig. 1-11.

1971a, *Über Ichnia vom Typ Helicoraphe und Helicodromites aus Gegenwart und geologischer Vergangenheit:* Deutsch. Akad. Wiss. Berlin, Monatsber., v. 13, no. 1, p. 72-79, 2 tables.

1971b, *Über Dictyodora liebeana (Ichnia invertebratorum), ein Beitrag zur Taxiologie und Ökologie sedimentfressender Endobionten:* Same, v. 13, no. 2, p. 136-151, text-fig. 1-11, pl. 1-3.

1971c, *Zur Ichnologie, Ökologie, und Phylogenetik der Tetrapoden des Karbon:* Same, Monatsber., v. 13, no. 7, p. 537-553, text-fig. 1-8, 2 pl.

1971d, *Bioturbation durch Decapoda (Crustacea) in Sandsteinen der sächsischen Oberkreide:* Same, Monatsber., v. 13, no. 9, p. 696-707, text-fig. 1, 2, 4 pl.

1971e, *Zur Kenntnis von Asterosoma (Vestigia invertebratorum):* Freiberger Forschungsh., C, v. 267, p. 7-17, text-fig. 1-4, 3 pl.

1971f, *Miscellanea aus dem limnisch-terrestrischen Unterperm (Rotliegendes) von Mitteleuropa; Teil 1:* Deutsch. Akad. Wiss. Berlin, Monatsber., v. 13, p. 937-948, text-fig. 1, 3 pl.

Müller, Gisela

1968, *Bohr-Röhren von unbekannten Anneliden und anderen Organismen in unterdevonischen Brachiopodenklappen aus der Eifel und dem Siegerland (Rheinisches Schiefergebirge):* 121 p., 35 p., 5 pl., Inaug. Diss. Univ. Köln (Köln).

Müller, Karl

1846, See under Rüppell: in Botan. Zeitung, v. 4, p. 79-83 (review of Rüppell, 1845).

Müller, K. J., & Nogami, Yasuo

1972, *Entöken und Bohrspuren bei den Conodontophorida:* Paläont. Zeitschr., v. 46, p. 68-86, text-fig. 1-11, pl. 14-16.

Murchison, R. I.

1839, *The Silurian system. Part 1. Founded on geological researches in the counties of Solop, Hereford, Radnor, Montgomery, Caermarthen, Brecon, Pembroke, Monmouth, Gloucester, Worcester, and Stafford; with descriptions of the coal-fields and overlying formations:* p. i-xxxiii, 1-578; Part II. *Organic remains,* p. 579-768, pl. 1-37, John Murray (London).

1850, *Memoria sulla struttura geologica delle*

Alpi, delle Apennini e dei Carpazi: 528 p., 2 pl., 2 tables, Stamperia granucale (Firenze).

1859, *On the succession of the older rocks in the northernmost counties of Scotland, with some observations on the Orkney and Shetland Islands*: Geol. Soc. London, Quart. Jour., v. 15, p. 353-418, pl. 12, 13.

1867, *Siluria*: 4th edit., xvii + 566 p., illus., 41 pl., John Murray (London).

Myannil [Männil], R. M.

1966, *O Vertikalnykh norkakh zaryvaniya v Ordovikskikh izvestiyakakh Pribaltiki*: in Organizm i sreda v geologischeskom proshlom, Akad. Nauk SSSR, Paleont. Inst., p. 200-207, text-fig. 1, 2, pl. 1, 2. [*A small vertically excavated cavity in Baltic Ordovician limestone.*]

Myers, A. C.

1970, *Some palaeoichnological observations of tube of Diopatra cuprea (Bosc)*: in Trace fossils, T. P. Crimes & J. C. Harper (eds.), Geol. Jour., spec. issue no. 3, p. 331-334, text-fig. 1, Seel House Press (Liverpool).

Nathorst, A. G.

1873, *Om några förmodade växtfossilier*: Översigt af K. Vetensk.-Akad. Förhandlingar, Stockholm, 1873, no. 9, p. 25-32, pl. 15-19.

1881a, *Om spår af nagra evertebrerade djur m. m. och deras palaeontologiska betydelse. (Mémoire sur quelques traces d'animaux sans vertèbres etc. et de leur portée paléontologique.)*: K. Svenska Vetenskapsakad., Handl., v. 18 (1880), no. 7, 104 p., text-fig. 1-32, 11 pl. (p. 61-104: abridged French transl. of Swedish text).

1881b, *Om aftryck af medusor i Sveriges kambriska lager*: Same, n. ser., v. 19, no. 1, p. 1-34, 6 pl.

1883a, *On the so-called "plant-fossils" from the Silurian of Central Wales*: Geol. Mag., ser. 2, v. 10, p. 33-34.

1883b, *Quelques remarques concernant les algues fossiles*: Soc. Géol. France, Bull., sér. 3, v. 11, p. 452-455.

1886, *Nouvelles observations sur des traces d'animaux et autres phénomènes d'origine purement mécanique décrits comme "Algues fossiles"*: K. Svenska Vetenskapsakad., Handl., v. 21, no. 14, 58 p., text-fig. 1-24, 5 pl.

Nestler, Helmut

1960, *Ein Bohrschwamm aus der weissen Schreibkreide (unt. Maastricht) der Insel Rügen (Ostsee)*: Geologie, v. 9, no. 6, p. 650-655, 1 pl.

Neuburg, M. F.

1934, *Issledovaniya po stratigrafii Uglenosnykh otlozheniy Kuznetskogo Basseina 1930-31 gg.*: Vses. Geol. Razved. Obed. SSSR, Trudy, vyp. 348, p. 1-44, text-fig. 1-2, pl. 1-4, 2 maps. (Eng. summ.). [*Explorations on the stratigraphy of the Carboniferous series of deposits of the Kusnetzk Basin carried out in 1930 and 1931.*]

Neuhaus, A.

1940, *Über die Erzführung des Kupfermergels der Haaseler und der Gröditzer Mulde in Schlesien*: Zeitschr. Angew. Mineralogie, v. 2, p. 304-343.

Neviani, Antonio

1925, *Di una nuova medusa fossile appartenente alle Aequoridae (Craspedotae) rinvenuta nelle argile classiche della Farnesina presso Roma*: Pont. Accad. Nuovi Lincei, Atti, v. 78, p. 148-153, 1 text-fig.

Newall, G.

1970, *A symbiotic relationship between Lingula and the coral Heliolites in the Silurian*: in Trace fossils, T. P. Crimes & J. C. Harper (eds.), Geol. Jour., spec. issue no. 3, p. 335-344, pl. 1, 2, Seel House Press (Liverpool).

Newberry, J. S.

1885, *Saporta's problematical organisms of the ancient seas*: Science, n. ser., v. 5, p. 507-508.

1888, *Fossil fishes and fossil plants of the Triassic rocks of New Jersey and the Connecticut Valley*: U. S. Geol. Survey, Mon., v. 14, 152 p., 26 pl.

1890, *On Dendrophycus triassicus Newb.*: Am. Naturalist, v. 24, p. 1068-1069.

Newman, W. A., Zullo, V. A., & Withers, T. H.

1969, *Cirripedia*: in Treatise on invertebrate paleontology, R. C. Moore (ed.), Part R, p. R206-R295, text-fig. 80-119, Geol. Soc. America & Univ. Kansas (Boulder, Colo.; Lawrence, Kans.).

Nicholson, H. A.

1873, *Contributions to the study of the errant annelides of the older Paleozoic rocks*: Royal Soc. London, Proc., v. 21, p. 288-290 (also Geol. Mag., v. 10, p. 309-310).

——, & Etheridge, Robert (jun.)

1880, *A monograph of the Silurian fossils of the Girvan District in Ayrshire. III. The Annelida and Echinodermata, with supplements on the Protozoa, Coelenterata, and Crustacea*: p. 237-341, pl. 16-24, W. Blackwood & Sons (Edinburgh, London).

——, & Hinde, G. J.

1875, *Notes on the fossils of the Clinton, Niagara, and Guelph formations of Ontario, with descriptions of new species:* Canad. Jour. Sci., Lit. History, n. ser., v. 14, p. 137-160.

Nielsen, Eigil

1949, *On some trails from the Triassic beds of East Greenland:* Meddel. Grønland, v. 149, no. 4, 44 p., text-fig. 1-27.

Niino, Hiroshi

1955, *Sand pipe from the sea floor off California:* Jour. Sed. Petrology, v. 25, p. 41-44.

Nitecki, M. H.

1968, *On the nature of the holotype of Nipterella paradoxica (Billings):* Fieldiana, Geology, v. 16, no. 11, p. 289-295, text-fig. 1-4.

————, & **Solem, Alan**

1973, *A problematic organism from the Mazon Creek (Pennsylvanian) of Illinois:* Jour. Paleontology, v. 47, p. 903-907, text-fig. 1,2, pl. 1,2.

Noel, Denise

1958, *Étude de coccolithes du Jurassique et du Crétacé inférieur:* Publ. Serv. Carte géol. Algérie (n. sér.), Bull., v. 20, p. 155-196, 11 pl.

Nopcsa, F. Baron

1923, *Die Familien der Reptilien:* Fortschr. Geologie, Paläontologie, no. 2, 210 p., 6 pl.

Norman, A. M.

1903, *New generic names for some Entomostraca and Cirripedia:* Ann. Mag. Nat. History, ser. 7, v. 11, p. 367-369.

Nowak, Wiesław

1957, *Kilka hieroglifów gwiaździstych z zewnętrznych Karpat fliszowych (Quelques hiéroglyphes étoilés des Karpates de Flysch extérieures):* Polsk. Towarzyst. Geol., Rocznik (Soc. Géol. Pologne, Ann.), v. 26 (1956), p. 187-224.

1959, *Palaeodictyum w. Karpatach fliszowych:* Kwartal. Geol., v. 3, p. 103-125, text-fig. 1-9, 6 pl. [*Palaeodictyum in the Carpathian flysch.*]

1961, *Z badań nad hieroglifami fliszu karpackiego 1. Niektóre hieroglify z warstw cieszynskich i grodziskich:* Spraw. Pos. Kom. Odd. Pan Krakówie p. 226-?. [*From research on hieroglyphs of the Carpathian flysch 1. Some hieroglyphs from the Cieszyn and Grodzisk beds.*] [Not seen by the editors.]

1970, *Spostrzezenia nad problematykami Belorhaphe i Sinusites z dolnokredowego i paleogeńskiego fliszu Karpat Polskich:* Kwartal. Geol., v. 14, p. 149-163, text-fig. 1-3, pl. 1-2. [*Problematic organic traces of Belorhaphe and Sinusites in the Carpathian Lower Cretaceous and Paleogene flysch deposits of Poland.*]

Oberhauser, Rudolf

1960, *Foraminiferen und Mikrofossilien "incertae sedis" der ladinischen und karnischen Stufe der Trias aus den Ostalpen und aus Persien:* Geol. Bundesanst. Wien, Jahrb., Sonderbd., v. 5, p. 5-46, pl. 1-6.

Obrhel, Jiři

1964, *Ein problematisches Mikrofossil aus dem Devon Böhmens:* Ústřed. Ústavu Geol., Vestnik, v. 39, no. 3, p. 217-218, pl. 1, 2.

Öpik, A. A.

1929, *Studien über das estnische Unterkambrium (Estonium). I-IV.:* Univ. Tartu., Acta Comment., ser. A, v. 15, no. 2, 56 p., text-fig. 1-7, 4 pl.

1933, *Über Scolithus aus Estland:* Same, Acta Comment., ser. A, v. 24, no. 3, 12 p., 2 pl.

1956, *Cambrian (Lower Cambrian) of Estonia:* 20th Internatl. Geol. Congress, El Sistema Cambrio. Symposium, Part I, John Rodgers (ed.), p. 97-126 (Mexico City).

1959, *Tumblagooda sandstone trails and their age:* Australia Bur. Min. Resources, Geology, Geophysics, Rept. 38, p. 3-20, text-fig. 1-19.

Oersted, A. S.

1843, *Annulatorum danicorum conspectus:* pt. 1, 52 p., 7 pl., Walianae (Maricolae-Hafniae).

Ohlson, Birger

1961, *Observations on Recent lake balls and ancient Corycium inclusions in Finland:* Commiss. Géol. Finlande, Bull., v. 96, p. 377-390, illus.

Okulitch, V. J.

1955, *Archaeocyatha:* in Treatise on invertebrate paleontology, R. C. Moore (ed.), Part E, p. E1-E20, text-fig. 1-13, Geol. Soc. America and Univ. Kansas Press (New York; Lawrence, Kans.).

Ooster, W. A.

1869, *Die organischen Reste der Zoophycos-Schichten der Schweizer Alpen:* in W. A. Ooster & C. Fischer-Ooster, Protozoa Helvetica, v. 1, p. 15-35, pl. 3-11.

Opler, P. A.

1973, *Fossil Lepidopteris leaf mines demonstrate the age of some insect-plant relationships:* Science, v. 179, p. 1321-1322, text-fig. 1.

Oppel, Albert

1862, *Ueber Fährten im lithographischen Schiefer (Ichnites lithographicus):* Museum Bayer. Staates, Paläont. Mittheil., v. 1, no. 2, p. 121-125, pl. 39.

d'Orbigny, Alcide

1835-47, *Voyage dans l'Amérique méridionale (le Brésil, la République oriental de l'Uruguay, la République Argentine, la Patagonie, la*

République du Chili, la République de Bolivia, la République du Péron) exécuté pendant les annees 1826, 1827, 1829, 1830, 1831, 1832, et 1833: v. 3, pt. 4 (Paléontologie), 188 p., 22 pl. (1842); atlas for part 8 (1847), Pitois-Levrault (Paris), Levrault (Strasbourg).

1849-52, *Cours élémentaire de paléontologie et de géologie stratigraphiques:* v. 1 (1849 or 1850), p. 1-299; v. 2 (1852), p. 1-382; v. 3 (1852), p. 383-847, V. Masson (Paris).

Orlowski, Stanislae, Radwański, Andrzej, & Roniewicz, Piotr

1970, *The trilobite ichnocoenoses in the Cambrian sequence of the Holy Cross Mountains:* in Trace fossils, T. P. Crimes & J. C. Harper (ed.), Geol. Jour., spec. issue no. 3, p. 345-360, text-fig. 1, 2, pl. 1-4, Seel House Press (Liverpool).

1971, *Ichnospecific variability of the Upper Cambrian Rusophycus from the Holy Cross Mts.:* Acta Geologica Polonica, v. 21, p. 341-348, 6 pl.

Osgood, R. G.

1970, *Trace fossils of the Cincinnati Area:* Palaeont. Americana, v. 6, no. 41, p. 281-444, text-fig. 1-29, pl. 57-83.

────, & Szmuc, E. J.

1972, *The trace fossil Zoophycos as an indicator of water depth:* Bull. Am. Paleontology, v. 62, no. 271, 22 p., text-fig. 1, 2, pl. 1, 2.

Otto, Ernst von

1852, *Additamente zur Flora des Quadergebirges in der Gegend um Dresden und Dippoldiswalde, enthaltend meist noch nicht oder wenig bekannte fossile Pflanzen:* 29 p., 7 pl., Meissen (Dippoldiswalde).

1854, *Additamente zur Flora des Quadergebirges in Sachsen:* part 2, 53 p., 9 pl., G. Mayer (Leipzig).

1855, *Fossile Würmer im Quadersandstein:* Allg. Deutsche Naturhist. Zeitg., n. ser., v. 1, p. 307-312, text-fig. 1-12.

Owen, Richard

1852, *Description of the impressions and footprints of the Protichnites from the Potsdam sandstone of Canada:* Geol. Soc. London, Quart. Jour., v. 8, p. 214-225, pl. 9-14A.

Ozaki, Kimihiko

1968, *Problematical fossils from the Permian limestone of Ahasaka, Gifu Prefecture:* Yokohama Nat'l Univ., Sci. Rept., v. 2, no. 14, p. 27-33, text-fig. 1-5, 3 pl.

Pabst, Wilhelm

1896, *Thierfährten aus dem Oberrothliegenden von Tambach in Thüringen:* Deutsch. Geol. Gesell., Zeitschr., v. 48, p. 638-643, pl. 14.

Packard, A. S.

1898, *A half-century of evolution with special reference to the effects on geological changes of animal life:* Am. Assoc. Advanc. Sci., Proc., 1898, p. 311-356. [text by Matthew, G. F., p. 323].

1900a, *View of the Carboniferous fauna of the Narrangansett Basin:* Am. Acad. Arts Sci., Proc., v. 35, p. 399-405, text-fig. 1.

1900b, *On supposed merostomatous and other Paleozoic arthropod trails, with notes on those of Limulus:* Same, Proc., v. 36, p. 61-71, text-fig. 1-5. (no. 4: 1900).

Pacltová, Blanka

1972, *Palaeocryptidium Deflandre from the Proterozoic of Bohemia:* Časopis pro Mineralogii a Geologii, v. 17, p. 357-364, text-fig. 1, pl. 1-4. (Czech. resumé.)

Papp, Adolf

1941, *Quergegliederte Röhren aus dem Oberkreide-Flysch der Alpen:* Palaeobiologica, v. 7, p. 314-317, text-fig. 1, 2.

1949, *Über Lebensspuren aus dem Jungtertiär des Wiener Beckens:* Österr. Akad. Wiss., Sitzungsber., math.-naturw. Kl., v. 158, p. 667-670.

Paréjas, Édouard

1935, *L'Organisme B de E. Joukowsky et J. Favre:* Soc. Phys. Histoire Nat. Genève, Comptes Rendus, v. 52, p. 221-224, text-fig. 1.

1948, *Sur quelques coprolithes de Crustacés:* Arch. Sci., v. 1, no. 3, p. 512-520, text-fig. 1-35.

Patteisky, Karl

1929, *Die Geologie und Fossilführung der mährischschlesischen Dachschiefer- und Grauwackenformation:* Naturwiss. Verein Troppau, 356 p., 26 pl.

Patterson, J. M.

1942, *Halymenites, a marine sandstone indicator:* Jour. Paleontology, v. 42, p. 271-273.

Paul, C. M.

1899, *Der Wienerwald. Ein Beitrag zur Kenntnis der nord-alpinen Flyschbildungen:* K. K. Geol. Reichsanst. Wien, Jahrb., v. 48 (1898), p. 53-178, text-fig. 1-27, pl. 2-6.

Paulus, Bruno

1957, *Das Spurenfossil Lennea schmidti im Devon der Eifel:* Senckenbergiana Lethaea, v. 38, p. 169-175, 1 pl.

Peck, R. E.

1974, *On the systematic position of the umbellids:* Jour. Paleontology, v. 48, p. 409-412, text-fig. 1.

Pelletier, B. R.

1958, *Pocono paleocurrents in Pennsylvania and Maryland:* Geol. Soc. America, Bull., v. 69, p. 1033-1064, text-fig. 1-19.

Péneau, J.

1946, *Étude sur l'Ordovicien Inférieur (Arénigien = Grès Armoricain) et sa faune (spéciale- ment en Anjou):* Soc. Études Sci. d'Angers, n. sér., Bull., v. 74-76 (1944-46), p. 37- 106, 8 pl.

Pequegnat, W. E., James, B. M., Bouma, A. H., Bryant, W. R., & Fredericks, A. D.

1972, *Photographic study of deep-sea environ- ments of the Gulf of Mexico:* in Contribu- tions on the geological and geophysical oceanography of the Gulf of Mexico, Richard Rezak & V. J. Henry (eds.), Texas A. & M. Univ. Oceanographic Studies, v. 3, p. 67-128.

Perry, J. B.

1872, *On the so-called Scolithi in the Potsdam:* Boston Soc. Nat. History, Proc., v. 14 (1870/71), p. 139.

Peruzzi, D. G.

1881, *Osservazioni sui generi Paleodictyon e Paleomeandron dei terreni cretacei ed eo- cenici dell'Apennino settentrionale e centrale:* Atti Soc. Toscana Sci. Nat., Mem., v. 5, p. 1-8, pl. 1.

Pettijohn, F. J., & Potter, P. E.

1964, *Atlas and glossary of primary sedimentary structures:* 370 p., 117 pl., Springer-Verlag (New York).

——, ——, & Siever, Raymond

1972, *Sand and sandstone:* 618 p., illus., Springer- Verlag (New York). (Biogenic structures, p. 127-131.)

Peyer, Bernhard

1945, *Über Algen und Pilze in tierischen Hartsub- stanzen:* Julius-Klaus Stiftung, Archiv, Erg.- Bd. 20, p. 496-546, text-fig. 1-48.

Pfefferkorn, H. W.

1971, *Note on Conostichus broadheadi Lesquereux (trace fossil: Pennsylvanian):* Jour. Paleon- tology, v. 45, p. 888-892, pl. 101.

Pfeiffer, Heinz

1959, *Über Dictyodora liebeana (Weiss):* Geolo- gie, v. 8, p. 425-433, text-fig. 1-5, pl. 1-3.

1965, *Volkichnium volki n. gen., n. sp. (Lebens- Spuren) aus den Phycoden-Schichten Thü- ringens:* Same, v. 14, p. 1266-1268, text- fig. 1, 2.

1967, *Der Magdeburg-Flechtinger Kulm und seine stratigraphische und regionale Stellung:* Same, v. 16, no. 7, p. 781-790, text-fig. 1-6.

1968, *Die Spurenfossilien des Kulms (Dinant)*

und Devons der Frankenwälder Querzone (Thüringen): Jahrb. Geologie, v. 2 (1966), p. 651-717, text-fig. 1-5, 10 pl. (1968).

Pflug, H. D.

1966, *Neue Fossilreste aus den Nama-Schichten in Südwest-Afrika:* Paläont. Zeitschr., v. 40, no. 1-2, p. 14-25, text-fig. 1-3, pl. 1, 2.

1970, *Zur Fauna der Nama Schichten in Südwest- Afrika I. Pteridinia, Bau und systematische Zugehörigkeit:* Palaeontographica, v. 134, A, no. 4-6, p. 226-262, text-fig. 1-14, pl. 20-23, 1 table.

——, & Strübel, Günter

1969, *Algen und Bakterien in präkambrischen Konkretionen:* Palaeontographica, v. 127, B, no. 1-6, p. 143-158, text-fig. 1-7, pl. 4, 5, tables.

Philipp, Hans

1904, *Paläontologisch-geologische Untersuchungen aus dem Gebiet von Predazzo:* Deutsch. Geol. Gesell., Zeitschr., v. 56, p. 1-98, pl. 1-6.

Pia, Julius

1927, *Thallophyta:* in M. Hirmer, Handbuch der paläobotanik, v. 1, p. 31-136, R. Oldenbourg (München, Berlin).

1936, *Algen und Pseudoalgen aus der spanischen Trias:* in M. Schmidt, Fossilien der span- ischen Trias, Heidelberger Akad. Wiss., math.-nat. Kl., Abhandl., v. 22, p. 9-17.

1937, *Review of A. Fucini: Problematica verru- cana. I. Pisa 1936:* Neues Jahrb. Mineral- ogie, Geologie, Paläontologie, 1937, Ref. III, p. 1094-1099.

1939, *Sammelbericht über fossile Algen: Soleno- poraceae 1930-1938, mis Nachträgen aus früheren Jahren:* Same, Ref., 1939, III, p. 731-760.

Picard, Leo

1942, *New Cambrian fossils and Palaeozoic prob- lematica from the Dead Sea and Arabia:* Hebrew Univ. Jerusalem, Geol. Dept., Bull., v. 4, no. 1, p. 1-18, 1 pl.

Pickett, J. W.

1972, *The ecology of worm populations in the Erins Vale Formation (Late Permian) south- ern Sydney Basin:* Geol. Soc. Australia, Jour., v. 19, p. 313-320, text-fig. 1, pl. 19, 20.

——, & Scheibnerova, Viera

1974, *The inorganic origin of "anellotubulates":* Micropaleontology, v. 21, p. 97-102, pl. 1, 2.

Pickett, T. E., Kraft, J. C., & Smith, Kenneth

1971, *Cretaceous burrows—Chesapeake and Dela- ware Canal, Delaware:* Jour. Paleontology, v. 45, p. 209-211, pl. 28.

Pietzsch, Kurt
1911, *Cruzianen aus dem Untersilur des Leipziger Kreises:* Deutsch. Geol. Gesell., Zeitschr., v. 62 (1910), p. 571-582, pl. 11-13.

Piper, D. J. W., & Marshall, N. F.
1969, *Bioturbation of Holocene sediments on La Jolla deep sea fan, California:* Jour. Sed. Petrology, v. 39, p. 601-606.

Piveteau, Jean
1955, *Ichnologie:* in Traité de Paléontologie, J. Piveteau (ed.), v. 5, p. 314-316.

Plessman, Werner
1966, *Diagenetische und kompressive Verformung in der Oberkreide des Harz-Nordrandes sowie im Flysch von San Remo:* Neues Jahrb. Geologie, Paläontologie, Monatsh., pt. 8, p. 480-493.

Plička, Miroslav
1962, *Rozšiření Palaeospirographis hraběi n. g. n. sp. (Chaetopoda, Polychaeta) v západní oblasti magurského flyše v ČSSR:* Ústřed. Ústavu Geol., Věstník, v. 37, no. 5, p. 359-364, text-fig. 1, pl. 1, 2. [*Distribution of Palaeospirographis hrabei n. g. n. sp. (Chaetopoda, Polychaeta) in the western region of the Magura Flysch in Czechoslovakia (prelim. report).*]

1965, *Nový rod fosilnich morškých sabellid z karpatského Flyše:* Vlastivédného Ústavu Volomonci, Zprávy, v. 122 (1963/64), p. 1-5, text-fig. 1, 2. [*New genus of fossil marine worms (Sabellidae) of Carpathian flysch (Czechoslovakia).*]

1968, *Zoophycos, and a proposed classification of sabellid worms:* Jour. Paleontology, v. 42, p. 836-849, text-fig. 1-11, pl. 107, 108.

1969, *Methods for the study of "Zoophycus" and similar fossils:* New Zealand Jour. Geology, Geophysics, v. 12, p. 551-573, text-fig. 1-18.

1970, *Zoophycos and similar fossils:* in Trace fossils, T. P. Crimes & J. C. Harper (eds.), Geol. Jour., spec. issue no. 3, p. 361-370, text-fig. 1-4, pl. 1, 2, Seel House Press (Liverpool).

Plieninger, W. H. Th. v.
1845, *(Reliefs im . . . Keupersandstein bei Stuttgart; Tubifex antiquus):* Ver. Vaterl. Naturkd. Württemberg, Jahresh., v. 1, p. 159, text-fig. 1-5, 1 pl. [Not seen by the editors.]

Plumstead, E. P.
1967, *General review of the Devonian fossil plants found in the Cape System of South Africa:* Palaeont. Africana, v. 10, p. 1-83, text-fig. 1, 2, 25 pl.

Pogue, J. B., & Parks, J. M., Jr.
1958, *Lower Permian occurrence of "amphibian tracks" (invertebrate burrows) in central Texas:* Geol. Soc. America, Bull., v. 69, no. 12, p. 1629 (abstr.).

Pohowsky, R. A.
1974, *Notes on the study and nomenclature of boring Bryozoa:* Jour. Paleontology, v. 48, p. 556-564, pl. 1.

Pomel, Auguste
1849, *Matériaux pour servir à la flore fossile des terrains jurassiques de la France:* Vers. Gesell. Deutsch. Naturf. Aerzte, Amtl. Ber. 25, Sept. 1847, p. 332-354.

Pompeckj, J. F.
1927, *Ein neues Zeugnis uralten Lebens:* Paläont. Zeitschr., v. 9, p. 287-313, pl. 5.

Portlock, J. E.
1843, *Report on the geology of the County of Londonderry, and of parts of Tyrone and Fermanagh:* 748 p., 37 pl., A. Milliken (Dublin, London).

Potonié, Henri
1893, *Die Flora des Rothliegenden von Thüringen:* K. Preuss. Geol. Landesanst., Abhandl., n. ser., v. 9, 298 p., 34 pl.

Poulsen, Valdemar
1963, *Notes on Hyolithellus Billings, 1871, class Pogonophora Johannson, 1937:* K. Danske Vidensk. Selskab, Biol. Meddel., v. 23, no. 12, 15 p.

Powers, Sidney
1922, *Gastropod trails in Pennsylvanian sandstones in Texas:* Am. Jour. Sci., v. 3, p. 103-107.

Poyarkov, B. V.
1966, *Devonskie kharofity Tyan'-Shanya:* in Iskopaemye kharofity SSSR, Akad. Nauk SSSR, Geol. Inst., Trudy, vyp. 143, p. 161-200, text-fig. 1-10, pl. 1, 2. [*Devonian charophytes from the Tien Shan.*]

Prantl, Ferdinand
1944, *O výskytu chondritů ve vápencich bránických -ga:* K. České Společnost. Nauk, Třída mat.-přir., Věstník, 1943, v. 5, 19 p., text-fig. 1, 2, 2 pl. [*The Chondrites beds of the Braniker Limestone-ga.*]

1946, *Two new problematic trails from the Ordovician of Bohemia:* Acad. Tchèque Sci., Bull. Internatl., Cl. Sci. math. nat. méd., v. 46 (1945), p. 49-59 (Reprint: p. 1-9), text-fig. 1-4, 2 pl.

Pratje, Otto
1922, *Fossile kalkbohrende Algen (Chaetophorites gomontoides) in Liaskalken:* Centralbl. Mineralogie, Geologie, Paläontologie, 1922, p. 299-301, text-fig. 1-3.

Prescher, Hans
1954, *Sedimentpetrographische Untersuchungen oberturoner Sandsteine im Elbstandsteingebirge:* Freiberger Forschungsh., C, v. 11, 96 p., text-fig. 1-28.

Pruvost, Pierre

1930, *La faune continentale du terrain houiller de la Belgique:* Muséum Royal Histoire Nat. Belgique, Mém., v. 44, p. 103-282, 14 pl.

Putzer, Hannfrit

1938, *Die Rhät-und Liasablagerungen am Seeberg bei Gotha, am Röhnbergrücken und bei Eisenach:* Jenaische Zeitschr. Naturwiss., v. 71, p. 327-444, text-fig. 1-13, pl. 6-15.

Quatrefages, M. A. de

1846, (not seen by the editors).

1849, *Note sur la Scolicia prisca (A. De Q.), an- nélide fossile de la craie:* Ann. Sci Nat., sér. 3, Zoologie, v. 12, p. 265-266.

Quenstedt, F. A.

1845-49, *Petrefactenkunde Deutschlands. 1. Abth., v. 1: Cephalopoden:* 580 p., 36 pl., L. F. Fues (Tübingen). (Atlas, 1849).

1879, *Petrefactenkunde Deutschlands. 1. Abth., v. 6: Korallen. Die Röhren-und Steinkorallen:* 1093 p., L. F. Fues (Leipzig). (Liefg. 7: 1879).

Quenstedt, Werner

1932a, *Die Geschichte der Chitonen und ihre allge- meine Bedeutung. (Mit Zusätzen):* Paläont. Zeitschr., v. 14, p. 77-96, text-fig. 1.

1932b, *Zufall, Gunst und Grenzen paläontologischer Überlieferung:* Gesell. Naturf. Freunde Ber- lin, Sitzungsber., 1932, p. 131-192.

Quilty, P. G.

1970, *Triangulina n. g. (Problematica) from the Tertiary of southern Australia:* Micropa- leontology, v. 16, p. 179-184, text-fig. 1-3.

Raciborski, Marian

1890, *Taonurus ultimus Sap. in Galizien:* K. K. Geol. Reichsanst. Wien, Verhandl., 1890, p. 265-266.

Radig, Franz

1964, *Die Lebensspur Tomaculum problematicum Groom 1902 im Llandeilo der Iberischen Halbinsel:* Neues Jahrb. Geologie, Paläon- tologie, Abhandl., v. 119, p. 12-18, text-fig. 1, 2, 1 pl.

Radoičić, Rajka

1959, *Some problematic microfossils from the Dinarian Cretaceous:* Zavod Geol. i Geofiz. Istraž., Vesnik, v. 17, p. 87-92.

1966, *O problematičnim mikrofosilima iz jure i krede Dinarida:* Zavod Geol. Geofiz. Istra- živanja, Vesnik, ser. A, v. 24-25, p. 269- 280, 7 pl. [*On problematic microfossils from the Jurassic and Cretaceous of the Dinarides.*] (French summ., p. 274.)

Radwański, Andrzej

1959, *Struktury litoralne w liasie w Dolince Smytniej:* Acta Geol. Polonica, v. 9, no. 2,

p. 231-280, text-fig. 1-9, pl. 19-24. (Russ. & Engl. summ.) [*Littoral structures (cliff, clastic dikes and veins, and borings of Po- tamilla) in the high-tatric Lias,* p. 270-278.]

1964, *Boring animals in Miocene littoral environ- ments of Sonkea Poland:* Acad. Polon. Sci., Bull., sér. sci. géol. géogr., v. 12, p. 57-62, 6 pl.

1965, *Additional notes on Miocene littoral struc- tures of southern Poland:* Same, Bull., sér. sci. géol. géogr., v. 13, p. 167-173, 4 pl.

1969, *Transgresja dolnego tortonu na południo- wych stokach Gór Świętokrzyskich (strefa zatok i ich przedpola):* Acta Geol. Polonica, v. 19, p. 1-164, text-fig. 1-37, 42 pl. [*Lower Tortonian transgression onto the southern slopes of the Holy Cross Mts.*] (Polish, with English resumé.)

1970, *Dependence of rock-borers and burrowers on the environmental conditions within the Tortonian littoral zone of Southern Poland:* in Trace fossils, T. P. Crimes & J. C. Harper (eds.), Geol. Jour., spec. issue no. 3, p. 371- 390, text-fig. 1-4, pl. 1-6, Seel House Press (Liverpool).

1972, *Remarks on the nature of belemnicolid bor- ings Dendrina:* Acta Geol. Polonica, v. 22, no. 2, p. 257-264, text-fig. 1-5. (Pol. summ.)

————, & Roniewicz, Piotr

1963, *Upper Cambrian trilobite ichnocoenosis from Wielka Wiśniówka (Holy Cross Mountains, Poland):* Acta Palaeont. Polonica, v. 8, p. 259-280, pl. 1-10. (English, with Pol. & Russ. summaries).

1967, *Trace fossils Aglaspidichnus sanctacrucensis n. gen., n. sp., a probable resting place of an aglaspid (Xiphosura):* Same, v. 12, p. 545- 552, text-fig. 1, pl. 1. (English, with Polish and Russ. summaries).

1970, *General remarks on the ichnocoenose con- cept:* Acad. Polon. Sci., Bull., v. 18, p. 51- 56 (with Russ. summ.).

1972, *A long trilobite-trackway, Cruziana semipli- cata Salter, from the Upper Cambrian of the Holy Cross Mts.:* Acta Geol. Polonica, v. 22, p. 439-447, text-fig. 1, 2, pl. 1. (Pol. resumé.)

Rafinesque, C. S.

1821, *Description of a fossil Medusa, forming a new genus, Trianisites Cliffordi:* Am. Jour. Sci. Arts, v. 3, p. 285-287.

Rankama, Kalervo

1948, *New evidence of the origin of pre-Cambrian carbon:* Geol. Soc. America, Bull., v. 59, p. 391-416, text-fig. 1-4, 6 pl.

1950, *Corycium resuscitatum: a discussion:* Jour. Geology, v. 58, p. 75-79.

Rasmussen, H. W.

1971, *Echinoid and crustacean burrows and their*

diagenetic significance in the Maastrichtian-Danian of Stevns Klint, Denmark: Lethaia, v. 4, p. 191-216, text-fig. 1-17.

Rauff, Hermann

1891, *Über Palaeospongia prisca Bornem., Eophyton z. Th., Chondrites antiquus, Haliserites z. Th. und ähnliche Gebilde:* Neues Jahrb. Mineralogie, Geologie, Paläontologie, 1891, II, p. 92-104.

1892, *Über Pseudoorganismen, besonders über Dictyodora and Crossopodia:* Deutsch. Geol. Gesell., Zeitschr., v. 44, p. 561-564.

1893, *Über Angebliche Spongien aus dem Archaicum:* Neues Jahrb. Mineralogie, Geologie, Paläontologie, 1893, v. 2, p. 57-67, text-fig. 1-3.

1895, *Über Porocystis pruniformis Cragin (=? Araucarites wordi Hill) aus der unteren Kreide in Texas:* Same, 1895, I, p. 1-15, 1 pl.

1896, *Über angebliche Organismen aus präkambrischen Schichten der Bretagne:* Same, 1896, v. 1, p. 117-138, text-fig. 1-17.

Raup, D. M., & Seilacher, Adolf

1969, *Fossil foraging behavior: computer simulation:* Science, v. 166, p. 994-995.

Raymond, P. E.

1920, *The appendages, anatomy and relationships of trilobites:* Connecticut Acad. Arts Sci., Mem., v. 7, 169 p., 11 pl.

1922, *Seaside notes:* Am. Jour. Sci., ser. 5, v. 3, p. 108-114.

1931a, *Notes on invertebrate fossils, with descriptions of new species. No. 4. Trails from the Silurian at Waterville, Maine:* Harvard Coll., Museum Comp. Zoology, Bull., v. 55, p. 184-194, 5 pl.

1931b, *Notes on invertebrate fossils, with descriptions of new species. No. 5. On the nature of Phytopsis bulbosum Hall:* Same, v. 55, p. 194-198.

1935, *Pre-Cambrian life:* Geol. Soc. America, Bull., v. 46, p. 375-392.

Rech-Frollo, M. -M.

1962, *Une nouvelle hypothèse sur l'origine des "Helminthoides":* Acad. Sci. [Paris], Comptes Rendus séanc., v. 254, p. 894-896.

Redini, R.

1938, *Sulla natura e sul significato chronologico di pseudofossili e fossili del Verrucano tipico del Monte Pisano:* Riv. Ital. Paleontologia, suppl., v. 40, p. 329-382, 6 pl.

Reineck, H. -E.

1955, *Marken, Spuren und Fährten in den Waderner Schichten (ro) bei Martinstein (Nahe):* Neues Jahrb. Geologie, Paläontologie, Abhandl., v. 101, p. 75-90, pl. 7-11.

1958a, *Wühlbau-Gefüge in Abhängigkeit von Sediment-Umlagerungen:* Senckenbergiana Lethaea, v. 39, p. 1-14, text-fig. 1-3, 5 pl.

1958b, *Über Gefüge von orientierten Grundproben aus der Nordsee:* Same, v. 39, p. 25-36, 54-56, text-fig. 1, 2, pl. 6-8.

1968, *Lebensspuren von Herzigeln:* Same, v. 49, p. 311-319, 3 pl.

1973, *Schichtung und Wühlgefüge in Grundproben vor der ostrafikanischen Küste:* "Meteor" Forsch.-Ergebnisse, ser. C, no. 16, p. 67-81, text-fig. 1-17.

———, Dörjes, Jürgen, Gadow, Sibylle, & Hertweck, Günther

1968, *Sedimentologie, Faunenzonierung and Faziesabfolge vor der Ostküste der inneren Deutschen Bucht:* Senckenbergiana Lethaea, v. 49, p. 261-309, text-fig. 1-15, 2 pl.

Reis, O. M.

1910a, *Beobachtungen über Schichtenfolge und Gesteinsausbildungen in der fränkischen Trias. I. Muschelkalk und untere Lettenkohle:* Geognost. Jahresh., v. 22, p. 1-285, 11 pl.

1910b, *Zur Fukoidenfrage:* K. K. Geol. Reichsanst., Jahrb., v. 59 (1909), p. 615-638, pl. 17.

1922, *Über Bohrröhren in fossilen Schalen und über Spongeliomorpha:* Deutsch. Geol. Gesell., Zeitschr., v. 73 (1921), p. 224-237, pl. 7.

Reish, D. J.

1952, *Discussion of the colonial tube-building polychaetous annelid Dodecaceria fistulicola Ehlers:* S. California Acad. Sci., Bull., v. 51, p. 103-107, pl. 20.

Reitlinger, E. A.

1957, *Sfery devonskikh otlozhenii Russkoi platformy:* Akad. Nauk SSSR, Doklady, v. 115, p. 774-776, 1 pl. [*Spheres in the Devonian deposits of the Russian Platform.*]

Renelt, F.

1943, *Asteriden aus der nordböhmischen Kreide:* Firgenwald, v. 13, pt. 2, p. 113.

Renz, Carl

1925, *Problematische Medusenabdrücke aus der Olonos—Pindos-Zone des Westpeleponnes:* Naturf. Gesell. Basel, Verhandl., v. 36, p. 220-223, text-fig. 1.

1930, *Ein Medusenvorkommen im Alttertiär der Insel Cypern (Cyprus):* Eclogae Geol. Helvetiae, v. 23, p. 295-300, text-fig. 1.

Resser, Ch. E.

1938, *Middle Cambrian fossils from Pend Oreille Lake, Idaho:* Smithson. Misc. Coll. v. 97, no. 3, 12 p., 1 pl.

———, & Howell, B. F.

1938, *Lower Cambrian Olenellus zone of the Ap-*

palachians: Geol. Soc. America, Bull., v. 49, p. 195-248, text-fig. 1, 13 pl.

Reynès, Pierre

1868, *Essai de géologie et de paléontologie avey-ronnaises:* 110 p., 7 pl., J. -B. Baillière et fils (Paris).

Rezak, Richard, & Henry, V. J. (eds.)

1972, *Contributions on the geological and geo-physical oceanography of the Gulf of Mex-ico:* Texas A. & M. Univ. Oceanographic Studies, v. 3, 303 p.

Rhoads, D. C.

1963, *Rates of sediment reworking for Yoldia limatula in Buzzards Bay, Massachusetts, and Long Island Sound:* Jour. Sed. Petrol-ogy, v. 33, p. 723-727, text-fig. 1-3.

1967, *Biogenic reworking of intertidal and sub-tidal sediments in Barnstable Harbor and Buzzards Bay, Massachusetts:* Jour. Geology, v. 75, p. 461-476.

1970, *Mass properties, stability and ecology of marine muds related to burrow activity:* in Trace fossils, T. P. Crimes & J. C. Harper (eds.), Geol. Jour., spec. issue no. 3, p. 391-406, text-fig. 1-4, pl. 1-3, Seel House Press (Liverpool).

————, & Stanley, D. J.

1965, *Biogenic graded bedding:* Jour. Sed. Petrol-ogy, v. 35, p. 956-963, text-fig. 1-7.

Rhumbler, Ludwig

1911-13, *Die Foraminiferen (Thalamophoren) der Plankton-Expedition:* Ergebnisse der Plank-ton-Exped. der Humboldt-Stiftung, V. Hen-sen (ed.), a) 1911, v. 3, Lief. c., p. 1-331, pl. 1-39 (1909); b) 1913, Pt. 2, *Systematik: Arrhabdammidia, Arammodisclidia und Arnodosammidia,* v. 3, Lief. c., p. 332-476, 65 fig., Lipsius & Fischer (Kiel).

Richardson, E. S., Jr.

1966, *Wormlike fossil from the Pennsylvanian of Illinois:* Science, v. 157, p. 75-76, text-fig. 1, 2.

Richardson, G., Gregory, D., & Pollard, J.

1973, *Anellotubulates are manufactured 'micro-fossils':* Nature, v. 246, no. 5432, p. 347-348, 1 text-fig.

Richter, Reinhard

1850, *Aus der thüringischen Grauwacke:* Deutsch. Geol. Gesell., Zeitschr., v. 2, p. 198-206, pl. 8, 9.

1851, *Über thüringische Graptolithen:* Same, Zeitschr. v. 3, p. 563-566.

1853a, *Thüringische Graptolithen:* Same, Zeitschr., v. 5, p. 439-464, pl. 12.

1853b, *Gaea von Salfeld:* Programm d. Realsch. Saalfeld, p. 3-32.

1856, *Beitrag zur Paläontologie des Thüringer*

Waldes. 1. Theil: Akad. Wiss. Wien, math.-nat. Kl., Denkschr., v. 11, p. 87-138, pl. 1-3.

1870, *Bemerkungen zu Ludwig's Abhandlung über paläozoische Pflanzenreste:* Neues Jahrb. Mineralogie, Geologie, Paläontologie, 1870, p. 207-209.

1871, *Aus dem Thüringischen Schiefergebirge:* Deutsch. Geol. Gesell., Zeitschr., v. 23, p. 231-256, pl. 5.

Richter, Rudolf

1920, *Ein devonischer "Pfeifenquarzit" verglichen mit der heutigen "Sandkoralle" (Sabellaria, Annelidae):* Senckenbergiana, v. 2, p. 215-235, text-fig. 1-6.

1921, *Scolithus, Sabellarifex und Geflechtquarz-ite:* Same, v. 3, p. 49-52.

1924, *Flachseebeobachtungen zur Paläontologie und Geologie. VII-XI:* Same, v. 6, p. 119-165, text-fig. 1-8.

1926, *Flachseebeobachtungen zur Paläontologie und Geologie. XII-XIV:* Same, v. 8, p. 200-224, pl. 3.

1927a, *Die fossilen Fährten und Bauten der Wür-mer, ein Überblick über ihre biologischen Grundformen und deren geologische Bedeu-tung:* Paläont. Zeitschr., v. 9, p. 193-240, pl. 1-4.

1927b, *Syringomorpha nilssoni (Torell) in nord-deutschen Geschieben des schwedischen Cambriums, ein glazialgeologisch verwend-bares Problematikum:* Senckenbergiana, v. 9, p. 260-268, text-fig. 1, 2.

1928, *Psychische Reaktionen fossiler Tiere:* Palaeo-biologica, v. 1, p. 225-244, text-fig. 4, pl. 23.

1931, *Tierwelt und Umwelt im Hunsrückschiefer; zur Entstehung eines schwarzen Schlamm-steins:* Senckenbergiana, v. 13, p. 299-342, text-fig. 1-16.

1937, *Marken und Spuren aus allen Zeiten. I-II:* Same, v. 19, p. 150-169, text-fig. 1-14.

1941, *Marken und Spuren im Hunsrückschiefer. 3. Fährten als Zeugnisse des Lebens auf dem Meeres-Grunde:* Same, v. 23, p. 218-260, text-fig. 1-17.

1954, *Fährte eines "Riesenkrebses" im Rheinischen Schiefergebirge:* Natur u. Volk, v. 84, p. 261-269.

1955a, *Die ältesten Fossilien Süd-Afrikas:* Senck-enbergiana Lethaea, v. 36, p. 243-289, 7 pl.

1955b, *Kunstform von Menschenhand oder ver-steinerte Tierwege?:* Natur u. Volk, v. 85, p. 337-344, text-fig. 1-5.

————, & Richter, Emma

1930, *Bemerkenswert erhaltene Conularien und ihre Gattungsgenossen im Hunsrückschiefer (Unterdevon) des Rheinlandes:* Sencken-bergiana, v. 12, p. 152-171, text-fig. 1-5.

1939a, *Marken und Spuren aus allen Zeiten. III. Eine Lebens-Spur (Syncoprulus pharma-*

ceus), gemeinsam dem rheinischen und böhmischen Ordovicium: Same, v. 21, p. 152-168, text-fig. 1-8.

1939b, *Marken und Spuren aus allen Zeiten. IV. Die Kot-Schnur Tomaculum Groom (= Syncoprulus Rud. & E. Richter), ähnliche Scheitel-Platten und beider stratigraphische Bedeutung:* Same, v. 21, p. 278-291, text-fig. 1-6.

1941, *Das stratigraphische Verhalten von Tomaculum als Beispiel für die Bedeutung von Lebensspuren:* Same, v. 23, p. 133-135.

1951, *Tetramerer Bau bei Tabulaten als Erklärung von "Brooksella rhenana":* Paläont. Zeitschr., v. 24, p. 146-164, pl. 11, 12.

Robison, R. A.
1969, *Annelids from the Middle Cambrian Spence Shale of Utah:* Jour. Paleontology, v. 43, p. 1169-1173, 1 pl.

Rodriguez, J., & Gutschick, R. C.
1970, *Late Devonian-Early Mississippian ichnofossils from western Montana and northern Utah:* in Trace fossils, T. P. Crimes & J. C. Harper (eds.), Geol. Jour., spec. issue no 3, p. 407-438, text-fig. 1-6, 10 pl., Seel House Press (Liverpool).

Roedel, Hugo
1926, *Ein kambrisches Geschiebe mit problematischen Spuren:* Zeitschr. Geschiebeforschg., v. 2, p. 22-26, 1 text-fig.

1929, *Ergänzung zu meiner Mitteilung über ein kambrisches Geschiebe mit problematischen Spuren:* Same, v. 5, p. 48-51, 1 text-fig.

Roemer, Ferdinand
1848, *Kritische Anzeige von James Hall's Paläontologie des Staates New York (Band I):* Neues Jahrb. Mineralogie, Geognosie, Petrefactenkd., 1848, p. 169-181.

1870, *Geologie von Oberschlesien:* 587 p., Atlas, L. Köhler (Breslau).

Rogers, H. D.
1838 [not seen by the editors].

Roniewicz, Piotr
1970, *Borings and burrows in the Eocene littoral deposits of the Tatra Mountains, Poland:* in Trace fossils, T. P. Crimes & J. C. Harper (eds.), Geol. Jour., spec. issue no. 3, p. 439-446, text-fig. 1-3, pl. 1, 2, Seel House Press (Liverpool).

Rooney, W. S., & Perkins, R. D.
1972, *Microboring organisms as environmental indicators and sediment tracers, Arlington Reef Complex, Australia:* Am. Assoc. Petroleum Geologists, Bull., v. 56, p. 650.

Rosenkranz, Dieter
1971, *Zur Sedimentologie und Ökologie von Echinodermen-Lagerstätten:* Neues Jahrb. Geologie, Paläontologie, Abhandl., v. 138, p. 221-258, text-fig. 1-10.

Ross, J. P.
1967, *Fossil problematica from Upper Ordovician, Ohio:* Jour. Paleontology, v. 41, p. 37-42, text-fig. 1, 2, 6 pl.

Rothpletz, August
1896, *Über die Flysch-Fukoiden und einige andere fossile Algen sowie über liassische, Diatomeen führende Hornschwämme:* Deutsch. Geol. Gesell., Zeitschr., v. 48 (1896), p. 854-914, pl. 22-24.

1913, *Über Kalkalgen, Spongiostromen und einige andere Fossilien:* Sveriges Geol. Undersök., Afhandl., Upps. ser. Ca, no. 10, 57 p., 9 pl.

1916, *Über die systematische Deutung und die stratigraphische Stellung der ältesten Versteinerungen Europas und Nordamerikas mit besonderer Berücksichtigung der Cryptozoen und Oolithe. II. Über Cryptozoon, Eozoon und Atikokania:* Bayer. Akad. Wiss., math.-phys. Kl., Abhandl., v. 28, text-fig. 1-4, 92 p., 8 pl.

Rouault, Marie
1850, *Note préliminaire sur une nouvelle formation découverte dans le terrain silurien inférieur de la Bretagne:* Soc. Géol. France, Bull., sér. 2, v. 7, p. 724-744.

Roux, Wilhelm
1887, *Über eine im Knochen lebende Gruppe von Fadenpilzen (Mycelites ossifragus):* Zeitschr. Wiss. Zoologie, v. 45, p. 227-254, pl. 14.

Rovereto, G.
1901, *Briozoi, Anellidi e spugne perforanti del Neogene Ligure:* Palaeont. Italica, v. 7, p. 219-234, text-fig. 1-5, pl. 28.

Rowell, A. J.
1971, *Supposed Pre-Cambrian brachiopods:* in Paleozoic perspectives: a paleontological tribute to G. Arthur Cooper, J. T. Dutro, Jr. (ed.), Smithson. Contrib. Paleobiology, no. 3, p. 71-79, pl. 1, Smithson. Inst. Press (Washington, D. C.).

Ruchholz, Kurt
1967, *Zur Ichnologie und Fazies des Devons und Unterkarbons im Harz:* Geologie, v. 16, p. 503-527, text-fig. 1-16, pl. 1-4.

Ruedemann, Rudolf
1916, *Paleontologic contribution from the New York State Museum:* New York State Museum, Bull., 189, 225 p., 46 text-fig., 36 pl.

1925, *The Utica and Lorraine Formations of New York. Pt. 2, no. 1. Plants, sponges, corals, graptolites, crinoids, worms, bryozoans, brachiopods:* Same, Bull., 262, 171 p., 75 text-fig., 13 pl.

1926, *The Utica and Lorraine Formations of New York. Part 2. Systematic paleontology. no. 2. mollusks, crustaceans and eurypterids:* Same, Bull., 272, 227 p., 26 text-fig., 28 pl.

1929, *Note on Oldhamia (Murchisonites) occidens (Walcott):* Same, Bull., 281, p. 47-51, 7 text-fig.

1934, *Palaeozoic plankton of North America:* Geol. Soc. America, Mem., v. 2, 141 p., 26 pl.

1942, *Oldhamia and the Rensselaer grit problem:* New York State Museum, Bull., v. 327, p. 5-12, 3 pl.

Rüger, L., & Rüger-Haas, P.

1925, *Palaeosemaeostoma geryonides v. Huene sp., eine sessile Meduse aus dem Dogger von Wehingen in Württemberg und Medusina liasica n. sp., eine coronatenähnliche Meduse aus dem mittleren Lias von Hechingen in Württemberg:* Heidelberger Akad. Wiss., math.-nat. Kl., Sitzungsber, 1925, no. 15, 22 p., text-fig. 1, 2, 2 pl.

Rüppell, Eduard

1845, *Oeffentliche Rede, gehalten am 22. November 1842 bei Gelegenheit des 25-jährigen Stiftungsfestes der Senckenbergischen Naturforschenden Gesellschaft:* Museum Senckenbergianum, v. 3, p. 197-214.

Rupp, A. W.

1966, *Origin, structure, and environmental significance of Recent and fossil calcispheres:* Geol. Soc. America, Program Ann. Mtg., p. 186 (San Francisco) (Nov.).

Rusconi, Carlos

1955, *Fósiles cámbricos y ordovícicos al oeste de San Isidro, Mendoza:* Museo Historia Nat. Mendoza, Revista, v. 8, no. 1-4, p. 3-64.

1956, *Oldhamias Ordovicicas de Mendoza:* Same, Revista, v. 9, no. 1-2, p. 47-53.

Russell, L. S.

1940, *Micrichnus tracks from the Paskapoo Formation of Alberta:* Royal Soc. Canada, Inst. Trans., v. 23, p. 67-74, text-fig. 1-4.

Rzehak, A.

1888, *Die Foraminiferen des kieseligen Kalkes von Nieder-Hollabrunn und des Melettamergels der Umgebung von Bruderndorf in Niederösterreich:* K. K. Naturhist. Hof. museum, Annalen, v. 3, 14 p., 1 pl.

Sacco, Federico

1888, *Note di Paleoicnologia Italiana:* Soc. Italiana Sci. Nat., Atti, v. 31, p. 151-192, 2 pl.

1939, *Palaeodictyon:* R. Accad. Sci. Torino, Mem., v. 69, p. 267-285, pl. 1, 2.

1940, *La "Sewardiella" Fuc. delle scisto verrucano del Monte Pisano:* Same, Atti, v. 76, p. 41-58, 1 pl.

Sahni, M. R.

1936, *Fermoria minima: a revised classification of the organic remains from the Vindhyans of India:* Geol. Survey India, Rec., v. 69, p. 458-468, pl. 43.

——, & Shrivastava, R. N.

1954, *New organic remains from the Vindhyan system and the probable systematic position of Fermoria Chapman:* Current Science, v. 23, p. 39-41, text-fig. 1-4.

Saint-Seine, Roseline de

1951, *Un Cirripède acrothoracique du Crétacé: Rogerella lecontrei n. g., n. sp.:* Acad. Sci. [Paris], Comptes Rendus, v. 233, p. 1051-1053, text-fig. 1-3.

1956, *Les Cirripèdes acrothoraciques échinocoles:* Soc. Géol. France, Bull., sér. 6, v. 5, p. 299-303, pl. 16, 17.

Salter, J. W.

1856, *On fossil remains in the Cambrian rocks of the Longmynd and North Wales:* Geol. Soc. London, Quart. Jour., v. 12, p. 246-252, text-fig. 1, 2, pl. 4.

1857, *On annelide-burrows and surface markings from the Cambrian rocks of the Longmynd:* Same, Quart. Jour., v. 13, p. 199-206, pl. 5.

1861, *On the fossils, from the High Andes, collected by David Forbes:* Same, Quart. Jour., v. 17, p. 62-73, pl. 4, 5.

1864a, *On some points in ancient physical geography, illustrated by fossils from a pebble-bed at Budleigh Salterton, Devonshire:* Geol. Mag., v. 1, p. 5-12, text-fig. 1-4.

1864b, *Notes on the fossils from Budleigh Salterton pebble-bed:* Geol. Soc. London, Quart. Jour., v. 20, p. 286-302, pl. 15-17.

1866, *Appendix. On the fossils of North Wales:* in A. C. Ramsay, The geology of North Wales, Geol. Survey Great Britain, Mem., v. 3, p. 239-363, pl. 1-28.

1873, *A catalogue of the collection of Cambrian and Silurian fossils contained in the Geological Museum of the University of Cambridge:* 204 p., University Press (Cambridge).

Saporta, Gaston de

1872-73, *Paléontologie française ou description des fossiles de la France [commencée par Alcide d'Orbigny et] continuée par une réunion de paléontologistes. 2 sér. Végétaux. Plantes jurassiques:* v. 1, 506 p., 70 pl., G. Masson (Paris). (p. 1-432, pl. 1-60 [1872]; p. 433-506, pl. 61-70 [1873]).

1878, *Sur une nouvelle découverte de plantes terrestres siluriennes, dans les schistes ardoisiers d'Angers, due à M. L. Crié:* Acad. Sci. [Paris], Comptes Rendus, v. 87, p. 767-771, text-fig. 1.

1882, *A propos des algues fossiles:* 82 p., 9 pl., Masson (Paris).

1884, *Les organismes problématiques des anciennes mers:* 100 p., 13 pl., Masson (Paris).

1887, *Nouveaux documents relatifs aux organismes problématiques des anciennes mers:* Soc. Géol. France, Bull., sér. 3, v. 15, p. 286-302, pl. 3-7.

1890 [1886-1891], *Paléontologie française ou description des fossiles de la France. 2 sér. Végétaux. Plantes jurassiques:* v. 4, 548 p., 74 pl., G. Masson (Paris). (p. 273-352, pl. 41-52: 1890).

————, & Marion, A. F.

1883, *Die paläontologische Entwicklung des Pflanzenreiches. Die Kryptogamen:* 250 p., 85 text-fig., Internatl. Wiss. Bibliothek (Leipzig).

Sardeson, F. W.

1896, *The Saint Peter sandstone:* Minnesota Acad. Sci., Bull., v. 4 (1892/1910), p. 64-88, pl. 2-6.

Sarjeant, W. A. S., & Kennedy, W. J.

1973, *Proposal of a code for the nomenclature of trace-fossils:* Canad. Jour. Earth Sci., v. 10, p. 460-475.

Sarle, C. J.

1906a, *Arthrophycus and Daedalus of burrow origin:* Rochester Acad. Sci., Proc., v. 4, p. 203-210, text-fig. 1-4.

1906b, *Preliminary note on the nature of Taonurus:* Same, Proc., v. 4, p. 211-214, text-fig. 1, 2.

Sauer, Walther

1955, *Coprinisphaera ecuadoriensis, un fosil singular del Pleistoceno:* Inst. Cienc. Nat. Univ. Central, Bol., v. 1, no. 2, p. 1-9, text-fig. 1-7.

1959, *Merkwürdige Kugeln in Tuffen Ecuadors und ihre Deutung:* Natur u. Volk, v. 89, p. 118-124, text-fig. 1-10.

Savage, N. M.

1971, *A varvite ichnocoenosis from the Dwyka Series of Natal:* Lethaia, v. 4, p. 217-233, text-fig. 1-17.

1972, *A preliminary note on arthropod trace fossils from the Dwyka Series in Natal:* Second Gondwana Symposium, South Africa (July to Aug. 1970), Proc. & Papers, p. 627-636, text-fig. 1, 2, 5 pl., Council Sci. Industr. Res. (Pretoria).

Savi, Paolo, & Meneghini, G. G.

1850, *Osservazioni stratigrafiche e paleontologiche concernanti la geologia della Toscana e dei paesi limitrofi. (Appendix to Murchison: Memoria sulla struttura geologica delle Alpi):* 246 p., 1 pl., Stamperia granducale (Firenze).

Scarabelli, G. G.

1890, *Necessità di accertare se le impronte cosi dette fisiche e fisiologiche provengno dalle superficie superiori o dalle inferiori degli strati. Osservazioni sopra il Nemertilites Strozzii Menegh.:* Soc. Geol. Italiana, Bol., v. 9, p. 349-358.

Schäfer, Wilhelm

1937, *Bau, Entwicklung und Farbenentstehung bei den Flitterzellen von Sepia officinalis:* Zeitschr. Zellforsch. u. Mikroskop. Anatomie, v. 27, p. 221-245.

1938a, *Über die Zeichnung in der Haut einer Sepia officinalis von Helgoland:* Zeitschr. Morphologie Ökologie der Tiere, v. 34, p. 129-134.

1938b, *Palökologische Beobachtungen an sessilen Tieren der Nordsee:* Senckenbergiana, v. 20, p. 323-331, text-fig. 1-10.

1939a, *Fossile und rezente Bohrmuschel-Besiedlung des Jadegebietes:* Same, v. 21, p. 227-254, text-fig. 1-14.

1939b, *Polypen-Kolonien im Watt.:* Natur u. Volk, v. 69, p. 408-411.

1941a, *Zur Fazieskunde des deutschen Wattenmeeres. 1. Dangast und die Ufersäume des Jadebusens:* Senckenbergiana Naturforsch. Gesell., Abhandl., v. 457, p. 1-33.

1941b, *Zur Fazieskunde des deutschen Wattenmeeres. 2. Mellum, eine Düneninsel der deutschen Nordsee-Küste:* Same, Abhandl., v. 457, p. 34-54.

1941c, *Assiminea und Bembideon, Fazies-Leitformen für MHW-Ablagerungen der Nordseemarsch:* Senckenbergiana, v. 23, p. 136-145, text-fig. 1-9.

1941d, *Fossilations-Bedingungen von Quallen und Laichen:* Same, v. 23, p. 189-216, text-fig. 1-19.

1943, *Weichköperbewegungen von Buccinum undatum:* Same, v. 26, p. 459-466, text-fig. 1-3.

1948, *Wuchsformen von Seepocken:* Natur u. Volk, v. 78, p. 74-78.

1949, *Sandkorallen:* Same, v. 79, p. 244-245.

1950a, *Über Nahrung und Wanderung im Biotop bei der Strandschnecke Littorina littorea:* Archiv Molluskenkunde, v. 79, p. 1-8.

1950b, *Nahrungsaufnahme und ernährungsphysiologische Umstimmung bei Aeolis papillosa:* Same, v. 79, p. 9-14.

1950c, *Der "Sipho" der Klaffmuschel (Mya arenaria):* Natur u. Volk, v. 80, p. 142-146.

1950d, *Klaffmuschel-Spülsäume am Wattenstrand:* Same, v. 80, p. 173-176.

1951a, *Der 'kritische Raum,' Masseinheit und Mass für die mögliche Bevölkerungsdichte innerhalb einer Art:* Deutsch. Zool. Gesell. Wilhelmshaven, Verhandl., v. 40, p. 391-395.

1951b, *Fossilisations-Bedingungen brachyurer Krebse:* Senckenberg. Naturforsch. Gesell., Abhandl., v. 485, p. 221-238.
1951c, *Erhabene Fährten:* Natur u. Volk, v. 81, p. 89-90.
1952a, *Biogene Sedimentation im Gefolge von Bioturbation:* Senckenbergiana, v. 33, p. 1-12, text-fig. 1-10.
1952b, *Biologische Bedeutung der Ortswahl bei Balaniden-Larven:* Same, v. 33, p. 235-246, text-fig. 1-4.
1953a, *Zur Fortpflanzung der Rochen:* Natur u. Volk, v. 83, p. 245-292.
1953b, *Zur Unterscheidung gleichförmiger Kot-Pillen meerischer Evertebraten:* Senckenbergiana, v. 34, p. 81-93, text-fig. 1-6.
1954a, *Form und Funktion der Brachyuren-Schere:* Senckenberg. Naturforsch. Gesell., Abhandl., v. 489, p. 1-65.
1954b, *Mellum: Inselentwicklung und Biotopwandel:* Naturwiss. Ver. Bremen, Abhandl., v. 33, p. 391-406.
1954c, *Modell-Versuch zur Formänderung der Mellum Plate:* Natur u. Volk, v. 84, p. 426-432.
1954d, *Über das Verhalten von Jungheringsschwärmen im Aquarium:* Archiv Fischereiwiss,. v. 64, p. 276-287.
1954e, *Dehnungsrisse unter Wasser im meerischen Sediment:* Senckenbergiana Lethaea, v. 35, p. 87-99, text-fig. 1-12.
1955a, *Über die Bildung der Laichballen der Wellhorn-Schnecken:* Natur u. Volk, v. 85, p. 92-97.
1955b, *Wale auf norwegischen Felsbildern, vom Meeresbiologen betrachtet:* Germania, v. 33, p. 333-339.
1955c, *Fossilisations-Bedingungen der Meeressäuger und Vögel:* Senckenbergiana Lethaea, v. 36, p. 1-25, text-fig. 1-3, pl. 1, 2.
1956a, *Wirkungen der Benthos-Organismen auf den jungen Schichtverband:* Same, v. 37, p. 183-263, text-fig. 1-35, pl. 1, 2.
1956b, *Gesteinsbildung im Flachseebecken am Beispiel der Jade:* Geol. Rundschau, v. 45, p. 71-84.
1956c, *Wale auf norwegischen Felsbildern im Lichte meerespaläontologischer Beobachtungen:* Natur u. Volk, v. 86, p. 233-240.
1956d, *Der kritische Raum und die kritische Situation in der tierischen Sozietät.* Senckenberg. Naturforsch. Gesell., Aufsätze u. Reden, no. 9, p. 1-38.
1957, *Aufgaben und Ziele der Meerespaläontologie:* Naturwissenschaften, v. 44, p. 294-299.
1959, *Gibt es eine Überspezialisierung im Laufe der stammesgeschichtlichen Entwicklung:* Natur u. Volk, v. 89, p. 65-73.
1962, *Actuo-Paläontologie nach Studien an der Nordsee:* 666 p., 277 text-fig., 36 pl., Waldemar Kramer (Frankfurt).
1965, *Aktuopaläontologische Beobachtungen: 4. Spiralfährten und "geführte Mäander":* Natur u. Museum, v. 95, p. 83-90.
1966, *Aktuopaläontologische Beobachtungen. 6. Otolithen-Anreicherungen:* Same, v. 96, p. 439-444.
1972, *Ecology and palaeoecology of marine environments:* G. Y. Craig (ed.), 568 p., 277 text-fig., 39 pl., Oliver & Boyd (Edinburgh).

Schaffer, F. X.
1928, *Hormosiroidea florentina n. g., n. sp., ein Fucus aus der Kreide der Umgebung von Florenz:* Paläont. Zeitschr., v. 10, p. 212-215, text-fig. 1-3.
1941, *Zur Frage der Sewardiellen:* Zentralbl. Mineralogie Geologie, Paläontologie, 1941, B, p. 358-361, text-fig. 1-4.
1942, *Der Wealden der Monti Pisani in der Toscana:* Reichsamt Bodenforschg., Bericht, 1942, p. 12-16.

Schafhäutl, K. E.
1851, *Geognostische Untersuchungen des Südbayrischen Alpengebirges:* 208 p., 45 pl., Literarisch-artistische Anstalt (München).

Schenk, August
1864, *Beiträge zur Flora der Vorwelt:* Palaeontographica, v. 11, p. 296-308, pl. 46-49.
1885, see Schimper, W. & Schenk, A. (1879-1890).

Schiller, W.
1930, *Die tektonische Natur von arthrophycus- und spirophyton-ähnlichen Gebilden im Altpaläozoikum der Provinz Buenos Aires (Argentinien):* Geol. Rundschau, v. 21, p. 145-151, text-fig. 1-4.

Schimper, W. Ph.
1846 [not seen by the editors].
1869-74, *Traité de Paléontologie végétale ou la flore du monde primitif:* v. 1, 740 p., 56 pl. (1869); v. 2, 522 p., pl. 57-84 (1870); p. 523-968, pl. 85-94 (1872); v. 3, p. 1-896, pl. 95-110 (1874). J. B. Baillière et fils (Paris).

————, & Schenk, August
1879-90, *Palaeophytologie:* in Handbuch der Palaeontologie, K. A. von Zittel (ed.), II. (Abth.): 958 p., 429 text-fig., Oldenbourg (München & Leipzig). (p. 1-152: 1879: p. 329-396: 1885).

Schindewolf, O. H.
1921, *Studien aus dem Marburger Buntsandstein. I, II:* Senckenbergiana, v. 3, p. 33-49.
1928, *Studien aus dem Marburger Buntsandstein. III-VII:* Same, v. 10, p. 16-54, text-fig. 1-14.
1956, *Über präkambrische Fossilien:* Geotekton. Symposium zu Ehren von H. Stille (F. Lotze, ed.), p. 455-480, pl. 31-34, Ferdinand Enke (Stuttgart).

1962, *Parasitäre Thallophyten in Ammoniten-Schalen:* Paläont. Zeitschr., H. Schmidt-Festbd., p. 206-215, text-fig. 1, pl. 21-23.

1963, *Pilze in oberjurassischen Ammoniten-Schalen:* Neues Jahrb. Geologie, Paläontologie, Abhandl., v. 118, p. 177-181, pl. 16.

Schlotheim, E. F. Baron von

1820, *Die Petrefactenkunde auf ihrem jetzigen Standpunkte durch die Beschreibung seiner Sammlung versteinerter und fossiler Überreste des Thier-und Pflanzenreiches der Vorwelt erläutert:* 437 p., 15 pl., Becker (Gotha).

1822, *Nachträge zur Petrefactenkunke. 1. Abt.:* 100 p., 21 pl., Becker (Gotha).

Schloz, Wilhelm

1968, *Über Beobachtungen zur Ichnofazies und über umgelagerte Rhizocorallien im Lias a Schwabens:* Neues Jahrb. Geologie, Paläontologie, Monatsh., 1968, p. 691-698, text-fig. 1, 2.

1972, *Zur Bildungsgeschichte der Oolithenbank (Hettangium) in Baden-Württemberg:* Arbeit. Inst. Geologie, Paläontologie, Univ. Stuttgart, n. ser., v. 67, p. 101-212, text-fig. 1-40, pl. 19-36.

Schmidt, Martin

1928, *Die Lebewelt unserer Trias:* 461 p., 1220 text-fig., Hohenlohe (Öhringen).

1934, *Cyclozoon philippi und verwandte Gebilde:* Heidelberger Akad. Wiss., math.-nat. Kl., Sitzungsber. 1934, pt. 6, 31 p., 4 pl.

Schmidt, W. J.

1954, *Über Bau und Entwicklung der Zähne des Knochenfisches Anarrhichas lupus L. und ihren Befall mit "Mycelites ossifragus":* Zeitschr. Zellforschg. u. Mikroskop. Anatomie, v. 40, p. 25-48. (See also Natur u. Volk, v. 85, p. 58-61, 1955).

Schmidtgen, O.

1927, *Tierfährten im oberen Rotliegenden bei Mainz:* Paläont. Zeitschr., v. 9, p. 101-107, text-fig. 1-7.

1928, *Eine neue Fährtenplatte aus dem Rotliegenden von Nierstein am Rhein:* Palaeobiologica, v. 1, p. 245-252, text-fig. 1, 2, 2 pl.

Schneid, Th.

1938, *Über eine interessante neue fossile Lebensspur aus dem mittleren Malm Frankens (Xenohelix suprajurassica n. sp.):* Zentralbl. Mineralogie, Geologie, Paläontologie, 1938, B, p. 312-315, 1 text-fig.

Schneider, Wilfried

1962, *Lebensspuren aus der Gräfenthaler Serie (Ordovizium) am Schwarzburger Sattel:* Geologie, v. 11, p. 954-960, text-fig. 1-8.

Schremmer, Fritz

1954, *Bohrschwammspuren in Actaeonellen aus der nordalpinen Gosau:* Österr. Akad. Wiss., math.-naturw. Kl., Sitzungsber., pt. 1, v. 163, p. 297-300, 1 pl.

Schroeder, P. C.

1968, *On the life history of Nereis grubei (Kingberg), a polychaete annelid from California:* Pacific Science, v. 22, p. 476-481, 1 text-fig.

Schröter, Carl

1894, *Notiz über ein Taenidium aus dem Flysch von Ganey bei Seewis:* Naturf. Gesell. Graubünden, Jahresber., n. ser., v. 37, p. 79-87, text-fig. 1, 2, 1 pl.

Schultz, C. B.

1942, *A review of the Daimonelix problem:* Univ. Nebraska Studies, Stud. Sci. Technol., v. 2, 30 p., text-fig. 1-17.

Scott, A. J.

1962, *Review: Treatise on invertebrate paleontology: Part W, Miscellanea (conodonts, conoidal shells of uncertain affinities, worms, trace fossils and problematica):* Jour. Paleontology, v. 36, 1398-1401.

Scott, D. H.

1900, *Studies in fossil botany:* 1st edit., 533 p., Adam and Charles Black (London).

Sederholm, J. J.

1911, *Geologisk öfversiktskarta öfver Finland. Sektionen B.2. Tammersfors, Beskrifning till bergartskartan:* Geologiska Kommissionen (Helsingfors). [Not seen by the editors].

1924, *Über die primäre Natur des Coryciums:* Centralbl. Mineralogie, Geologie, Paläontologie, 1924, p. 717-718.

1925, *Nochmals das Corycium:* Same, 1925, B, p. 360-363, text-fig. 1-4.

Sedgwick, Adam

1848, *On the organic remains found in the Skiddaw slate, with some remarks on the classification of the older rocks of Cumberland and Westmoreland:* Geol. Soc. London, Quart. Jour., v. 4, p. 216-225, text-fig. 1, 2.

Seebach, K. A. L. von

1876 [not seen by the editors].

Seilacher, Adolf

1953a, *Studien zur Palichnologie. I. Über die Methoden der Palichnologie:* Neues Jahrb. Geologie, Paläontologie, Abhandl., v. 96, p. 421-452, text-fig. 1-10, pl. 14.

1953b, *Studien zur Palichnologie. II. Die fossilen Ruhespuren (Cubichnia):* Same, Abhandl., v. 98, p. 87-124, text-fig. 1-5, 7 pl.

1953c, *Der Brandungssand als Lebensraum in Gegenwart und Vorzeit:* Natur u. Volk, v. 83, p. 263-272, text-fig. 1-9.

1954, *Die geologische Bedeutung fossiler Lebens-spuren:* Deutsch. Geol. Gesell., Zeitschr., v. 105, p. 213-227, text-fig. 1-3, pl. 7, 8.

1955, *Spuren und Fazies im Unterkambrium:* in O. H. Schindewolf & A. Seilacher, Beiträge zur Kenntnis des Kambriums in der Salt Range (Pakistan), Akad. Wiss. Lit. Mainz, math.-nat. Kl., Abhandl., no. 10, 1955, p. 11-143, text-fig. 1-6, pl. 22-27.

1956a, *Der Beginn des Kambriums als biologische Wende:* Neues Jahrb. Geologie, Paläonto-logie, Abhandl., v. 103, p. 155-180, text-fig. 1, 2, pl. 8, 9.

1956b, *Ichnocumulis n. g., eine weitere Ruhespur des schwäbischen Jura:* Same, Monatsh., 1956, p. 153-159, text-fig. 1-5.

1957, *An-aktualistisches Wattenmeer?:* Paläont. Zeitschr., v. 31, p. 198-206, text-fig. 1, 2, pl. 22, 23.

1959, *Zur ökologischen Charakteristik von Flysch und Molasse:* Eclogae Geol. Helvetiae, v. 51 (1958), p. 1062-1078, text-fig. 1, 3 pl.

1960, *Lebensspuren als Leitfossilien:* Geol. Rund-schau, v. 49, p. 41-50, text-fig. 1-3, 2 pl.

1962, *Paleontological studies on turbidite sedi-mentation and erosion:* Jour. Geology, v. 70, p. 227-234.

1963, *Lebensspuren und Salinitätsfazies:* Fortschr. Geol. Rheinld. u. Westfal., v. 10, p. 81-94, text-fig. 1-6, 1 table.

1964a, *Sedimentological classification and nomen-clature of trace fossils:* Sedimentology, v. 3, p. 253-256.

1964b, *Review of Häntzschel, W.: Trace fossils and problematica.—Treatise on Invertebrate Pa-leontology (Herausgeber, R. C. Moore), Part W, 177-245, text-fig. 109-149, Law-rence, Kansas:* Zentralbl. Geologie, Paläon-tologie, Teil II, p. 875.

1964c, *Biogenic sedimentary structures:* in Ap-proaches to paleoecology, J. Imbrie & N. D. Newell (eds.), p. 296-316, John Wiley & Sons, Inc. (New York).

1967a, *Vorzeitliche Mäanderspuren:* in Die Strassen der Tiere, H. Hediger (ed.), p. 294-306, text-fig. 1-8, Verl. Vieweg (Braunschweig).

1967b, *Bathymetry of trace fossils:* Marine Geol-ogy, v. 5, p. 413-428, text-fig. 1-4, 2 pl.

1967c, *Fossil behavior:* Scientific American, v. 217, no. 2, p. 72-80.

1968, *Sedimentationsprozesse in Ammonitenge-häusen:* Akad. Wiss. Lit., Math.-nat. Kl., Abhandl., Jahrg. 1967, no. 9, p. 191-203, text-fig. 1-5, 1 pl.

1969a, *Sedimentary rhythms and trace fossils in Paleozoic sandstones of Libya:* in Geology, archaeology and prehistory of the southwest-ern Fezzan, W. H. Kanes (ed.), Libya, 11th ann. field conf. 1969, p. 117-123, text-fig. 1, pl. 1, 2 (1970).

1969b, *Paleoecology of boring barnacles:* Am. Zoologist, v. 9, p. 705-719, text-fig. 1-8, 6 pl.

1970, *Cruziana stratigraphy of "nonfossiliferous" Paleozoic sandstones:* in Trace fossils, T. P. Crimes & J. C. Harper (eds.), Geol. Jour., spec. issue no. 3, p. 447-476, text-fig. 1-11, pl. 1, Seel House Press (Liverpool).

——, & **Crimes, T. P.**

1969, *"European" species of trilobite burrows in eastern Newfoundland:* in North Atlantic— geology and continental drift, Marshall Kay (ed.), Am. Assoc. Petroleum Geologists, Mem. 12, p. 145-148, text-fig. 1, 1 pl.

——, & **Hemleben, Christoph**

1966, *Beiträge zur Sedimentation and Fossil-führung des Hunsrückschiefers, Teil 14, Spurenfauna und Bildungstiefe des Huns-rückschiefers:* Hess. Landesamt Boden-forsch., Notizblatt, v. 94, p. 40-53, text-fig. 1-5, pl. 2-4.

——, & **Meischner, D.**

1965, *Fazies-Analyse im Paläozoikum des Oslo-Gebietes:* Geol. Rundschau, v. 54, p. 596-619, text-fig. 1-13, 1 table.

Selley, R. C.

1970, *Ichnology of Paleozoic sandstones in the southern desert of Jordan: a study of trace fossils in their sedimentological context:* in Trace fossils, T. P. Crimes & J. C. Harper (eds.), Geol. Jour., spec. issue no. 3, p. 477-488, text-fig. 1-5, pl. 1, Seel House Press (Liverpool).

Sellwood, B. W.

1970, *The relation of trace fossils to small sedi-mentary cycles in the British Lias:* in Trace fossils, T. P. Crimes and J. C. Harper (eds.), Geol. Jour., spec. issue no. 3, p. 489-504, text-fig. 1-6, pl. 1, Seel House Press (Liver-pool).

1971, *A Thalassinoides burrow containing the crustacean Glyphaea undressieri (Meyer) from the Bathonian of Oxfordshire:* Pa-laeontology, v. 14, p. 589-591, pl. 108.

1972, *Regional environmental changes across a Lower Jurassic stage-boundary in Britain:* Same, v. 15, p. 125-157, text-fig. 1-14, pl. 28, 29.

Selwyn, A. R. C.

1890, *Tracks of organic origin in rocks of the Animikie group:* Am. Jour. Sci., ser. 3, v. 39, p. 145-147.

Serres, Marcel de

1840, *Description de quelques mollusques fossiles nouveaux des terrains infra-jurassiques et de craie compacte inférieure du Midi de la*

France: Ann Sci. Nat. Paris (Zool.), sér. 2, v. 14, p. 5-25, pl. 1, 2.

Seward, A. C.

1894, *Catalogue of the Mesozoic plants in the Department of Geology, British Museum. The Wealden flora. Pt. 1. Thallophyta—Pteridophyta:* 179 p., 11 pl., Brit. Museum (Nat. History) (London).

1898, *Fossil plants for students of botany and geology. I:* 452 p., Cambridge Univ. Press (Cambridge).

1903, *Fossil floras of the Cape Colony:* South Afr. Museum, Ann., v. 4, pt. 1, p. 1-122, text-fig. 1-8, 14 pl. (Ganzer Bd.: 1903-08).

1931, *Plant life through the ages; a geological and botanical retrospect:* 601 p., illus., The University Press (Cambridge, Eng.).

Sharpe, C. F. S.

1932, *Eurypterid trail from the Ordovician:* Am. Jour. Sci., ser. 5, v. 24, p. 355-361, text-fig. 1, 2.

Shaw, A. B.

1955, *Paleontology of northwestern Vermont. V. The Lower Cambrian fauna:* Jour. Paleontology, v. 29, p. 775-805, pl. 73-76.

Sheldon, R. W.

1968, *Probable gastropod tracks from the Kinderscout Grit of Soyland Moor, Yorkshire:* Geol. Mag., v. 105, p. 365-366, pl. 12.

Shepard, C. U.

1867, *On the supposed tadpole nests, or imprints made by the Batrachoides nidificans Hitchcock, in the red shale of the New Red Sandstone of South Hadley, Mass.:* Am. Jour. Sci., ser. 2, v. 43, p. 99-104.

Shinn, E. A.

1968, *Burrowing in Recent lime sediments of Florida and the Bahamas:* Jour. Paleontology, v. 42, p. 879-894, pl. 109-112.

1972, *Worm and algal-built columnar stromatolites in the Persian Gulf:* Jour. Sed. Petrology, v. 42, p. 837-840, text-fig. 1-3.

Sieber, R.

1937, *Neue Untersuchung über die Stratigraphie und Ökologie der alpinen Triasfaunen:* Neues Jahrb. Mineralogie, Geologie, Paläontologie, Beil.-Bd., v. 78, p. 123-188.

Silén, Lars

1946, *On two new groups of Bryozoa living in shells of molluscs:* Arkiv Zoologi, v. 38B, no. 1, p. 1-7.

1947, *On the anatomy and biology of Penetrantidae and Immergentiidae (Bryozoa):* Same, v. 40A, no. 4, p. 11-48, text-fig. 1-70.

Silliman, Benjamin, Jr.

1851, *On the origin of a curious spheroidal struc-* *ture in certain sedimentary rocks:* Am. Assoc. Adv. Sci., Proc., v. 4, p. 10-12.

Silva, S. de Oliveira

1952, *Siluriano no Rio Tapajos:* Engenharia, Mineraçao e Metallurgia, v. 16, p. 380, text-fig. 1.

Silvestri, A.

1911, *Sulla vera natura dei "Paleodictyon":* Soc. Geol. Italiana, Boll., v. 30, p. 85-106, text-fig. 1, 2, pl. 6, 7.

Simonelli, Vittorio

1905, *Intorno ad alcune singolari paleoicniti del Flysch appenninico:* R. Accad. Sci. Bologna, Mem., ser. 6, v. 2, p. 91-96 (263-268), text-fig.

Simpson, Frank

1967, *O niektórych róznicach w morfologii Palaeodictyon Meneghini:* Polsk. Towarzyst. Geol., Rocznik, v. 37, p. 509-514, pl. 35, 36. [*Some morphological variants of Palaeodictyon Meneghini.*]

1969, *Rotamedusa roztocensis gen. et sp. nov., meduza z Eoceńskiego fliszu Karpackiego:* Same, Rocznik, v. 39, no. 4, p. 697-703, text-fig. 1, 2, pl. 114, 115. [*Rotamedusa roztocensis gen. et sp. nov., a medusa from the Eocene flysch of the Carpathians.*] [Pol., with Eng. summ.]

1970, *O sedymentacji środkowego eocenu serii magurskiej w Polskich Karpatach zachodnich:* Same, Rocznik, v. 40, p. 209-286, text-fig. 1-18, pl. 6-11. [*Sedimentation of the middle Eocene of the Magura Series, Polish western Carpathians.*]

Simpson, Scott

1957, *On the trace-fossil Chondrites:* Geol. Soc. London, Quart. Jour., v. 112, p. 475-499, text-fig. 1, 2, pl. 21-24.

1970, *Notes on Zoophycos and Spirophyton:* in Trace fossils, T. P. Crimes & J. C. Harper (eds.), Geol. Jour., spec. issue no. 3, p. 505-514, text-fig. 1-4, Seel House Press (Liverpool).

Sinclair, G. W.

1951, *The generic name Bilobites:* Jour. Paleontology, v. 25, p. 228-231.

Ślączka, Andrzej

1964, *Meduza z fliszu karpackiego—Kirklandia multiloba, n. sp.:* Polsk. Towarzyst. Geol., Rocznik, v. 34, no. 3, p. 479-486, text-fig. 1-3, pl. 19-20. [*Kirklandia multiloba, n. sp. —a jellyfish from the Carpathian flysch.*]

1965, *Nowe problematyki radialne z fliszu karpackiego:* Spraw. Pos. Kom. Odd. Pan Krakowie, 1965, p. 470-471 (Pol., with Eng. summ.). [*New star-shaped problematica from the Carpathian flysch.*]

Smith, John
1893, *Peculiar U-shaped tubes in sandstone near Crawfurdland Castle and in Gowkha Quarry, near Kilwinning:* Geol. Soc. Glasgow, Trans., v. 9, p. 289-292, pl. 10.

Smith, N. D., & Hein, F. J.
1971, *Biogenic reworking of fluvial sediments by staphylinid beetles:* Jour. Sed. Petrology, v. 41, p. 598-602.

Sohl, N. F.
1969, *The fossil record of shell boring by snails:* Am. Zoologist, v. 9, p. 725-734, text-fig. 1-15, table 1.

Sokolov, B. S.
1972, *The Vendian Stage in earth history:* 24th Internatl. Geol. Congress, sec. 1, p. 78-84 (Montreal).
1973, *Vendian of northern Eurasia:* in Arctic geology, M. G. Pitcher (ed.), Am. Assoc. Petroleum Geologists, Mem. 19, p. 204-218, text-fig. 1-6.

Sollas, W. J.
1893, *The geology of Dublin and its neighbourhood:* Geologists' Assoc., Proc., v. 13, p. 91-122, text-fig. 1-16, pl. 3, 4.
1895, *Pucksia Mac Henryi, a new fossil from the Cambrian rocks of Howth:* Royal Dublin Soc., Sci. Proc., n. ser., v. 8, pt. 4, p. 297-303.
1900, *Ichnium Wattsii, a worm track from the slate of Bray Head, with observations on the genus Oldhamia:* Geol. Soc. London, Quart. Jour., v. 56, p. 273-286, pl. 17-19.

Solle, Gerhard
1938, *Die ersten Bohrspongien im europäischen Devon und einige andere Spuren:* Senckenbergiana, v. 20, p. 154-178, text-fig. 1-22.

Soot-Ryen, Tron
1969, *Mytilacea:* in Treatise on invertebrate paleontology, R. C. Moore (ed.), Part N, p. N271-N281, text-fig. C16-C22, Geol. Soc. America & Univ. Kansas (Boulder, Colo.; Lawrence, Kans.).

Sordelli, Ferdinando
1873, *Descrizione di alcuni avanzi vegetali delle argille plioceniche Lombarde, coll'aggiunta di un Elenco delle piante fossili finora conosciute in Lombardia:* Soc. Italiana Sci. Nat., Atti, v. 16, p. 350-429, pl. 4a-7a.

Spandel, Erich
1909, *Der Rupelton des Mainzer Beckens, seine Abteilungen und deren Foraminiferenfauna, sowie einige weitere geologisch-paläontologische Mitteilungen über das Mainzer Becken:* Offenbacher Ver. Naturk., Bericht. Tätigkeit, p. 57-230, 2 pl.

Speck, Josef
1945, *Fährtenfunde aus dem subalpinen Burdigalien und ihre Bedeutung für Fazies und Paläogeographie der oberen Meeresmolasse:* Eclogae Geol. Helvetiae, v. 38, no. 2 (1945), p. 411-416, text-fig. 1, pl. 15.

Spjeldnaes, Nils
1963, *A new fossil (Papillomembrana sp.) from the Upper Precambrian of Norway:* Nature, v. 200, no. 4901, p. 63-64, text-fig. 1-3, 1 pl. (London).

Sprigg, R. C.
1947, *Early Cambrian (?) jellyfishes from the Flinders Range, South Australia:* Royal Soc. South Australia, Trans., v. 71, p. 212-224, text-fig. 1-7, pl. 5-8.
1949, *Early Cambrian "jellyfishes" of Ediacara, South Australia and Mount John, Kimberley District, Western Australia:* Same, v. 73, p. 72-99, text-fig. 1-10, pl. 9-21.

Squinabol, Senofonte
1887, *Contribuzione alle flora fossile dei terreni terziarii della Liguria. I. Fucoidi ed Elmintoidea:* Soc. Geol. Italiana, Boll., v. 6, p. 545-561, pl. 14-19.
1890, *Alghe e Pseudoalghe italiane:* Atti Soc. Ligustica Sci. Nat. Geogr., v. 1, p. 29-49, p. 166-199, pl. 5-12.
1891, *Contribuzioni alla flora fossile dei terreni terziarii della Liguria. I. Alghe:* p. i-xxv, pl. A-E, Sordomuti (Genova).

Squire, A. D.
1973, *Discovery of Late Precambrian trace fossils in Jersey:* Geol. Mag., v. 110. p. 223-226, pl. 1.

Stanley, D. J.
1971, *Fish-produced markings on the outer continental margin east of the middle Atlantic states:* Jour. Sed. Petrology, v. 41, p. 159-170, text-fig. 1-8.

Stanley, S. M.
1969, *Bivalve mollusk burrowing aided by discordant shell ornamentation:* Science, v. 166, p. 634-635, text-fig. 1, 2.

Stanton, R. J., Jr.
1966, *Paleoecologic and stratigraphic value of radiosphaerid calcispheres in North America, and the significant variables in calcisphere classification:* Geol. Soc. America, Program Ann. Mtg., p. 211 (San Francisco) (Nov.).

Staub, Móric
1899, *Über die "Chondrites" benannten fossilen Algen:* Földtani Közlony, v. 29, p. 110-121, text-fig. 1-4 (Hungar. text: p. 16-32).

Stefani, Carlo de
1879, *La Montagnola Senese, studio geologico. VI.*

Delle Eufotidi e delle altre rocce appartenenti all'Eoceno superiore: R. Comit. Geol. Italia, Boll., v. 10, p. 431-460.

1885, *Studi paleozoologici sulle creta superiore e media dell'Apennino settentrionale:* Atti R. Accad. Lincei, Mem., ser. 4, v. 1, p. 73-121, 2 pl.

————, Major, C. J. Forsyth, &
Barbey, William

1895, *Karpathos. Étude géologique, paléontologique et botanique:* 180 p., 15 pl., Bridel & Cie (Lausanne).

Stehmann, Erich

1934, *Das Unterkambrium und die Tektonik des Paläozoikums auf Bornholm:* Greifswald Univ., Geol.-Pal. Inst., Abhandl., v. 14, 63 p., 10 pl.

1935, *Über Wurmröhren im Nexösandstein auf Bornholm:* Frankfurter Beitr. Geschiebeforsch., Beih. Zeitschr. Geschiebeforsch., 1935, p. 28-33.

Steinmann, Gustav

1907, *Einführung in die Paläontologie:* 2nd edit., 542 p., 902 text-fig., W. Engelmann (Leipzig).

Stephenson, L. W.

1941, *The larger invertebrate fossils of the Navarro Group of Texas:* Univ. Texas, Publ. no. 4101, 641 p., 10 text-fig., 95 pl.

1952, *Larger invertebrate fossils of the Woodbine formation (Cenomanian) of Texas:* U. S. Geol. Survey, Prof. Paper 242, p. 1-226, pl. 8-59.

Sternberg, K. M. Graf. von

1820-38, *Versuch einer geognostisch-botanischen Darstellung der Flora der Vorwelt:* v. 1-8, 364 p., 136 pl., Fr. Fleischer (Leipzig, Prague) [v. 1, pt. 1, p. 1-24 (1820); pt. 2, p. 1-33 (1822); pt. 3, p. 1-39 (1823); pt. 4, p. 1-48 (1825); pt. 5, 6, p. 1-80 (1833); v. 7, 8, p. 81-220 (1838)].

Stevens, G. R.

1968, *The Amuri fucoid:* New Zealand Jour. Geology, Geophysics, v. 11, p. 253-261.

Stiehler, A. W.

1857, *Beiträge zur Kenntnis der vorweltlichen Flora des Kreidegebirges im Harze:* Palaeontographica, v. 5, p. 45-80, pl. 9-11.

**Stöcklin, Jovan, Ruttner, A., &
Nabavi, M. H.**

1964, *New data on the lower Paleozoic and Pre-Cambrian of north Iran:* Iran, Geol. Survey, Rept. no. 1, p. 1-29.

Størmer, Leif

1934, *Downtonian Merostomata from Spitsbergen, with remarks on the suborder Synziphosura:*

Norske Vidensk Akad. Oslo, Skrifter, 1934, 2. Bind, mat.-nat. Kl., no. 3, 26 p., text-fig. 1-4, 2 pl.

Stoneley, H. M. M.

1958, *The Upper Permian flora of England:* Brit. Museum (Nat. History), Bull., v. 3, no. 9, p. 295-337, text-fig. 1-16, 5 pl.

Stopes, M. C.

1913, *Catalogue of the Mesozoic plants in the British Museum (Nat. Hist.). The Cretaceous flora. Pt. 1. Bibliography, algae and fungi:* 285 p., 2 pl., Brit. Museum (Nat. History) (London).

Stoppani, Antonio

1857, *Studii geologici e paleontologici sulla Lombardia del sacerdote prof. Antonio Stoppani, colla, Descrizione de alcune nuove specie di pesci fossili di Perledo e di altre località lombarde, studii di Cristoforo Bellotti:* 461 p., C. Turati (Milano).

Stout, Wilber

1956, *The fossil Conostichus:* Ohio Jour. Sci., v. 56, no. 1, p. 30-32, text-fig. 1, 2.

Straaten, L. M. J. U. van

1949, *Occurrence in Finland of structures due to subaqueous sliding of sediments:* Commiss. Géol. Finlande, Bull., v. 144, p. 9-18, text-fig. 1-11.

1954, *Sedimentology of Recent tidal flat deposits and the Psammites du Condroz (Devonian):* Geologie en Mijnbouw, v. 16, p. 25-47, text-fig. 1-15, 2 pl.

Stradner, H.

1961, *Über fossile Vorkommen von Nannofossilien im Mesozoikum und Alttertiär:* Erdöl-Zeitschr., v. 77, p. 77-88, 99 text-fig.

Strigel, Adolf

1929, *Das süddeutsche Buntsandsteinbecken:* Naturhist-med. Ver. Heidelberg, Verhandl., n. ser., v. 16, p. 80-465.

Stur, Dionys

1877, *Die Culm-Flora. I. Die Culm-Flora des mährisch-schlesischen Dachschiefers. II. Die Culm-Flora der Ostrauer und Waldenburger Schichten:* K. K. Geol. Reichsanst. Wien, Abhandl., v. 8, 472 p., 44 pl. (I: Wien 1875).

Sullivan, C. J., & Öpik, A. A.

1951, *Ochre deposits, Rumbalora, Northern Territory:* Australia Bureau Min. Res., Geology Geophysics, Bull., v. 8, 27 p., 9 pl.

Summerson, C. H.

1951, *Cambrian tracks in the Lamotte sandstone:* Jour. Paleontology, v. 25, p. 533-536, text-fig. 1, 2.

Sun, Y. C.

1924, *Contributions to the Cambrian faunas of North China:* Palaeont. Sinica, ser. B, v. 1, no. 4, 90 p., text-fig. 1, pl. 1-5.

Surlyk, F., Bromley, R. G., Asgaard, U., & Pedersen, K. R.

1971, *Preliminary account of the mapping of the Mesozoic formations of south-east Jameson Land:* Grønlands Geol. Unders., Rapport, no. 37, p. 24-32, text-fig. 1-11.

Świdziński. H.

1934, *Uwagi o budowie Karpat fliszowych—Remarques sur la structure des Karpates flyscheuses:* Państ. Inst. Geol., Biul. (Serv. Géol. Pologne), v. 8, no. 1, p. 75-139 (Polish), p. 141-199 (French), text-fig. 1-4, pl. 7-10.

Szczechura, Janina

1969, *Problematic microfossils from the upper Eocene of Poland:* Rev. Espan. Micropaleontologia, v. 1, no. 1, p. 81-94, text-fig. 1, 2, 4 pl.

Sze, H. C.

1951, *Über einen problematischen Fossilrest aus der Wealdenformation der südlichen Mandschurei:* Sci. Rec. Chunking, v. 4, p. 81-83, 1 pl.

Taljaard, M. S.

1962, *On the paleogeography of the Table Mountain Sandstone series:* South Afr. Geogr. Jour., v. 44, p. 25-27, pl. 1, 2.

Tanaka, Keisaku

1970, *Sedimentation of the Cretaceous flysch sequence in the Ikushumbetsu area, Hokkaido, Japan:* Geol. Survey Japan, Rept. 236, 102 p., text-fig. 1-48, 12 pl.

1971, *Trace fossils from the Cretaceous flysch of the Ikushumbetsu area, Hokkaido, Japan:* Same, Rept. no. 242, 31 p., text-fig. 1, 11 pl.

Tappan, Helen, & Loeblich, A. R., Jr.

1968, *Lorica composition of modern and fossil Tintinnida (ciliate Protozoa), systematics, geologic distribution, and some new Tertiary taxa:* Jour. Paleontology, v. 42, p. 1378-1394, text-fig. 1, pl. 165-171.

Tasch, Paul

1968, *A Permian trace fossil from the Antarctic Ohio Range:* Kansas Acad. Sci., Trans., v. 71, p. 33-37, text-fig. 1, 2.

Tate, George

1859, *The geology of Breadnell, in the county of Northumberland, with a description of some annelids of the Carboniferous formation:* The Geologist, 1859, p. 59-70, pl. 2.

Tate, Ralph

1876, in R. Tate & J. F. Blake, *The Yorkshire Lias:* 475 p., 23 pl., J. Van Voorst (London).

Tauber, A. F.

1944, *Über prämortalen Befall von rezenten und fossilen Molluskenschalen durch tubikole Polychaeten (Spionidae):* Palaeobiologica, v. 8, no. 1/2, p. 154-172, illus.

1949, *Paläobiologische Analyse von Chondrites furcatus Sternberg:* Geol. Bundesanst. Wien, Jahrb., v. 92, no. 3-4, p. 141-154, text-fig. 1-3.

Taylor, B. J.

1967, *Trace fossils from the Fossil Bluff Series of Alexander Island:* Brit. Antarctic Survey, Bull., v. 13, p. 1-30, text-fig. 1-9.

Taylor, J. D.

1970, *Feeding by predatory gastropods in a Tertiary (Eocene) molluscan assemblage:* Palaeontology, v. 13, p. 254-260, pl. 46.

Taylor, M. E.

1966, *Precambrian mollusc-like fossils from Inyo County, California:* Science, v. 153, p. 198-201, text-fig. 1-4.

Teichert, Curt

1934, *Inlandeis und Gletscher Ostgrönlands:* Natur u. Volk, v. 64, p. 140-151, text-fig. 1-13.

1945, *Parasitic worms in Permian brachiopod and pelecypod shells in Western Australia:* Am. Jour. Sci., v. 243, p. 197-206, text-fig. 1, 3 pl.

1964a, *Recent German work on the Cambrian and saline series of the Salt Range, West Pakistan:* Geol. Survey Pakistan, Records, v. 11, p. 1-8.

1964b, *Doubtful taxa:* in Treatise on invertebrate paleontology, R. C. Moore (ed.), Part K, p. K484-K490, text-fig. 348-351, Geol. Soc. America & Univ. Kansas Press (New York; Lawrence, Kans.).

1965, *Devonian rocks and paleogeography of Central Arizona:* U. S. Geol. Survey, Prof. Paper 464, 181 p., 40 text-fig., 21 pl.

1970, *Runzelmarken (wrinkle marks):* Jour. Sed. Petrology, v. 40, p. 1056.

1972, [Discussion] in Walter Häntzschel & O. Kraus, Names based on trace fossils (ichnotaxa): request for a recommendation. Z. N. (S.) 1973: Bull. Zool. Nomenclature, v. 29, p. 140-141.

1973, *Clay rolls as pseudofossils:* Geotimes, v. 18, no. 5, p. 11-12 (May).

Teilhard, de Chardin, P. P.

1931, *On an enigmatic pteropod-like fossil from the Lower Cambrian of Southern Shansi, Biconulites grabaui, nov. gen., nov. sp.:*

Geol. Soc. China, Bull., v. 10, p. 179-188, text-fig. 1, 2, 2 pl.

Termier, Henri, & Termier, Geneviève

1947, *Un organisme récifal du Cambrien marocain: Anzalia cerebriformis nov. gen. nov. sp.:* Soc. Géol. France, Bull., ser. 5, v. 17, p. 61-66, text-fig. 1-4.

1951, *Sur deux formes énigmatiques de l'Ordovicien marocain: Leckwyckia et Khemisina:* Serv. géol. Maroc, Div. Mines et Géol., Notes et Mém., v. 85 (=Notes Serv. géol., 5), p. 187-198, text-fig. 1-6.

1964, *Les couches à Anzalia du Cambrien inférieur du Haut Atlas:* Same, Notes et Mém., v. 172 (1963), p. 7-9, text-fig. 1-16.

Terquem, Olry, & Berthelin, G.

1875, *Étude microscopique des marnes du lias moyen d' Essey-lès-Nancy, zone inférieure de l'assise à Ammonites Margaritatus:* 126 p., text-fig., pl., F. Savy (Paris).

Ters, Mireille, & Deflandre, Georges

1966, *Sur l'age cambro-silurien des terrains anciens de la Vendée littorale (ex-Briovérien):* Acad. Sci. [Paris], sér. D, Comptes Rendus, 262, p. 339-342, 1 pl.

Thiollière, Victor

1858, [Talk on the Oxford clay, Great Oolite, Lower Oolite, and Lias]: Soc. Géol. France, Bull., sér. 2, v. 15, p. 710-720. (Réunion extraordinaire, Seance Sept. 1858.)

Thomas, D. H.

1935, *On Dinocochlea ingens B. B. Woodward, and other spiral concretions:* Geologists' Assoc., Proc., v. 46, p. 1-17, text-fig. 1, 2, 2 pl.

1961, *Skylonia mirabilis gen. et. sp. nov., a problematical fossil from the Miocene of Kenya:* Ann. Mag. Nat. History, ser. 13, v. 4, no. 42, p. 359-363, text-fig. 1, pl. 13.

Thomasset, J. -J.

1932, *Sur un champignon fossile: Mycelites ossifragus (Roux):* Soc. Géol. France, Bull., sér. 5, v. 1 (1931), p. 597-603, text-fig. 1-4.

Thomson, P. W.

1940, *Beitrag zur Kenntnis der fossilen Flora des Mitteldevons in Estland:* Loodusuur. Seltsi Aruanded, v. 45 (1938), p. 1-24, 7 pl.

Thusu, B.

1972, *Depositional environments of the Rochester Formation (Middle Silurian) in southern Ontario:* Jour. Sed. Petrology, v. 42, p. 930-934, text-fig. 1, 2.

Tillyard, R. J.

1936, *Description of the fossils:* in T. W. E. David, & R. J. Tillyard, Memoir on fossils of the Late Pre-Cambrian (newer Proterozoic) from the Adelaide Series, South Australia, p. 63-84, pl. 1-10, Angus & Robertson (Sydney).

Toepelman, W. C., & Rodeck, H. G.

1936, *Footprints in late Paleozoic red beds near Boulder, Colorado:* Jour. Paleontology, v. 10, p. 660-662, text-fig. 1, 2.

Tomlinson, J. T.

1963, *Acrothoracican barnacles in Paleozoic myalinids:* Jour. Paleontology, v. 37, p. 164-166, text-fig. 1, 1 pl.

Toots, Heinrich

1963, *Helical burrows as fossil movement patterns:* Wyoming Univ., Contrib. Geology, v. 2, no. 2, p. 129-134, text-fig. 1-4.

1967, *Invertebrate burrows in the non-marine Miocene of Wyoming:* Same, v. 6, no. 2, p. 93-96.

Torell, O. M.

1868, *Bidrag till Sparagmitetagens geognosi och paleontologi:* Acta Univ. Lundensis, Lunds Univ. Årsskr., v. 4, pt. 2, 40 p., 3 pl.

1869, *Om Sparagmitetagens fauna og flora:* Skandinaviske Naturforsk. Forhandl., v. 10, Møde Christiania 1868, p. LXVI-LXVII.

1870, *Petrificata Suecana Formationis Cambricae:* Lunds Univ. Årsskr., v. 6, pt. 2, no. 8, p. 1-14.

Toula, Franz

1900, *Lehrbuch der Geologie:* 410 p., 367 text-fig., 30 pl., A. Hölder (Wien).

Trautschold, H. A.

1867, *Einige Crinoiden und andere Thierreste des jüngeren Bergkalks im Gouvernement Moskau:* Soc. Impér. Nat. Moscou, Bull., v. 40 (1867), p. 1-49, pl. 1-5.

Tromelin, Gaston de

1877, *Étude de la faune du grès silurien de May, Jurques, Campandré, Mont-Roberts, etc. (Calvados):* Soc. Linnéenne Normandie, Bull., sér. 3, v. 1 (1876-77), p. 5-82.

1878, *Étude des terrains paléozoiques de la Basse-Normandie, particulièrement dans les départments de l'Orne et du Calvados:* Assoc. franç. Avanc. Sci., Comptes Rendus, v. 6, sess. Le Harve, p. 493-501.

———, & Lebesconte, Paul

1876, *Essai d'un catalogue raisonné des fossiles siluriens des départements de Maine-et-Loire, de la Loire-Inférieure et du Morbihan:* Assoc. Franç. Avanc. Sci., v. 4, sess. Nantes, p. 601-661.

Trusheim, Ferdinand

1934, *Ein neuer Leithorizont im Hauptmuschelkalk von Unterfranken:* Neues Jahrb. Mineralogie, Geologie, Paläontologie, Beil.-Bd.

71, Abt. B, no. 3, p. 407-421, text-fig. 1, 2, 1 pl.

Turner, R. D.

1969, *Superfamily Pholadacea Lamarck, 1809:* in Treatise on invertebrate paleontology, R. C. Moore (ed.), Part N(2), p. N702-N741, text-fig. E162-E212, Geol. Soc. America & Univ. Kansas (Boulder, Colo.; Lawrence, Kans.).

Twenhofel, W. H.

1919, *Pre-Cambrian and Carboniferous algal deposits:* Am. Jour. Sci., ser. 4, v. 48, p. 339-353.

1924, *The geology and invertebrate paleontology of the Comanchean and "Dakota" formations of Kansas:* Kansas State Geol. Survey, Bull., v. 9, 135 p., 23 pl.

1928, *Geology of Anticosti Island:* Geol. Survey Canada, Mem., v. 154, 481 p., text-fig. 1, 60 pl.

Udden, J. A.

1898, *Fucoids or coprolites:* Jour. Geology, v. 6, p. 193-198, pl. 7, 8.

Ulrich, E. O.

1880, *Catalogue of fossils occurring in the Cincinnati Group of Ohio, Indiana, and Kentucky:* 26 p., Barclay (Cincinnati).

1889, *Preliminary description of new Lower Silurian sponges:* Am. Geologist, v. 3, p. 233-248.

1904, *Fossils and age of the Yakutat Formation. Description of collections made chiefly near Kodiak, Alaska:* Harriman Alaska Exped., v. 4, Geol., Paleont., p. 125-146, pl. 11-21 (Washington).

Umbgrove, J. H. F.

1925, *Eenige problematische Fossilen uit het Limburgsche Krijt:* Natuurhist. Maandbl., v. 14, p. 99-100, 1 pl.

Valeton, Ida

1971, *Tubular fossils in bauxite and the underlying sediments of Surinam and Guyana:* Geologie en Mijnbouw, v. 50, p. 733-741.

Van Gundy, C. E.

1951, *Nankoweap Group of the Grand Canyon Algonkian of Arizona:* Geol. Soc. America, Bull., v. 62, p. 953-959, text-fig. 1, 3 pl.

Van Straelen, Victor

1938, *Sur des restes de crustacés fouisseurs du Viséen inférieur du Nord de la France:* Musée Royal Histoire Nat. Belgique, Bull., v. 14, no. 30, 6 p., text-fig. 1-5.

Van Tuyl, F. M., & Berckhemer, Fritz

1914, *A problematic fossil from Catskill Formation:* Am. Jour. Sci., ser. 4, v. 38, p. 275-276, text-fig. 1.

Vanuxem, Lardner

1842, *Geology of New York, pt. III, comprising the survey of the 3d geological district:* 306 p., W. & A. White and J. Visscher (Albany).

Vassoevich [Vassojevič], N. B.

1932, *O nekotorykh priznakakh pozvolyayushchikh otlicht'oprokinutoe polozhenie flishevykh obrazovaniy ot normalynogo:* Akad. Nauk SSSR, Geol. Inst., Trudy, v. 2, p. 47-64, text-fig. 1-6, 3 pl. [*Some data allowing us to distinguish the overturned position of Flysch sedimentary formations from normal ones.*]

1951, *Usloviya obrazovaniya flisha:* 240 p., 86 text-fig., 10 pl., Gostoptekhizdat (Leningrad). [*The conditions of the formation of flysch.*]

1953, *O některych fliševych těksturach (znakach):* Lwowskogo Geol. Obščestva, Trudy, geol. ser., v. 3, p. 17-85, 10 pl. [*On some flysch textures (traces).*]

Vecchio, Celeste del

1919, *Su alcuni riliefi e impronte del Senoniana della Brianza:* Natura (Riv. Sci. Nat.), v. 10, p. 73-83, text-fig. 1-5.

Veevers, J. J.

1962, *Rhizocorallium in the Lower Cretaceous rocks of Australia:* Australia Bur. Min. Res., Geology, Geophysics, Bull., no. 62, p. 1-21, pl. 1-3.

1970, *Upper Devonian and Lower Carboniferous calcareous algae from the Bonaparte Gulf Basin, Northwestern Australia:* Same, Bull. 116, Paleont. Papers 1968, p. 173-188, text-fig. 1, 2, pl. 25-47.

Verma, K. K., & Prasad, K. N.

1968, *On the occurrence of some trace fossils in the Bhander Limestone (Upper Vindhyans) of Rewa district, M. P.:* Current Science, v. 37, no. 19, p. 557-558.

Villa, A. F. da

1844, *Sulla constituzione geologica e geognostica della Brianza, e segnatamente sul terreno cretaceo: memoria di Antonio e Giovanni Battista Villa:* 46 p., Presso gli editori dello Spettatore industriale (Milano).

Vinassa da Regny, P. E.

1904, *Fossili ed impronte del Montenegro:* Soc. Geol. Italiana, Boll., v. 23, p. 307-322, pl. 9.

Vita-Finzi, C., & Cornelius, P. F. S.

1973, *Cliff sapping by molluscs in Oman:* Jour. Sed. Petrology, v. 43, p. 31-32, text-fig. 1, 2.

Vitális, Sándor

1961, *Életnyomok a salgótarjáni barnakőszénmedencében:* Foldtani Közlony, v. 91, p. 3-19, text-fig. 1, 2, 15 pl. [*Trace fossils from the northern Salgótarján (Hungary) coal basin.*] (French summ.)

Viviani, Domenico

1805, *Phosphorescentia maris, quatuordecim lucescentium animalculorum novis speciebus illustrata a Domenico Viviani, . . . Accedit novi cujusdam generis e Molluscorum familia descriptio et anatomes . . .:* 17 p., 5 pl., J. Giossi (Genua).

Vlcek, V.

1902, *O některých problematických zkamenělinách českého cambria a spodniho siluro:* Palaeontogr. Bohemiae, v. 6, 9 p., 2 pl. (=Česká Akad. Cisare Frant. Josefa, tr. II). [*On some problematical fossils from the Bohemian Cambrian and Lower Silurian.*]

Voigt, Ehrhard

1957, *Harmeriella ? cretacea n. sp., ein fragliches parasitisches Bryozoon aus der Schreibkreide von Rügen:* Senckenbergiana Lethaea, v. 38, no. 5-6, p. 345-357, text-fig. 1-6, pl. 1.

1959, *Endosacculus moltkiae n. g. n. sp., ein vermutlicher fossiler Ascothoracide (Entomostr.) als Cystenbildner bei der Oktokoralle Moltkia minuta:* Paläont. Zeitschr., v. 33, p. 211-223, text-fig. 1, 2, pl. 25, 26.

1962, *Verkhnemelovye mshanki Evropeyskoi chasti SSSR i nekotorykh sopredelnykh oblastey:* p. 1-65, 28 pl., Moskov. Univ. (Moskva). [*Upper Cretaceous bryozoans of the European part of the USSR and some adjacent regions.*] (Ger. abstr.).

1965, *Über parasitische Polychaeten in Kreide-Austern sowie einige andere in Muschelschalen bohrende Würmer:* Paläont. Zeitschr., v. 39, p. 193-211, text-fig. 1-3, 3 pl.

1970, *Endolithische Wurm-Tunnelbauten (Lapispecus cuniculus n. g. n. sp. und Dodecaceria [?] sp.) in Brandungsgeröllen der oberen Kreide im nördlichen Harzvorland:* Geol. Rundschau, v. 60, pt. 1, p. 355-380, text-fig. 1-7. (with Engl., Fr., Russ. summ.).

1971, *Fremdskulpturen an Steinkernen von Polychaeten-Bohrgängen aus der Maastrichter Tuffkreide:* Paläont. Zeitschr., v. 45, p. 144-153, text-fig. 1, 2, pl. 15, 16.

1972a, *Tonrollen als potentielle Pseudofossilien:* Natur u. Museum, v. 102, p. 401-410, text-fig. 1-10.

1972b, *Über Talpina ramosa v. Hagenow 1840, ein wahrscheinlich zu den Phoronoidea gehöriger Bohrorganismus aus der Oberen Kreide:* Akad. Wiss. Göttingen, II. math.-phys. Kl., Nachricht., no. 7, p. 93-126, pl. 1-5.

1973, *Comments on the application concerning trace fossils. Z. N. (S.) 1973:* Bull. Zool. Nomenclature, v. 30, pt. 2, p. 69-70.

————, & Häntzschel, Walter

1956, *Die grauen Bänder in der Schreibkreide Nordwest-Deutschlands und ihre Deutung als Lebensspuren:* Geol. Staatsinst. Hamburg, Mitteil., v. 25, p. 104-122, text-fig. 1, 2, pl. 15, 16.

————, & Soule, J. D.

1973, *Cretaceous burrowing bryozoans:* Jour. Paleontology, p. 21-23, text-fig. 1, pl. 1-4.

Vokes, H. E.

1941, *Fossil imprints of unknown origin:* Am. Jour. Sci., v. 239, p. 451-453, 1 pl.

Volk, Max

1960, *Bifasciculus radiatus n. g. n. sp., eine Lebensspur aus dem Griffelschiefer des thüringischen Ordoviziums:* Geol. Blätter Nordost-Bayern, v. 10, p. 152-156, text-fig. 1-4.

1961, *Protovirgularia nereitarum (Reinhard Richter), eine Lebensspur aus dem Devon Thüringens:* Senckenbergiana Lethaea, v. 42, p. 69-75, 2 pl.

1967, *Tigillites (Rouault 1850) ähnliche Spuren (Röhrentunnel) aus dem tieferen Kulm von Steinach:* Hallesches Jahrb. Mitteldtsch. Erdgesch., v. 8 (1966), p. 97-99, text-fig. 1.

Vonderbank, Klaus

1970, *Geologie und Fauna der tertiären Ablagerungen Zentral-Spitzbergens:* Norsk. Polarinst. Skrifter, no. 153, 119 p., text-fig. 1-31, 21 pl.

Voorthuysen, J. H. van

1949, *Lagena-x:* Micropaleontologist, v. 3, no. 2, p. 31, text-fig. 1-4.

Vyalov [Vialov], O. S.

1962, *Problematica of the Beacon Sandstone at Beacon Height West, Antarctica:* New Zealand Jour. Geology & Geophysics, v. 5, p. 718-732, text-fig. 1-11.

1964a, *Zvezdchatye ieroglify iz Triasa severovostoka Sibiri:* Akad. Nauk SSSR, Sibirsk. Otdel. Geol. i Geofiz., no. 5, p. 112-115, text-fig. 1-3. [*Star-shaped hieroglyphs from the Triassic of northeastern Siberia.*]

1964b, *O prirode Cylindrites tuberosus Eichwald iz paleogena priaralya:* Moskov. Obshch. Ispyt. Prirody, Byull., Otdel. geol., v. 39, p. 163-167, text-fig. 1-4. [*On the nature of Cylindrites tuberosus Eichwald in the Paleogene of the Lake Aral area.*]

1964c, *Yavleniya prizhiznennogo zamurovaniya (immuratsii) v prirode:* in Voprosy zakonomernostey i form razvitiya organicheskogo mira, Vses. Paleont. Obshch., Trudy, 7 sess., p. 193-194. [*Phenomena of vital immuration in nature.*]

1964d, *Network structures similar to those made*

by tadpoles: Jour. Sed. Petrology, v. 34, p. 664-666, text-fig. 1.

1966, *Sledy zhiznedeyatelnosti organizmov i ikh paleontologicheskoe znacheni:* Akad. Nauk SSSR, Inst. Geol. Geokhim. Goryuch. Iskopaem., Lvov. Geol. Obshch., 219 p., 51 text-fig., 53 pl. [*Trace fossils and their paleontologic significance.*]

1968a, *O zvezdchatykh problematikakh:* Vses. Paleont. Obshch., Ezhegodnik, v. 5, p. 326-343, text-fig. 1-4, pl. 1, 2. [*On star-shaped problematica.*]

1968b, *Materialy k klassifikatsii iskopaemykh sledov i sledov zhiznedeyatelnosti organizmov:* Paleont. Sbornik, v. 1, no. 5, p. 125-129. Russ., with Engl. summ.) [*Materials to classification of fossil traces and vital activities of organisms.*]

1968c, *Nakhodka krupnogo paleodiktiona v Karpatakh:* Akad. Nauk SSSR, Sbornik, Geol. i Geokhim. Goruch. Iskopaem., Kiev, p. 46-49, text-fig. 1, 2. [*The discovery of a large Paleodictyon in the Carpathians.*]

1969, *Vintoobraznyy khod chlenistonogogo iz Kryma:* Paleont. Sbornik, Izdatel. Lvov. Univ., Vyp. 1, no. 6, p. 105-109, 1 text-fig. (Eng. summ.) [*Screw-like motion of Arthropoda from Cretaceous deposits of the Crimea.*]

1971, *Redkie problematiki iz mesozoya Pamira i Kavkaza:* Same, Vyp. vtoroy no. 7, p. 85-93, 2 pl. [*Rare Mesozoic problematica from the Pamir and Caucasus.*]

1972a, *The classification of the fossil traces of life:* 24th Internatl. Geol. Congress, sec. 7, p. 639-644, 2 text-fig. (Montreal). [*Klassifikatsiya iskopaemykh sledov zhizni:* Russ. text printed in Akad. Nauk SSSR, Mezhdunarodnyy Geol. Kongress, 24th sess., p. 20-29.]

1972b, *Printsipy klassifikatsii sledov zhizni:* Paleont. Sbornik, Izdatel. Lvov. Univ., Vyp. 1, no. 9, p. 60-66. (Eng. summ.) [*The principles of the classification of traces of life.*]

1972c, *Bioglify iz paleogena Dagestana:* Same, Izdatel. Lvov. Univ., Vyp. 2, no. 9, p. 75-80, text-fig. 1, 2, pl. 1-4. (Eng. summ.) [*Bioglyphs from the Paleogene of the Daghestan.*]

————, & Golev [Wiałow, Golew], B. T.

1960, *K sistematike Paleodictyon:* Akad. Nauk USSR, Doklady, v. 134, p. 175-178, 1 text-fig. [*On the systematics of Paleodictyon.*]

1962, *Paleodictyonidae iz fliša Jugoslavije:* Zavod za Geološka i Geofiz. Istrazivanja, Sedimentol., Book 2/3 (1962-63), p. 5-19, text-fig. 1-4. (Eng. summ.). [*Paleodictyonidae from the flysch of Yugoslavia.*]

1964, *Printsipy podrazdeleniya Paleodictyon:* Geo-

logija i Razvedka, Izvestiya vyssh. Uchebn. Zavedeniy, 1964, no. 1, p. 37-48, text-fig. 1-3. [*Principles for the subdivision of Paleodictyon.*]

1965, *O drobnom podrazdeleni gruppy Paleodictyonidae:* Moskov. Obshch. Ispyt. Prirody, Byull., v. 40, no. 2, p. 93-114. [*On the detailed subdivision of the Paleodictyonidae.*]

1966a, *Krytyczny przeglad nowych albo mało znanych form Paleodictyonidae:* Polsk. Towarzyst. Geol., Rocznik, v. 36, no. 2, p. 181-198, text-fig. 1-3, pl. 10, 11. (Pol., with Fr. resumé; Russ., p. 184-198). [*Critical review of new or slightly recognizable forms of Paleodictyonidae.*]

1966b, *O paleodiktionakh iz flisha Bolgarii:* Spisanie na B'lgarskoto Geologich. Druzhestvo, v. 27, pt. 2, p. 173-178, text-fig. 1-3. [*On the flysch Paleodictyons in Bulgaria.*]

————, Gorbach, L. P., & Dobrovolska, T. I.

1964, *Vikopni zirkopodibni sliduzhittediyalnosti morskikh organizmiv iz Skhidnogo Krimu:* Akad. Nauk Ukrainskoi RSR, Geol. Zhurnal, v. 24, no. 4, p. 92-97, pl. 1, 2. [*Star-shaped trace fossils of marine organisms from the eastern Crimea.*]

————, & Kantolinskaya, I. I.

1968, *Sledy sverleniy khishchnykh gastropod v rakovinakh Miotsenovykh Foraminifer:* Paleont. Sbornik, Izdatel. Lvov. Univ., Vyp. 2, no. 5, p. 88-94, 1 text-fig. (Eng. summ.). [*Boring traces of predatory gastropods on shells of Miocene Foraminifera.*]

————, & Ulyanova, A. G.

1968, *Sledy sverleniy na rakovinakh Miotsenovykh ostrakod:* Same, Izdatel. Lvov. Univ., Vyp. 2, no. 5, p. 81-87, 1 text-fig. (Eng. summ.) [*The drilling traces on tests of Miocene Ostracoda.*]

————, & Zenkevich, N. L.

1961, *Sled polzayushchego zhivotnogo na dne tikhogo okeana:* Akad. Nauk SSR, Izvestiya, ser. geol., 1961, no. 1, p. 52-58, text-fig. 1-3. [*Trail of a crawling animal on the floor of the Pacific Ocean.*]

Wade, Mary

1968, *Preservation of soft-bodied animals in Precambrian sandstones in Ediacara, South Australia:* Lethaia, v. 1, p. 238-267.

1970, *The stratigraphic distribution of the Ediacara fauna in Australia:* Royal Soc. S. Australia, Trans., v. 94, p. 87-104.

Wähner, Franz

1903, *Das Sonnwendgebirge in Unterinntal, ein Typus alpinen Gebirgsbaues . . .:* 272 p., F. Deuticke (Leipzig, Wien).

Wagner, Georg

1932, *Beobachtungen am Meeresstrand und ihre Bedeutung für die Geographie der Vorzeit:* Aus der Heimat, Öhringen, v. 45, p. 161-173, pl. 17-38.

Walcott, C. D.

1883, *Fossils of the Utica slate:* Albany Inst., Trans., v. 10, p. 18-38, pl. 1, 2.

1890, *The fauna of the Lower Cambrian or Olenellus Zone:* U. S. Geol. Survey, Ann. Rept., v. 10, pt. 1, p. 509-774, pl. 43-98.

1896, *Fossil jelly fishes from the Middle Cambrian terrane:* U. S. Natl. Museum, Proc., v. 18, p. 611-614, pl. 31-32.

1898, *Fossil medusae:* U. S. Geol. Survey, Mon., v. 30, 201 p., 47 pl.

1899, *The Pre-Cambrian fossiliferous formations:* Geol. Soc. America, Bull., v. 10, p. 199-244, pl. 22-28.

1912a, *Notes on fossils from limestone of Steeprock series, Ontario, Canada:* Canada Dept. Mines, Geol. Survey Branch, Mem., v. 28, p. 16-23, pl. 1, 2.

1912b, *Cambrian geology and paleontology. II. No. 9. New York Potsdam-Hoyt Fauna:* Smithson. Misc. Coll., v. 57, no. 1, p. 249-279, pl. 37-49.

1914, *Cambrian geology and paleontology. III. No. 2. Pre-Cambrian algal flora:* Same, v. 64, p. 77-156, pl. 4-23.

1918, *Cambrian geology and paleontology, IV. No. 4. Appendages of trilobites:* Same, v. 67, no. 4, p. 115-216, pl. 14-42.

1931, *Addenda to descriptions of Burgess shale fossils:* Same, v. 85, no. 3, 46 p., 23 pl.

Walker, M. V.

1938, *Evidence of Triassic insects in the Petrified Forest National Monument, Arizona:* U. S. Natl. Museum, Proc., v. 85, p. 137-141, 4 pl.

Walter, —.

1903, *Über Nemertites Sudeticus Roem., sein Vorkommen und seine Entstehung:* Centralbl. Mineralogie, Geologie, Paläontologie, 1903, p. 76-78.

Walther, Joh.

1904, *Die Fauna der Solnhofener Plattenkalke, bionomisch betrachtet:* Festschrift f. E. Haeckel, Med.-naturwiss. Gesell. Jena, Denkschr., v. 1, 81 p., 1 pl.

Wanner, Johannes

1938, *Beiträge zur Paläontologie des Ostindischen Archipels. XV. Balanocrinus sundaicus n. sp. und sein Epöke aus dem Altmiocän der Insel Madura:* Neues Jahrb. Mineralogie, Geologie, Paläontologie, Beil.-Bd. 79, B, p. 385-402, pl. 10, 11.

1940, *Gesteinsbildende Foraminiferen aus Malm und Unterkreide des östlichen Ostindischen Archipels:* Paläont. Zeitschr., v. 22, p. 75-99.

1949, *Lebensspuren aus der Obertrias von Seran (Molukken) und der Alpen:* Eclogae Geol. Helvetiae, v. 42, p. 183-195, text-fig. 1-5.

Warme, J. E.

1967, *Graded bedding in the Recent sediments of Mugu Lagoon, California:* Jour. Sed. Petrology, v. 37, p. 540-547.

1970, *Traces and significance of marine rock borers:* in Trace fossils, T. P. Crimes, & J. C. Harper (eds.), Geol. Jour., spec. issue no. 3, p. 515-526, pl. 1-4, Seel House Press (Liverpool).

————, **& Marshall, N. F.**

1969, *Marine borers in calcareous terrigenous rocks of the Pacific Coast:* Am. Zoologist, v. 9, p. 765-774.

————, **Scanland, T. B., & Marshall, N. F.**

1971, *Submarine canyon erosion: Contribution of marine rock burrowers:* Science, v. 173, p. 1127-1129, 1 text-fig.

Wasmund, Erich

1936, *Relief-Fährten am Winterstrand der Insel Usedom:* Geol. Rundschau, v. 27, p. 492-498, text-fig. 1.

Webby, B. D.

1969a, *Trace fossils (Pascichnia) from the Silurian of New South Wales, Australia:* Paläont. Zeitschr., v. 43, p. 81-94, text-fig. 1-5, pl. 10.

1969b, *Trace fossils Zoophycos and Chondrites from the Tertiary of New Zealand:* New Zealand Jour. Geology & Geophysics, v. 12, p. 208-214, text-fig. 1-3.

1970a, *Brookvalichnus, a new trace fossil from the Triassic of the Sydney Basin, Australia:* in Trace fossils, T. P. Crimes, & J. C. Harper (eds.), Geol. Jour., spec. issue no. 3, p. 527-530, 1 pl., Seel House Press (Liverpool).

1970b, *Late Precambrian trace fossils from New South Wales:* Lethaia, v. 3, p. 79-109, 21 text-fig.

1970c, *Problematical disk-like structure from the Late Precambrian of western New South Wales:* Linnean Soc. New South Wales, Proc., v. 95, p. 191-193, pl. 10.

Webster, C. L.

1920, *Observations on some marine plants of the Iowa Devonian, with descriptions of new genera and species:* Am. Midland Naturalist, v. 6, p. 286-289 (no. 11: 1920).

Weigelt, Johannes

1927, *Rezente Wirbeltierleichen und ihre paläo-*

biologische Bedeutung: 227 p., text-fig. 1-28, 37 pl., M. Weg (Leipzig).

1929, *Fossile Grabschächte brachyurer Decapoden als Lokalgeschriebe in Pommern und das Rhizocoralliumproblem:* Zeitschrift f. Geschiebeforsch., v. 5, no. 1-2, p. 1-42, pl. 1-4.

Weimer, R. J., & Hoyt, J. H.

1964, *Burrows of Callianassa major Say, geologic indicators of littoral and shallow neritic environments:* Jour. Paleontology, v. 38, p. 761-767, text-fig. 1, 2, pl. 123, 124.

Weiss, Ernst

1884a, *Vorlegung des Dictyophytum Liebeanum Gein. aus der Gegend von Gera:* Gesellsch. Naturf. Freunde Berlin, Sitzungsber., 1884, p. 17.

1884b, *Beitrag zur Culm-Flora von Thüringen:* Preuss. Geol. Landesanst., Jahrb. 1883, p. 81-100, pl. 11-15.

Weiss, Willi

1940, *Beobachtungen an Zopfplatten:* Deutsch. Geol. Gesell., Zeitschr., v. 92, p. 333-349.

1941, *Die Entstehung der "Zöpfe" im schwarzen und braunen Jura:* Natur u. Volk, v. 71, p. 179-184, text-fig. 1-7.

Weissenbach, R. N. P.

1931, *Ein neues Problematikum aus den devonischen Knollenkalken ga(Gg₁)¹:* Státn. Geol. Úst. C. S. R., Sbornik, v. 9 (1930), p. 57-82, text-fig. 1, pl. 1, 2.

Weller, Stuart

1899, *Kinderhook faunal studies. I. The fauna of the vermicular sandstone at Northview, Webster County, Missouri:* Acad. Sci. St. Louis, Trans., v. 9, p. 9-51, pl. 2-6.

Wells, J. W., & Hill, Dorothy

1956, *Zoanthiniaria, Corallimorpharia, and Actiniaria:* in Treatise on invertebrate paleontology, R. C. Moore (ed.), Part F, p. F232-F233, text-fig. 163, 164, Geol. Soc. America & Univ. Kansas Press (New York; Lawrence, Kans.).

Westergård, A. H.

1931, *Diplocraterion, Monocraterion and Scolithus from the Lower Cambrian of Sweden:* Sver. Geol. Undersök., ser. C, Avh. och Upps., no. 372 (=Årsbok. 25, no. 5), 25 p., 10 pl.

Wetzel, Otto

1961, *New microfossils from Baltic Cretaceous flintstones:* Micropaleontology, v. 7, p. 337-350, 3 pl.

1967, *Rätselhafte Mikrofossilien des Oberlias (E): neue Funde von "Anellotubulaten":* Neues Jahrb. Geologie, Paläontologie, Abhandl., v. 128, p. 341-352, 4 pl.

Weyland, Hermann, & Budde, Ernst

1932, *Fährten aus dem Mitteldevon von Elberfeld:* Senckenbergiana, v. 14, p. 259-273, text-fig. 1-21.

Wheeler, H. E., & Quinlan, J. J.

1951, *Precambrian sinuous mud cracks from Idaho and Montana:* Jour. Sed. Petrology, v. 21, p. 141-146.

White, C. D.

1901, *Two n. sp. of algae of the genus Buthotrephis, from the Upper Silurian of Indiana:* U. S. Natl. Museum, Proc., v. 24, p. 265-270, 3 pl.

1928, *Algal deposits of Unkar Proterozoic age in the Grand Canyon, Arizona:* Nat. Acad. Sci., Proc., v. 14, p. 597-600.

1929, *Flora of the Hermit shale, Grand Canyon, Arizona:* Carnegie Inst. Washington, Publ., v. 405, 221 p., 51 pl.

Whiteaves, J. F.

1883, *On some supposed annelid tracks from the Gaspé sandstone:* Royal Soc. Canada, Proc., & Trans., v. 4 (1882/83), p. 109-111, pl. 11, 12.

Whitehouse, F. W.

1934, *A large spiral structure from the Cretaceous beds of Western Queensland:* Queensland Museum, Mem., v. 10, p. 203-210, text-fig. 1-4, pl. 32.

Wilckens, Otto

1947, *Paläontologische und geologische Ergebnisse der Reise von Kohl-Larsen (1928-29) nach Süd-Georgien:* Senckenberg. Naturforsch. Gesell., Abhandl., v. 474, 75 p., 9 pl.

Willard, Bradford

1935, *Chemung tracks and trails from Pennsylvania:* Jour. Paleontology, v. 9, p. 43-56, pl. 10, 11.

———, & Cleaves, A. B.

1930, *Amphibian footprints from the Pennsylvanian of the Narragansett Basin:* Geol. Soc. America, Bull., v. 41, p. 321-327, pl. 4.

Williamson, I. A., & Williamson, R. I. H.

1968, *Trace fossils from Namurian sandstone, Yorkshire:* Geol. Mag., v. 105, p. 562.

Williamson, W. C.

1881, *On the organisation of the fossil plants of the Coal-Measures, Part 10:* Royal Soc. London, Philos. Trans., v. 171 (1880), p. 493-539, pl. 14-21.

1887, *On some undescribed tracks of invertebrate animals from the Yoredale rocks, and on some inorganic phenomena, produced on tidal shores, simulating plant-remains:* Manchester Lit. Philos. Soc., Mem., ser. 3, v. 10, p. 19-29, pl. 1-3.

Wills, L. J.

1970, *The Bunter Formation at the Bellington pumping station of the East Worcestershire Waterworks Company:* The Mercian Geologist, v. 3, p. 387-398.

————, & Sarjeant, W. A. S.

1970, *Fossil vertebrate and invertebrate tracks from boreholes through the Bunter Series (Triassic) of Worcestershire:* The Mercian Geologist, v. 3, p. 399-414, pl. 31-33.

Wilson, A. E.

1948, *Miscellaneous classes of fossils, Ottawa Formation, Ottawa-St. Lawrence Valley:* Geol. Survey Canada, Dept. Mines Res., Bull., v. 11, 116 p., 4 text-fig., 28 pl.

Wilson, M. E.

1931, *Life in the pre-Cambrian of the Canadian Shield:* Royal Soc. Canada, Proc. & Trans., ser. 3, v. 25, sec. 4, p. 119-126, 1 pl.

Winkler, T. V.

1886, *Histoire de l'Ichnologie. Étude ichnologique sur les empreintes de pas d'animaux fossiles:* Arch. Musée-Teyler, sér. 2, v. 2, p. 241-440, pl. 8-19.

Wolfe, M. J.

1969, *A trace fossil from the lower Dalradian, Co. Donegal, Eire:* Geol. Mag., v. 106, p. 274-276, pl. 14, 15.

Wood, Alan

1955, *The origin of the structure known as Guilielmites:* Geol. Mag., v. 72, p. 241-245.

————, & Smith, A. J.

1959, *The sedimentation and sedimentary history of the Aberystwyth Grits (Upper Llandoverian):* Geol. Soc. London, Quart. Jour., v. 114, pt. 2, no. 454, p. 163-195, text-fig. 1-10, pl. 6-8.

Woodward, B. B.

1922, *On Dinocochlea ingens n. g. n. sp., a gigantic gastropod from the Wealden Beds near Hastings:* Geol. Mag., v. 59, p. 242-248, pl. 10, 11, text-fig. 1.

Wray, J. L.

1967, *Upper Devonian calcareous algae from the Canning Basin, Western Australia:* Colorado School Mines, Prof. Contrib., no. 3, p. 1-76, text-fig. 1-18, pl. 1-11.

Wright, T. S.

1856, *Description of two tubicolar animals:* Roy. Phys. Soc. Edinburgh, v. 1, p. 165-167, 1 text-fig.

Wunderlich, Friedrich

1972a, *Nordseeküste, Umwelt im ökonomischen Fortschritt:* Natur u. Museum, v. 102, p. 326-371, text-fig. 1-7.

1972b, *Georgia coastal region, Sapelo Island, U. S. A.: sedimentology and biology: III. Beach dynamics and beach development:* Seckenbergiana Maritima, v. 4, p. 47-79, text-fig. 1-9, 7 pl.

Wurm, A.

1912, *Untersuchungen über den geologischen Bau und die Trias von Aragonien:* Deutsch. Geol. Gesell., Zeitschr., v. 63 (1911), p. 38-174, text-fig. 1-26, pl. 5-7.

Yabe, Hisakatsu

1939, *Note on a Pre-Cambrian fossil from Lyôtô (Liautung) Peninsula:* Japan. Jour. Geology, Geography, v. 16, p. 205-207, pl. 10, 11.

1949, *Problematical fossils on the stratification plane of some older rocks from Japan and Manchuria:* Japan Acad., Proc., v. 25, p. 116-121, text-fig. 1-4.

1950, *Taonurus from the Lower Permian of the Eastern Hills of Taiyuan, Shansi, China:* Same, Proc., v. 26, p. 36-39.

Yochelson, E. L.

1973, *Comments on the application concerning trace fossils. Z. N. (S.) 1973:* Bull. Zool. Nomenclature, v. 30, pt. 2, p. 70-71.

Young, F. G.

1972, *Early Cambrian and older trace fossils from the Southern Cordillera of Canada:* Canad. Jour. Earth Sci., v. 9, p. 1-17.

Young, G. M.

1967, *Possible organic structures in early Proterozoic (Huronian) rocks of Ontario:* Canad. Jour. Earth Cci., v. 4, p. 565-568, text-fig. 1, 1 pl.

Zahálka, Břetislav

1957, *Nalez medusovité formy v Kride beskydské:* Ústřed. Ústavu Geol., Vestnik, v. 32, p. 234-296, 1 pl. [*On the occurrence of a medusae-like form in the Cretaceous of the Beskydy.*]

Zenker, J. C.

1836, *Historisch-topographisches Taschenbuch von Jena und seiner Umgebung besonders in naturwissenschaftlicher und medicinischer Beziehung:* J. C. Zenker (ed.), 338 p., 1 map, Wackenhoder (Jena).

Zigno, Baron Achille de

1856-68, *Flora fossilis formationis oolithicae. Le piante fossili dell'oolite:* v. 1, livr. 1, p. 1-32, pl. 1-6 (1856); livr. 2, p. 33-64, pl. 7-12 (1858); livr. 3-5, p. 65-223, pl. 13-25 (1867); v. 2, livr. 1, p. 1-48, pl. 26-29 (1873); livr. 2-3, pl. 49-120, pl. 30-37 (1881); livr. 4-5, p. 121-203, pl. 38-42 (1885), Seminario (Padua).

Zimmermann, E.

1889, *Über die Gattung Dictyodora:* Deutsch. Geol. Gesell., Zeitschr., v. 41, p. 165-167.

1891, *Neue Beobachtungen an Dictyodora:* Same, v. 43, p. 551-555.

1892, *Dictyodora Liebeana (Weiss) und ihre Beziehungen zu Vexillum (Rouault), Palaeochorda marina (Gein.) und Crossopodia henrici (Gein.):* Gesellsch. Freunde Naturwiss. Gera, Jahresber., v. 32-35, p. 28-63, 1 pl.

1894, *Weiteres über angezweifelte Versteinerungen (Spirophyten und Chondrites):* Naturwiss. Wochenschr., v. 9, p. 361-366, text-fig. 1-11.

Zittel, K. A. von (ed.)

1876-90, *Handbuch der Palaeontologie. Abt. 1. Palaeozoologie:* v. 1, viii + 765 p., 557 fig. (1876-80); Abt. 2. *Palaeophytologie* (begun by W. Ph. Schimper, cont'd by A. Schenk): v. 5, 958 p., 429 fig. (1879-90), Oldenbourg (München & Leipzig).

Zittel, K. A. von

1913, *Vermes:* in Text-book of paleontology, C. R. Eastman (ed.), p. 135-142, text-fig. 213-225, Macmillan & Co., Ltd. (London).

1924, *Grundzüge der Paläontologie (Paläozoologie), I. Abt.: Invertebrata:* Neubearbeitet von Dr. F. Broili, 6th edit., 733 p., 1467 text-fig., R. Oldenbourg (München, Berlin).

ADDENDUM (MICROPROBLEMATICA)

When this volume was in page proof, the following publication came to our knowledge (courtesy of H. Kozur):

Kozur, H., & Mostler, H., 1972, *Mikroproblematika aus Lösungsrückständen triassischer Kalke und deren stratigraphische Bedeutung:* Gesellsch. Geol. Bergbaustudien, Innsbruck, Mitt., v. 21, p. 989-1012, 6 pl.

Here, the following 14 new genera of Microproblematica are described:

Hollow, conical tubes

Argonevis Kozur & Mostler, 1972, p. 992 [*A. nuda*; M]. *U.Trias.(up.Nor., Hallstätter Kalk)*, Eu.(N.Aus.).

Limolepis Kozur & Mostler, 1972, p. 993 [*L. interruptus*; M]. *M.Trias.(low.Ladin.)-U.Trias.(Rhaet.)*, Eu.(Hung.-N. Italy-E. Alps).

Erinea Kozur & Mostler, 1972, p. 994 [*E. triassica*; M]. *U.Trias.(low.Carn.-Rhaet.)*, Eu. (E. Aus.-Hung.-Czech.).

Venerella Kozur & Mostler, 1972, p. 994. Two species described, no type species indicated. *U.Trias.(Nor.)-M.Jur.(Malm)*, Eu.(Aus.-Hung.-Czech.-Yugosl.-N. Italy).

Nemotapis Kozur & Mostler, 1972, p. 996. Two species described, no type species indicated. *M.Trias.(mid.Anis.)-Jur.(Malm)*, Eu. (Aus.-Hung.-Italy-Greece).

Antler-like skeletal elements (possibly holothurian sclerites)

Cornuvacites Kozur & Mostler, 1972, p. 997. Two species described, no type species indicated. *M.Trias.(up. Anis.)-U.Trias.(up.Carn.)*, Eu.(Aus.-Hung.-N. Italy).

Concavo-convex perforated plates (?echinoderms)

Irinella Kozur & Mostler, 1972, p. 999 [*I. canalifera*; M]. *U.Trias.(Carn.)*, Eu.(Aus.-Hung.).

Hook-shaped forms (?echinoderms)

Bogschites Kozur & Mostler, 1972, p. 1000 [*B. carnicus*; M]. *U.Trias.(mid.Carn.)*, Eu. (Aus.-Hung.).

Havinellites Kozur & Mostler, 1972, p. 1000 [*H. spinosus*; M]. *U.Trias.(mid.Carn.)*, Eu. (Aus.-Hung.).

Miscellaneous stalked forms

Strechoritina Kozur & Mostler, 1972, p. 1001 [*S. radiata*; M]. *U.Trias.(up.Nor.)-L.Jur. (Lias.)*, Eu.(Aus.-Hung.).

Uvanogelia Kozur & Mostler, 1972, p. 1001 [*U. incurvata*; M]. *M.Trias.(up.Ladin.)-U.Trias.(low.Rhaet.)*, Eu.(Aus.-Hung.-N.Italy).

Radimonis Kozur & Mostler, 1972, p. 1002 [*R. foliacea*; M]. *U.Trias.(low.Carn.)*, Eu. (Aus.-Hung.).

Placerotapis Kozur & Mostler, 1972, p. 1002 [*P. subplanus*; M]. *M.Trias.(low.Ladin.)-U.Trias.(up.Carn.)*, Eu.(Aus.-Hung.).

Fanerocoelia Kozur & Mostler, 1972, p. 1003 [*F. pennata*; M]. *U.Trias.(mid.Nor.)*, Eu. (N.Aus.).

All described fossils consist of high magnesium calcite. Most of them are extremely abundant and some are of stratigraphic importance.

[Descriptions supplied by Curt Teichert]

INDEX

Italicized names in the following index are considered to be invalid, with exception of foreign phrases; those printed in roman type, including morphological terms, are accepted as valid. Suprafamilial names are distinguished by the use of full capitals and author's names are set in small capitals with an initial large capital. Page references having chief importance are in boldface type (as **W100**).